AF251677

Institution of Civil Engineers

Bridge design, construction and maintenance

Proceedings of the two day international conference organised by the Institution of Civil Engineers and held in Beijing on 17–18 September 2007

Edited by Dr Robert Lark

Co-sponsored by:

Organising committee:

Dr Robert Lark, Cardiff University, UK (*Chairman*)
Professor Joan Casas, Technical University of Catalonia, Spain
Dr Stuart Davis, Mott Macdonald, UK
Professor Dan Frangopol, University of Colorado, USA
Mr Naeem Hussain, Ove Arup & Partners, Hong Kong
Professor Liu Xila, Chinese Civil Engineering Society, China
Mr Barry Mawson, Structures and Buildings Board ICE, UK
Professor Andrezej Nowak, University of Nebraska, USA
Mr Gandhi Suppiah, GS Technology Consultants Ltd, UK

Technical Advisory Group:

Professor Ben Barr, Cardiff University, UK
Mr Mike Chubb, Atkins Global, UK
Professor Aarne Jutila, Helsinki University of Technology, Finland
Dr Donald Pearson-Kirk, AccordMP, UK
Professor Lin Shaopei, Shanghai Jiao Tong University, China
Professor Wojech Radomski, Warsaw University of Technology, Poland
Professor Shanmugam, Universiti Kebangsaan, Malaysia
Mr Ian Sloane, Connell Wagner Limited, New Zealand
Professor Jiri Strasky, Technical University of Brno, Czech Republic

Published for the Organising Committee by Thomas Telford Publishing, Thomas Telford Ltd, 1 Heron Quay, London E14 4JD. www.thomastelford.com

Distributors for Thomas Telford books are
USA: ASCE Press, 1801 Alexander Bell Drive, Reston, VA 20191-4400, USA
Japan: Maruzen Co. Ltd, Book Department, 3–10 Nihonbashi 2-chome, Chuo-ku, Tokyo 103
Australia: DA Books and Journals, 648 Whitehorse Road, Mitcham 3132, Victoria

First Published 2007

A catalogue record for this book is available from the British Library
ISBN 978-07277-3593-5

Flash drive manufactured by Gold Products Ltd, Telford, Shropshire

Foreword

Bridge engineering not only requires a comprehensive knowledge and understanding of design and construction techniques, but also demands that the challenge of the long-term behaviour and maintenance of such structures is met. This requires an innovative, multi-disciplinary approach built on a sound foundation of experience and research.

The Institution of Civil Engineers organised its fifth international conference on current and future trends in bridge design, construction and maintenance, in conjunction with a team of world-renowned bridge engineers. The conference was held in Beijing on 17 and 18 September 2007 and these are the proceedings of that conference. The event has a history of attracting high quality presentations and delegates from around the world. This year was no exception with presentations from twenty-five countries.

The conference provided a forum for bridge practitioners and researchers to share their experiences and understanding, to highlight notable successes and to ensure that progress is maintained. The programme was designed to be as wide-ranging as possible so that all might benefit.

Dr Robert Lark
(Chairman of the organising committee)
Cardiff University, UK

Contents:

Theme two: Advances in the understanding of the structural behaviour of bridges

Theme three: Durability, structural reliability, risk management and health monitoring

Theme six: The future of condition monitoring, assessment and maintenance

Author index

Opening plenary session:

State-of-the-art in long-span bridges

Incheon Bridge – Fast track design check of the cable stayed bridge

Cecilia Cheong, Arup, Hong Kong
Chris Conroy, Halcrow Group Ltd, Swindon, UK
Matt Carter, Arup, Hong Kong
Dr Innes D Flett, Halcrow Group Ltd, Swindon, UK

Abstract

A 12.3 km long sea crossing is currently under construction at Incheon in South Korea. At a cost of US$1.4 billion, the crossing will link the new Incheon International Airport on Yeongjong island to Songdo (New City) and the new International Free Enterprise Zone (IFEZ) which are both currently under construction. A cable stayed bridge will cross the 625.5m wide by 74m high navigation channel leading to the Port of Incheon. With an 800m long main span, this will be the longest spanning bridge in South Korea and will form part of one of the longest sea crossings in the world.

A joint venture team comprising Halcrow, Arup and local consultant Dasan was appointed by design and construct contractor Samsung Construction JV (SCJV) as the Contractor's Checking Engineer (CCE) in March 2005. This paper will introduce the role of the CCE and describe the procurement of the cable stayed bridge, illustrating the part that the CCE plays in the fast track delivery of the project.

Introduction

The Incheon Bridge, illustrated in Fig. 1, will carry six lanes of traffic across the straits between Yeongjong island and the Korean peninsula. The project is being procured by the Korea Expressway Corporation (KEC). The Incheon Bridge Company Ltd, the AMEC led Concessionaire, in joint venture with the City of Incheon, will finance and manage the toll-bridge for 30 years before returning the project to the Korean government.

Figure 1. Incheon Bridge

Bridge design, construction and maintenance 2007, Thomas Telford, London

Construction of the bridge was let on a Design & Build basis and construction works began in June 2005. SCJV, a group of seven Korean construction companies are carrying out the works with design services provided by consultants Seoyeong Engineering (Korea), Chodai (Japan) and others.

The majority of the length of the bridge is constructed as low level viaduct structures with pretensioned precast 50m long concrete box girder spans. Where the alignment rises to cross the navigation channel, precast segmental balanced cantilever approach bridges with 145m spans link the viaducts to the cable stayed bridge which provides the 800m long navigation span itself (Figure 2).

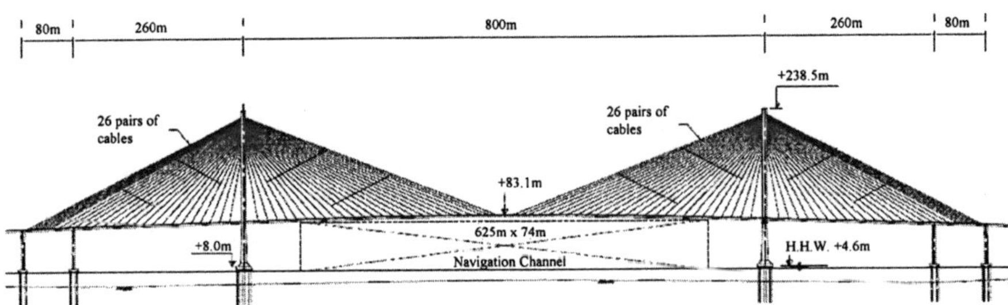

Figure 2. General arrangement of cable stayed bridge

The bridge is constructed over tidal flats and in up to 20m depth of water. Marine deposits overly rock strata. All of the foundations are large diameter cast in place concrete piles socketed in the weathered or soft rock strata.

Korea is in a region of moderate seismicity and the bridge is designed for a 1,000 year return period event which governs the design of the substructures. In addition the bridge can be subject to typhoon wind loading which is significant for the high level structures. In particular wind buffeting loads and aerodynamic stability has been important to the design of the cable stayed bridge.

Crossing the main navigation route into Incheon port the bridge is designed to withstand ship impacts of up to 100,000 DWT. Protection is provided in the form of sacrificial dolphin structures which are configured around the piers close to the navigation channel.

The role of the Contractor's Checking Engineer

The CCE's role was to perform an independent check of the permanent works and major temporary works to confirm that they were in accordance with the basis of design and carry out an independent review of the remaining temporary works as well as review a number of technical reports. The independent checks were deemed to be a higher order check than the independent review as only the drawings were received for the permanent works requiring a complete re-analysis of the structure with corresponding stress checks and calculations. Both drawings and calculations were received for the independent review of the temporary works which did not necessarily require any further analysis.

As a fast track project, the design was prepared as a sequence of packages in accordance with the demands of the construction schedule. Managing and organising the CCE's joint venture team required communicating and coordinating with the checking teams, which were located in different offices worldwide, as and when the design packages became available. Working closely with the design team in the Incheon site office was essential for the effective delivery of design packages to the checkers and dealing with the day to day issues arising from during the procurement of the design.

Site offices were provided by SCJV and senior representatives from the CCE maintained a full time presence there during the design phase of the project. A project specific management plan and quality procedures were developed to help manage the coordination of inputs from all parties and keep track of the checkers comments and the certification process.

The contract duration for the CCE's checking role was initially 2 years but this was extended in March 2007 to keep a CCE presence on site further into the construction period.

Basis of Design

Design Standards

The design basis for the Incheon Bridge was originally set out in two key documents, the Project Performance Requirements (PPR) written by the Ministry of Construction and Transportation and the Concessionaire's Supplementary Requirements (CSR) introduced by the Incheon Bridge Company Ltd.

The PPR references the AASHTO LRFD Bridge Design Specifications as the key standard for structural design. However, in order to ensure a consistent performance with other Korean bridges the PPR also wrote out a full set of loads and load combinations which were extracted from the Korea Bridge Design Standards. The bridge had to be designed to cover both the LRFD and the PPR loading conditions although in both cases analysis, verification and detailing requirements were in accordance with LRFD. In almost all cases the PPR loads proved to be the governing design condition.

Development of the Design Manual

Before award of the contract, SCJV entered into negotiations with the Concessionaire and KEC concerning interpretation of the design basis. Issues involved clarification of the design basis, interpretation of ambiguous or conflicting requirements and requests for relaxation of certain requirements which were perceived as unnecessarily onerous. The CCE took an active role in assisting with these negotiations and wrote a number of Technical Reports providing an independent opinion making reference to projects and standards from around the world.

SCJV produced a Design Manual for the project which consolidated the requirements of the PPR and CSR and introduced further clauses as a result of their negotiations with the Concessionaire and KEC. The CCE was required to review and certify that the Design Manual complied with the PPR and CSR and that it was appropriate for the Incheon Bridge project. The review resulted in various comments being raised which were passed on to SCJV for information and action where necessary.

A number of key changes proposed by the CCE had to be introduced before the early design packages could be certified. In order to avoid delay whilst the revised Design Manual was being approved, SCJV produced a number of Design Manual Addenda as separate documents which were referenced by the check certificates. This allowed KEC approval of the relevant Design Manual Addendum and the certified drawings to be carried out in parallel.

Cable Stayed Bridge

Design Units

The project was split into a number of design units to define the packages in which the detailed design would progressively be finalised and checked. For the Cable Stayed Bridge, the main structure comprised of 7 design units which are described in Table 1 and Figure 3. There were a number of additional design packages for ancillary or miscellaneous items.

Table 1. Definition of design units

No.	Main Division	Second Division	Notes
1	Substructure A	CSB 1	Piles and pilecap (pylon)
2		CSB 2	Piles and pilecap (piers)
3	Substructure B	CSB 1	Pylon and associated bearings
4		CSB 2	Piers and associated bearings
5	Superstructure	Cables	-
6		Sidespan Girder	Large segments to be erected by floating crane
7		Mainspan Girder	Small segments to be erected by derrick

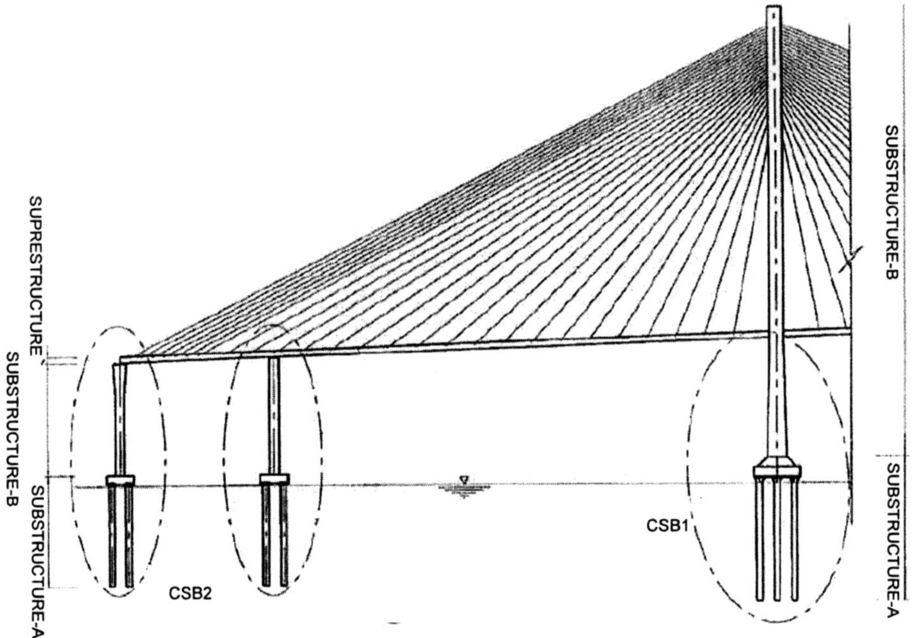

Figure 3. Illustration of design units

Figure 4 shows the schedule of checking for the cable stayed bridge and illustrates the significant overlap between the design, checking and construction which is the consequence of the fast track procurement approach.

Design Check

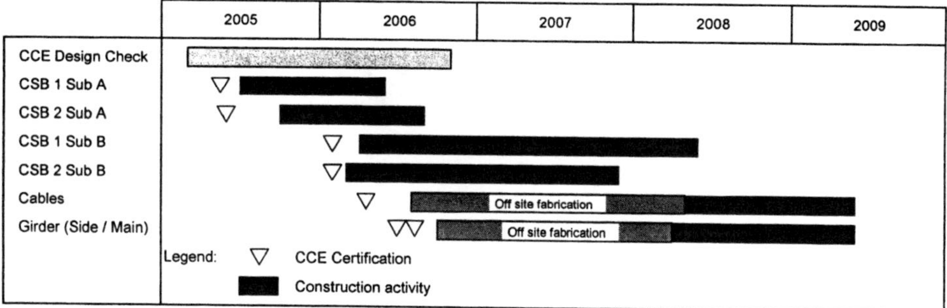

Figure 4. Design check schedule

Meeting the fast track requirements of the project was particularly demanding for the cable stayed bridge. Only 12 weeks were available between the CCE starting work and the scheduled date for certification of the pylon piles and pilecap (CSB1 Sub A).

As well as geotechnical capacity and reinforcement checks for the substructure, this initial certification required review of the design basis and a full global analysis of the structure including both wind buffeting analyses carried out using TDV RM2000 and response spectrum seismic analyses with mode specific damping carried out using Oasys GSA. Furthermore the feasibility of the complete structure had to be reviewed in order to provide a reasonable degree of security that the foundation loads would not be increased as design of the pylon, deck and articulation progressed.

The CCE approached this demanding schedule by working with a high degree of interaction with the designer. Key design data was checked and agreed upon and foundation loads compared prior to production of the reinforcement drawings by the designer for checking and certification.

In order to accurately capture the dynamic interaction between the cable stayed bridge and the approach bridges, the global analysis model included a complete representation of the approach bridges (Figure 5). The approach bridge elements were activated for all modal analyses (seismic loads and wind buffeting loads). However, to reduce computational time these additional elements were deactivated for live load optimisation analyses.

The analysis package LARSA was also used to check the pylon and backspan piers. Although both the LARSA model and the RM2000 model were created using common geometry and modelling assumptions derived from the GSA model, the use of three different analysis packages operated by three different engineers provided a degree of internal checking of analysis output and allowed direct comparison of results. For example, mean wind analysis results from RM2000 could be compared with pseudo static wind load results from GSA as a check of the application of wind load (which differs conceptually between the packages).

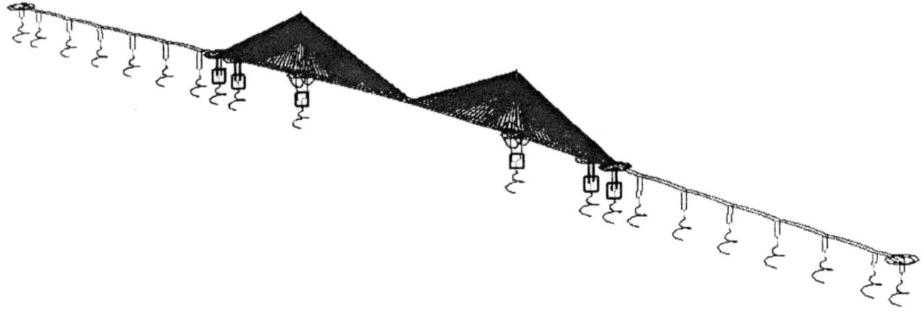

Figure 5. Global analysis model established in Oasys GSA

Foundations

The foundations consist of 3.0m diameter cast in place bored piles end bearing in rock sockets. Permanent steel casings cantilever above the mud line to allow the pilecap to be constructed above water.

A key issue was whether the piles could reasonably be designed to behave elastically during the design seismic event which would allow a reduction in the highly congested transverse reinforcement albeit at the expense of additional longitudinal reinforcement. The CCE carried out a thorough review of the implications making reference to a seismic hazard assessment independently carried out for the Seoul metropolitan area and concluded that there was sufficient conservatism in the design event and sufficient ductility in the piles with ordinary reinforcement detailing that the structures would be able to withstand a 1 in 2,500 year event without collapse if designed elastically for the nominal 1 in 1,000 year design event.

Construction of the foundations began whilst design and checking of the substructure and superstructure was still progressing. Piles were installed by marine equipment and for each foundation a precast structure (PC House) was used to provide a working platform and permanent formwork (Figure 6).

Figure 6. Installation of the pylon PC house

Pylon & Piers

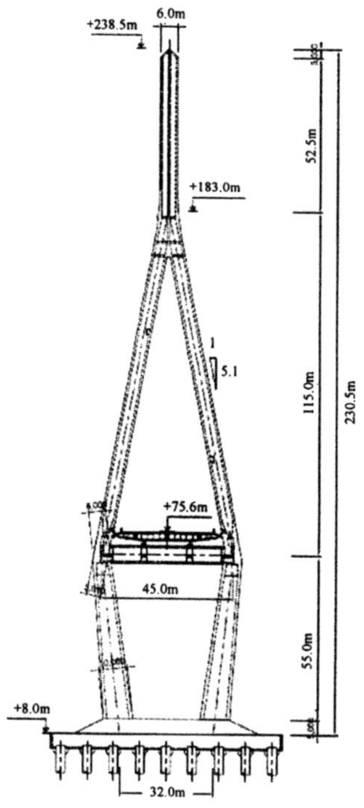

Figure 7. Pylon

The 230m tall pylon is a reinforced concrete hollow section in a diamond configuration (Figure 7) which provides torsional stability to the main span and minimises the size of foundation which must be protected from ship impacts. A steel anchor box inside the upper part of the pylon transfers load from the stay cables into the concrete structure.

The pylon legs are constructed with an autoclimbing jump form. The pylon cross beam was precast and installed by floating crane. Construction of the pylon above the cross beam is now ongoing. (Figure 8)

Figure 8. Recent construction progress

Two reinforced concrete hollow anchor piers are provided in each 340 m sidespan dividing the total length into 260m + 80m. Uplift of the sidespan girder is prevented by concrete counterweight cast inside the steel box as well as vertical tie down cables at each anchor pier.

Steel Superstructure

The stay cables are preformed parallel wire strand supplied by Nippon Steel. The largest cables consist of 301 no. 7mm wires with an ultimate tensile strength of 1770 MPa. A dimple pattern on the HDPE sheathing prevents rain-wind induced vibration. External friction dampers will also be provided to raise the structural damping of the cables above a specified minimum of 0.5% of critical damping.

The deck is a 33.4 m wide orthotropic steel box girder suspended up to 83.1m above mean sea level. The stay cables are anchored in tubes running in the plane of the external webs. Internal webs provide additional strength and stiffness. The typical segment length is 15.0m in the main span and the field splice consists of a full penetration butt weld in the deck plate together with bolting of the webs and bottom flange plates as well as bolted splices for stiffeners.

Construction Method

The SCJV have two large floating cranes dedicated to the construction of the Incheon Bridge with the larger having a lifting capacity of 3,000 tonnes. Given the availability of these cranes the construction methods have made maximum use of them. For the cable stayed bridge, two temporary piers will be installed to allow the side span girder to be fully erected by floating crane prior to installation of any stay cables. Cantilever construction of the main span will then follow traditional practice with a derrick crane on the deck (Figure 9).

As well as having the advantage of speed of installation, this method is more aerodynamically robust since it avoids having 260m long balanced cantilevers as would be required if cantilever construction progressed from the pylon towards the permanent sidespan piers.

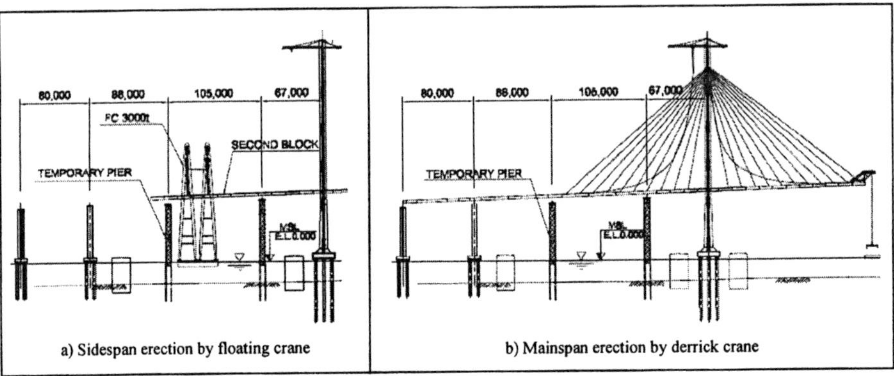

Figure 9. Construction Method

Temporary Works

The temporary works were classified in 3 categories namely, Major Temporary Works (MTW), Temporary Works (TW) and Method Statements (MS) all of which required checking to different levels. The TW's and the MS's required only a detailed review where both drawings and calculations were provided and these were certified accordingly. However, the MTW's, which were identified as temporary works that could impose significant loads on the permanent works or temporary works which represented a significant safety risk, required a complete independent design check as was required for the permanent works.

The cable stayed bridge included two items classified as MTW's. The temporary pylon struts are trusses which will prop apart the inclined upper legs of the pylons to prevent overstress of the legs during construction as well as avoiding locked in bending moments. The temporary back span piers are required to support the sidespan prior to installation of the stay cables.

Both MTW's were checked by a thorough evaluation of the interaction between the permanent and temporary works, incorporating the MTW into the global analysis model. For the temporary sidespan bents, the CCE carried out a wind buffeting analysis to confirm that the proposed pseudo-static wind load with a gust factor of 1.9 would be a conservative evaluation.

Tests & Investigations

A number of studies were carried out by SCJV during the design development period which were independently reviewed and certified by the CCE. These included:

- Probabilistic Seismic Hazard Assessment
- Oceanographic Investigations
- Ground Investigations
- Pile Load Tests
- Wind Tunnel Testing
- Ship Impact Protection Test Programme

Each of these investigations were summarised by SCJV in a Technical Report which the CCE reviewed and certified with regard to the appropriateness of the content, methodology and general principles. Two key sets of studies relevant to the cable stayed bridge are described in more detail below.

Wind Tunnel Testing

Wind tunnel testing was carried out to determine the aerodynamic force coefficients for the deck and tower and to investigate flutter and vortex shedding stability.

Sectional model tests were carried out at 1:100 scale for a variety of cross sections investigating different configurations of the leading edge and a 1:50 scale confirmatory test was carried out for the preferred cross section. The deck section was shown to be stable.

A confirmatory aeroelastic full bridge model test at 1:150 scale also showed the bridge to be aerodynamically stable. In addition an aeroelastic model test was carried out for the free standing tower which did not indicate any unacceptable aerodynamic effects.

In addition to the wind tunnel testing carried out by SCJV, a wind climate analysis was provided by the Concessionaire which confirmed that the PPR design wind speed was conservative and also recommended turbulence characteristics for use in wind buffeting analyses.

Ship Impact Protection

Ship impact protection is provided in the form of circular sheet piled dolphins filled with crushed rock and tied together with a reinforced concrete cap. The dolphins were designed to provide both deterministic and probabilistic protection, the former being to stop a 100,000 DWT design vessel travelling at 4.5 m/s directly towards the cable stayed bridge pylon and the latter being to reduce the annual collapse frequency to less than 1 in 10,000 when considering a distribution of design vessels heading towards any point on the bridge axis in any direction. The probabilistic design was carried out in accordance with the method detailed in AASHTO LRFD based on an initial ship velocity of 10 knots.

The dolphins work by dissipating energy through various mechanisms; crushing of the ships bow, local deformation of the dolphin, passive resistance of the soil and friction between ship and dolphin. A large portion of the energy dissipation comes from mobilising the passive resistance of the soil. A reliable way to estimate impact dissipation in soil structures is through testing of a physical model in a centrifuge which allows earth pressures to be correctly modelled at a reduced scale. However, due to the time and expense required for centrifugal model testing it is preferred to use the results to calibrate a non-linear finite element analysis which will then allow analysis of different configurations. This method, which had previously been adopted for Stonecutters Bridge [1], was followed for the design of the Incheon Bridge ship impact protection.

After establishing the behaviour of an individual dolphin by this method, a series of impact simulations were carried out to determine the behaviour of the dolphin configuration. These analyses were independently checked by the CCE using the Arup in house programme IMPS. In certifying the ship impact protection, the CCE reviewed and certified all of the documents related to the test programme and impact simulations which demonstrated the stopping capacity of the dolphins. The design drawings were also checked and certified to confirm the construction details and the ability of the dolphins to withstand gravity and hydraulic loads as well as a short return period seismic event.

Conclusion

For a bridge of this scale a full independent check including independent analysis and verification is essential for ensuring safety. However, introduction of an Independent Checking Engineer in a traditional role would inevitably lead to delay in a fast track project.

By working within the contractor's organisation, the CCE is able to work in parallel with the design team rather than waiting until completion of the design before commencing work. This allows an interactive checking process whereby analytical results can be compared prior to production of all detailed drawings. The CCE is also able to take a proactive role providing technical advice and helping developing solutions to design problems which inevitably arise during the design process.

The cable stayed bridge provides an illustration of the value that can be added to the project by the CCE role. The overlap between design and construction is clear with certification of the superstructure taking place after substructure construction was well advanced. With this being the most complex structure on the project it naturally falls onto the critical path and early certification of the pylon piles was essential to ensure that the project remained on schedule. A highly interactive approach was required in order to ensure that the target date was met.

The CCE is playing a vital role in ensuring the delivery of this complex project safely and on time.

Completion of construction is planned for October 2009.

Acknowledgement

This paper has been published with the permission of the Samsung Construction Joint Venture and Incheon Bridge Company Limited. Figures 1 to 3 and 6 to 9 were provided by the SCJV.

References

[1]   LEE D.M., and PEIRIS N., "Modelling of Ship Impact on a Bridge Foundation", *IABSE Symposium Shanghai 2004: Metropolitan Habitats and Infrastructure*, IABSE Report, Volume 88, 2004.

Stonecutters Bridge
Detailed Design and Construction of Superstructure

Steve KITE, Associate, Arup, Hong Kong SAR, China
Ngai YEUNG, Chief Technical Officer, Arup, Hong Kong SAR, China
Stephen KWOK, Resident Engineer, Arup, Hong Kong SAR, China

Abstract

Stonecutters Bridge will span 1018m across Rambler Channel in Hong Kong providing 73.5m high clearance above the busy shipping channel. Due to the typhoon prone environment, detailed design was dominated by the dynamic effects of the turbulent wind loading. Considerable wind tunnel testing was carried out as part of the design process. The high expected percentage of heavy goods vehicles also influenced the design considerably.

High level prestressed concrete decks act as anchor spans to support the steel main span. The concrete back span twin box girders have been constructed in-situ and require a substantial temporary works structure to remain in place until the stay cables are installed. The main span steelwork boxes are fabricated and assembled in China. A 4000T heavy lift scheme for the deck over land precedes main span cantilevering which will take place with 18m long erection segments. Stay cables are prefabricated parallel wire cables up to 540m long.

Keywords: cable-stayed bridge; twin-box deck; wind tunnel testing; construction method;

Introduction

Stonecutters Bridge will form the centrepiece of Hong Kong's latest infrastructure improvement, the new Route 8 - an alternative road to the international airport link, also providing enhanced access connections into the new container terminal on Tsing Yi Island. The bridge crosses the Rambler Channel at the entrance to Kwai Chung Container Port, one of the busiest in the world. Minimum disruption to the operations of the port was a high priority in the design and construction planning.

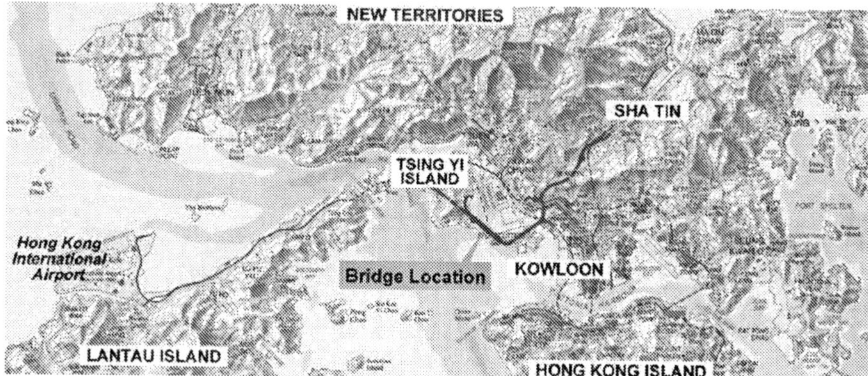

Fig. 1 Location of Stonecutters Bridge

Detailed design by Arup and Cowi took place in 2002 and 2003. The design is dominated by the effects of extreme loading conditions associated with the site. The construction contract was awarded to the Maeda-Hitachi-Yokogawa-Hsin Chong Joint Venture, and construction began in April 2004 with completion scheduled for 2009. This paper describes some of the interesting design problems tackled and how the challenges of erecting this major bridge are being overcome.

The Bridge
The 1596m long cable-stayed bridge has a steel main span of 1018m, with prestressed concrete back spans each side of 79.75m, 70m, 70m and 69.25m. The circular tapered mono-column towers stand on land at the bridge centre line between the two longitudinal boxes of the twin girder deck. The towers are formed of concrete to +175m, are of composite construction with an outer stainless steel skin to +293m and are topped by a lighting feature to +298m. Stay cables are in 2 planes arranged in a modified fan layout and attached to the outside edges of the deck. The deck girders are connected with cross girders spaced at 18m in the main span, coinciding with the stay anchorage spacing, and 20m in the back spans where the stays anchorages are spaced at 10m. The concrete back spans are monolithic with the piers Three intermediate piers have single rectangular tapered column shafts, while the end piers at the interfaces to the adjoining viaducts are twin column portal structures. Both sides of the bridge are on reclaimed land and foundations are large diameter bored piles to rock. Pile lengths are between 50m and 110m and shaft diameters are between 2.2m and 2.8m.

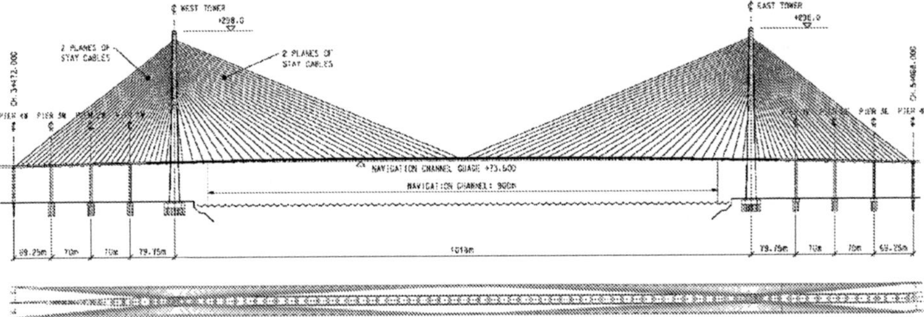

Fig. 2 Elevation and plan

Design Criteria
The design criteria are summarised in a Design Memorandum, based on the Hong Kong Structures Design Manual for Highways and Railways (SDM), supplemented by BS5400 and other relevant codes. In a number of areas it has been necessary to establish site specific design criteria. For example from study of the wind climate, parameters are derived for the dynamic wind loading. Risk analysis gave a rationale for the assessment of seismic effects and dynamic model tests and analysis were used for deriving accidental ship impact effects, [1]. The Design Memorandum was updated continuously during the design process as improved information was finalised. The design life of the bridge is 120 years.

Wind Loading
The wind loading applied reflected the wind climate of Hong Kong as described in the SDM. It was supplemented with data from other sources to create a comprehensive wind model suitable for dynamic wind load assessments of the bridge.

Ocean exposure is characterised by high wind velocities and relatively low turbulence. The characteristic 1-hour mean wind speed at deck level equals 52 m/s for ocean exposure (south westerly directions). When the wind is approaching over mountainous/urban terrain the wind velocity is lower but the level of turbulence is high. The 1-hour mean wind speed at deck level equals 42 m/s for mountainous/urban exposure (westerly to south easterly directions). The high turbulence wind has generally governed the design.

Global Analysis

FE Model

The global analysis included modelling the construction process, as a sequence of phases. A time indication is linked to each phase allowing calculation of effects such as creep, shrinkage and relaxation. In addition to dead load, live load, wind, temperature, seismic load etc. the bridge has been analysed for a number of accidental load cases, including stay rupture and ship impact.

Displacements

The displacements of the upwind girder at mid-span are listed in Table 1, which shows that the total transverse displacements of the girder are similar for ocean and for mountainous/urban exposure. However, vertical girder displacements are greater for the more turbulent wind regime despite the lower mean wind speed. The same observation is made for the twist angle of the deck.

		Upwind girder displacement at mid-span		
Exposure	**Load case**	**Transverse (m)**	**Vertical (m)**	**Twist (deg)**
-	Traffic load	-	-1.81	-
Mountainous/ urban exposure	Mean wind	0.47	-0.03	0.20
	Buffeting	±0.76	±1.64	±1.04
	Total wind	1.23	-1.67	1.24
Ocean exposure	Mean wind	0.66	-0.07	0.26
	Buffeting	±0.56	±1.03	±0.70
	Total wind	1.22	-1.10	0.96

Table 1 Displacements for live load and wind.

Dynamic Performance

Fundamental eigenfrequencies of the girder for the completed bridge are listed in Table 2. The Design Memorandum requirement that the ratio between the natural frequency of the lowest pure torsion mode and the lowest pure vertical bending mode shall exceed 1.2 is fulfilled.

Eigenfrequency (Hz)	**Mode shape description**
0.143	1st symmetric horizontal, girder
0.195	1st symmetric vertical, girder
0.236	1st asymmetric vertical, girder
0.350	1st asymmetric horizontal, girder
0.428	1st symmetric torsion, girder

Table 2 Fundamental eigenfrequencies for girder vibration in completed bridge.

Wind Tunnel Testing

During the initial phases of the design, before results from wind tunnel testing and on-site measurements were available, simulations of the aerodynamic behaviour of the bridge were carried out by means of the computer program DVMFLOW. This is a two-dimensional version of the discrete vortex method programmed for simulation of flow around bluff bodies, stationary, elastically suspended or in prescribed motion. As the design progressed substantial section model wind tunnel testing was carried out to verify the aerodynamic stability and to supply refined information on the steady state wind load coefficients for the global analysis.

The results of the section model tests show that with the chosen cross section the bridge is stable against divergent instability with the critical wind speed far exceeding the 95m/s limiting acceptance criteria. Confirmatory 3D aeroelastic Full Bridge Model Tests at scale 1:200 have also been carried out.

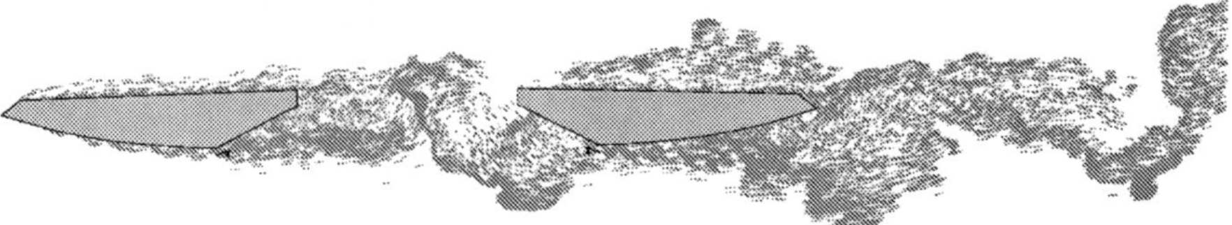

Fig. 3 Flow plot from DVMFLOW model, twin box girder including guide vanes.

Concrete Back Spans

In the back spans the steel deck extending 49.75m behind each tower, and the remainder is in concrete. The concrete decks have varying geometry to account for the changing road width on the west side to accommodate the on and off-ramps, and the start of a change in horizontal alignment and superelevation on the east side.

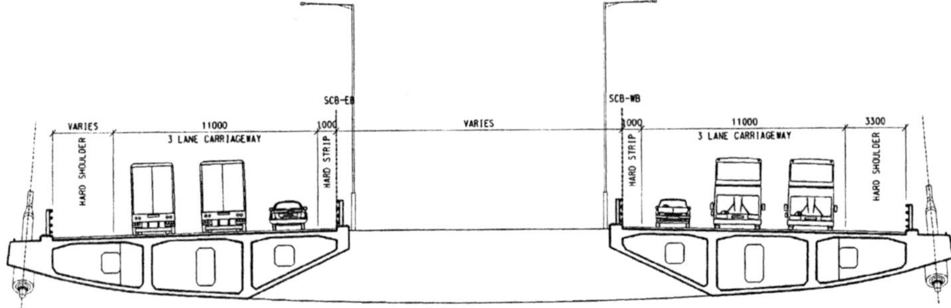

Fig 4. Concrete deck – cross section in East Back Spans.

The concrete spans act as stiff anchor spans. Due to the layout with the stay cables on the outer edge of the twin box girders, the load path for the stay forces is a combination of bending and torsion of the longitudinal structure, and bending of the cross girders. In the permanent load condition, the balance between longitudinal box torsion and cross girder bending is dependent on the methods of construction, as initial stressing of the stay cables could apply large torsions to the boxes. This is overcome by applying temporary prestress at the top of the cross girders during construction, prior to building them in to the pier cross heads. This temporary prestress is applied as a bowstring, which in effect mimics the action of the stay cables. When the stays are later stressed the bowstring is released. The combined effect is a much reduced dead load torsion in the boxes.

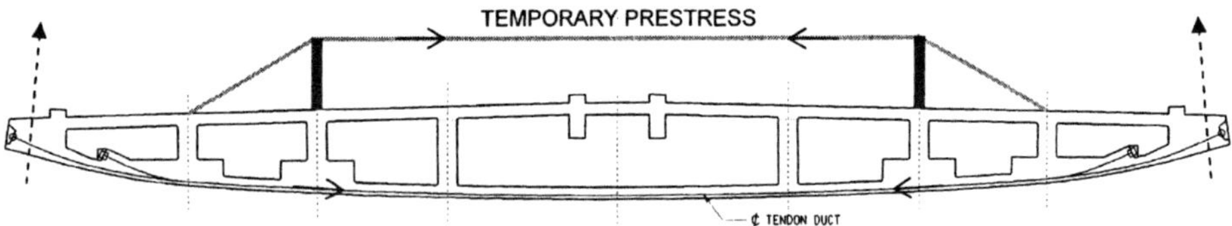

Fig. 5 Schematic prestress of cross girders.

The resistance to the remaining torsion forces from permanent loads and from the dominant live load effects of buffeting wind on the main span is provided by using 60MPa concrete, and thickening the typically 250mm thick slabs to 500mm thick in the sections near the piers. This additional material also provides the required weight to give a reasonable balance in permanent vertical loads between the large main span and shorter back spans, thereby minimising the vertical bending forces that the back span decks must carry. However, the resulting cross sectional area requires a large amount of longitudinal prestress to carry the live load vertical bending effects. The prestressing tendons are mostly external which avoids a reduction in the section available to carry the torsion and shear stresses, and provides for the option of future replacement. Some internal tendons are also included where there is insufficient space within the girder for all the tendons to be external. Up to seventeen 6-37 tendons are required per girder in the end span, where the axial compression from the inclined stay cables is small. As the compression increases towards the tower, the requirement for prestress force reduces correspondingly.

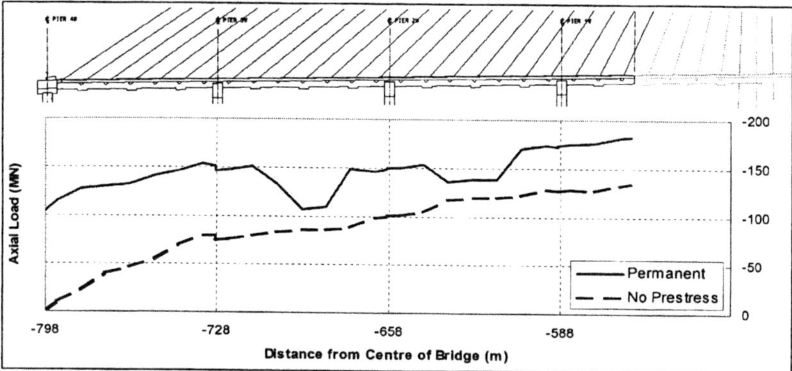

Fig. 6 Permanent Axial Load in Back Span Girder

The intermediate pier shafts are single columns on the bridge centre line. They are between 60 and 65m in height and sufficiently flexible in the longitudinal direction for bearings not to be required. They were constructed with 60MPa concrete using a hydraulic climbing form system from VSL. Four pockets were cast into the outer face at each pour to provide the support points for the climbform. With typical pour heights of 4m, a cycle time for construction of each lift of 6 days was achieved, with concrete finishing works undertaken from trailing platforms hanging below the main working platforms.

The monolithic connections with the deck cater have a complex geometry. Cantilever type cross heads are 9m deep at the root, and incorporate the curved deck soffit slab to maintain continuity for the deck axial compression. The pours were split into manageable sizes. A temporary works truss cantilevering from the pier shaft provided the support in the temporary condition before the concrete gained the required strength.

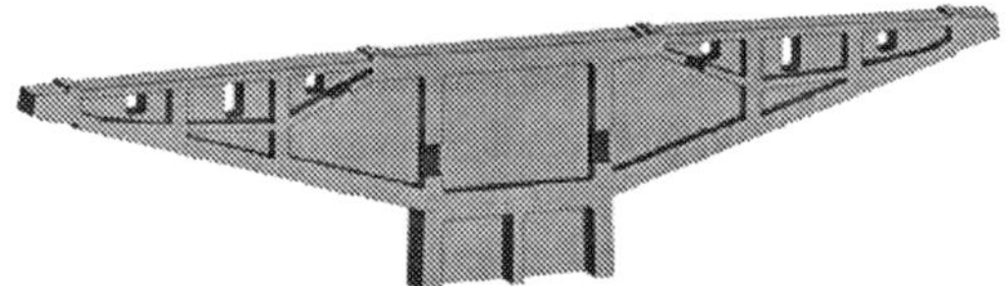

Fig. 7 Typical Cross Head – cut away section.

The end piers are portal frames to provide additional resistance for lateral stability and provide a suitable platform to accommodate the bearings for the adjacent viaduct structures. The piled foundations here are arranged in a single line rather than the more typical group of 3 × 2. This creates a more flexible framing in the longitudinal direction and thereby reduces the bending moments required to be carried due to thermal movements.

Deck Construction Sequence

In each span the three cross girders are cast first as independent units. Once the first stage of transverse prestressing is applied, the two longitudinal bays between these cross girders are constructed. After the remaining transverse prestressing, the final deck pours stitch the span concrete to the pier cross heads. When a continuous deck was formed, the longitudinal prestress was then applied, with stressing taking place at the ends of the deck where there is adequate access.

This sequence allows independent components of the deck to be constructed and adjusted to the correct geometry prior to forming an increasingly complex non-determinate structure.

Falsework System

The contractor elected to use precast columns on bored pile foundations to provide the required temporary support system which carries the weight of the concrete decks until the stay cables are installed. Two columns per cross girder up to 60m tall are spaced at 35.8m to suit the web locations. This system provides a relatively stiff support, which minimises the deflections that occur as the deck segments are cast, and also minimises uncertainty of those deflections. This reduces the adjustments required to correct the levels. The 50MPa concrete columns are hollow sections, 2.5m square, with wall thicknesses up to 250mm. Vertical prestressing bars are coupled together and stressed down onto the segments at each of three 20m high modules. Steel truss members brace the columns together in both the longitudinal and transverse directions, and plan bracing also is introduced at the top of each module to form structures rigid enough to carry the heavy deck even in the full typhoon condition.

Level adjustment is provided at the top of each temporary column, with four 500T jacks, used to move elements of the deck up or down as necessary, to ensure the correct alignment of the independent cross girders, prior to connecting them to each other and then to the pier cross heads.

Fig. 8 Back Span Falsework during erection

Cross girders

The 4m wide concrete cross girders each contain 10 internal prestressing tendons in the permanent works which pass from stay cable to stay cable through the bottom slab. To accommodate the anchorages at the edges of the deck and to avoid a clash with the stay cable anchorage, the tendons fan out. Therefore, in this region, an 8m wide section is cast as part of the cross girder construction.

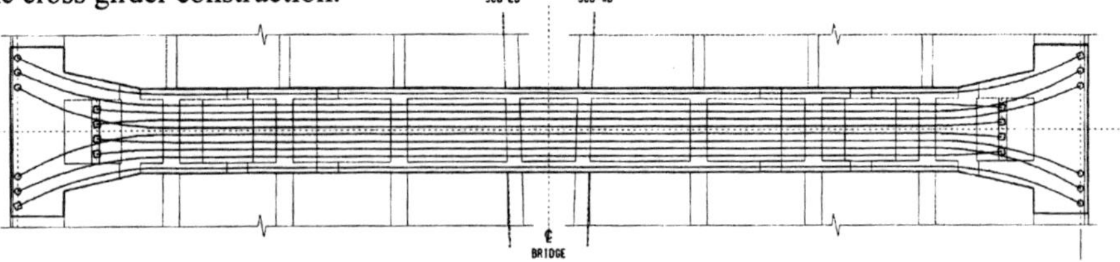

Fig. 9 Transverse Prestress at Cross Girder

Temporary support during casting was provided by stiff steel trusses which span the 35.8m between the falsework columns, and wing trusses for the areas outboard of the columns. The area on top of the trusses was decked out, with scaffolding on top of this forming the curved shape of the soffit. A two stage pour – firstly the bottom slab and webs, then the top slab – minimised the additional stresses locked into the girders due to the weight of the second pour being partially carried by the first stage.

The contractor modified the envisaged arrangement of temporary transverse prestress slightly to give better clearance for access along the top of the deck. The pairs of horizontal cables, providing a force of 5MN, are located 4.5m above the top slab. Additional transverse prestress tendons are provided in the region above the falsework columns to counteract the hogging moment.

Longitudinal Girders

Steel falsework trusses were hung from the constructed cross girders and cross heads to support casting the longitudinal box girders in each bay. Similar decking as used for the cross girders was placed on these trusses to form the working platform, and then scaffolding was provided to form curved soffit shape. Due to the larger pours and more complex shapes, a three staged pour is used – bottom slab first, then webs, lastly the top slab – with careful evaluation of the locked-in stresses this causes to ensure that they can be accommodated by the design.

Longitudinal prestressing is a combination of internal and external tendons of varying lengths, running from the end of the bridge into the girder, or from near the steel-concrete deck interface. In this way, the areas where stressing must take place are limited, and handling operations for the 2¼T jack for the 6-37 tendons are limited. Small temporary holes were left in the top slab to allow easier access for stressing the tendons which cross the steel-concrete interface rather than transporting the jack along the inside of the deck.

Steel Deck

The 53.3m wide steel bridge deck consists of twin box girders connected by cross girders. The total length of the steel superstructure is 1117.5m.

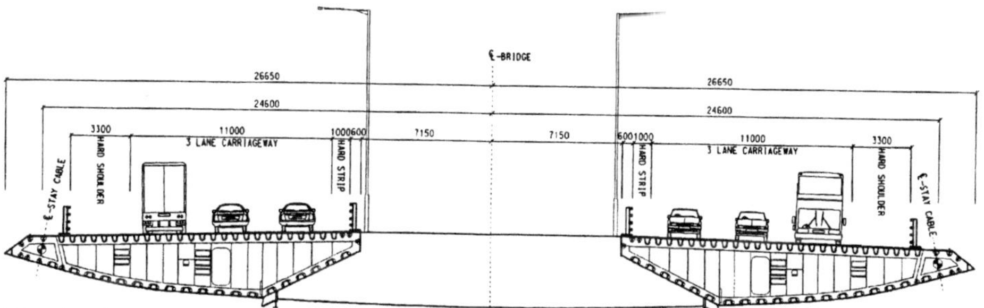

Fig. 10 Steel superstructure - cross section in the main span.

For the longitudinal girders the maximum plate thickness applied is 40mm. Central webs are introduced in the back spans to provide continuity with the central webs in the concrete deck. For the cross girders the maximum plate thickness is 50mm (adjacent to the towers). The internal surfaces of the steel superstructure are corrosion protected by dehumidification. The quantity of structural steel amounts to 32,000 tonnes in total.

Steel grade S420 M and S420 ML have been specified. Whilst normalised steel has been used for the previous major steel bridges in Hong Kong, thermo-mechanical steel has been specified for Stonecutters Bridge due to the increased weldability associated with lower alloying content, which is an important advantage considering the thickness of plates requiring full penetration butt welding.

Live Load Effects

The typhoon wind loading has governed the design, and in particular the dynamic response of the structure to turbulent wind was found to be a significant loading. The dominance of the typhoon wind loading on the steel deck is illustrated in Figure 11, which shows Ultimate Limit State vertical bending moment envelopes in one of the longitudinal girders.

For global highway loading the moment envelope is as expected for a typical cable-stayed bridge with sagging moment greatest in the middle of the main span and lesser hogging moment lobes to either side of the centre point. As would be expected the mean wind (0° incidence angle) has little effect in the vertical direction. However, when the effect of buffeting response is considered the load effect is dramatically altered with both the sagging and the hogging moment governed by the buffeting response.

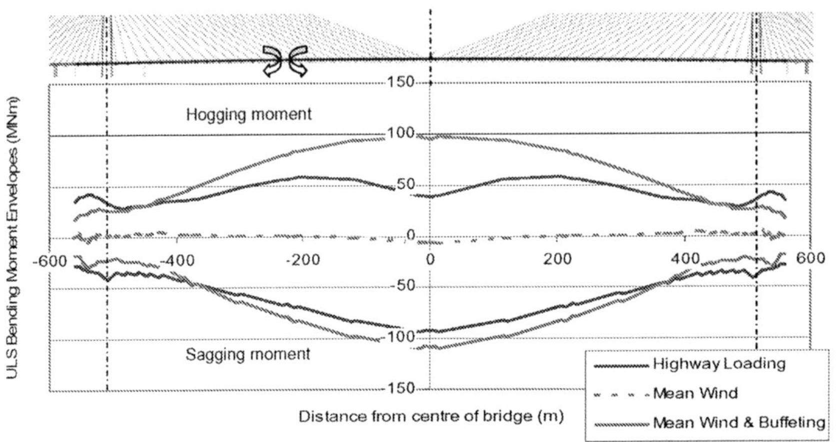

Fig. 11   Ultimate Limit State vertical bending moment envelopes per girder.

The typhoon loading also leads to in-plane bending of the deck. This is combined with an axial compression due to the stay forces, which increases towards the towers. The governing stress state is compression, which is most severe towards the outer edges of the heavily stiffened girders. The cross sectional area increases by approximately 70% from mid span towards the towers (see Table 3).

The cross girders are subject to biaxial bending due to transverse bending of the deck between the stay anchorages and in-plane vierendeel action of the twin box girder. The maximum bending occurs in the cross girders located between the towers and quarter span.

Location	Area (m^2)	Moment of inertia, horizontal axis (m^4)	Moment of inertia, vertical axis (m^4)
Mid span	1.04	1.41	30.1
At tower	1.78	2.60	53.3

Table 3 Section properties of steel girder (one girder).

Fatigue Design of Orthotropic Steel Deck

The bridge is located in a sub-tropical climate with afternoon summer time temperatures frequently above 30°C. The reduction in stiffness of asphalt surfacing at high temperatures means that the benefit of the surfacing in acting compositely with the deck plate to reduce local stresses is limited.

Furthermore, the traffic crossing the bridge is expected to contain an unusually high number of heavy goods vehicles due to the bridge linking to the Kwai Chung Container Port. It is estimated that 42% of the total traffic will be heavy goods vehicles. The resulting commercial traffic is compared with typical values from other sources in Table 4.

Location	Eurocode BS EN 1991 Part 3	Hong Kong Structures Design Manual	Stonecutters Bridge
Slow lane	$0.5 - 2.0 \times 10^6$	2.0×10^6	3.9×10^6
Adjacent lane	-	1.5×10^6	2.9×10^6

Table 4 Annual flow of commercial traffic.

To cope with this intense fatigue loading, without beneficial composite action with the surfacing, the orthotropic steel deck has been designed with an 18mm thick deck plate and 325mm deep, 9mm thick trough stiffeners. The distance between the diaphragms is 3.8m in general, and 2.8m at the cross girders. This results in a relatively stiff top flange, which is also advantageous in terms of an extended lifetime of the 60mm thick mastic asphalt surfacing.

Verification

The steel deck verification has been based upon a two-staged analysis. The first stage involved a space frame global analysis from which sectional forces were found at critical sections. The corresponding stresses are then checked against the relevant clauses of BS 5400, Part 3. The longitudinal girders are designed as compression members with longitudinal stiffeners whereas the cross girders are designed as beams, also with longitudinal stiffeners.

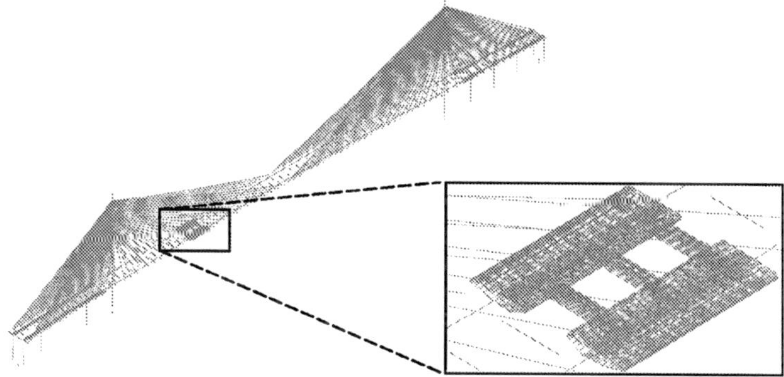

Fig. 12 Semi-local model of steel deck included in global model

The second stage involved a confirmatory semi-local model of the steel deck to investigate in detail the connection between the longitudinal girders and the cross girders and also to investigate shear lag effects associated with the introduction of the stay force towards the outer edge of the deck.

The traditional approach applied during a detailed design is to select a limited part of the structure, for instance a single erection section, and make a separate detailed model. However, some problems arise from this approach, especially regarding boundary conditions and loading on the detailed model. In order to avoid such uncertainties and to take advantage of the fact that all loads are generally already included in the global beam model, the semi-local model has been included in the global model of Stonecutters Bridge.

Fabrication

Steel deck panels are fabricated at a facility in North Eastern China. The plates are cut to shape and stiffeners formed and welded on to create sub-assemblies such as deck plates, bottom panels, diaphragms and cross girder panels. These are shipped to a yard in Guangdong province, Southern China, where assembly takes place on two production lines. Match fabrication to ensure a consistent cross section shape and the correct segment alignment is crucial to ensure site welding the segments together in Hong Kong proceeds without problems.

Erection of Deck around Towers

A 4000T heavy lift scheme erects the deck which is over land. An 88m length of deck formed from 6 sections of each of the longitudinal boxes was welded at ground level and strand jacked into place. The jacking operation uses a large bracket arrangement on the tower to carry most of the load, with additional load being carried on the tip of the concrete deck.

Main Span Erection

18m long segments will be lifted from dynamic positioning barges as the main span deck cantilevers out over the Rambler Channel. A 200m square work zone, and an 8 hour window for lifting operations will ensure minimum disruption to shipping. Local jacking to match the face geometry of the new segment supported on lifting strands to the existing geometry of the cantilever, supported at its outer edges by the stay cables, will be required to allow the welding to take place.

Stay Cables

Prefabricated PWS stay cables are supplying by Nippon Steel with Ø7mm galvanised wires and extruded outer HDPE sheathing. The tensile strength of the wires is 1770Mpa and allowable stress at service load is 769MPa. Stay cable outer diameters vary from 113mm (163 wires) near the towers to 192mm (499 wires) towards the end of the back spans. The longest cable is 540m long and weighs about 70 tonnes. Stressing will take place at deck level, with the jacking equipment transported along on an underslung gantry. Temporary measures to control cable vibrations will be used until installation of the internal hydraulic dampers to achieve the required level of 4% damping.

Conclusion

Stonecutters Bridge was the first cable-stayed bridge with a span over 1km for which detailed design was completed. This presented many challenges, including defining project specific design loadings and criteria to extend the standard codes of practice. Dynamic effects of turbulent wind dominated the design.

Construction has been underway since April 2004, and the Contractor has overcome many challenges to build the high level concrete back spans. Erection of the steel deck segments, fabricated in China, will continue to narrow the gap between the towers, with completion of the bridge scheduled for early 2009.

Fig. 13 Stonecutters Bridge – a new icon for Hong Kong in 2009

Acknowledgement

This paper is published with permission from Highways Department, HKSAR Government.

References

[1]   KITE S, FALBE-HANSEN K., VEJRUM T., HUSSAIN N., PAPPIN J., "Stonecutters Bridge – Design for Extreme Events", *IABSE Symposium Lisbon 2005: Structures and Extreme Events*, IABSE Report Volume 90, 2005.

Stonecutters Bridge
Design and Construction of the Towers

Jonathan CHEUNG, Senior Engineer, Arup, Hong Kong SAR, China
Peter THOMPSON, Associate Director, Arup, Hong Kong SAR, China

Abstract

Stonecutters Bridge in Hong Kong, across Rambler Channel at the entrance to Kwai Chung Container Port, will become one of the longest cable-stayed bridges in the world. Total length of the bridge is 1596m. The most prominent feature of the bridge is the two single tapered cylindrical towers each 295m tall. These support a high level road deck providing 73.5m navigation headroom to the busy container port. The towers stand on shore. The 1018m main span is in steel, while the back spans on each side are in concrete.

This paper describes the development from the initial design for the towers at Reference Scheme to the final configuration, and the detailed design and construction of the final tower including its foundation.

Introduction

Stonecutters Bridge is part of Route 8, providing a strategic route between the Hong Kong International Airport and the Eastern part of New Territories. The bridge will be highly visible from the harbour and the urban areas. Because of its prominent location, Highways Department of Hong Kong SAR decided to build a landmark structure and to conduct an international design competition in 2000. The wining bridge concept of the competition was selected as the Reference Scheme (RS) for further design development [1].

The contract for the detailed design and supervision for the bridge was awarded to Arup and Cowi in March 2002. Construction of the bridge by the Maeda-Hitachi-Yokogawa-Hsin Chong JV started in April 2004, and is anticipated to be complete in early 2009.

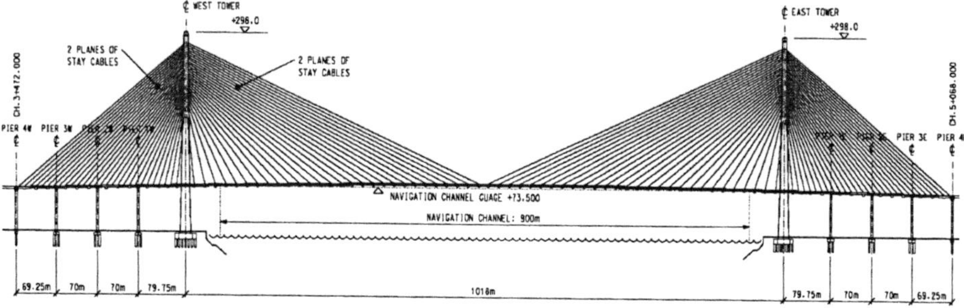

Fig. 1 Elevation and Plan of Stonecutters Bridge

One of the main features of the Reference Scheme bridge is the adoption of the 295m tall mono-column towers with the upper part of metallic look. Major design and construction aspects for the Stonecutters Bridge towers include:-

Bridge design, construction and maintenance 2007, Thomas Telford, London

i.   Technical review of the RS towers;
ii. Design and construction of the lower concrete part of the final tower;
iii. Design of the composite structure for the upper part of the final tower; and
iv. Fabrication and erection of stainless steel skins and cable anchor boxes.

Reference Scheme Tower

The RS tower is of reinforced concrete for the 167m lower part and all structural steel for the 115m upper part. Before commencing detailed design, a technical review was carried out to assess structural performance, constructability and durability of the RS towers. The review proposed two significant modifications.

Composite structure for the upper tower

A circular tower is known to be susceptible to vortex shedding vibration. In service, the peak amplitude of the oscillation at the top of the RS tower was estimated to be 0.45m. The 1st lateral frequency of the RS tower coincided with the natural frequencies of the stay cables, hence excitation of cables due to linear resonance was possible. As a result, the upper part of the tower was changed to a composite structure of stainless steel skin, reinforced concrete and steel cable anchor boxes at the core. The greater mass, stiffness and damping of the composite tower improved its response to the vortex shedding, and thereby avoiding the risk of the cable excitation.

Articulated joint at Tower

A monolithic connection between the tower and the decks, as proposed in the Reference Scheme, was abandoned and a joint introduced, as shown in figure 2. Longitudinal hydraulic buffers are not only designed to minimise the effects of the long term loadings (i.e. traffic loads, temperature, creep and shrinkage, etc.) on the tower by allowing the movement of the decks relative to the tower, but also are used to restrain the deck movement due to the short term loadings such as seismic and buffeting wind. Bearings provide lateral support to the decks. Advantages of introducing the joint are identified below.

Fig. 2 Articulated Joint

i.   Reduction of restraining forces on the tower due to temperature effects;
ii. Reduction of torsion in the tower during full cantilevering and in-service stages; and
iii. Elimination of a "Hard Point" in the deck so that the combination of high bi-axial bending and compression is reduced.

Final tower configuration

The general configuration of the finalised tower is shown in figure 3. The lower part of the tower is in concrete from the top of the pile cap to the level +175mPD. The tower section tapers linearly from an elongated circular section of 24m x 18m at the base to a true circular section of diameter 10.909m at the level +175mPD. The wall thickness reduces from 2m to 1.4m over this lower concrete part. At the level +77.75mPD, there is a diaphragm slab of 2.5m thick. This receives the longitudinal and transverse horizontal loads imposed by the deck via the hydraulic buffers and the lateral bearings.

The upper part of the tower is a composite concrete-steel structure from the level +175mPD to +293mPD. The circular tower section tapers linearly from 10.909m to 7.159m over the upper part. The external stainless steel skin is 20mm thick. The concrete wall thickness decreases from 1.4m to 0.8m. There are 32 stainless steel skin segments and 25 cable anchor boxes. The lowermost 3 sets of stay cables are anchored at the concrete corbels while the remaining 25 sets are installed at steel anchorages of the anchor boxes. A pendulum mass damper installed at the level +293mPD, is designed to mitigate the vibration of the tower in the service condition. A 5m tall glazing structure of diameter 7.2m with aesthetic lighting tops off each tower, making 295m tall structures which will be amongst the tallest bridge towers in the world.

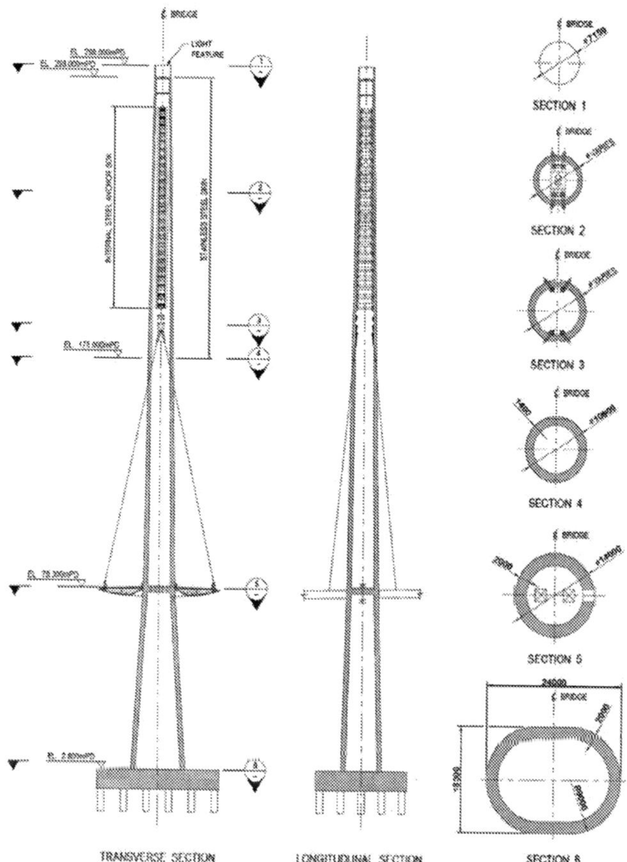

Fig. 3 General Arrangement of Tower

Foundation

Piles

Detailed ground investigations and geophysical survey were undertaken to develop a geological model, as shown in figure 4. The ground conditions were found to include sand fill over alluvium with thin layers of marine deposits. The alluvium is typically underlain by Saprolites before reaching the bedrock.

Due to high loading, piled foundations are adopted. Tower foundations consist of a large pile cap and 29 end-bearing bored piles founded on rock. The pile is 2.8m shaft diameter and 4.2m bell-out at the base, and the pile lengths vary from 60m to 110m. Concrete strength for the pile is 45MPa. Bearing capacity of the pile is provided by a combination of end bearing and socket stress within the bedrock. Negative skin friction loads resulting from potential liquefaction and long term settlement of the newly reclaimed land have been considered in the pile design. In accordance with the performance requirements, the piles are designed to behave elastically under the application of seismic loading effects.

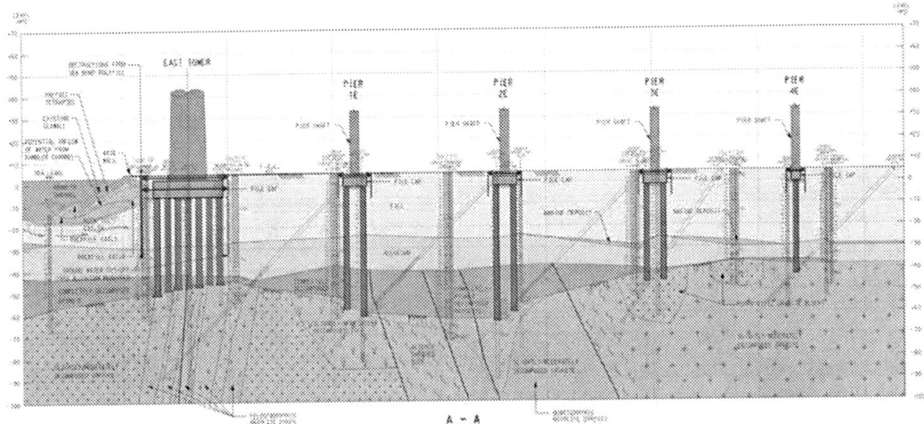

Fig. 4 Geological Model

The piles were constructed by the conventional method of using temporary casing, oscillator, reverse circulating drill and expanding bit to form the bell out. The main challenges for the piling works were to meet the specified pile verticality and to extract the temporary casings, for the piles deeper than 80m. Consequently, telescoping casing method together with left-in permanent liners were adopted to overcome the problems for construting such deep piles.

Tower Pile Cap

Dimensions of the tower pile cap are 47.4m × 36.4m × 8m deep and concrete strength is 45MPa. The finite element software "Strand 7" was used for analysis. Due to the flexibility of the cap, inner piles near the tower received higher loads than outer piles. Figure 5 shows the details of the finite element model where part of 18m tall tower is modelled by brick elements; the pile cap is modelled by 8 nodal shell elements; each pile is modelled by a beam element and its top 2m part is modelled by brick elements to provide area support.

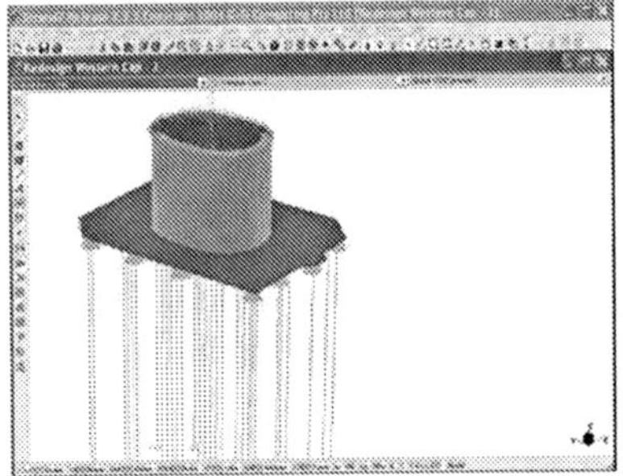

Fig. 5 FEM for Tower Pile cap

Dewatering system and massive concrete pouring are the most important issues for constructing a large pile cap in the permeable sand adjacent to the sea. A sheet pile cofferdam with deployment of 200mm diameter pumps inside the cofferdam was used. An automatic monitoring system was established to monitor the ground water level at the specified locations on a 24-hour basis. This monitoring ensured that there was no excessive settlement induced on the nearby structures.

Fig. 6 Concreting Works

Total concrete volume for the pile cap is approximately 12,000m³. Such large concrete pouring requires careful considerations on the control of temperature. A combination of the use of 60% ground granulated blastfurnace slag (GGBS), typically 1m thick individual concrete pours to form the whole cap, addition of flaked ice in the concrete and insulation to concrete surface, was successful to control the maximum allowable temperature below 80°C and the maximum differential temperature less than 30°C. It was noted that the early tensile strength of the concrete containing GGBS is low and that some shrinkage cracks were found at the top surface of each concrete pour. Additional reinforcement bars was placed in the intermediate layers of the cap to prevent the formation of such cracks.

Lower Tower

Due to high slenderness, a second order analysis was performed to determine the amplification to be applied to the tower forces from the global analysis. The amplification fully accounts for the material non-linearity effect.

Second order effects are more critical in the transverse direction, because higher restraint is provided by the stay cables in the longitudinal direction. The governing design load case is the permanent dead load plus transverse wind loads from the

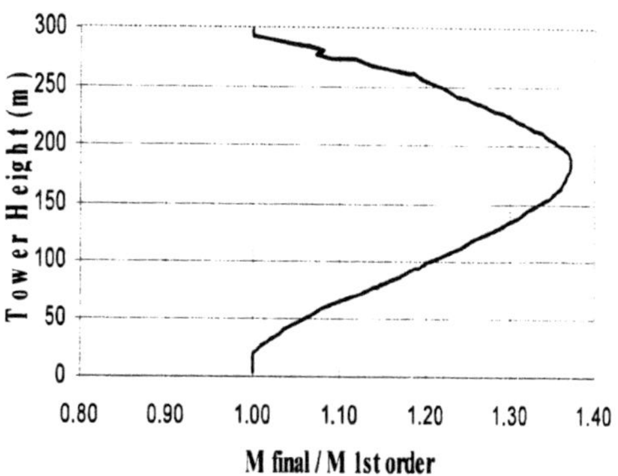

Fig. 7 Second Order Effects

south without traffic loads, under the ultimate limit state. The resultant forces obtained from the summation of the first and second order effects were used for the reinforcement design of the towers. Figure 7 gives the amplification factors applied to the first order effects. The peak amplification of almost 40% occurs around +175mPD.

A semi-local finite element model was developed to check the stability of the cracked tower. The result showed that a softened tower attracts less forces resulting from the wind effects. Lateral wind loads are shed away from the tower to the deck. Tower instability is therefore not occurred. Also, the model was used to assess the effects of initial imperfections and concluded that the tower is still structurally adequate.

Figure 8 shows typical reinforcement arrangement for the lower part of the tower. The tower is of 60MPa in compressive strength, and φ50mm high yield reinforcement (460MPa) is used. Reinforcement ratio varies from 2% at the tower base to 4% at +175mPD. The outer horizontal links (32mmφ) resists both shear and torsion while the inner horizontal links (25mmφ) only resists shear.

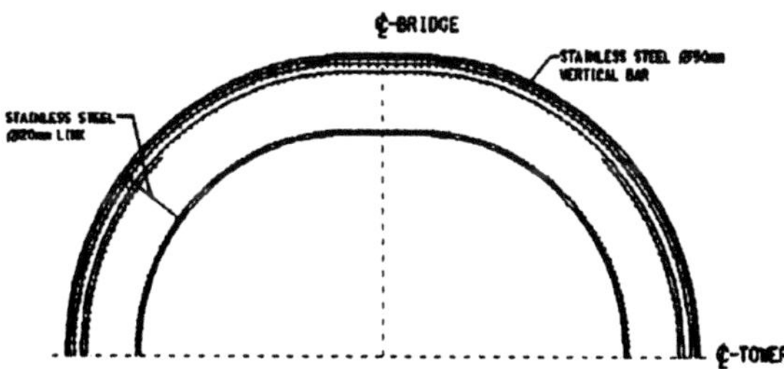

Fig. 8 Lower Tower Reinforcement

To ensure durability of the tower for the 120 years design life, not only was Grade 60 concrete used with a water/cement ratio 0.35 employing 35% PFA and 8% micro-silica for reducing the permeability, but also the concrete cover was increased to 60mm and the design crack width set to not exceed 0.25mm for good chloride resistance. In addition, the outermost reinforcement and links are grade 1.4301 stainless steel from the tower base to the level +175mPD.

Climbing Formwork System

Fig. 9 Lower Tower Construction

The lower part of the tower was cast in 44 construction lifts. Each lift is approximately 4m in height. A climbing formwork system as shown in figure 9 was used to form the complex tower shape. There are three major components in the formwork system: a steel framing structure with 10 pairs of screw jacks providing working platforms necessary for rebar fixing, formwork erection and dismantle, concrete pouring, curing and silane application; a shutter formwork consisting of outer 24mm thick plywood and inner steel mould; and a trailing platform hanging below the steel framing structure. The form was raised by 4m for every lift using the screw jacks during the climbing operation and then supported on 10 sets of shear cones which are inserted in the previous cast concrete section. The external plywood is flexible to be adjusted to suit the tower geometry at different levels and is durable for the repeated concreting operation. Cycle time for the construction of a typical tower lift is about 7 days.

UPPER TOWER

Stainless Steel Skin

A stainless steel skin forms the external face of the composite tower between the levels +175mPD and +293mPD, which is considered to be a visually sensitive area. At the start of the detailed design, a study was undertaken to identify a grade of stainless steel and determine an appropriate surface finish for the stainless skin. Also, a prototype was produced representing a section of the skin to examine fabrication feasibility and investigate any detailing problems, as shown in figure 10.

Fig. 10 Prototype

Austenitic and duplex grade of stainless steels had been under consideration for the skin. Duplex stainless steel was found to be superior, in terms of strength and corrosion resistance. The austenitic stainless steel is relatively more susceptible to the staining and pitting corrosion in an aggressive marine environment. Typical austenitic stainless steel has a low yield strength (~220MPa). Therefore, the grade 1.4462 duplex stainless steel to EN 10088 with yield strength of 500MPa, was selected.

Fig. 11 Shot-Peened Finish

The skin surface needs finishing treatment to give a slightly textured non-directional appearance. This produced a uniform low reflective surface with good dirt shedding ability. In the prototype, the plates of the polished 1K finish received two stages finishing process to achieve the desired surface finish. The surface was blasted with aluminium oxide and then shot-peened with a glass bead media to achieve a uniform surface roughness in a range of 1 to 1.25 microns. The contractor proposed an alternative one-stage process in which the skin was shot-peened with the mixture of aluminium oxide and glass beads to achieve the same surface finish. The resulting finish can be seen in figure 11.

Composite Design

Composite action amongst out stainless steel skin, the steel cable anchor box at the core and the reinforced concrete, contributes to the structural capacity of the upper tower section. Shear studs of ϕ16mm up to 300mm long, are utilised to transfer loads between the concrete and the skin or the anchor box. The studs are distributed evenly at a spacing of 300mm approximately so that the skin plates and the anchor box plates can reach their yield strength before local buckling.

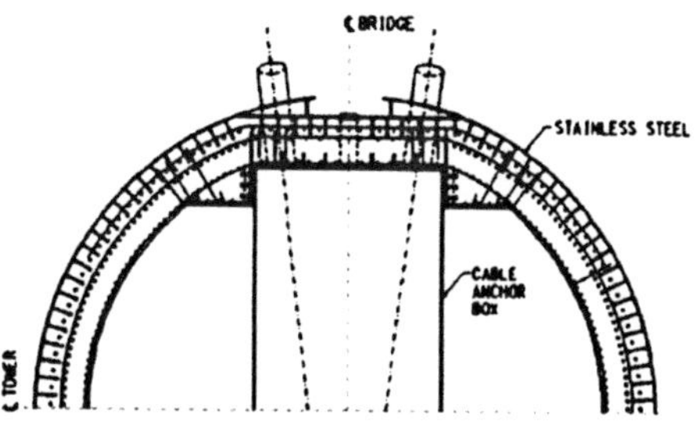

Fig. 12 Half section of upper tower

The horizontal bolted splice between the stainless skin segments or the anchor box segments is designed to resist compression only. Therefore, the skin and the anchor box are treated as a compression element in deriving the tower sectional capacity and their tensile strength is not taken to control the crack width of the concrete section. The skin is designed like the horizontal reinforcement to resist the shear and torsion for the tower section above the level +220mPD where the skin is not fully utilized in sustaining vertical bending moments and axial forces. Webs and flanges of the anchor box make no contribution to resist the shear and torsion. As the upper tower is primarily in compression, vertical reinforcement does not yield in tension for the critical loading cases and is actually utilised to enhance the compressive capacity of the tower section. Vertical reinforcement ratio decreases linearly from 3.6% at +175mPD to 1.2% at +293mPD.

To enable the lifting operation of the skin be more manageable, each skin segment is fabricated in two halves. High strength friction grip vertical bolted splice joins the two halves. ϕ22mm duplex steel bolts and vertical duplex steel splice plates form the connection. The contact surface of the connecting plates requires treatment to ensure that the friction factor is greater than 0.2. The bolts are preloaded to achieve the shank tension of 165KN. It is designed that there is no slip of the plates under the serviceability limit state, but the connection is allowed to slip and then remaining applied loads are resisted by the bearing of the bolts under the ultimate limit state.

Stay Cable Anchorages

There are two anchoring types for the stay cables in the upper tower. The lowest three sets of stay cables are anchored at the concrete corbels (figure 13) from the level +175mPD to +195.65mPD, while the remaining stay cables are anchored at the steel anchorages of the internal anchor boxes (figure 14) from the level +195.65mPD to +280.5mPD.

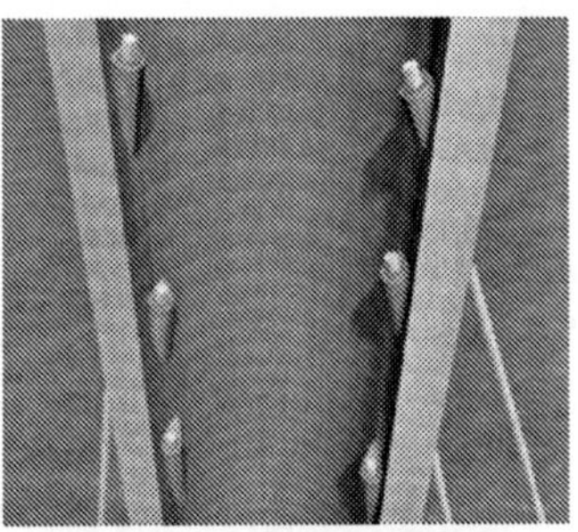

Fig. 13 Concrete Corbel

At the corbels, the angle of the stay cables is close to vertical so that the induced splitting force can be accommodated by the reinforced concrete. The cable force is transmitted to the tower wall, through a 70mm thick anchor plate and a ribbed steel guide pipe embedded in the concrete. Flat bar rings welded to the pipe transfer the cable force in shear to the surrounding concrete. Reinforcement for the concrete corbel is detailed as a prestressing blister.

In the anchor boxes, the stay cable anchorage acts as a beam to transfer the cable force in shear to the side walls. The anchorage comprises anchorage webs, anchor plates and parts of the front wall used as a tension flange. The vertical component of the cable force is transferred to the concrete walls via shear studs.

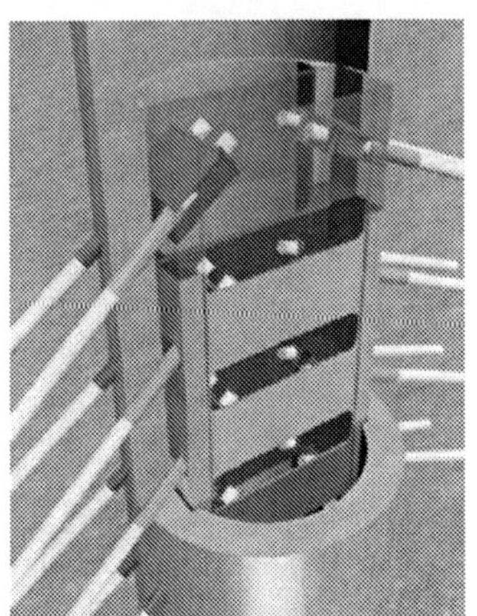

Fig. 14 Cable Anchor Box

Because of the access openings in the side walls of the anchor box, discrete tension ties are formed to resist the opposing components of horizontal cable stay forces. Approximate 90% of the balanced horizontal cable force is taken by the tension tie, while the remaining 10% is carried through the concrete section. The unbalanced horizontal cable force is delivered in shear to the concrete through the shear studs on the 600mm wide strips of the side walls cast in the concrete. Shear studs on these strips carry approximately 2/3 of the vertical cable force into the concrete wall. Shear studs on the front wall of the anchor box carry the remaining 1/3 of the vertical cable force.

Steelwork Fabrication

Stainless Steel Skins

Grade 1.4462 stainless steel plates with 1K polished surface finish were supplied by Outokumpu from Sweden to the fabrication yard at Zhongshan, in Guangdong province, China.

The skin segment height varies from 5.6m to 3.2m. The heaviest half segment is approximate 27 tons. Each segment is fabricated in 2 halves from 20mm thick plates, with 25mm thick stiffening flanges, and 25mm thick intermediate stiffening rings. Vertical bolted splices join the halves, and horizontal bolted splice connect segments at the flanges. For the typical half segments, 4 flat plates are connected using full strength butt welds and then the joined plate is rolled into the correct shape. However, for the first 4 skin segments which are of 5.6m in height, the plates were rolled to the correct curvature before welding due to the width limitations of the bending roller machine. Matching fabrication of two successive skin segments was carried out to ensure correct levels, alignment, orientation and radii of the matching segments.

The geometry of the upper tower is controlled by the geometry of the steel skin and anchor box. The relative geometry of the skin and anchor box is therefore important and must be controlled carefully. Therefore, a rolling trial assembly of 3 sets of stainless steel skin segments and anchor box segments are carried out in the fabrication yard (figure 15) to ensure the precise alignment of stay cable anchorages and guide pipes, and to correct cumulative errors. The top segment from each trial assembly is carried forward to the next trial assembly. This fit-up process is repeated for all segments.

Fig. 15 Rolling Assembly of 3 Segments

Anchor Box

Steel plates up to 75mm thick are used to fabricate the cable anchor box segments. Segment heights vary from 4.35m to 2.9m, widths from 3m to 2m, and lengths taper from 7.5m to 5.6m with the increasing elevation. The heaviest segment is about 20 tons. Careful planning of the assembly sequence along with control and correction of weld distortions were important aspects of fabricating such large scale and complex structural steelworks.

Fig. 16 Fabricated Anchor Box

Matching fabrication process for the skin is adopted in manufacturing the anchor box segments. Trimming of a 15mm green area at the top of each segment and inserting of shim plates at the splice joint, are used to correct any misalignment due to the weld shrinkage and distortion.

Construction Sequence

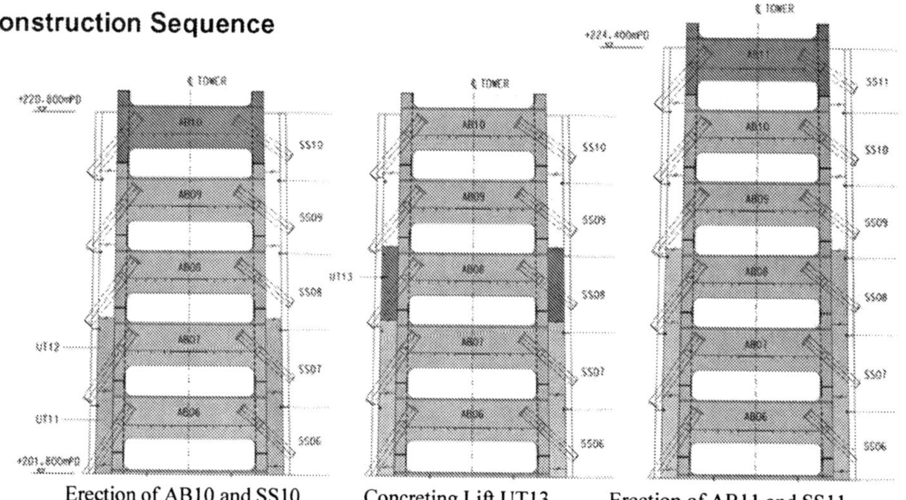

Fig. 17 Construction Sequence for Upper Tower

Figure 17 demonstrates the construction sequence for the upper tower construction. The skin and anchor boxes are used as permanent formwork for the concreting. Erection of the skin and anchor box segments advances two full cycles ahead of the concreting. This sequence facilitates the assembly of the segments on site with the interlock between the skin segments, to minimize the segment deformation due to the concreting works. Reinforcement fixing particularly for fitting the bars through the shear studs of both the skin and anchor box segments, is a relatively difficult operation because of small working area available.

Site erection of the tower steelwork on top of the lower tower reproduces the same geometry as achieved in the trial assembly. Adjustment to the prefabricated segments is not possible on site. Therefore, verticality of the first 2 skins was crucial in determining the alignment of the remainder of the upper tower. Extensive surveys were carried out to ensure the correct positioning of the first 2 skins, prior to the first concrete pour.

The large sail area makes the half skin prone to the wind effects. There was concern over controlling the behavior of the skin when the lifting operations take place in windy conditions. Several lifting trials were undertaken to verify that the skin was under control when the gust wind speed was below 40km/hr. Erection of the anchor box segments is much easier as its shape suffers from relatively less wind effects.

Fig. 18 Erection of 1st Stainless Steel Skin

Conclusion

The elegant tower is the main aesthetic feature of the Stonecutters Bridge, which will be an attraction in Hong Kong. In an attempt to respect the intent and appearance of the reference scheme as selected in the international design competition, innovative design ideas and advanced construction methods have been exploited to create an aesthetic pleasing and structural efficiency tower structure.

Acknowledgement

This paper is submitted with the permission of the Highways Department, the Government of the Hong Kong Special Administrative Region.

References

[1] FALBE-HANSEN, K.,HAUGE, L. and WONG, C., 2006, Stonecutters Bridge – International Design Competition and Reference Scheme Reviewing, International Conference on Bridge Engineering – Challenges in the 21st Century, Hong Kong SAR 2006, Proceedings.

Advancements in Long Span Bridge Construction: the Stonecutters Bridge Experience

M. J. Tapley, Chief Engineer MHYH JV, Hong Kong
B. W. West, Project Manager MHYH JV, Hong Kong
Dr. S. H. R. Sham, Executive Director, Maunsell AECOM, Hong Kong

Abstract
Stonecutters Bridge will be a landmark cable stayed bridge crossing the entrance to the Hong Kong container terminal. The construction of the bridge has been far from straightforward, demanding tough decisions and innovative solutions. The construction of the bridge can be divided into roughly five areas:
- Construction of the concrete backspans;
- Fabrication of the deck and tower steelwork;
- Construction of the two towers;
- Lifting of the steel backspan decks; and
- Main span erection and stay cable installation.

This paper gives an overview of the construction of the bridge with particular focus on areas where the erection process has moved beyond usual construction practice.

Introduction

Figure 1: Stonecutters Bridge

Stonecutters Bridge is one of a new generation of long span cable stayed bridges. With a main span of 1018m and towers 298m high, the bridge has presented joint venture Maeda – Hitachi – Yokogawa – Hsin Chong (JV), with many challenges. With Maunsell AECOM supporting the JV as consultant for the construction engineering, the JV has adopted robust and innovative solutions to the problems faced.

Bridge design, construction and maintenance 2007, Thomas Telford, London

Figure 1 shows a rendering of the bridge at completion. The main span and part of the backspans comprise twin orthotropic steel deck girders connected by cross girders at 18m centres. The remainder of the backspan decks is prestressed concrete cast monolithically with the supporting piers. The towers take the form of single tapering hollow concrete sections. The upper tower steel anchor boxes at the core act compositely with the concrete and outer stainless steel skins.

Deck Erection Analysis and Wind Tunnel Testing

Bridge geometry control methods and procedures have been developed by Maunsell, for use in a prediction, survey, re-analysis and possible adjustment cycle in the bridge erection. The overall objectives are to ensure that the target geometry of the completed structure is achieved and that the structure has adequate strength and performance at all construction stages.

Thee framework for bridge geometry consists of a co-ordinated set of activities in the construction planning phase, the fabrication phase and the erection phase. All activities in these three phases are robustly integrated to support the prediction, survey, re-analysis and possible adjustment cycle. Erection analyses underpin many of the activities in the framework and some details are given in this paper.

The erection analyses and structural verification of the erection stages have been supported by a full programme of wind tunnel testing of the proposed construction sequence and methods. Maunsell advises JV in bridge aerodynamics and has been responsible for the planning, management and supervision of the wind tunnel investigations for Stonecutters Bridge construction.

The emphasis of the Contractor's wind tunnel testing is on the behaviour of the bridge structure during the erection stages, notably its response to vortex shedding, buffeting and divergent amplitude effects.

The principal aspects of the wind tunnel testing included deck section model testing, aeroelastic freestanding tower model testing on different erection heights, and aeroelastic bridge model testing to investigate the bridge erection stages. Furthermore wind tunnel tests on stay cables have been carried out to determine cable drag and to examine susceptibility to rain-wind induced oscillations. The wind tunnel testing has also been augmented by Maunsell's computational buffeting analyses.

Typhoon events are critical for the construction of Stonecutters Bridge with major activities planned to take place in typhoon windows. Contingencies are also set in place for overruns and in case typhoons develop rapidly.

Construction of the Concrete Backspans

The concrete backspans comprise four piers cast monolithically with an in-situ deck. Without stay cables the concrete decks are unable to support themselves and therefore it has been necessary to construct a falsework system which acts both as a platform for casting the decks and also as support for the deck until such time as the stay cables have been installed. Figure 2a shows an elevation of the falsework system and Figure 2b an aerial view of the falsework during deck construction. The 60m-high falsework system is one of the most substantial ground-supported systems erected to this height.

The concrete backspans comprise the following key components:
1)   Pier shafts: each pier (approximately 60m high) has been constructed by a self climbing jump form in lift heights of 4m;
2)   Pier crossheads: multi-stage concrete pours on large trusses cantilevering from the piers;
3)   Deck cross-girders: constructed on transverse trusses supported on precast concrete temporary towers.   Once complete the load from the girders is transferred to elastomeric bearings which are able to accommodate movements of the structure during the remainder of construction.
4)   Deck longitudinal girders: longitudinal trusses suspended by hangers from the previously constructed cross girders. Initially the intermediate decks are cast and then subsequently the "stitch" decks connecting to the permanent piers are cast.

Top Figure 2a: Elevation of Falsework System.
Bottom Figure 2b: Aerial View of the Backspan Falsework System following Construction of the Cross Girders

The designer for Stonecutters Bridge, Arup, envisaged the concrete deck would be supported by a falsework system combined with computer linked jacks enabling compensation for any differential settlement of supports. Rather than provide such a jacking arrangement the JV has founded the falsework on in situ concrete end bearing piles and precast concrete towers. In this manner deflection of the falsework has been restricted to the relatively small and predictable elastic shortening of the piles and towers. Although provision has also been made for a simple system of compensation jacking, it has not been deemed necessary as the falsework has behaved very much as anticipated and the deck levels have satisfied targets.

Pulverized-fuel ash (pfa) has been used in the concrete mix to give a concrete that meets strict chloride resistance requirements, produces low heat of hydration and exhibits good early strength gain. The same mix design has been adopted for both the concrete backspans and the towers. Uniform appearance and colour have also been a requirement of the Engineer. These features have been achieved by adopting a common mix design throughout and using a single source of aggregate.

One unusual feature on Stonecutters Bridge has been the use of bowstring PT mounted 4.5m above each of the cross girders. The design requires the concrete deck in the temporary state to be in a similar condition to that after stay cable installation. Hence a PT system has been provided which forces the cross girders into sagging as if the deck were supported by the stays at its edges. The PT is stressed prior to the casting of the "stitch" deck when the decks become monolithic with the permanent piers.

Fabrication and Testing of Stay Cables

The parallel wire stay cable strand (PWS) system has been adopted for Stonecutters Bridge. PWS cables minimize the diameter and thus the blockage to wind and for this reason were the preferred choice of the designer. Nippon Steel are supplying the cables which are being manufactured at Jiangyin China.

Considerable testing has been required on the cables prior to the commencement of manufacture. Initial testing was carried out in the wind tunnel to establish their susceptibility to wind-rain induced vibration. To minimize this effect the cables have a dimpled surface texture which then required further testing to verify the drag coefficient.

Tests have also been carried out to prove the tensile strength and the fatigue resistance on the cables. This programme of tests has been carried out by CTL near Chicago USA, which is the only test facility able to handle cables of the size being used on the bridge. CTL also performed a watertightness test of a full-size cable and socket assembly in accordance with the CIP Recommendations of 2002. The arrangement for the fatigue test is shown in Figure 3.

Figure 3: Stay Cable Fatigue Test Rig

Following the successful completion of the test programme manufacture of the cables has commenced at Jiangyin with cable installation starting Summer 2007.

Fabrication of Deck and Tower Steelwork

Fabrication of the steelwork has taken place at three locations in China:

- Fabrication of deck panels   Shanhaiguan, Northern China
- Assembly of deck girder segments   Dongguan, Pearl River Delta
- Fabrication of tower anchor boxes and skins   Zhongshan, Pearl River Delta

It has been necessary to use the three different sites as each fabrication yard has particular skills which could not be found at any single site.

The segments for the deck, the tower anchor boxes and the tower stainless steel skins are being carefully match fabricated to ensure the geometry will be within tolerance following erection on site. The deck segments are fabricated on two production lines of seven or eight segments with trial assemblies being carried out on three segments at a time. Once the trial assembly is complete the segments are sent to the paint shop prior to dispatch to Hong Kong.

Stainless steel skins and anchor boxes are fabricated separately in pairs. Once the match fabrication is complete, trial assemblies are carried out where three anchor boxes and three stainless steel skins are combined. Only at this stage are the anchor tubes welded to the stainless steel skins to ensure compatibility with the anchor boxes. It is critical in this process that the skins and anchor boxes fit perfectly together, if the trial assembly cannot be repeated on site the geometry of the upper tower will soon become impossible to achieve. Figure 4a shows the match fabrication of two stainless steel skins.

Once the trial assembly of the skins and anchor boxes is complete the skins are sent for shot peening. If the skins were polished stainless steel the upper tower would be susceptible to dazzling reflection in bright sunshine. To avoid this effect the duplex stainless steel skins are shot peened - the application of light abrasive to give a non-reflective finish. Achieving uniform colour and even appearance has required selection of suitable abrasive materials and careful workmanship.

Construction of the Towers

Pre-drilling prior to the commencement of the piling work on the West Tower identified a fault line where the bedrock was considerably deeper than originally anticipated. Even with the relocation of some piles it was still necessary to excavate piles over 100m in depth. Although not unique, piles of such length are unusual and particular care was required during the excavation and concreting processes.

The construction of the pile caps for the towers required the construction of a large cofferdam 50 x 39 x 10m deep. With this excavation immediately behind the seawall extensive dewatering was required using up to 20 no. pumps. The pile caps were cast in seven lifts.

Above Figure 4a: Fabrication of Stainless Steel Skins
Right Figure 4b : East Tower with brackets for the backspan steel deck heavy lift operation

Construction of the lower towers (see Figure 4b) has been carried out by means of a conventional jump form system. The tapered shape of the tower did however require modification of the jump form after every lift, with the timber forms being cut and reshaped to achieve a curved surface without obvious lines between formwork panels. The jump form comprises a self contained 'bird cage' incorporating all access platforms necessary for rebar fixing, formwork fixing and striking, concrete pouring, curing and the application of the Silane protection coating. At level +175m, the outer part of the jumpform was removed as the stainless skins from this point onwards act as the outer form of the concrete. Concrete operations are by means of a skip lifted by tower crane; the height of the tower has prohibited the use of concrete pumps. The concrete is placed evenly in layers around the circumference of the tower. Although the tower wall is up to 2m thick the rise in temperature of the concrete following placement has not been excessive primarily due to the use of pfa in the concrete mix design.

The erection of the first stainless steel skin presented a considerable challenge. The large surface area made it susceptible to the effects of wind: careful control was required during its lift to 170m above ground level. Once installed the skin had to positioned very accurately as any errors at this stage would be compounded for the remainder of the upper tower. Consequently hydraulic jacks were used to level the skin to an accuracy of ±1mm. With skins and anchor boxes being match fabricated there should not be need to carry out further adjustment to the top of the tower.

Erection of the Steel Backspans

The JV originally considered using a large scale falsework system to erect the steel deck segments for the backspan. To withstand the strong typhoon winds in Hong Kong the size of the steel members became considerable, which combined with the world shortage of steel, made the cost of the system excessive. As a consequence the JV formed an alliance with VSL to develop a heavy lift scheme using strand jacks mounted on the already completed tower and concrete backspan. In this alliance both parties share in any cost savings due to changes in methods and temporary works. Maunsell is primarily responsible for the construction engineering aspects of the Heavy Lift Scheme, including geometry control and structural verification of the permanent works during the heavy lift operation.

The back span steel deck segments are delivered by barge to purpose built jetties immediately in front of the two towers and then off-loaded by a 54m long unloading frame attached to the tower. The segments are fixed to sledges and moved along a series of rails by hydraulic jacks to the correct location for assembly. The longitudinal deck segments are then welded together at ground level. Once the welding of the longitudinal deck girders is complete the heavy lift operation can be carried out.

The sequence for the heavy lift is as follows:
- The two longitudinal girders are lifted simultaneously (2000t) each by strand bundles from brackets on the tower and backspan concrete deck (see figures 5a and 5b).
- Hydraulic jacks on the deck and on the backspan lifting frame are used to horizontally shift the deck girders to their final position. It is not possible to lift directly into position due to the tapering shape of the towers.
- Cross girders are lifted from ground level to position by strand jacks, fixed by keeper plates and welded to the longitudinal girders.
- The stitch between the steel and concrete decks is cast, with props to eliminate movement during the operation.
- The permanent lateral bearings and hydraulic buffers between the deck and tower are activated.
- Stay cables are installed, initially only taking part of the load. Once all the backspan steel deck stay cables are installed the strand jacks are de-stressed, thus lowering the deck to final position and effectively stressing the stay cables to full load.

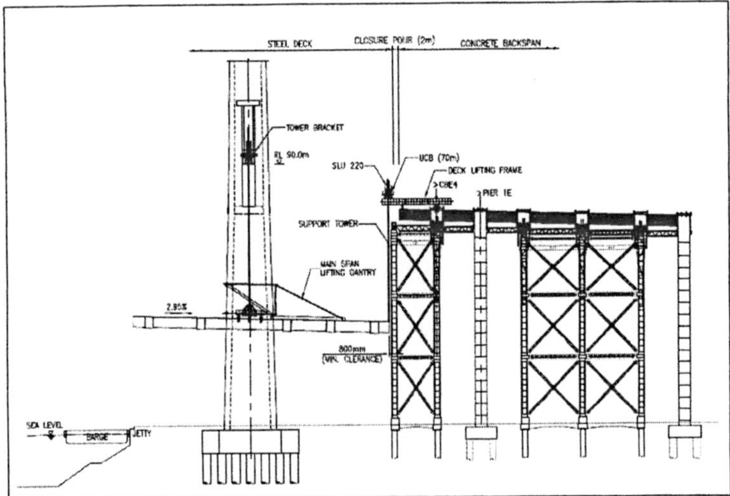

Figure 5a: Elevation of Heavy Lift Operation

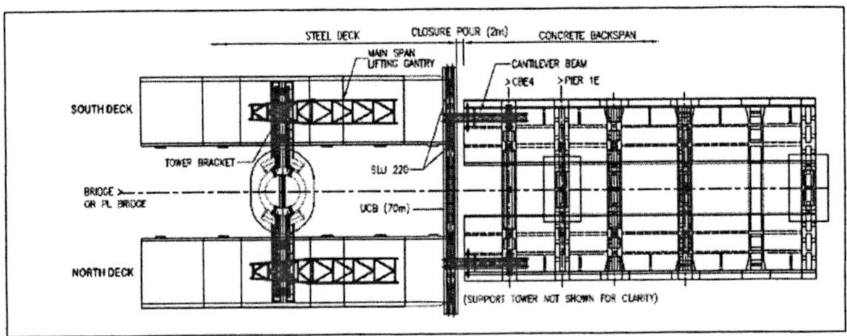

Figure 5b: Plan of Heavy Lift Operation

In developing the scheme one of the most significant criteria has been the need to design the system to withstand typhoon loading. Although the period during which the deck is supported by strand is only two months it is during the typhoon season. It has therefore been necessary to prepare contingency measures for approaching typhoons. Ties and struts are provided as connections to the backspan concrete decks and temporary lateral bearings restrict movements at the towers.

Erection of the Main Span and Installation of Stay Cables

In determining the construction methods the erection of the main span for Stonecutters Bridge the JV has sought to minimize the impact on the shipping lanes passing beneath the bridge. With Hong Kong container terminals operating 24 hours a day and ships continually passing beneath the bridge, it is a clear requirement that there shall be no delays to shipping.

The segments are delivered by barge and lifted to position by a gantry at the extremity of the deck cantilever. Figure 6 shows the arrangement.

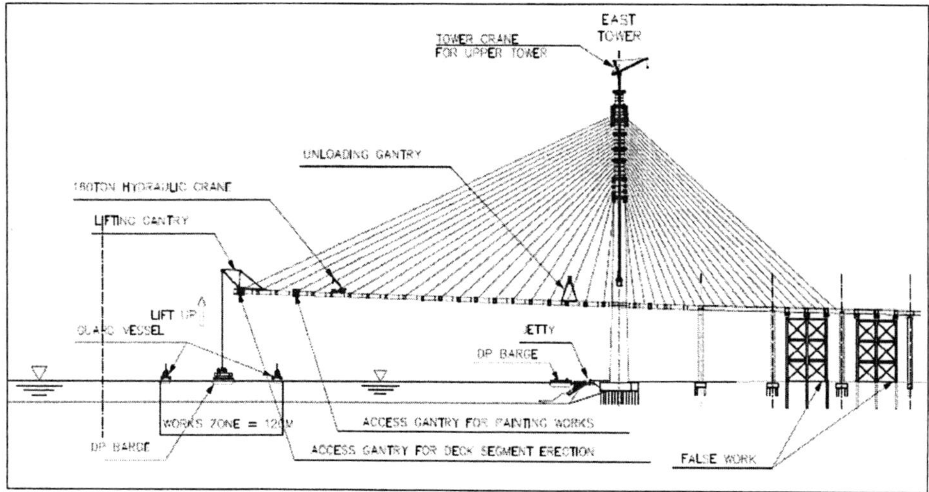

Figure 6: Main Span Segment Erection

In order to minimize the time of impact to the channel a dynamically positioned (DP) barge is used to deliver the segment. Using GPS the barge is able to position itself accurately and then using its DP thrusters it is able to hold position for the duration of the deck lifting operation. The lifting gantry uses winches rather than strand jacks, again as a means to reducing the time for the lift. A deck lift should usually take one hour from the time the lifting equipment is attached to the segment.

One aspect of the lift operation which has required careful consideration has been matching the geometry of the lifted segment with that of the already erected cantilever. The end of the cantilever is deformed transversely by the weight of the lifting gantry whereas the segment being lifted has no such deformation. In order to avoid this geometry mismatch the JV forces the lifted segment to the same geometry as the cantilever end by means of applying temporary transverse external prestress. The prestress is placed externally as a bow-string above the segment thus causing the cross-girder to deflect in the same manner as previously erected segment. The arrangement for PT on the lifted segment is shown in Figure 7. Minor adjustments are applied to the lifted segment by means of small hydraulic jacks.

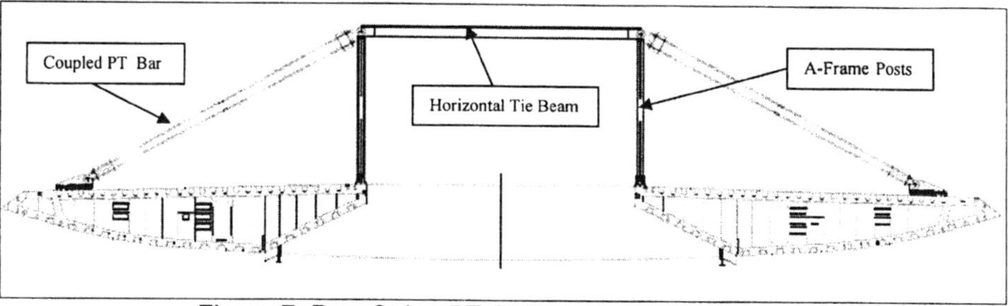

Figure 7: Bow String PT System for Lifted Segment

As stated above the PWS system has been adopted for the stay cables which consequently means that the cables arrive on site pre-fabricated. The longest cables are 540m long, thus presenting a major challenge in terms of lifting and handling. The tower cranes used for the construction of the towers have insufficient capacity to lift the tower anchorage socket of the longer cables and as such an additional erection crane has to be installed on top of the tower. On deck the cables are extended to the cable anchorage by means of strand jacks.

Construction Progress

At the end of April 2007 construction is progressing well. The East Backspan concrete deck is complete and the West Backspan nearing completion. The East Tower is nearing 200m high with the first three levels of stainless steel skins erected. The West Tower is about 50m lower, with the first installation of skins in late Spring 2007. The Heavy Lift of the East Backspan steel deck is imminent, with the welding of the deck segments at ground level almost complete in preparation for the lifting operation.

Main span closure of the bridge is currently planned for late Summer 2008 with overall completion following in early 2009.

Summary

Stonecutters Bridge has many unique features due to its prominent location, proximity to the Container Terminal and its span length. In order to construct the bridge it has been necessary to develop methods and technology accordingly. The concrete backspans have required the erection of a massive falsework structure, while erection of the backspan steel decks above land require a 4000t heavy lift. With main span erection soon to commence the project is nearing completion, adding one more to Hong Kong's list of prestigious bridges.

Acknowledgements

This paper is submitted with the permission of Highways Department, the Government of the Hong Kong Special Administrative Region.

Theme one:

Lessons that can be learnt from the past

Arnodin and Structural Durability

B.R. Mawson   Cardiff School of Engineering, Cardiff University.
R.J. Lark   Cardiff School of Engineering, Cardiff University.

Abstract

On September the 12th 1906, the Newport Transporter Bridge was opened to the public. Taking the form of a 200 metre steel, suspension bridge structure carrying a moving aerial ferry, it carries up to six cars at a time across the River Usk in Newport South Wales. The writers were involved in the structures £3.5m restoration to full working order in 1995 and during the work, were grateful for the foresight shown by the designer, the French Engineer Ferdinand Arnodin, for the future maintenance and ensured longevity of the 'bridge'.

Arnodin was a successful designer of suspension bridges in France through the latter part of the 19th century. He was concerned to extend the life of bridges as much as possible and published several treatise on the replaceability of components within a structure. This he put into practice in his designs and in particular the bridge at Newport, one of his later structures.

This paper discusses the way his work greatly assisted the restoration of the Newport Transporter Bridge and highlights the lessons that can be learnt from such an approach. However, it also identifies how current knowledge and experience in relation to the detailing required to minimise corrosion and aspects relating to the behaviour and longevity of the cables would have enabled his design to be improved. Such observations highlight measures that can be taken to maintain and prolong the life of cable structures.

Introduction

The Newport Transporter Bridge has a clear span of 196.6m. and is described in detail in a comprehensive article published in The Engineer to coincide with the bridge's opening in 1906 (ref. 1). Essentially an aerial ferry, it is one of only seven believed still to be in existence and is unusual because of the hybrid suspension/cable-stayed arrangement supporting the high-level beam from which the ferry or 'gondola' is suspended (Fig. 1).

The gondola, measuring 10m x 12.2m, is suspended by vertical and raking cables from a 32m-long travelling frame. (It is interesting to note that, because of its French design, all dimensions for the bridge were specified in metric units.) The frame runs on pairs of rails attached to the bottom flanges of the rail-bearing girders, and these girders form part of the stiffening boom which, at a clear height of 50m above the waterway, gave sufficient clearance for the passage of the high-mast sailing ships that commonly plied the River Usk when the bridge was constructed. The boom, comprising two, 236m-long, 4m-deep parallel trusses,

Bridge design, construction and maintenance 2007, Thomas Telford, London

was designed to distribute the load as evenly as possible to the main suspension system that supports it.

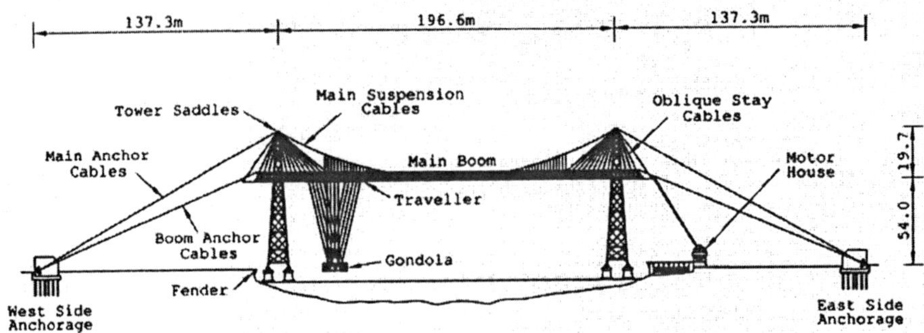

Fig. 1   General arrangement elevation

Vertical members are laced double channels, and diagonal bracing in the vertical plane is achieved by means of cables connected to U-bolts secured around pins at the panel points (Fig 2). The same pins connect the boom girder to the suspension system by vertical wrought-iron hangers over the central portion of the span and by the oblique stay-cables near the towers. The top chords of the stiffening girder also have spring devices at approximately the centre and quarter points, and although the purpose of these unusual devices is not immediately obvious, as will be discussed later, their significance should not be underestimated.

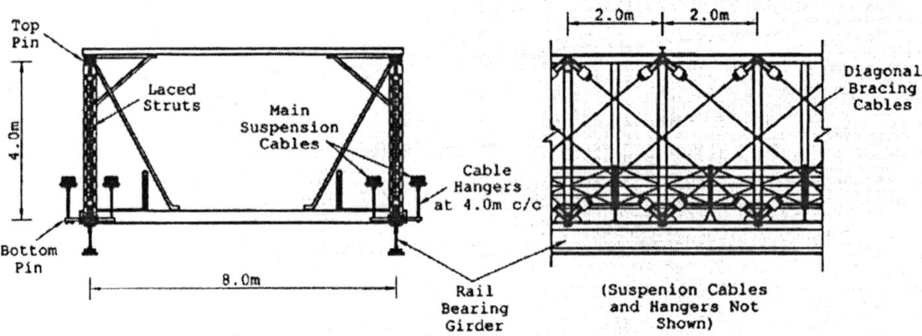

Fig. 2   Details of stiffening girders

The main suspension and anchor cables are grouped in two pairs of four on each tower leg (i.e. 16 no. total), and they are finished at their ends with cast steel sockets. These are connected to the tower saddles by large, 'U'- shaped shackles and to the anchorages by steel bolts which are seated on granite blocks incorporated at the rear of the masonry anchorages. More correctly described as spiral strand, the 200m-long, 68mm diameter, main-span cables are composed of 127 no. 5mm diameter wires in a 36/30/24/18/12/6/1 configuration in section, while the similarly grouped, 70mm diameter, anchor span cables are composed of 5.28mm wire and are 157m long. The results of tensile tests performed on wires from a sample of the original cable indicated that its tensile strength was of the order of 65-70ton/in2 (1000-1080N/mm2). Quality was generally inferior to the most basic grade of wire that would be used today, but the strand did compare well with the original specification which stipulated that all wires should be of the same size and with a minimum tensile strength of

63.5ton/in2. On removal from the bridge, however, the strands were found to contain not only corroded but also broken wires and evidence of fatigue damage, possibly because, although ultimate wire strength is a carefully controlled variable related directly to cable strength, wire fatigue strength is highly variable and sensitive to both environmental and processing defects. At the top of the tower legs, each tower saddle is supported on what have been described as 'expansion carriages' composed of two 2.75 x 1.2 x 0.02m thick steel plates separated by nine no. 80mm diameter steel rollers. Even the original designer, Ferdinand Arnodin, commented that these carriages would be of limited value because of the configuration of the stay cables and, given their badly corroded and seized state prior to refurbishment, it seems probable that they have experienced little, if any, movement since construction.

Fig. 3   General view from south

Following its opening, the bridge operated from dawn to dusk every day (Fig 3), with Sunday mornings being reserved for routine maintenance, but, although tolls were charged, the bridge never paid its way. Maintenance therefore became increasingly difficult to finance and, with the bridge's declining use following the opening of the new George Street Bridge in 1964, its future was a matter for much local debate.

Broken wires on the cables were first noticed in the late 1950s, and in 1961 Sir Alfred Pugsley was called in to inspect them. He advised replacement of the boom anchor cables (eventually undertaken in 1969) and commented on the slackening of the main cables. The oblique stay-cables were replaced in 1979, and in 1981 the structure saw its 75th anniversary celebrated in style. Following this it was given historic building status Grade II*, (UK classifications are III, II, II*, I, the latter being the highest). By 1985, however, so many wire breakages were apparent that the integrity of the bridge could no longer be assured. Gwent County Council (the Highway Authority at the time) was therefore left with no option but to close it while attempts were made to quantify what capacity remained and to consider further actions. The investigations into condition, the repairs needed and the way those repairs were carried out is fully reported by Lark et al. (ref. 4) and Mawson and Lark (Ref. 5).

Once funding had been sought and found, the £3.5million restoration project was undertaken in three main phases commencing in 1993 and completed for a re-opening in December 1995.

The main work comprised:

- Phase I: Extensive repairs to the towers, access stairs, and walkways.
- Phase II: Replacement of the main suspension and anchor cables, with refurbishment of the anchorages.
- Phase III: Repairs to the main boom, refurbishment of the saddles, and various minor works, including refurbishment of the gondola itself.

This sequence of work was decided by following the structure's strength from its foundations upwards. After considerable analysis, the decision was taken to, wherever possible, restore the ninety year old structure to a condition approaching that as originally built in 1906. Fortunately, there existed a comprehensive archive for the bridge, held mostly in the local museum. This included the designer's drawings and calculations, together with specifications, design commentaries and various other documents prepared by Arnodin, which effectively represented a maintenance manual for the bridge. During the restoration, extensive reference was made to this archive and the accompanying progress photographs. The unusual nature of this historic structure and the difficulties posed by its restoration led the authors to research the works of Arnodin and his contemporaries, and this generated a tremendous respect for those Engineers who were undoubtedly working close to the edge of their engineering knowledge.

Ferdinand Arnodin, the designer

Ferdinand Arnodin was born in 1845 and lived in Chateau Neuf sur Loire, where his father worked for Marc Seguin and his brothers. He studied at Orleans and then Paris, before joining his father as an inspector in the Companie Seguin.

His experience with the Seguin brothers and the recent knowledge of suspension bridge failures of the 1850s, which had subsequently led to a moratorium on suspension bridge construction in Europe, encouraged him to keenly promote the comeback of suspension bridges for major river crossings in the latter part of the 19th century. He developed and used a combination of a suspended and cable stayed support to the deck which enabled a simplified analysis of the bridge suspension. This combination is similar to that used by Roebling in America. In 1872, Arnodin set up his own business with a view to reviving suspension bridge construction in France and developed his Systeme Arnodin. This was first employed on the Pont de Saint Ilpize (70 m main span) in 1879 and with it he pioneered the use of spirally wound steel cables arranged so that the central portion of the main span was hung from groups of parabolic cables while the outer portions were supported directly from the towers by oblique stay cables. This enabled him to feel much more confident about assuming that the main cables were carrying a uniformly distributed load.

In 1886 the theoretical basis of the system proposed by Arnodin was examined by Professor M. Levy, details of which are set out in his treatise *La Statique Graphique et ses Applications aux Constructions*.(ref. 2&3). Initially, Arnodin used timber for the decks of his structures but later turned to steel, experimenting with various forms. The boom or stiffening girder used at Newport is a double-intersection Whipple truss.

Arnodin's involvement with the more mechanically complex transporter bridge form came through collaboration with the Spanish engineer and Architect, Alberto del Palacio. In 1888, Palacio had been preparing a scheme to bridge the river Nervion at Portugalete near Bilbao. The navigation requirements led him to propose a transporter bridge for the site. He considered that a suspension system for the high level beams would be the most economical, but his experience was in heavy steel and masonry. So he enquired around and was advised to consult Ferdinand Arnodin in France as one of the then leading experts in suspension bridge design. This he did and they worked together on the design, which exists today.

As a businessman, Arnodin was always concerned to promote his firm and it was probably through such advertising that R.Haynes, the Borough Engineer of Newport heard of this work and that led to the appointment of Haynes and Arnodin as Joint Engineers for the Newport Transporter Bridge in 1901. In fact, the bulk of the design and drawings were produced in France at Arnodin's works in Chateau Neuf sur Loire. These are the drawings held in the Museum at Newport, but they are not accurate as the steel used was manufactured in the UK to imperial dimensions, which are generally slightly larger than the metric dimensions required.

Arnodin's Thesis

Among the documents held in the archive are two, which lead readers to appreciate the thinking and foresight of Arnodin the designer. They aid considerably understanding of the design approach and detailing present at Newport. Many of the features incorporated at Arnodin's insistence, directly aided the work of restoration some 90 years later!

One paper *("Note sur LES CABLES TEMOINS- systeme F.Arnodin" par M.F.Arnodin)* outlines the measurement of loads and stresses within a complex structure and proposes the use of 'witness' cables to measure cable forces. Here a small wire of identical material to the cable being measured is hung alongside the cable. Load is adjusted with a spring balance until the wire lies in the same vertical curve as the cable. The load reading then gives a direct means of calculating the cable load. At Newport, tests to prove the load carrying capacity of the bridge were carried out on the completed structure during the commissioning trials between 29 August and 5 September 1906. Supervised by Arnodin, they consisted of applying a series of test loads to the gondola while the stresses in the cables were then measured using this technique.

However, perhaps the most significant document is one entitled *'De l'amovibilite des pieces dans les ponts suspendus' (The removability of components in suspension bridges)*, in which Arnodin expounds the principle of replaceability and discusses at length the way structures of significant scale should be designed for maintenance and hence good durability. The various aspects of structural design and material specification are covered and these are now discussed with particular reference to the structure at Newport. Written in 1890, this document makes impressive reading today (Fig 4).

Fig. 4   Sketch and signature from Arnodin's paper

Structural layout

Arnodin stresses the advantage of simple or clear structural action which can be understood and interpreted giving good predictability of induced stress levels. He advocated the advantages of suspension systems with their direct tensile stresses over the more complex stress distribution found in plate girders.

At Newport, he used his well-developed combination suspension system, which functioned in a manner similar to that to used by Roebling in the USA.

To ensure good loaded distribution, he employed a long traveller to apply the load to the stiffening girder and suspension system; and adjusted the stiffness of the main stiffening girder to avoid full beam action. Thus at three positions in the girder's compression flange there are spring devices to limit the load in that member and thus the maximum moment, that can be carried by the stiffening girder. At Newport, these 'springs' compress and extend significantly as the load travels across the bridge, and this undoubtedly does lead to increased longitudinal movement at the towers, but an improved and controlled stress distribution in the stiffening girder and main cables.

Structural details

Arnodin continued the principle of load distribution predictability through to the details, at the same time as incorporating facilities for changing structural components. He recognised the desirability of undertaking such maintenance and renewal without taking the structure out of commission and so he considered the temporary stages of renewal and, again, the predictability of load/stress paths.

At Newport, the main cables are grouped in two sets of four at each side (Fig 5). This hanging system allows for the renewal and removal of one cable at a time, using a detail which re-balances the load and ensures that the capacity is temporarily reduced by a little more than one eighth. His document of 1890 contains an illustration of such a detail, but for 5 cables (Fig 4). A similar principle is applied to the details suspending the gondola.

Fig. 5 Cables and hangers in groups of 4

Materials and maintenance

Arnodin stressed the need to understand materials; not only for their initial physical properties, but also the way those properties would deteriorate with time. He divided the deterioration process into two types. In his texts he refers to one 'enemy' with and one 'enemy' without advance warning.

Firstly he discusses those that are readily visible and thus give due warning, for example, corrosion of metal which is expansive. This would be visible as corrosion products (rust) or, where internal, corrosion would cause visible swelling, especially in the cables. He was probably unaware of the non-expansive 'black rust' that can occur in steel with limited oxygen availability.

The second category of deterioration that he recognises is that which remains invisible. Arnodin refers to complex stress fields that change with load application and the development of molecular change with crystallisation within the metal. Probably what we know today as embrittlement and fatigue. But he was writing 120 years ago! This, he suggests, is less of a problem in trusses and suspension bridges, where stresses are simple axial tension or compression, than in the more complex arrangement of plate girders.

He considered it essential to recognise these differences in devising a maintenance programme for the longevity of a structure. Visible deterioration could be monitored and components changed as and when appropriate. As for the non-visible deterioration, items subject to this should be changed on a regular basis safely within their expected durability.

Assuming a component durability of 100 years, he proposed a maintenance schedule which required replacement of 10 per cent of the structure every 10 years, thus a rolling programme of maintenance could be established with predictable annual costs ensuring an indefinite life, or at least as long as those of the non-replaceable parts such as the masonry.

He admitted that, at the first 10 year maintenance cycle, the decision to discard components with a potential remaining life of 90 years would be politically difficult. He does also admit that this approach is for significant structures and would not be appropriate for small bridges.

While not specific to Newport, Arnodin's paper on maintenance is included in Newport's documents and is effectively a maintenance manual for the bridge. Facilities for changing components at Newport were incorporated in the suspension system and stiffening girder, but not in the towers. The permanent masonry and concrete works in the anchorages and foundations where obviously undertaken with great care, no doubt on strict instruction from Arnodin himself, recognising that under his approach, these items should control the life of the bridge.

There is no doubt that Arnodin's details greatly assisted with the cable replacement, which formed part of the recent work, although, the authors sought reassurance of the temporary redistribution of loads during cable changing by strain gauging. Perhaps we lacked Arnodin's confidence.

Inspectability and testing

Arnodin argued that inspectability was key to ensuring structural longevity. At Newport we see a number of features that facilitate this. Access stairways and walkways on both sides of the stiffening girder provide a high level crossing for pedestrians, but also assist with the visual inspection of a large proportion of the structure. Above this level, vertical ladders lead to a platform at the Tower Top. However, while the principle of this high level platform was sound, it failed to meet modern safety standards and had to be replaced during the restoration work.

As noted earlier, testing formed an important part of Arnodin's work, both for the quality of materials to be used, for which he wrote tight specifications, and for the performance of the structure. The full record of the commissioning tests and measurements are on file and signed by Arnodin himself. They provide a useful reference for comparing the behaviour of the structure today, with the inevitable redistribution of loading that has taken place within a cable structure of this type over its lifetime.

One interesting aside that is recorded in the commissioning notes is the confusion that arose over the use of metric measurements on the drawings for use in the UK where imperial measurements were still the order of the day. The rounding up from centimetres to inches lead to a considerable increase in the dead load and a consequential slight decrease in the live load capacity.

Restoration and structural deficiencies

Arnodin applied these principles of structural redundancy, replaceability and maintainability to Newport wherever he could. Thus components that were recognised as having a fixed life, such as the cables, were provided with facilities for replacement. Where components could not be replaced, it is very apparent that great care was taken, such as that with the masonry. However, Arnodin's hoped for maintenance schedule was never followed, possibly because all his papers are in French, but mainly because of costs. To operate the bridge requires a considerable workforce, the cost of which leaves little for maintenance. The argument for replacing part-used components, therefore fails.

Arnodin's provisions certainly helped considerably the work of restoration, but nearly half the cost was for work that he had not identified clearly. In particular, the towers with their form somewhat similar to the Eiffel Tower in Paris contained many details that form rust traps. With repainting access difficult, this is a problem.

Additionally, insufficient care was taken to ensure lines of action met correctly at intersections. This is a particular problem at the portal frame tower tops, where the 'St. Andrews Cross' bracing meets the legs. Severe fatigue cracking had developed and this had to be bridged by the addition of further bracing. In fairness to the designer, this problem was exaggerated by the switch from metric sections to imperial.

The view 100 years on

As the restoration work progressed, those involved, especially the authors, quickly began to appreciate the stature and foresight of Ferdinand Arnodin, a name largely unknown to British Engineers. He lived from 1845 to 1924, a period that saw great developments in the field of bridge engineering, and he was a significant contributor. As well as being an excellent designer, he seems ahead of his time in thinking about maintenance, durability and whole life costing. We are sure that he would not feel out of place today. A number of highway bridges in France and the Transporter Bridges at Matrou in France, Bilbao in Spain and Newport in the UK are his lasting monument.

References

1. The Transporter Bridge at Newport. *The Engineer*, 14 Sept, 1906, 263–265.

2. LEVY M., Mémoires sur le Calcul des Ponts Suspendus Rigides, 2nd edn., Annales de Ponts et Chaussés, Paris, 1886.

3. LEVY M. *La Statique Graphique et ses Applications aux Constructions—III Partie - Arcs. Ponts Suspendus* 3rd edn. , Corps de Révolution, Paris, 1918.

4. LARK, R. J., MAWSON B. R., SMITH A.K.,The refurbishment of Newport transporter bridge. *The Structural Engineer*, 1999, 77, No. 16, 15-21 August.

5. MAWSON B. R. and LARK, R. J., Newport Transporter Bridge- an historical perspective. *Civil Engineering*, Institution of Civil Engineers 2000, **138**, 40-48 February.

Newcastle High Level Bridge – Refurbishment and Strengthening of a Historic Road/Rail Bridge

Dr S G Davis, Mott MacDonald Ltd, Croydon, UK
T C Abbott, Mott MacDonald Ltd, Croydon, UK

Abstract

The High Level Bridge is a historic crossing of the River Tyne in Newcastle, UK, carrying both road and rail traffic. A principal inspection and structural assessment was carried out by Mott MacDonald in 1999. Since then major strengthening and refurbishment works have been designed and are currently being implemented. Works have included the complete removal of the original paint system and visual inspection of some 57,000m^2 of metalwork and repair of all critical defects found; complete refurbishment of the lower road deck; strengthening of cast iron rail deck cross girders and a detailed programme of cast iron testing and research into its fatigue performance and durability.

Figure 1: The High Level Bridge in its historic setting on the River Tyne

History

The famous 'High Level Bridge' has carried trains, road vehicles and pedestrians across the river Tyne between Gateshead and Newcastle in the UK for more than 155 years. The bridge was built as the final link in the London to Edinburgh Railway and being the only high level crossing of the river, incorporated a tolled road deck to accommodate the increasing number of road vehicles. The bridge was designed and constructed by Robert Stephenson and Thomas Harrison at the end of the cast iron era and was opened to rail traffic in 1849. Just two years earlier another Stephenson designed bridge, the Dee Railway Bridge, collapsed and there were grave concerns within the bridge world about the continued use of cast iron. Reports in the ICE Proceedings at the time highlight the great lengths that Stephenson and

Bridge design, construction and maintenance 2007, Thomas Telford, London

Harrison went to in order that the quality of the cast iron used on the High Level Bridge was optimised, this included careful control of the mix constituents, methods of casting through to the proof testing of each component. Notable failures of further cast iron bridges including the Tay Rail Bridge (1878), Inverythan Bridge (1882) and Norwood Junction (1891) lead to decisions to remove or strengthen all cast iron bridges subjected to live load bending. These failures have been attributed to the presence of major internal casting flaws. The client and asset owner, Network Rail still have a number of cast iron arch underbridges but these are working predominantly in compression, the High Level Bridge is a rare example of an underbridge supporting a live railway with cast iron members in tension bending. The bridge is now a 'Grade 1 Listed' structure in the UK, which recognises its place in engineering history and places strict controls on the works that can be carried out, in particular with respect to appearance and authenticity.

Structural Form

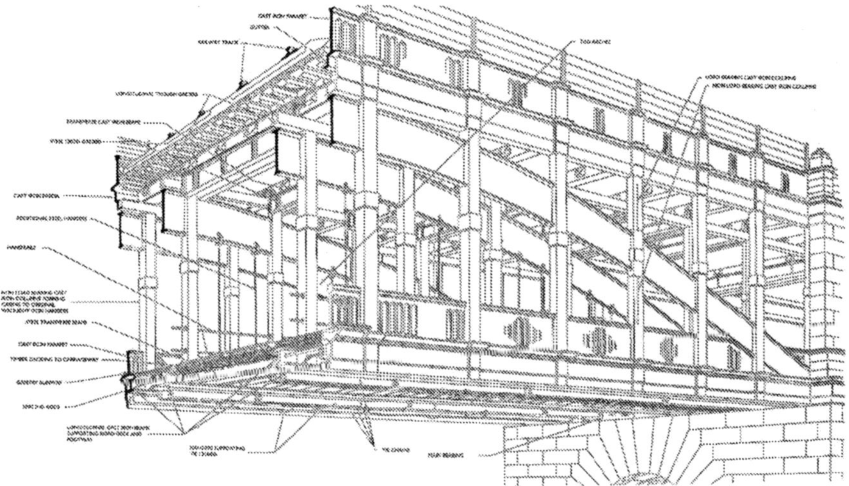

Figure 2: Complex structural form of the High Level Bridge

As illustrated in Figure 2, the bridge has a highly complex structural form, employing tied arches, tension hangers, compression struts and cross members, and includes cast iron, wrought iron, steel and timber components. Each of the six 39m main spans comprises four cast iron arch ribs restrained by wrought iron tie chains anchored at the bearings. The road deck is suspended beneath the arch ribs by wrought iron bars, which support cast iron longitudinal girders and cross girders, which were originally timber. The rail deck is carried by cast iron U trough girders which are supported above each arch rib by cast iron columns, timber waybeams supported in cast pockets in the cross girders then directly supported the rails of the three tracks. Including the approaches, there are a total of 136 cast iron rail deck cross girders across the 412m long bridge.

When the bridge was designed, locomotives of 25 tons were the norm but as the railway rapidly expanded loads soon increased to 40 tons. Initially the bridge with its three tracks carried an average of one train per hour. By 1890 locos weighed over 80 tons and there were 800 train movements per day. By the 1920's the 160-ton Pacific Class locos were common. Various modifications have been made to the bridge over its lifetime to accommodate these changes. The rail deck cross girders were strengthened in 1890 by the addition of steel 'swan neck' girders. In 1922 the road-deck was strengthened by replacing the timber cross beams with steel cross girders to take double-track electric tramcars, previously horse-drawn trams had used the bridge. In the 1950's the timber waybeams and rail decking were replaced by a heavier ballasted track and in 1990 the three train tracks were reduced to two.

Assessment

Although much research has been undertaken into the fatigue performance of steel, very little is understood about the fatigue performance of cast iron and the assessment codes treat cast iron with appropriate conservatism. An initial assessment of the cast iron cross girders highlighted that they were theoretically life expired/overstressed. A more detailed assessment adopting bridge specific real train loadings demonstrated a significant reduction in overstress, nevertheless the cross girders remained theoretically overstressed. A feasibility study was undertaken to determine what options were available. The following options were highlighted:

- Strengthening of individual cross girders
- Slewing of the two existing tracks
- Implementing a network change to limit the traffic using the bridge
- Testing of the cast iron to obtain improvements on the current code limits

Slewing of the tracks to reinstate the original design concept and implementation of a network change to reduce rail loads, both improved the situation but did not remove the theoretical overstress. The owner, Network Rail, decided to pursue both detailed design of a strengthening option and a research programme into the fatigue performance of the cast iron. As an interim measure load restrictions were placed on the bridge.

Strengthening Option

The cast iron rail deck cross girders were highlighted within the assessment as being below strength. There are a total of 136 cast iron rail deck cross girders within the bridge and a high percentage required strengthening. The chosen strengthening solution was an innovative post tensioning system which required steel fabricated anchorages to be bonded and dowelled onto/into the cast iron of the cross girders such that the strands could be tensioned to reduce the tensile bending stresses. The solution was however not cheap.

Cast Iron Testing

In parallel with the detailed design of the strengthening solution it was decided to undertake a study into the structure specific fatigue performance of cast iron and in particular the cast iron rail deck cross girders. In order to do this, three cross girders were removed from an un-trafficked area of the bridge for full scale testing. The testing was carried out in stages, each stage being more intrusive than the last but providing greater understanding of the material. The intension was to justify an improvement on the limiting stress approach stipulated within the assessment code and thereby reduce, or even eliminate the need for the expensive strengthening.

The stages of testing are summarised below:

Stage 1 – Non destructive in-situ testing

This was necessary to determine the integrity and quality of the cast iron material and utilised Ultrasonic Testing methods (UT) to investigate for internal defects. Hardness testing and replication/ spectrographic examination were also carried out to provide information on the physical and chemical composition of the material.

Extensive use was made of Ultrasonic Testing

Stage 2 – Small scale fatigue testing

This was necessary to establish a statistically reliable fatigue performance spectrum (SN) curve for the material as well as its Ultimate Tensile Strength and Modulus of Elasticity.

Machined Ø12.5mm tensile specimen

Fatigue testing of machined test specimens

Stage 3 – Large scale fatigue testing

Due to size and scale effects, it was considered that the only reliable way of determining the fatigue characteristics of the cast iron cross girders was to undertake full scale testing. This option was only made possible due to the fact that there are un-trafficked areas of the rail deck from which full length cross girders could be extracted. 'Listed Building Consent' was obtained on the condition that the cross girders would be repaired and returned to their original location on completion.

Three full length cross girders were therefore lifted out of the North Approach. In order to do this, the concrete deck was first removed using hydro-demolition and associated cast iron elements such as balustrades, fascia panels, drainage etc were carefully dismantled and removed for storage.

Deck furniture had to be dismantled before cross girders could be extracted

The three cross girders were blast cleaned, inspected and cut into two halves to provide six fatigue test specimens. The six test beams were transported to the University of Manchester where they were placed into a specially designed fatigue test frame and subjected to cyclic loads of predetermined range to a maximum of 2 million cycles.

The fatigue damage history was determined by detailed examination of historical records. Account was taken of modifications such as change in dynamic effects resulting from ballasting of the tracks and the relocation and reduction of tracks. The three extracted beams had the advantage of having little or no fatigue history or 'damage' as the area from which they were removed had not been subjected to live loading.

Each full scale test piece was instrumented and provided defined points on an SN curve. Points correlated well with the small sample fatigue test results when size effects were taken into account. Also taken into account were stress raising details. These were analysed using finite element modelling to determine stress concentration factors related to a variety of details including holes and casting details.

A second phase of testing was carried out utilising the quarter sections which were mounted within a smaller test rig and subjected to lower stress ranges to a higher number of cycles (up to 5 million) such that a non propagation stress limit could be established.

Lifting out of the 12m long cross girders during a night time road closure

Full scale fatigue testing at the University of Manchester

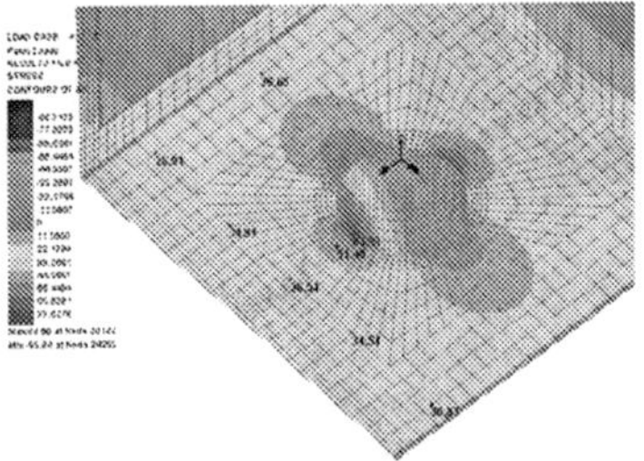

Finite element analysis was used to determine stress concentration factors

Cross girder after failure

From the test data, SN curves were derived for the various stress concentration factors, the curves were then utilised to determine the theoretical cumulative fatigue damage for individual details and more importantly the remaining useful life. The remaining useful life of the cross girders can be established for any combination of loads thereby enabling Network Rail, to make informed decisions on the future management of the structure.

Repair and Reinstatement

In order that legal requirements associated with the Grade 1 Listing ('Listed Building Consent' conditions) could be met, the chosen repair solution utilised a novel combination of 'Metalock' mechanical stitching and use of high strength carbon fibre bonded plates to reinstate tensile flexural capacity. This method maximised re-use of original fabric, whilst minimising any external change of appearance.

Conclusions

A common desire of all parties was that the High Level Bridge remained a 'working structure', however when subjected to modern assessment codes this meant that wholesale strengthening was necessary. The fatigue testing has validated an alternative assessment approach, moving away from limiting stress theory, to a bridge specific fatigue damage approach enabling residual fatigue life to be determined.

The cast iron testing of the High Level Bridge has also greatly enhanced the knowledge of the fatigue performance of cast iron which will allow industry in general to make more informed decisions on the conservation of historic cast iron structures subjected to cyclic loading.

Acknowledgements

The authors would like to thank the following parties:
> Network Rail
> May Gurney Ltd
> University of Manchester
> English Heritage

Rebuilding the Alexander Hamilton Bridge - The Future of Bridge Construction

Martin H. Kendall Edwards and Kelcey Inc, New York, NY, USA.
Tariq Bashir New York State DOT, New York, NY, USA

Abstract

The Alexander Hamilton Bridge carries the Cross Bronx Expressway, a section of Interstate Route I-95, over the Harlem River in the middle of New York City, as well as serving as the intersection with Interstate Route I-87. The deck is in bad condition and requires total replacement. Significant deterioration of the structure, and structural vulnerability due to lack of redundancy, needed to be addressed. The major concern was how to rehabilitate the bridge, and keep 188,000 vehicles per day moving with the minimum of disruption. The paper describes the development of the proposed rehabilitation solution for the mainline bridge, which will accommodate the mandated requirement of maintaining all existing lanes during peak hours and also provide for meeting FHWA standards for shoulders on the bridge. The solution included replacement of two bridges that span over the Cross Bronx Expressway, with piers located in the median of the highway. In order to permit the MPT on the AHB, these bridges had to be replaced in order to eliminate these piers.

Introduction

The Alexander Hamilton Bridge carries Interstate Highway I-95 over the Harlem River between the island of Manhattan and the South Bronx in the heart of New York City.

Figure 1 Aerial View of Alexander Hamilton Bridge Complex

The Alexander Hamilton Bridge complex includes a major interchange between I-95 and I-87. This last highway is designated as the Major Deegan Expressway inside New York City, and is the primary Interstate highway going north out of New York City, converting into the New York State Thruway feeding Albany and Buffalo in upstate New York, and

Montreal in Canada. The bridge was constructed in 1964 and was the last link from the Cross Bronx Expressway to the top end of Manhattan, and from there to the connection to the George Washington Bridge over the Hudson River between New York State and New Jersey. The Alexander Hamilton Bridge carries a peak volume of 188,000 vehicles per day (vpd) on I-95, with 40,000 vpd using the interchange with I-87.

Bridge Structures

There are nine (9) bridge structures as a part of the overall Alexander Hamilton Bridge complex to be reconstructed in this project. These comprise three groups of structures. They are the Alexander Hamilton Bridge mainline structure; two bridges (Ramp TE and Undercliff Avenue) which carry roadways over I-95 (Cross Bronx Expressway), and have piers in the center median of the Cross Bronx Expressway; and the Highbridge Interchange, comprised of six (6) bridge structures and associated on-grade roadways.

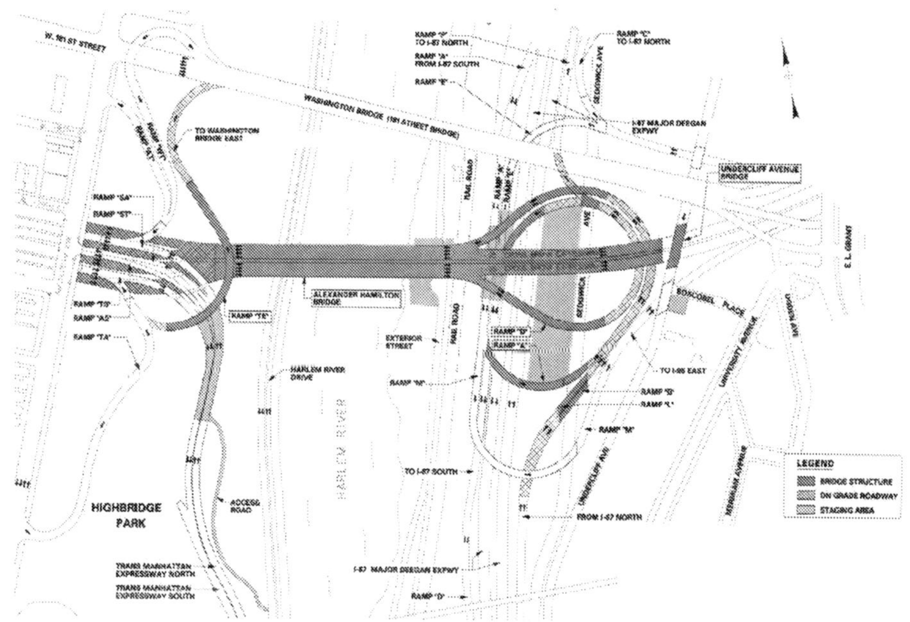

Figure 2 - Layout Plan

Alexander Hamilton Bridge Mainline Structure

The Alexander Hamilton Bridge is a ten (10) span bridge structure approximately 452.7m (1485-ft) long between abutments, and is designated as Bridge 1. It comprises a River Span bridge over the Harlem River, with West and East Approach viaducts. The River Span has a deck length of 166.3 m (545.7-ft), and a supporting under-deck main arch clear span of 153.9 m (505-ft.). The West and East Approaches are 51.6 m (169-ft) and 247.2 m (811-ft) long, respectively. The roadway is on a tangent from the West Abutment through the West

Approach spans, the main river span and two of the East Approach spans. Thereafter the roadway has a 608m (1995ft) radius curve through to the east abutment. The layout plan and elevation are shown in Figures 2 and 3.

Superstructure

Each set of lanes (eastbound and westbound) is supported at present on its own individual structure. Four different structure types comprise this bridge. These are the West Approach (2 girder spans), the Main Arch Bridge (1 Span), the East Approach Girder Spans (6 spans) and the East Approach stringer span (1 span). The Main Arch Span consists of a total of four (4) parallel hingeless steel box section arches supporting spandrel columns that support four parallel longitudinal spandrel girders. The arches are founded on concrete anchorages ("skewbacks") at each side of the river. The spandrel girders support the transverse floorbeams and cantilevered brackets. The deck stringers are framed into the floorbeams and brackets. The stringers support a 180mm (7") thick non-composite concrete deck slab, and a minimum thickness of 65mm (2½") asphalt wearing surface.

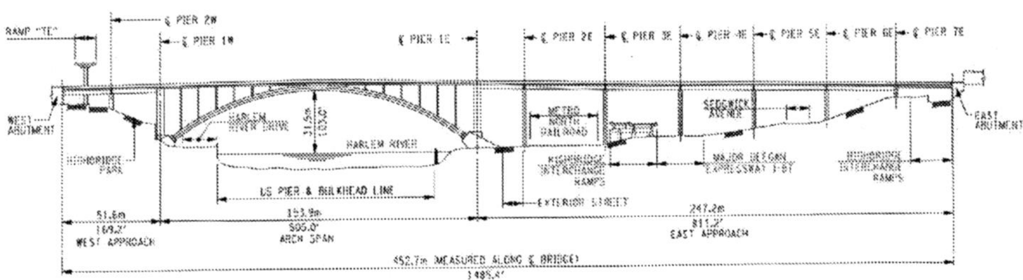

Figure 3 - Elevation on Alexander Hamilton Bridge

Eight of the nine approach spans each consist of pairs of simply supported, non-redundant steel girders, transverse floorbeams, and longitudinal stringers supporting a deck structure similar to that on the arch span. The stringers are either seated on the floorbeams or framed into the floorbeams. Only the last span on the east end of the east approach is comprised of multiple girder stringers supporting a similar deck structure as before (See Figure 4).

Substructure

The West and East anchorages for the arch span consist of reinforced concrete pedestals on stepped spread footings founded on bedrock. The substructures for the Approach Spans generally are made up of large diameter circular reinforced concrete column pier shafts, one shaft supporting each girder, with lightly reinforced concrete tie beams connecting the tops of the column pier shafts below bearing level. The column shafts are supported on pile caps which themselves are supported by multiple reinforced concrete cast-in-place piles. The last pier on the east end of the East Approach is a concrete wall supported on 1.525m (5ft.) diameter reinforced concrete cast-in-place shafts.

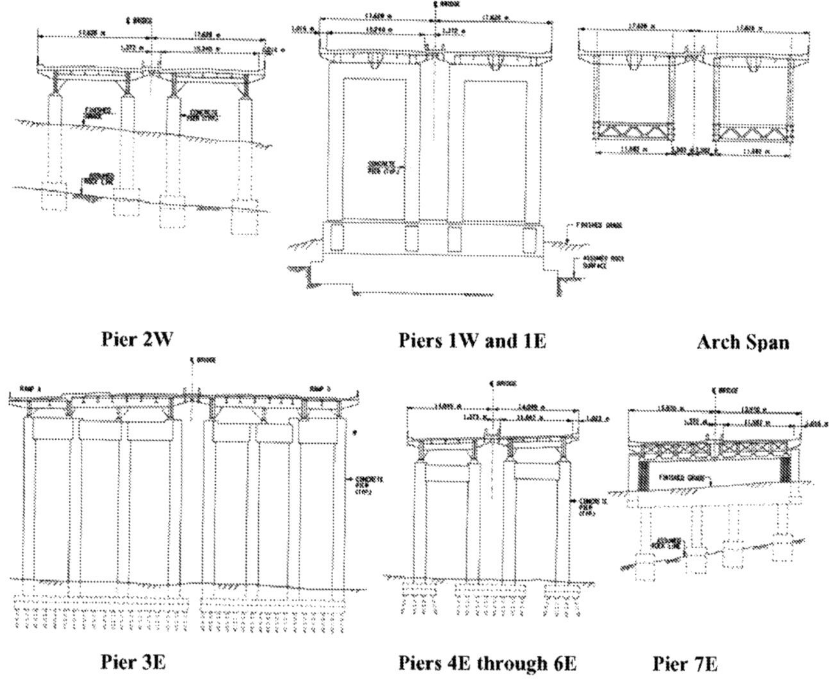

Pier 2W Piers 1W and 1E Arch Span

Pier 3E Piers 4E through 6E Pier 7E

Figure 4 - Existing Cross Sections

Ramp TE and Undercliff Avenue

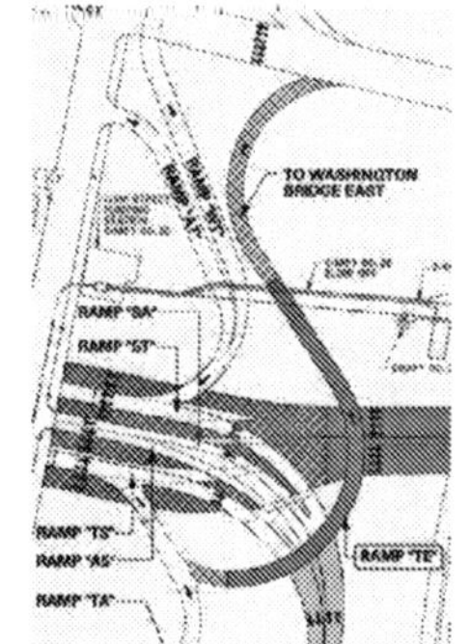

Figure 5 Ramp TE

Ramp TE (Figure 5) is a one lane roadway (originally designed for two lanes) which carries traffic from the eastbound Trans Manhattan Expressway to the original bridge crossing of the Harlem River, known as the Washington Bridge. It is supported by a ten (10) span concrete box structure, designated as Bridge 2, which is on a very tight radius (64.2m; or 210 ft., on centerline). The bridge has ten equal spans of 20.1m (66 ft.), and has a superelevation of approximately 6% throughout the majority of its length. The piers are comprised of single large diameter (1.98m or 6'-6" diameter), ½" thick steel pipes filled with reinforced concrete, supported on spread footings on rock. One of the piers is located in the center median of the Cross Bronx Expressway, in the middle of Span 2W, which is the westernmost span of the mainline Alexander Hamilton Bridge.

Undercliff Avenue Bridge (Figure 6) is a two span reinforced concrete rigid frame bridge, designated as Bridge 9, carrying a local street consisting of one traffic lane, two parking lanes, and two pedestrian sidewalks, spanning over the Cross Bronx Expressway immediately to the east of the East Abutment of the Alexander Hamilton Bridge. Foundations are on spread footings on rock.

Highbridge Interchange
The connecting ramps between the Alexander Hamilton Bridge and the Major Deegan Expressway (I-87) are tightly curved roadways supported on multiple bridge structures (see Figure 7).

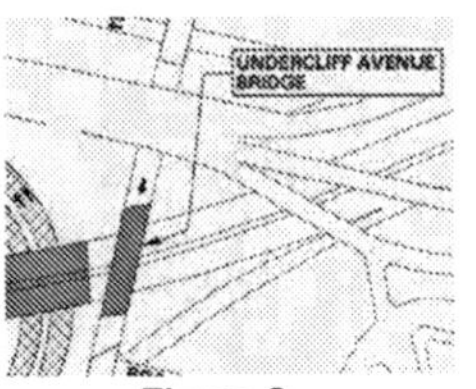

Figure 6
Undercliff Avenue

Exit Ramps from Eastbound Cross Bronx Expressway
Leading down from the eastbound Cross Bronx Expressway is a two lane wide roadway ramp, designated Ramp D. This ramp passes underneath the most easterly span (Span 7E) of the Alexander Hamilton Bridge; and then splits into two single lane roadways, one serving the southbound lanes of I-87 and the other the northbound lanes. The two lane roadway is supported on an elevated bridge structure, designated Bridge 3, passing over I-87 and a local street (Sedgwick Avenue) then on grade under the Alexander Hamilton Bridge. The bridge structure originally consisted of twin steel plate girders supporting transverse steel floor beams and longitudinal steel stringers supporting a reinforced concrete deck. The piers are comprised of single large diameter (1.98m/6'-6") diameter reinforced concrete shafts, supported on spread footings on rock. This structure was modified in 1988 with the addition of an additional two steel plate girders located underneath the floor beams and between the original plate girders as a means of eliminating the fracture critical nature of the original design. The split of the roadway is on a short simple span steel stringer structure, with non-composite concrete deck, over Sedgwick Avenue, designated as Bridge 4. The ramp to the northbound lanes of I-87 is on grade. The ramp to the southbound lanes of I-87 is initially on grade, followed by an elevated bridge structure, designated Bridge 5, which passes over both sets of lanes of I-87 and then terminates at the ramp structure that feeds the southbound lanes of I-87. Bridge 5 has a structure similar to that used for Bridge 3.

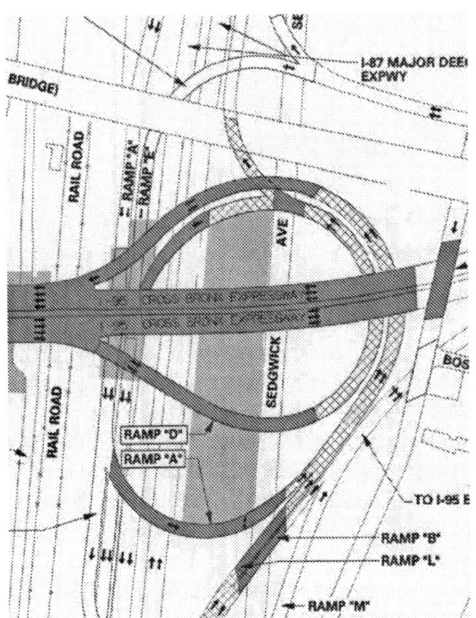

Figure 7 Highbridge Interchange Ramps

Entrance Ramps to Westbound Cross Bronx Expressway
Leading up to the westbound Alexander Hamilton Bridge is a roadway ramp designated as Ramp A. This ramp commences in two locations. The first is from the southbound lanes of I-87, and is a one lane roadway ramp, designated Ramp A. This ramp is on elevated structure, designated Bridge 6, which passes over both sets of lanes of I-87 and Sedgwick Avenue. The

bridge structure is similar to that structure described for Bridge 3. The second is the two-lane roadway ramp leading from the northbound lanes of I-87, which is on grade up to the point where it passes over Sedgwick Avenue. Immediately before this point, the roadway splits into two separate one lane roadways, designated Ramps B and L, which are carried over Sedgwick Avenue on a simply supported steel stringer and non-composite concrete deck bridge, designated Bridge 8. Thereafter, Ramp L is on grade, branches off to the right and feeds the eastbound lanes of I-95. Ramp B is also a roadway on grade, and merges with Ramp A. The combined roadway is also designated as Ramp A. This two-lane ramp passes underneath the Alexander Hamilton Bridge immediately to the east and adjacent to Ramp D The roadway is on grade until it reaches Sedgwick Avenue. It passes over Sedgwick Avenue on an elevated bridge structure, designated Bridge 7, and continues on over both lanes of I-87 before joining the westbound lanes of the Alexander Hamilton Bridge. The bridge structure is similar to that described for Bridge 3. Immediately before the junction with the Alexander Hamilton Bridge, the two lanes are merged into one. This one lane essentially is added as an extra lane to the three lanes of the westbound lanes of the East Approach of the main bridge.

Rehabilitation Requirements

The reinforced concrete deck of the Alexander Hamilton Bridge is not in good condition. The liberal use of roadway de-icing material over the years has resulted in significant concrete spalling and cracking of the concrete. The configuration of the structure of the approach spans (simply supported twin girders) was also deemed to be fracture critical, necessitating multiple additional hands-on inspections. The thermal movement performance of the bridge was also giving cause for concern with rocker bearings exhibiting tilts of over 50mm (2") at normal mean temperatures (20°C/68°F).

However the bridge complex is an extremely important node on the US Interstate Highway system, particularly with respect to traffic entering and leaving New York City. The New York State Department of Transportation was adamant that any scheme for the rehabilitation of the bridge complex allow for the maintenance of the full number of traffic lanes at all times. No permanent closure of lanes would be permitted.

Rehabilitation Scheme

The scheme that was developed to meet the criteria defined above not only meets the need to maintain all lanes of traffic, but also ultimately restores the roadway on the bridge to meet the Federal Highway Administration (FHWA) highway design standards with respect to safety shoulders

Proposed Structural Changes – Alexander Hamilton Bridge

There are primarily three major structural changes proposed for the Alexander Hamilton Bridge. These are widening the deck by 3.35m (11ft) to each side, eliminating half of the transverse deck joints and the whole of the longitudinal deck joint, and the introduction of additional girders in the Approach Spans.

3.35 m (11-ft) Deck Widening to Each Side

It is proposed to extend the width of the deck on both the north and south edges of the bridge by 3.35 m (11-ft). This change is proposed in order to maintain a 4-lane traffic pattern in each direction on the West Approach Spans and the Main River Bridge, and a 3-lane traffic pattern in each direction for the East Approach Spans, during construction. In addition this will also provide for full shoulders to each set of lanes after completion of the bridge rehabilitation. (See Figure 8 for typical cross sections).

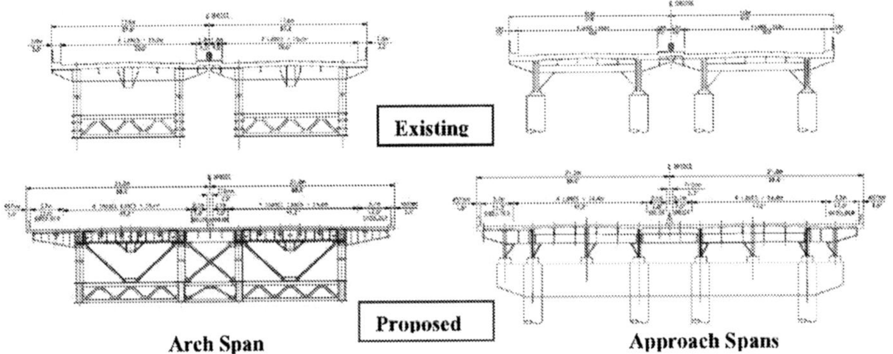

Figure 8 - Deck Widening Cross Sections

Deck Joint Elimination

The transverse deck joints to be eliminated include the joints at Piers 2W, 2E, 4E, and 6E for the approach spans, and at the arch panel points 3 and 3' for the arch span. The longitudinal deck joint is the joint located along the centerline of the entire bridge, which divides the bridge into two independent superstructures. The elimination of these joints is proposed due to the significant corrosion observed during inspection for the structural members adjacent to all of the deck joints. Further, the elimination of the longitudinal joint facilitates the various construction stages, as well as increasing the overall stiffness of the bridge structure. Median floorbeams are proposed throughout to connect the two existing superstructures. (See Figure 8). In addition, the elimination of the transverse deck joints at Piers 2W, 2E, 4E, and 6E will allow the conversion of the individual simply supported girder spans to new two-span continuous girder structures, thus increasing the redundancy of the Approach Spans structures.

In addition to the median floorbeams in the arch spans, new median horizontal struts and sway frames are proposed from PP7 though 7'. The outer arch ribs require additional top cover plates being added due to the corrosion that has occurred to the top flange plates of the ribs, and to the additional dead load being introduced with the widening. No additional stiffening is required to the inner arch ribs. The spandrel columns were originally all designed with the same section, so as a result the only strengthening required to these columns is to those columns at Panel Points 8 and 8', which are the longest elements.

Introduction of Additional Girders in Approach Spans.

It is proposed to reconfigure the structural arrangement of the cross section of the Approach

Spans, by converting from twin plate girder arrangements for each of the north and south superstructures, to a homogeneous multiple girder cross section. This is accomplished by adding a total of four (4) retrofit girders to the overall cross section. Two (2) new exterior girders are to be added to the structure, one each on the north and south sides of the bridge. These girders are introduced to support the new cantilevers for the extended width of deck. A further two (2) new interior truss girders are to be added between the existing plate girders, one to each of the north and south superstructures, in order to eliminate the fracture criticality of the existing main girders, and to balance the structural framing within the cross section.. (See Figure 8 for typical cross section). The interior girders have been designed as trussed girders in order to facilitate the installation of these elements around the existing floor beams.

Modifications to Substructure

The existing substructure is to be modified from its existing arrangement of independent column shafts each supporting a steel girder to full reinforced concrete rigid frames. This is to be accomplished by the construction of reinforced concrete cap beams to the tops of the existing column shafts, with cantilevers to support the new exterior girders and the replacement of all existing bearings with new pot bearings. The capacity of the existing foundations is only exceeded at Piers 2E and 4E, specifically in the piles supporting the outer column shafts. The solution adopted is to install new drilled Cast-In-Place piles between the existing pile caps, to preload the new piles to share the existing dead loads from the existing column shafts, and then to construct new pile caps to join up all the existing pile caps. A detailed seismic analysis was undertaken of the entire structure, and the only location where there was any concern was at the base of the column shafts of Pier 5E, where a shear overload under seismic loading was indicated. This will be accommodated by the introduction of shear collar reinforcement at the base of these column shafts. The seismic analysis was undertaken on an iterative basis to establish the most effective distribution of fixed and expansion bearing arrangements at the tops of the column shafts to reduce the seismic loading of the column structures to a minimum, with the new two-span continuous girder arrangement of the Approach Span superstructures.

Proposed Structural Reconfiguration – Ramp TE and Undercliff Avenue

In order to provide the necessary lateral clearances to permit the widening of the Alexander Hamilton Bridge, two of the piers of the Ramp TE bridge structure needed to be relocated as they are presently positioned immediately adjacent to the edge of the roadway deck of Span 2W. One of the piers supporting Ramp TE is located in the center median of the Alexander Hamilton Bridge in the middle of Span 2W, and effectively prohibits the relocation of traffic lanes during staged construction for the mainline bridge at the west side of the Harlem River. On the east side of the river, just to the east of the east abutment, the center pier of the Undercliff Avenue Bridge also prohibits the relocation of traffic lanes during staged construction for the eastern approach. The condition of the Ramp TE bridge structure is not good, with deteriorated concrete evident in a number of places, including failures in the main support joints for the spans over the Cross Bronx Expressway. The condition of the rigid concrete frame bridge for Undercliff Avenue is fair, but the need to eliminate the center pier dictated the replacement of this bridge.

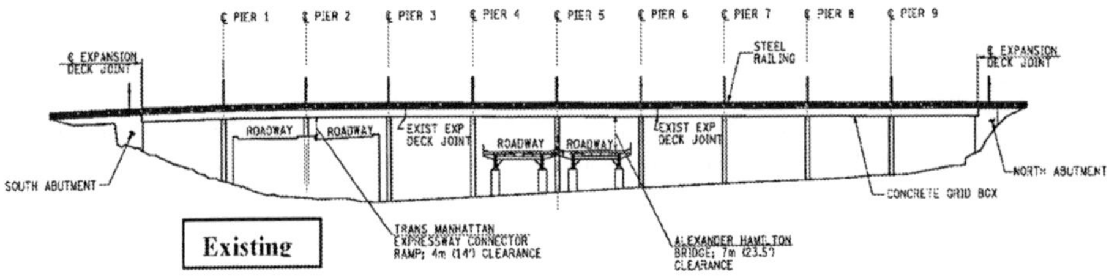

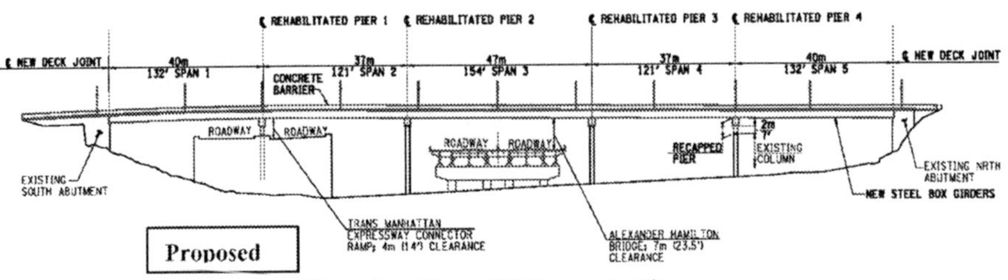

Figure 9 - Ramp TE Reconstruction

For these primary reasons it was decided to replace these two bridge structures in their entirety, with designs that eliminated the piers in the central median of the Cross Bronx Expressway. The new bridge structure for Ramp TE will be a twin steel tub girder structure supporting a reinforced concrete composite deck. The bridge will be reconstructed in the same location as the existing, and remains on a very tight centerline radius of 64.1m (210ft), with a 6% superelevation. The number of spans is being reduced from 10 equal spans of 20.1m (66ft) to a five span arrangement of varying centerline lengths, with the piers positioned to suit the existing features. The abutments are retained, as are two pier shafts and

foundations. New cap beams will be constructed for these shafts. The other two piers will have foundations that use the existing spread footings as a part of the overall foundation but will have new shafts completely. Undercliff Avenue Bridge will be replaced with a multiple steel stringer bridge with a composite concrete deck with a single clear span of 39.5m (129'-6"). Because of clearance requirements the girders are very shallow for the span and have had to be designed with a significantly uneconomic cross section to provide the necessary structural capacity.

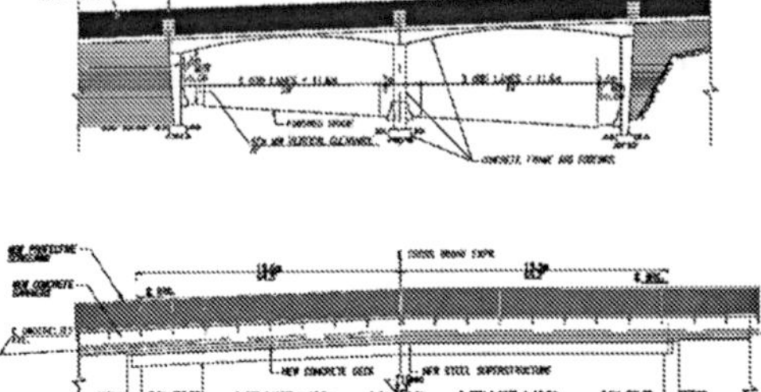

Figure 10 - Undercliff Avenue Bridge Reconstruction

Proposed Structural Changes - Highbridge Interchange

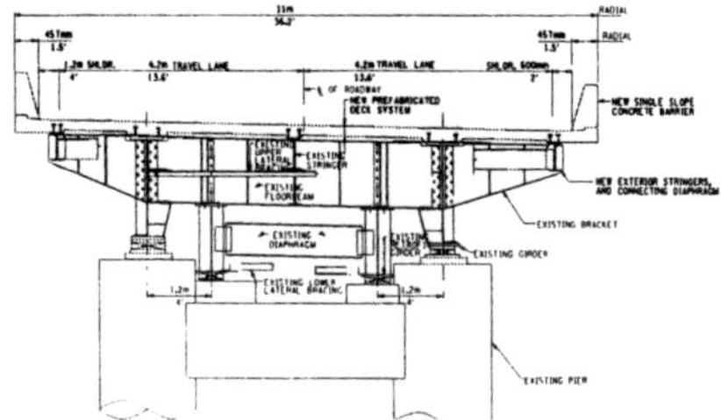

Figure 11 - Cross Section
Highbridge Interchange Ramp Bridges

The Highbridge Interchange Ramps are essentially being reconstructed as a replacement in kind. The longer bridge structures (Bridges 3, 5, 6, and 7) were retrofitted in 1990 with additional plate girders to increase the redundancy of the structures. The rehabilitation to be undertaken as a part of this project consists of replacing the existing reinforced concrete decks with new composite cast-in-place reinforced concrete decks. It had originally been anticipated that the replacement would have been undertaken using precast deck units, but in order to undertake such construction, the ramps would have had to have been closed to traffic for full weekend closures of approximately 48 hour durations. In November 2004 an emergency patch repair and resurfacing contract was undertaken for Ramp A using the traffic schemes developed for such closures, and the traffic delays that were experienced were of such magnitude that the New York City DOT (who are responsible for all traffic management inside the boundaries of New York City) requested that alternative methods of traffic management be investigated for this work. The finally adopted scheme is to construct temporary bridges and roadways alongside each of the ramps to carry the full traffic volume, whilst the decks of the ramp bridges and the roadways on grade are reconstructed. This will result in an additional cost of reconstruction estimated as more than $15 million, but the saving in travel time costs are estimated to be well in excess of this figure.

Summary
The future of bridge construction is epitomized in the reconstruction needs of the Alexander Hamilton Bridge Complex. A design for highways, bridges and highway interchanges that was state-of-the-art nearly 50 years ago has become overloaded in terms of traffic volume carried, and has become comparatively higher risk in terms of structural safety, as a result of a more in-depth knowledge of structural performance. In order to bring the bridge and interchange complex up to modern day standards, major reconstruction is required, but the need to maintain traffic flow for the benefit of the traveling public is a matter of overwhelming importance. As a result, the needs of the rehabilitation of the Alexander Hamilton Bridge Complex have been driven by the need to maintain all lanes of traffic at all times, on both the mainline bridge and the interchange ramps. This has necessitated the reconstruction of two additional bridges to permit this maintenance of traffic flow, as well as a solution for reconstruction of the interchange ramps that requires the construction of temporary bridges to permit the ramp reconstruction to proceed without affecting the traffic capacity of the interchange. This has resulted in a significant increase in rehabilitation cost over and above that which would normally have been expected in normal bridge construction activities.

Experiences with the Application of Various Slab-on-girder Bridge Systems in Slovakia

Jaroslav Halvonik – Associate Professor*
Ľudovít Fillo – Professor*
Viktor Borzovič – Assistant Professor*
* Slovak University of Technology in Bratislava, Slovakia
Vladimír Benko – Associate Professor - University of Technology in Vienna, Austria

Abstract

In the last ten years, the most significant development in the transportation infrastructure in Slovakia has been the construction of the national highway system. In order to speed up the process and minimize the cost, the effective design choice is the slab-on-girder bridges. This paper summarizes experiences regarding the application and behavior of precast prestressed girder bridges with various types of cast-in-place deck, which are produced in Slovakia. A wide range of completed structures including simply supported and semi-continuous bridges with reinforced concrete pier diaphragms are discussed.

Introduction

The Slovak highway network is still under development, but the major routes, which are part of the Trans-European Transport Network (TEN-T), are either under construction or in planning. The priority axis No 25, which runs from Gdansk (Poland) via Brno (the Czech Republic) and Bratislava (Slovakia) to Vienna (Austria) includes the longest elevated part of the highway in Slovakia, a 1.76 km long bridge. The other three TEN-T roadway routes running west – east and south – north require elevated highways due to the rough landscape and are a national priority. Concrete slab-on-girder bridges with up to 42 m long spans are the most economical option for elevated highways, overpasses, underpasses and overheads. Slovakia has a half-century long tradition of developing prestressed precast concrete girders, which are designed as simply supported structures with a flexible link over the piers. The obvious drawbacks, such as the large number of bearing pads, lateral instability during construction, costly maintenance and time-consuming rehabilitation with traffic disruptions has led to the design of semi-continuous bridges with diaphragms over piers.

Overview of precast bridge girders in Slovakia

The types of precast bridge girders produced in Slovakia are detailed in the table below.
Unlike other countries, design manuals including detailed drawings have been prepared for each type and length of girder. A designer is not required to analyze and detail a structure provided that the design is within the limits on use defined in the design manual. The use of girders not covered by a design manual is subject to detailed structural analyses and design.

Bridge design, construction and maintenance 2007, Thomas Telford, London

Type	Shape	Type of Prestressing	Length [m]	Producer
I-96	I	Post-tensioning	24, 27, 30, 42	Doprastav
DPS-VP I97	I	Pre-tensioning	12, 15, 18, 21	Doprastav
DPS-VP I04	I	Pre-tensioning	24, 27, 30	Doprastav
DZ-97	T	Pre-tensioning	21, 24, 27, 30	ZIPP
DZ – 97	double T	Pre-tensioning	12, 15, 18	ZIPP
IST-97	inverted T	Pre-tensioning	9, 12, 15,18, 21, 24	Inžinierske stavby
IST-97	I	Pre-tensioning	27 and 30	Inžinierske stavby
IPM-I	I	Pre-tensioning	12,15,18, 21, 24, 27, 30	Sučany

Post-tensioned precast beams I-96

Bridge girders I-96 ranging up to 42 m are the most frequently used girders in Slovakia. The girders are produced in four lengths (24, 27, 30 and 42 m) with 1.10, 1.25, 1.4 and 2.0 m depths. They consist of three segments tied together by post-tensioned tendons at the construction site before their erection. The maximum girder spacing in the transverse direction is 1.50 m (structural analysis and design are required to allow for a bigger spacing). The number of tendons varies in accordance with girder length. A 42 m long girder is prestressed using 10 tendons. All other shorter girders are prestressed by 7 tendons.
Materials:
Girder: Concrete: C45/55

  Tendon: 4xϕ15.5 mm/1800 MPa (ϕ15.5 mm is 7-wire strands)
Slab: Thickness: 200 mm

  Concrete: C30/37

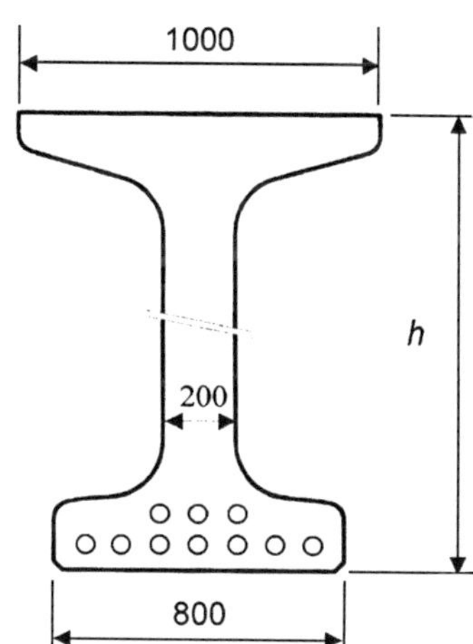

Figure 1. Girder I-96 – Section and front view

Early after the introduction of the I-96 girders, two problems were encountered. The 42 m long girders had:

- Excessive cambers from 90 mm to 120 mm (including creep), which were 1.5 -2 times the predicted values based on a time dependant analysis;
- Lateral instability of the girders during the construction stage (before and during the deck pour).

Excessive cambers result in changes in elevations and screeds during construction and significantly varying slab depths along the spans, thereby challenging the economical, structural and technological efficiency of the system. To find the cause, detailed monitoring was performed by the company "Projstar Ltd." and material testing [1] was carried out. The monitoring included:
- A detailed measurement of segment dimensions;
- Measurement of the camber during post-tensioning, and before and after the deck pour;
- Measurement of the prestressing force in selected tendons using magnetic sensors (see Fig. 2b)

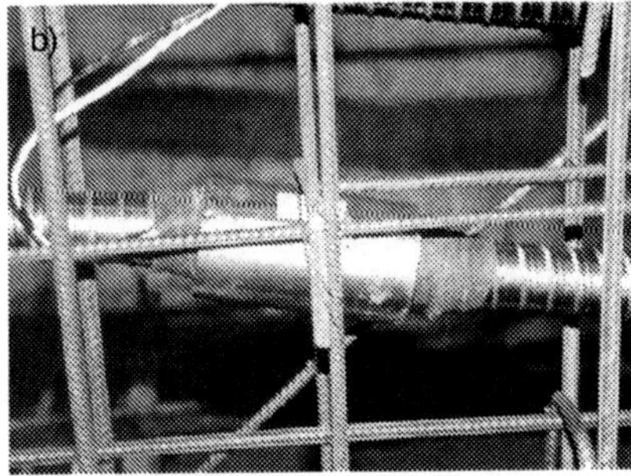

Figure 2. Girder I-96, 42 m long - Excessive camber (Photo Projstar)

The following reasons for the excessive cambers were determined:
1. Low modulus of elasticity ranging from 28 to 31 GPa due to there being only two aggregate fractions in the C35/45 concrete;
2. High compressive stresses during tensioning, leading to non-linear creep and causing excessive creep deformations;
3. Inclined segment ends due to non-uniform shrinkage over the girder depth and thermal effects (technological procedures were reviewed and excluded as a possible reason for the end inclination).

The following measures resulted from the monitoring and the above observations:
1. The class of concrete was changed from C35/45 to C45/55
2. The composition of the aggregates was changed to increase the modulus of elasticity to over 42 GPa
3. Introduce the recommendation that designers assume a deck depth of 240 mm when assessing the permanent loads and bill of quantities for spans ranging from 30 to 40 m.

The second problem, of the lateral instability of the precast girders during erection and deck pour, occurred during construction of the elevated highway at Beckov, completed in 1997. This elevated highway consists of two expansion units with four and five spans, together 152 simply supported composite girders (Fig.3). Continuity of the deck slab over the intermediate piers was ensured by a flexible link to allow for rotational flexibility of the girders. Each precast girder is supported by two pot bearings that allow rotations around both axes and movement in the expected direction. However, the girders in several spans rotated slightly around the longitudinal axis during the deck pour. The rotation caused the sealing steel rings of upper and lower part of bearings to come into contact on one side and open up a gap on the other side (V effect). This defect threatened the structural performance of the bearings in the bridge structure.

Figure 3. Highway D1 – Elevated highway at Beckov (Photo Cemos)

This unfavourable experience with the performance of slab-on-girder structures with flexible links over intermediate supports led to continuous systems with reinforced concrete diaphragms over piers. Two construction methods are currently used. Both are based on the use of precast diaphragms of width depending on the type and length of the precast system (measured in the longitudinal direction of the bridge). For longer spans of over 40 m a width of 2.8 m is used, for shorter spans (up to 33 m) the width is 2.3 m or less. The depth of precast diaphragms depends on the construction method used.

Figure 4. Highway D1 - Section Ladce-Sverepec (Photo Doprastav)

The first method is based on the use of precast diaphragms placed on hydraulic jacks that are located on the top of the piers (Fig.4). Subsequently the precast girders are placed on these diaphragms and the gap is filled with concrete (Fig.5). This method requires deeper precast diaphragms (0.55 to 0.65 m) depending on the space between bearings and the length of the cantilevers. The major advantage is an elimination of scaffolding during construction.

The shortcoming of this method is in increased amount of reinforcement in the precast diaphragms, which are required to carry their self-weight, the precast girders and the cast-in-place deck. Therefore, this method cannot be used for construction of bridges composed of longer precast girders, e.g. 40 m long I-96 girders.

The second method requires scaffolding to support the precast diaphragms (Fig.6, 7). Scaffolding supports are usually spaced by 2 m on a pier foundation. The depth of these precast diaphragms can be less (0.40 to 0.50 m) in comparison to the system described above because the diaphragms are continuously supported throughout their length.

Figure 5. Highway D1 - Section Ladce-Sverepec, 30 mlong precast I-96 girders (Phote Doprastav)

Figure 6. Highway D1- Section Sverepec - Vrtižer, precast diaphragm

Figure 7. Highway D1 - Section Sverepec - Vrtižer, elevated highway under construction (I-96 precast girders of 40 m)

Pre-tensioned precast girders DPS-VP I04 and DPS-VP I97

One of the most used slab-on-girder systems for construction of elevated highways in Slovakia consists of the precast girders DPS-VP I04 with lengths of up to 30 m and cast-in-place slabs with thickness ranging from 0.16 to 0.20 m. The depth of the precast girders is 1.40 m. The girders were originally designed as pre-tensioned members spaced at 1.50 m. Due to the new concept of construction the prestressing system had to be updated and some of the pre-tensioned strands were replaced by three draped 4-strand tendons anchored in the upper part of the girder ends (Fig.8). The girders are now also cast from high performance concrete C55/67. More than 2,000 girders of this type have been produced in the precast plant Senec within last two years. A typical bridge cross-section used for the construction of elevated highways from DPS-VP I04 precast girders can be seen in Fig.9.

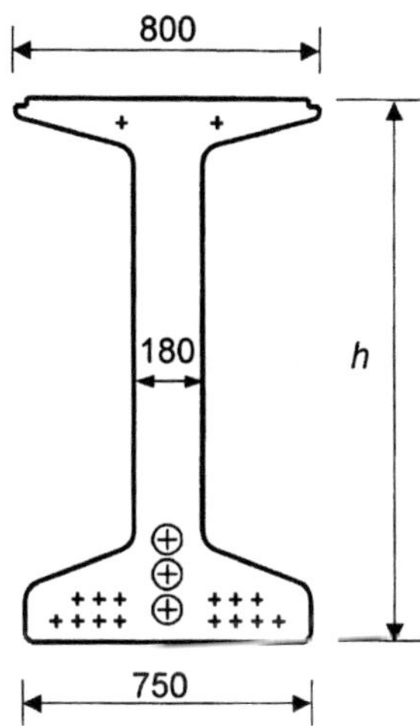

Figure 8. Girders DPS-VP I04 at a storage yard

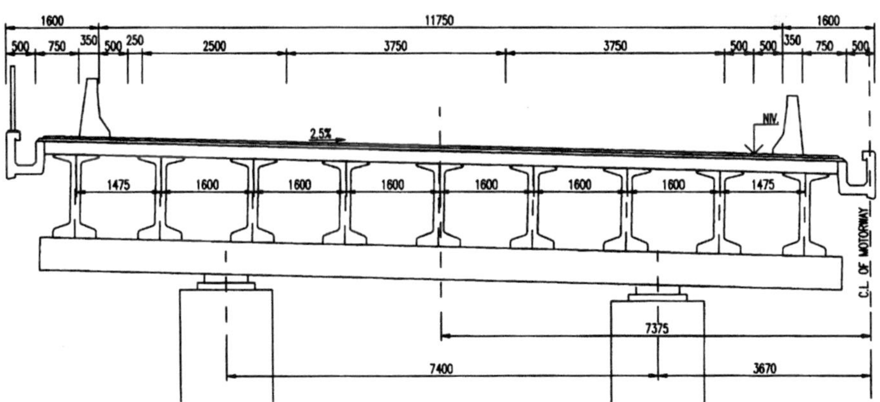

Figure 9. Typical bridge cross-section of DPS-VP I04

The precast girders DPS-VP I97 are pre-tensioned members designed in four length variants 12 – 15 – 18 and 21 m (Fig.10). The depth of girders varies from 0.60 to 0.95 m. The maximum spacing between the girders is 1.05 m. These girders were used only for construction of single span bridges [2].

Materials:
Girder: Concrete: C55/67 (high performance concrete)
Strands: 18xϕ15.5 mm/1800 MPa (21 m long girder)
Slab: Thickness: 200 mm

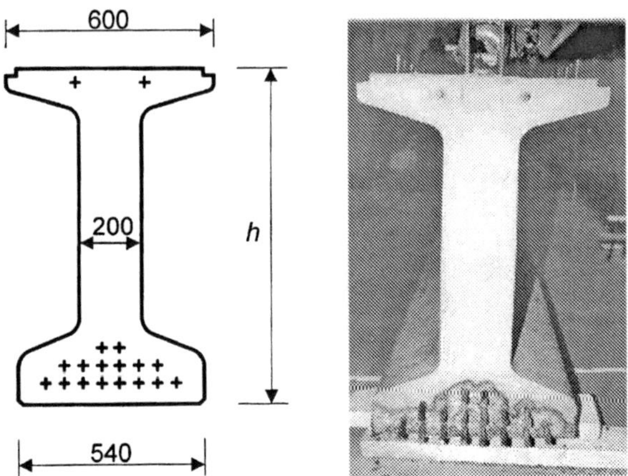

Figure 10. Precast girder DPS-VP I97, 21 m long

Precast girders DZ 97

The precast girders DZ 97 were designed by Dopravoprojekt for the construction company Zipp in 1997. The precast girders were originally designed as pre-tensioned members in eight length variants. Shorter girders were of double T shape in 12 - 15 - 18 m lengths and with depth varying from 0.76 to 0.93 m. The longer girders of 21 - 24 - 27 and 30 m have T cross-section with depths varying from 1.00 m to 1.40 m.

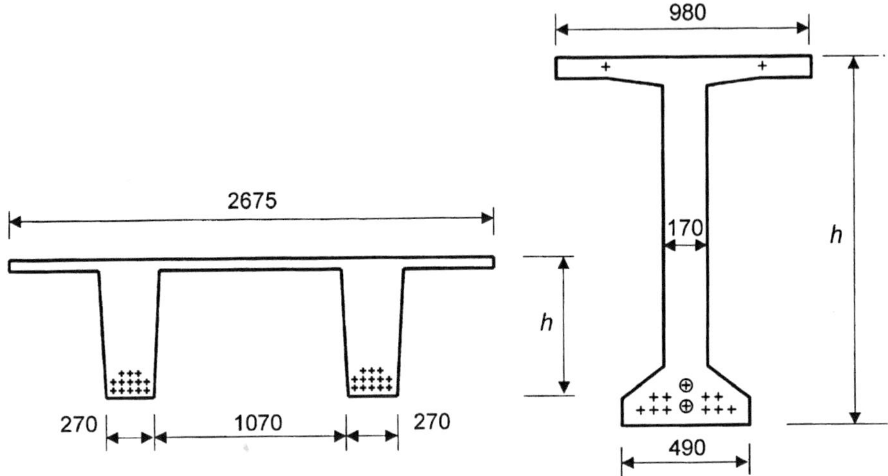

Figure 11. Precast girders DZ 97, 18 m and 30 m long

They are spaced at 1 m and were intended to be placed side by side avoiding the use of formwork for the deck. The system did not allow sufficiently for lateral girder deviations.

The thickness of the slab is 0.16 m. The girders DZ 97 of 30 m length were also used for the construction of semi-continuous elevated highways. This continuity allowed increased girder spacings up to 2.6 m. Eight strands were replaced by two 4 - strand draped tendons (Fig.11). A completed structure is shown in Fig.12. For more information see reference [5].

Figure 12. Elevated highway from precast girders DZ 97 on D1 highway Ladce – Sverepec (Doprastav)

Precast girders IST 97

The precast girders IST 97 were designed by the Faculty of Civil Engineering at SUT for a company Inzinierske stavby in 1997. These precast girders of an inverted T shape are produced in lengths up to 24 m. For girders of 27 and 30 m length an upper flange was required due to the excessive interface shear between the girder and the cast-in-place slab. The depth of these girders varies from 0.70 m (length of 9 m) to 1.40 m (length of 30 m).

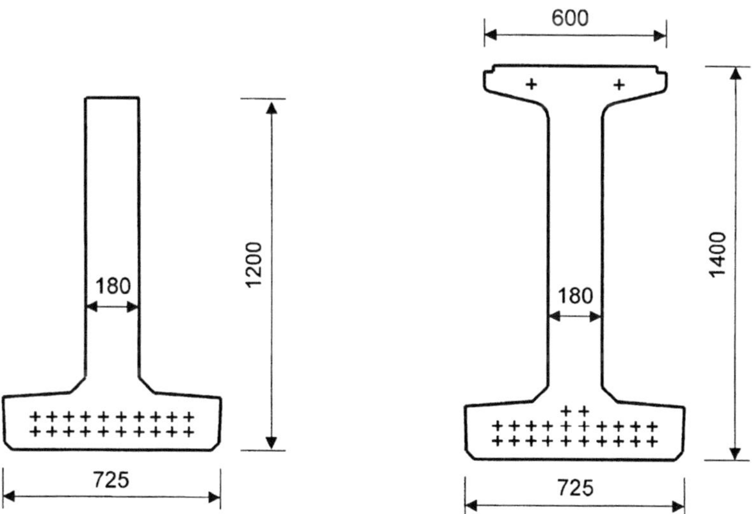

Figure 13. Precast girders IST 97 of 24 m and 30 m length

The girders are pre-tensioned by 7-wire strands of ϕ15,5 mm/1800 MPa from C45/55concrete. A maximum space between the girders is 1.25 m. Deck thickness is 0.20 m of C30/37 concrete. These girders were used for the construction of single span bridges only.

Structural analysis

Structural analysis of slab-on-girder structures is more complex than analysis of standard prestressed members, where all parts of the section are cast at the same time. Postponed deck pours lead to different concrete ages of the precast beam and the cast-in-place slab and consequently to different shrinkage and creep enhanced by significantly higher stresses in the prestressed girders in comparison with stresses in the slab. Composite behaviour of a member means compatibility of deformations over the depth of a girder which leads to the development of supplementary axial stresses that may adversely influence girder stresses and have an impact on the durability of a structure.

Complexity of the analysis grows by the assumption of continuity over the intermediate supports. Free deformation (rotation) at both ends of a composite girder due to differential shrinkage and creep is restrained by massive diaphragms which causes the development of secondary internal forces. In normal construction schedules, the age of the precast girders is less than one year at the time of a deck pour and so tensile stresses will develop at the bottom of the girders due to the positive, secondary bending moments. These secondary moments can be even larger than the negative moments due to the permanent actions at the intermediate supports. Therefore, it is necessary to properly anchor the precast girders into the diaphragms, e.g. by extension of the pre-tensioned strands or by reinforcement. The precast girders should also be sufficiently embedded in the diaphragms - the common length used in Slovakia being 0.70 m (Fig.14).

Figure 14. Reinforced concrete diaphragm under construction

Non-prestressed, reinforced concrete diaphragms make the analysis even more complex. Cracking over an intermediate support reduces the stiffness of this area and it leads to a redistribution of the internal forces.

The following assumptions were used for taking into account diaphragm cracking on the flexural stiffness of a member.

- Tension stiffening according to [6] was taken into account;
- Two layers of reinforcement were assumed:
  - $\phi22/100$ mm and $\phi16/200$ mm, which represents a minimum reinforcement determined for diaphragms of 2.1m depth. The deck is composed of precast girders DPS-VP I04 or DZ 97 with a depth of 1.40 m and a span length between 30 and 33m;
  - $\phi25/100$ mm and $\phi16/200$ mm, which is a minimum reinforcement determined for diaphragms of 2.6m depth. The deck is composed of precast beams I-96 with a depth of 2.0 m, and a span length between 40 and 42 m (Fig.14);
- Maximum stress in the longitudinal reinforcement is 280 MPa, which is the allowable stress for the assumed type of reinforcement according to the Slovak Code [7].

The computer analysis of the model with reduced flexural stiffness showed an increase in the sagging moments of 12 %.

Conclusions

Slab-on-girder systems are one of the most effective construction methods for elevated highway bridges of up to 42 m long spans. It speeds up the construction of bridges, particularly in mountain regions with a long winter season. Cold weather with subzero temperatures usually limits construction works using cast-in-place concrete. To the contrary, precast girders can be produced in stationary precast plants in advance, stored in a storage yard, transported to the construction site and finally erected. Furthermore, the quality of precast elements is higher than that of cast-in-place concrete members. The problems with the large amount of bearings required and the lateral instability of precast girders during their erection and the deck pour can be overcome by providing semi-precast diaphragms that ensure full continuity over the intermediate support. Fewer bearings, only two under each diaphragm, allows for easier maintenance and extends the lifetime of a bridge.

Acknowledgement

The authors gratefully acknowledge the financial support of the Scientific Grant Agency of the Ministry of Education of the Slovak Republic and the Slovak Academy of Sciences (VEGA 733).

References

[1] CHANDOGA, M.- HALVONÍK, J.- JAROŠEVIČ, P.: *The monitoring of DPS VP I97 Bridge beam.* Projstar PK Ltd., Bratislava, 2002. 80 p.

[2] CHANDOGA, M.- HALVONÍK, J.: *Interactive Design of Prestressed Precast Bridge Beams from High Performance Concrete.* In: Proceeding of Symposium *fib,* Concrete Structures the Challenge of Creativity. Avignon, June 2004. p 272-274 .

[3] HALVONÍK, J.- BORZOVIČ, V.- FILLO, Ľ.: *An Experimental Investigation of Composite Continuous Girders.* In: 2nd *fib* Congress. Naples, Jun 2006. p.354-355.

[4] *Precast Concrete Bridges, fib* Bulletin 29. State-of-art report prepared by Task Group 6.4, November 2004. 81p.

[5] RAKOVSKÝ, R.: *Composite Concrete Bridges on Motorway D1 Ladce – Sverepec.* In: National Report of Slovak republic for 2nd *fib* Congress. Naples, Jun 2006. p. 67-73.

[6] EN 1992-1-1: *Eurocode 2: Design of concrete structures. Part 1: General rules and rules for buildings.* December 2004. 250 p.

[7] STN 73 6206-71: *Design of concrete bridge structures.* Praha, March 1971. 50 p.

Condition assessment and Retrofitting of an Old Reinforced Concrete Arch Bridge

Dr Y. F. Fan, Associate Prof., Dalian University of Technology, Dalian, P.R.China
J. Zhou, Professor, Dalian University of Technology, P.R.China
X. Z. Zhang, Engineer, Jilin Highway Management Bureau, Changchun, P.R.China
T. Zhu, Associate Prof., Dalian University of Technology, Dalian, P.R.China

Abstract

In this study, the structural behavior of an old reinforced concrete arch bridge was studied, and retrofitting of the bridge was carried out. Firstly, field inspection (inclusive of test of material properties, static and dynamic tests of the bridge) was performed on the old reinforced concrete bridge. Based on the field experimental results, the present state of the bridge was evaluated. A retrofitting scheme was determined, and the bridge was retrofitted. Finally, a field inspection was executed on the retrofitted bridge. Field experimental results achieved from the un-retrofitted and retrofitted bridge are compared. It is concluded that the structural behavior of the retrofitted bridge has been improved significantly.

Introduction

Due to a wide variety of unforeseen conditions, it will never be possible or practical to design and build a bridge that has a zero percent probability of failure. Normal operation, overloading and environmental attack will lead to the early deterioration of large amounts of in-service bridges. In the USA, according to a 2005 report by the National Material Advisory Board, the number of bridges that are deteriorating has risen steadily from 34.6% in 1992 to 27.1% in 2003(ASCE, 2005). It will cost $9.4 billion over 20 years to eliminate all bridge deficiencies. In China, it is reported that more than 30% of the nation's highway bridges are structurally deficient, and many bridges are being added to this list every year. These early deficiencies potentially endanger the safety of the in-service bridges, and accidents have been caused in practice. Nowadays, the safety of bridges has become to be a critical issue and attracted more and more attention by scientists and engineers worldwide.

Field inspection of the old bridge

Damage description of the old bridge

The 6 span reinforced concrete arch bridge was constructed in 1960s. The total length of the bridge is 246m (Figure.1).

Figure 1. Sketch of the reinforced concrete arch bridge

Bridge design, construction and maintenance 2007, Thomas Telford, London

From the field survey, it was discovered that much damage had occurred on the bridge members including both side of the arch ribs, columns and beams, the bridge deck. Large amounts of concrete cover had spalled and the reinforcement in the concrete had been rusted severely. The 2 principal reasons responsible for the bridge deterioration were:

(1) Interactive action of the environment and loading: over 40 years, the interactive action of the environment and load will lead to the aging and cracking of the concrete, which will accelerate the corrosion process of the reinforcement and expansion of the concrete. With the deterioration growing gradually, the concrete cover will be spalled, strength and stiffness of the material will be reduced.

(2) Overloading of the bridge in the service: frequent overloading of the bridge will result in the cracking of the bridge members, and the strength and stiffness of the members will be deteriorated. Some small sections of the columns were crushed.

Field inspection of the old bridge

To investigate the present state of the bridge, static and dynamic field tests were performed on the old arch bridge respectively.

Static inspection of the bridge

According to the 'General code for design of highway bridges and culverts (JTG D60-2004)', the present code of the Communications Ministry of P.R. China, half of the traffic load for Grade-15 should be applied to the bridge and the deformations of the bridge were recorded.

Form the field survey, it is observed that maximum deformation of the bridge is 6-8mm.

Dynamic survey of the bridge

A digital Signal Processor System (DSPS) and AR-5F accelerometer sensors were applied for the dynamic survey of the bridge. Since the weakening of a structure can be characterized by a reduction in stiffness, which in turn will lower the frequencies and influence other modal properties, the dynamic properties, which will not change unless the mass or stiffness is altered, may be regarded as the vibrational signature or 'fingerprint' of the bridge. To acknowledge the current dynamic characteristics of the bridge, field dynamic tests were carried out. Layout of the measuring points is plotted in Figure.2. The ambient vibration method is the most practical means for an automated system to determine a bridge's vibrational signature, and was employed in this paper. From the frequency response, which can be measured directly from a vibrating structure using digital signal processing equipment and fast Fourier transform routines, resonant frequencies, mode shapes and system damping corresponding to each resonance were achieved.

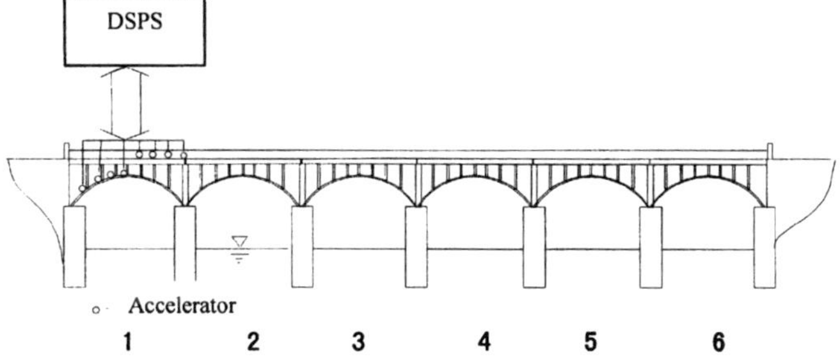

Figure 2 Layout of the accelerator measuring point

Though the six spans of the bridge are the same, different damages occurred on every span. Natural frequencies for the 6 spans were monitored respectively, which are listed in Table.3. The damping ratio derived from the 1st mode shape is 2.5-4.6%. From the field survey, it can be concluded that the more severe the damage, the lower the frequency of the bridge span.

Condition assessment of the old bridge

According to the structural assessment method recommended in the "Technical code of maintenance for Highway Bridges (JTGH11-2004)", the condition scale and weighting were determined. The structural condition score for the old bridge was 52.8, which is less than 60. Therefore, the bridge was in need of repair.

Considering that the present condition of the damaged bridge cannot satisfy the current traffic, retrofitting is essential. Based on the current codes for the Ministry of P.R. China, the retrofitting designs were executed.

Retrofitting scheme for crucial load-bearing members

Arch ribs

The arch rib is the crucial load-bearing member for an arch bridge. Strengthening the arch ribs is the universal method used in the retrofitting design of arch bridges. Concrete C40 was used to strengthen the reinforced concrete arch ribs. The designed section is plotted in Figure.3.

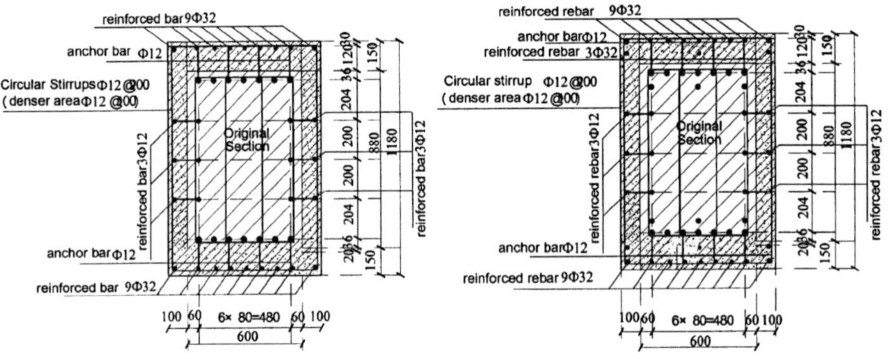

(a)arch rib section in compressive district   (b)arch rib section in tensile district

Figure 3. Retrofitted arch section (unit:mm)

Column

Concrete C40 was used to strengthen the reinforced concrete columns. The section is plotted in Figure.4.

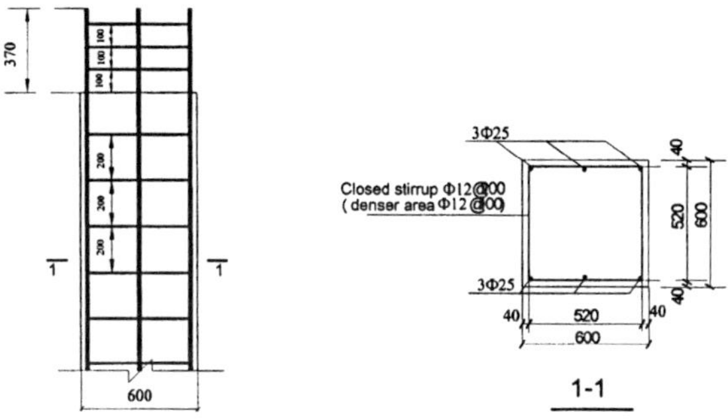

Figure 4. Retrofitted column section (unit:mm)

Cross beam

Concrete C40 was used to strengthen the reinforced concrete cross beams. The section is plotted in Figure.5.

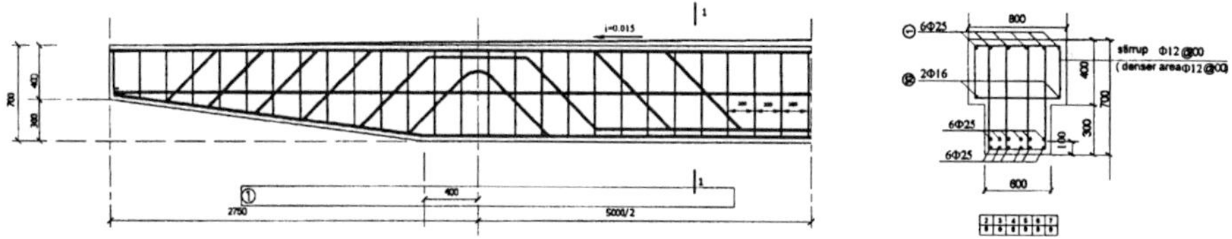

Figure 5. Retrofitted beam section (unit:mm)

Collar beam

Concrete C40 was used to strengthen the reinforced concrete collar beam. The section is plotted in Figure.6.

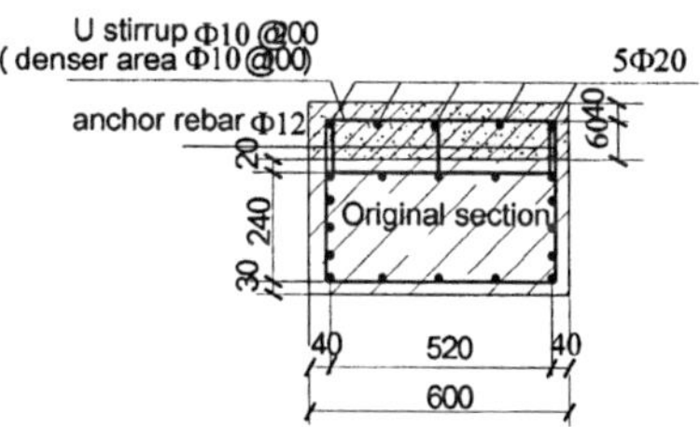

Figure 6. Retrofitted beam section (unit:mm)

Comparison of the response of un-retrofitted and retrofitted bridge

Static response

Using the layout of the loading vehicles specified in the Chinese codes, the static response (strain, deflection, etc) of the structure was determined for the 3 load cases respectively. Considering that the old bridge was severely damaged, half of the traffic load specified in the code was applied as the traffic load in case 1. For the retrofitted bridge, 2 traffic-loading cases (case 2 and case 3) were applied to the bridge respectively (Table 1). For all 3 load cases, the vehicles were placed symmetrically on two lines side by side. The layout of the vehicles for a 500kN load (including a standard vehicle weighing 200kN and a heavy vehicle weighing 300kN) is shown in Figure 7.

Both the un-retrofitted results and retrofitted results of these analyses are listed in Table 2.

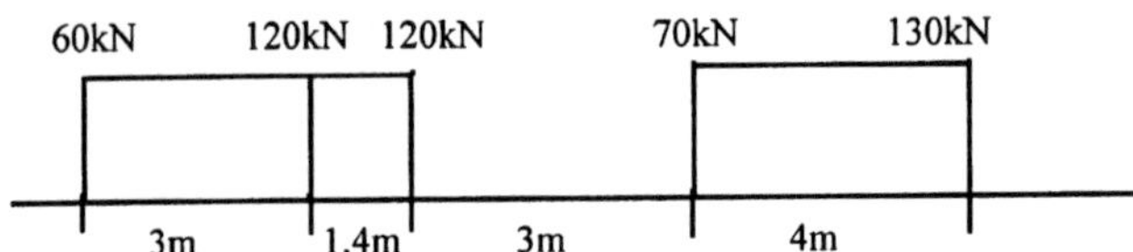

Figure 7. Layout of the traffic load

Table 1. Loading cases

	Case 1	Case2	Case3
	Traffic Load	Traffic Load	Traffic Load
Un-retrofitted bridge	1 vehicle weighing 155kN +1 vehicle weighing 194kN	-	-
Retrofitted bridge	-	2 vehicles weighing 200kN+ 2 vehicle weighing 300kN	6 vehicles weighing 200kN +2 vehicles weighing 300kN

Table 2. Deflection of the bridge under different loading cases

Location of the measuring point	Deflection /mm		
	Un-retrofitted	Retrofitted	
	Case1	Case 2	Case 3
midspan of span 1#	3.0	3.356	3.967
midspan of span 2#	4.0	3.347	3.789
midspan of span 3#	4.0	3.352	3.834
midspan of span 4#	3.0	3.343	3.742
midspan of span 5#	4.0	3.421	3.882
midspan of span 6#	4.0	3.033	3.181

From Table 2, it can be seen that under a traffic load of 1800kN (Load case 3), the maximum displacement of the retrofitted bridge is 3-4mm, which is much the same as the maximum displacement under a traffic load of 500kN for the un-retrofitted bridge. That is, although the truck weight increases 360%, the maximum displacement shows only a little increase for span 1#, while for span 2# to span 6#, the maximum displacements decrease. Therefore, the retrofitting effect is efficient.

Dynamic response

To identify the dynamic characteristics of the retrofitted bridge, field dynamic tests were carried out. The ambient vibration method was the most practical means for an automated system to determine a bridge's vibrational signature, and was employed in this paper. From the frequency response, which can be measured directly from a vibrating structure using digital signal processing equipment and fast Fourier transform routines, resonant frequencies, mode shapes and system damping corresponding to each resonance were achieved. Natural frequencies for the 6 spans were monitored respectively, which are listed in Table.3.

Table 3. Comparison of natural frequencies between un-retrofitted and retrofitted bridge

Span number	Un-retrofitted		Retrofitted	
	1st frequency / Hz	2nd frequency / Hz	1st frequency / Hz	2nd frequency / Hz
1#	/	4.75	3.66	5.61
2#	/	5.23	3.66	5.61
3#	/	5.23	4.08	5.61
4#	/	4.97	3.66	5.61
5#	/	3.76	3.90	5.81
6#	/	4.52	4.10	5.85

From Table.3, it can be seen that for the damaged bridge before being retrofitted, the 2nd frequency of the 6 spans of the bridge are between 3.76-5.23Hz, while for the retrofitted bridge, the 2nd frequency of the 6 spans of the bridge have significantly increased(7%-50%), reaching values between 5.61-5.85Hz. On the other hand, though the six spans of the bridge are the same as each other, different damages occur in every span. Natural frequencies for the 6 spans of the un-retrofitted bridge are quite different. However, after being retrofitted, the natural frequencies for the 6 spans are much closer to each other. Therefore, the global stiffness of the retrofitted bridge has been obviously improved, and the structural stiff nesses for the 6 spans are more uniform.

Impact response

Sole timber with height of 10cm was placed on the midspan of span 4# of the bridge. Trucks with weights of 20kN and 30kN were used as excitation for the un-retrofitted and retrofitted bridges respectively. Impact responses for both the un-retrofitted and retrofitted bridge are recorded in Table 4.

Table 4. Comparison of impact response between un-retrofitted and retrofitted bridge

Span number	un-retrofitted		retrofitted	
	acceleration /g	maximum displacement /mm	acceleration /g	maximum displacement/mm
1#	/	/	0.22	2.5
4#	0. 36	3.65	0.25	3. 0

From Table.4, it can be seen that impact accelerations were 0.36g and 0.25g for the un-retrofitted and retrofitted bridge respectively. That is, although the truck weight increases by 50%, the vibrational amplitude decreases by 18%. Therefore, the retrofitting effect is efficient.

Conclusions

In this study, the structural behavior of an old reinforced concrete arch bridge was studied, and retrofitting of the bridge was performed. Based on field inspections (including material property tests and static and dynamic tests of the bridge) executed on an old arch reinforced concrete bridge, the present state of the bridge was evaluated. A retrofitting scheme was determined, and the bridge was retrofitted. Field inspections were executed on the retrofitted bridge and the static and dynamic properties of the un-retrofitted and retrofitted bridges were recorded and compared. It is shown that the structural behavior of the retrofitted bridge has been significantly improved.

Acknowledgements

This paper presents some of the results of a project Funded by Jilin Highway Management Bureau of China, and is partially funded by a Liaoning Provincial Funded project (Grant No. 1050259).

References

[1] Code for Communications Ministry of P.R. China (1989), "General code for design of highway bridges and culverts (JTJ 021-89) ", Beijing: China Communications Press

[2] Code for Communications Ministry of P.R. China (1992), "Technical specification for strengthening concrete structures (CECS 25: 90)", Beijing: China Planning Press

[3] Code for Communications Ministry of P.R. China (2004), "Code for design of highway reinforced concrete and prestressed concrete bridges and culverts", Beijing: China Communications Press

[4] Code for Communications Ministry of P.R. China (2004), "Technical code of maintenance for highway bridge (JTGH11-2004)", Beijing: China Communications Press

Theme two:

Advances in the understanding of the structural behaviour of bridges

Investigation of mechanical properties of corroded reinforcement using simulated corrosion

Dr. Y. G. Du, The University of Birmingham, Birmingham, UK
C. W. Tang, formerly The University of Birmingham, Birmingham, UK
Y. Xu, Mott MacDonald, Bristol, formerly The University of Birmingham, UK,

Abstract

This paper reports the experimental results of simulated corrosion performed on reinforcement specimen to investigate the effect of corrosion on their mechanical properties, and particularly, to clarify some confusions caused by the contradictory results reported by different researchers who used the reinforcements corroded electrochemically. The removal of metal from a reinforcement surface caused by corrosion was simulated by mechanically machining the reinforcement with specified attack penetration and localization length, before it was subjected to a tension test to examine the variation of its mechanical properties. On basis of the experimental results, it has been found that the reduction of strength of corroded reinforcement was caused by the maximum penetration that was beyond the average penetration determined by weight loss, while the greater decrease of reinforcement elongation was due to the distribution of different residual sections of corroded reinforcement along its length.

Introduction

The premature deterioration of infrastructures, such as bridges, buildings and tunnels, is mainly caused by corrosion of reinforcement in concrete, which impairs the safety and serviceability of the structures and has ever been a major concern to their users and owners.

Over the past decade, considerable amount of literature has been published on aspects of corrosion mechanism, measurement and protection. However, relatively a few of papers were published on the mechanical properties of corroded reinforcement[1-10]. In all reported researches, reinforcement specimens were exposed to either an outdoor chloride environment or indoor accelerated corrosion or were removed from actual structures which had suffered from chloride intrusion or concrete carbonation, before they were subject to tension tests to examine the variation of their mechanical properties.

Bridge design, construction and maintenance 2007, Thomas Telford, London

From the experimental results, Maslehuddin et al reported that up to 1.1% corrosion hardly changed the strength and elongation of reinforcement[1]. By measuring the smallest sectional area of corroded reinforcement, Palssom et al also reported that the average yield and ultimate strengths of reinforcement with less than 10% loss of cross-sectional area are similar to those with more than 30% sectional loss, and even a slight increase in the yield strength was noted in pitted specimens[2]. However, Andrade et al, Lee et al, Morinaga, Zhang et al, Du et al and Almusallam argued that significant corrosion dramatically decreased the strength and elongation of reinforcement[3-9]. Regarding the ratio of ultimate to yield strengths, Zhang et al reported that corrosion reduced its value from 1.5 to 1.1[6], while Maslehuddin et al, Andrade et al, Palssom et al and Du et al reported that corrosion hardly affects this strength ratio [2,3, 8,9]

It is apparent that although the above works are valuable, the results reported by different researchers were not always consistent and some even contradictory. As a result, the experimental works were conducted on the simulated corrosion reinforcements to study the effect of corrosion on the mechanical properties of the reinforcements, and, particularly, to clarify some confusions caused by the above contradictory results reported by different researchers who used the actually corroded reinforcements. The reinforcement specimens were first mechanically machined with specified attack penetration and localization length to simulate the removal of metal from a reinforcement surface caused by corrosion, before they were subjected to a tension test to examine the variation of its mechanical properties.

Experimental Programme

Test Specimens

A total of 29 plain reinforcement specimens with the same length of 450*mm* and the same diameter of 16mm were manufactured, 23 of which were mechanically machined, as shown in **Figure 1,** to simulate the irregular attack penetration on the reinforcement surface caused by corrosion of reinforcement embedded in concrete.

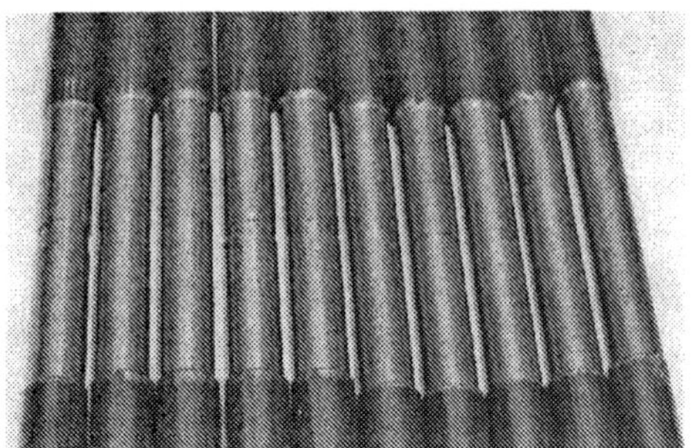

Figure 1 Test Specimen

The reinforcement specimens were first uniformly machined with minimum penetration x_{min} varying from 0.98 to 1.20mm along a length of 150mm, prior to a further localised machining in their middle of the length with both maximum penetration x_{max} and localization length $L_{smim,}$ ranging from 1.02 to 1.60mm and from 2.27 to 9.39mm, respectively, and with the residual areas of all machined reinforcements almost perfect circular ones, as schematically shown in Figure 2. Consequently, both average penetration $x = (x_{min} + x_{max})/2$ and the equivalent amount of corrosion $Q_{cor}=(4x/d_0)$ vary from 1.02 to 1.34mm and from 6.4 to 8.4%, respectively.

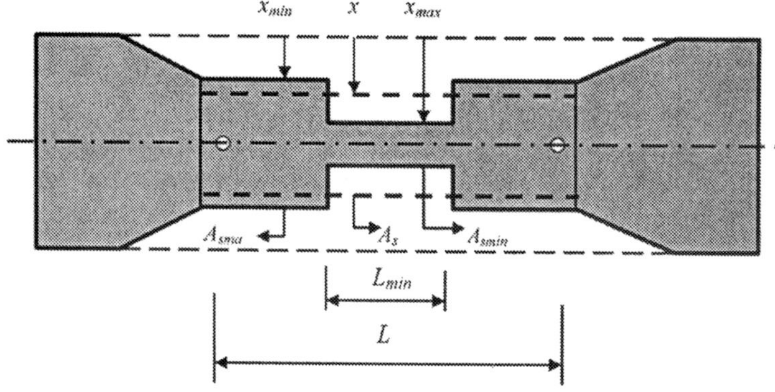

Figure 2 Research Variables

The research variables of the experimental work are the average penetration x, the ratio of maximum to average penetrations $\eta = x_{max}/x$ and localization length L_{smin}.

Uniaxial Tension Tests of Reinforcement

A Denision testing machine was used to perform the uniaxial tension tests on the reinforcement specimens, which was equipped with a data-processing computer and an electrical extensometer. The programmed computer was used to control the test and to record the applied load. The electrical extensometer has a maximum stroke of 0.3mm over its 50mm gauge length and was used to measure the linear elastic elongation of reinforcement to determine its elasticity, prior to the yielding of the reinforcement. In addition, two marks with initial distance of $L=80mm$, 5 times of the bar diameter, were made on the reinforcement surface prior to the tension test. After reinforcement fractured, the extension between these two marks was measured and was taken as the reinforcement elongation.

Prior to and after the tension test, the residual and deformed diameter of reinforcement specimens were carefully measured using a vernier caliper and instantly recorded for the calculation of attack penetration and the analysis of experimental results.

Results and Discussions

Fracture of Reinforcement Specimen

The experimental results show that, under an increasing tension force, all of the machined reinforcement specimens always fractured at their minimum residual section, A_{smin}, that has a maximum penetration, x_{max}, as shown in Figure 3, following the occurrence of an obvious yielding plateau on their

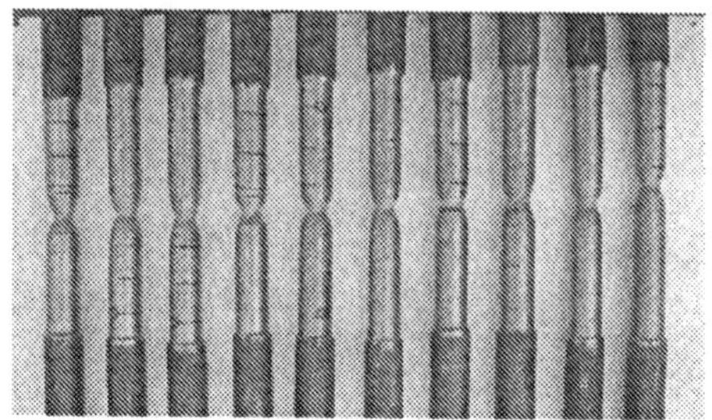

Figure 3 Fracture of Reinforcement

load-elongation curves at their yield forces.

Hence, it can be inferred that, although some uncertain factors due to the chemical composition, manufacturing process and quality control of reinforcement production might probably cause some random variation in the mechanical properties of reinforcement bars, a significant amount of corrosion would be able to cause corroded reinforcements always to yield and, particularly, to fracture first at their minimum residual area with maximum corrosion penetration.

Effect of Corrosion on Reinforcement Forces

The variation of yield force and ultimate force of reinforcement specimens with the attack penetration is shown in Figure 4. Here, the yiled force is the tension load under which reinfoceemnt specimen underwent substantial elongation without an increase of applied load. The ultimate force is the maximum tention force that the reinforcement is able to carry.

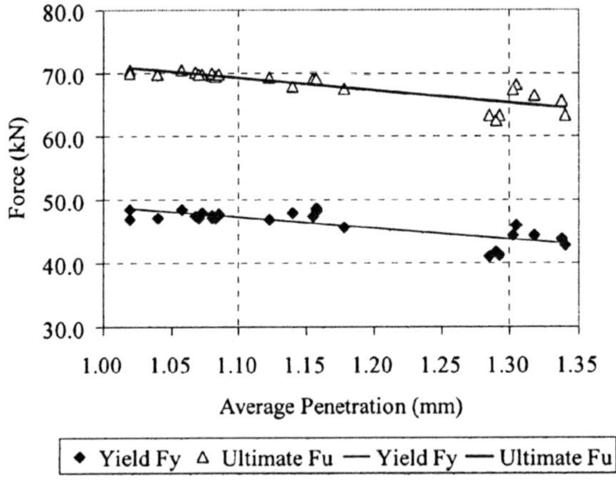

Figure 4. Effect of Corrosion on Yield and Ultimate Forces of Reinforcements

It is clear that both yield and ultimate forces of machined reinforcements reduces as an increase of attack penetration, due to the fact that the residual area of the machined reinforcement decreases as a result of attack penetration on reinforcement surface. In addition, Figure 4 indicates that the regressed trend-line of yield force F_y of machined reinforcements is almost parallel to that of their ultimate force F_u. In other words, attack penetration has little influence on the ratio of ultimate to yield forces of machined reinforcements, which is consistent with those reported on accelerated corroded reinforcements[7] and well confirms that corrosion does decrease both yield and ultimate forces of reinforcement, but hardly affect their ratio F_u/F_y of ultimate to yield forces.

Effect of Corrosion on Reinforcement Strength

The effect of corrosion on reinforcement strength is shown in Figure 5. Here, the yield strength of reinforcement specimen were calculated by dividing the yield force F_y by using four different types of cross sectional areas, that is, the nominal area A_{s0} of corresponding non-machined reinforcement, the minimum, maximum and average residual areas A_{smin}, A_{smax} and A_s with the maximum, minimum and average attack penetration x_{max}, x_{min} and x, respectively, of machined reinforcement.

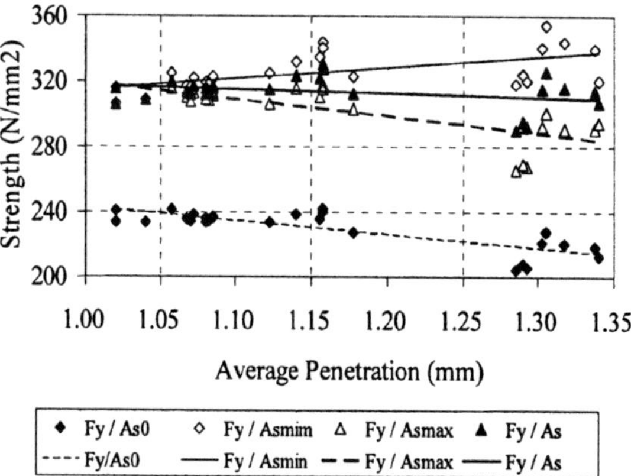

Figure 5 Effect of Corrosion on Reinforcement Strength

Figure 5 shows that the effect of corrosion on the yield strength of a machined reinforcement greatly depends on the type of the section used in its calculation. Using the nominal area A_{s0} of non-machined reinforcement significantly under-estimated the yield strength of machined reinforcement, due to actual reduction of its effective cross-section. Using the minimum area A_{smin} of machined reinforcement caused a slight increase of its yield strength, because of the fact that the steel material of reinforcement may not be uniform along its length and the weakest point of steel material of the machined reinforcement unlikely coincides with the point where the residual section of the machined reinforcement is the minimum one with the limited localization length L_{smin}. While the use of both average and maximum residual sections resulted in a reduction of yield strength of machined reinforcement as an increase of the attack penetration, since the both of them are greater than the actual residual section, i.e. minimum section A_{smin} of the machined reinforcement.

The above experimental results clarify the confusion regarding the effect of corrosion on reinforcement strength[1, 2, 3-9]. In addition the different type of reinforcement and techniques used in their experimental works, the contradictory results were mainly caused by the different type of the cross sections of corroded reinforcements used by the different researchers.

Furthermore, it should be noted that, due to the very irregular surface of actually corroded reinforcement, a direct measurement of their real residual area A_{smin} becomes very difficult. As a result, either an amount of corrosion calculated using the weight loss of corroded reinforcement or average penetration determined using the corrosion current measured in-site using polarization techniques are often used to measure the residual effective area of corroded reinforcements and calculate their yield strength accordingly in the assessment of practical corrosion-damaged structures. As a result, care should be taken when the average section is used to estimate the residual area of corroded reinforcement with pitting corrosion, which may over-estimate the bearing forces of corroded reinforcement.

The effect of the ratio of maximum to average penetration x_{max}/x on the reinforcement strength by using its average section is shown in Figure 6. It is clear that the reinforcement strength does not change too much with the penetration ratio, since it mainly depends on the properties of steel material of reinforcement and the minimum residual area of corroded reinforcement

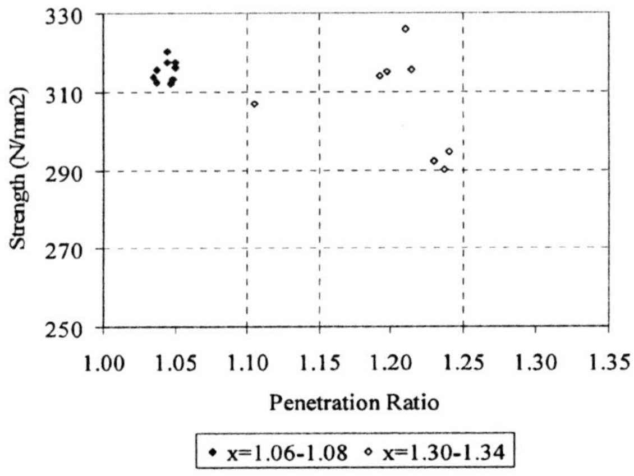

Figure 6 Effect of Penetration Ratio on Reinforcement

Effect of Corrosion on Reinforcement Elongation

The variation of reinforcement elongation with an increase of average penetration x under different values of localization length L_{smin} are shown in Figures 7.

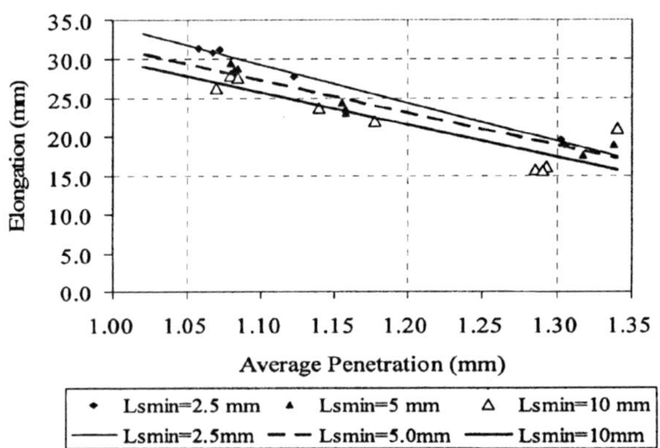

Figure 7 Effect of Attack Penetration on Reinforcement Elongation

It is clear that, compared with those shown in Figure 5, an increase of attack penetration caused the reinforcement elongation to decrease much more rapidly than its average yield strength. For example, as average penetration increased from 1.05 to 1.30mm, the yield strength of machined reinforcements decreased from 317 to 310 N/mm^2, only by 2.2%, while the value of their elongation with a localization length of L_{smin}=5.0mm reduced from 29.5 to 18.9%, by more than 35%. In other words, attack penetration has much more substantial impair on reinforcement elongation than on its strength, which is well consistent with those reported from experimental results of the corroded reinforcement [3,8,9].

Figure 7 also shows that an increase of localisation length results in a reduction of reinforcement elongation. For the same value of average penetration, the longer the localization length, the smaller the reinforcement elongation. The reason behind this result is that as an increase of localization length, the possibility of the localization of the weakest point of the steel material within the minimum residual section increases. In other words, for a longer localization length, the weakest point of steel material more likely coincides with the minimum residual section which promotes the failure of corroded reinforcement under a tension force and therefore decrease its total elongation. Hence, in practice, the elongation of corroded reinforcement with more pitting corrosion sections is likely smaller than those of corroded reinforcement with a single pitting, for the same type of reinforcement and same residual sections and penetration ratio.

Figure 8 shows the effect of the ratio $\eta=x_{max}/x$ of maximum to average penetration on reinforcement elongation under similar values of attack penetration x.

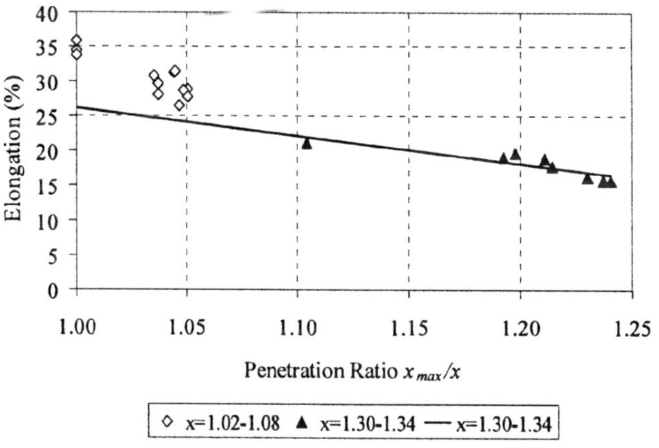

Figure 8 Effect of Penetration Ratio on Reinforcement Elongation

It is apparent that, apart from the attack penetration, the penetration ratio impairs reinforcement elongation considerably. For about 1.30mm of attack penetration, an increase of penetration ratio from 1.10 to 1.24 decreased the reinforcement elongation from 22 to 16.5%, by more than 33%.

As qualitatively discussed on the basis of experimental results of corroded reinforcement[9] and analytically investigated using the newly developed numerical model[10], the penetration ratio affects reinforcement elongation, because it dominates the distributions of the effective area of residual sections and, as a result, of the different levels of tension stress along the length of a corroded reinforcement when it fractures at its minimum residual section under a tension force. A small penetration ratio, to say, less than 1.05, means that the effective areas of residual section of corroded reinforcement along its length are almost the same and, therefore, the tension stresses over its residual sections are nearly similar each other and able to beyond its yield strength when the corroded reinforcement fractures at it's the minimum residual section. As a result, its deformation capacity is able to be fully developed before its fracture. However, if the penetration ratio is too large, to say, 1.25, the effective areas of residual sections and corresponding tension stresses/strains of a corroded reinforcement vary greatly along its length. When the corroded reinforcement fractures at its minimum residual section, the tension stresses/strains within the rest of residual section may likely be less than its yield stress and as a result, its deformation potential is unable to be fully developed. Hence, an increase of penetration ratio of corroded reinforcement decreases its elongation substantially.

After reinforcement specimens fracture at their minimum residual section, their deformed and residual diameters were carefully measured using a vernier caliper, as shown in the Figure 9.

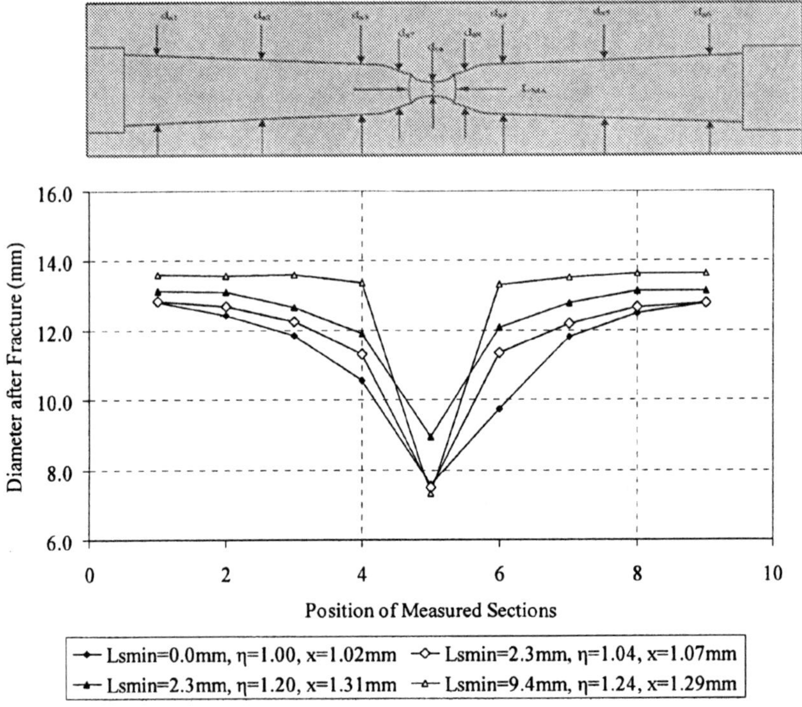

Figure 9 Deformed Diameter of Reinforcement after Fracture

Figure 9 show that attack penetration, localization length and penetration do have substantial influence on reinforcement elongation. Among all of the measured diameters, the deformed diameter of the uniformly machined specimen with L_{smin}=0.0mm, η=1.00 and x=1.02mm is the smallest and changes most gradually from fracture point, which indicates that uniformly machined specimen has a substantial longitudinal elongation along its length. An increase of either penetration ratio from 1.00 to 1.24 or localization length from 0.00 to 9.4mm, however, limit the transverse shrinkages and, of course, longitudinal elongation of machined specimens, which results a large values of deformed residual diameter and rapid change of radiometer as an increase of the distance from the fracture point of the reinforcement. Hence, the distribution of the different residual section along the length of the corroded reinforcement should be responsible for the huge reduction of reinforcement elongation

Conclusion

1. Corrosion decreases reinforcement forces, but does not affect the ratio of its ultimate to yield forces significantly.
2. Corrosion does not decrease the yield strength, if calculated using the minimum residual area of corroded reinforcement.
3. Corrosion does decrease the strength, if calculated using the sectional area other than minimum residual area, of corroded reinforcement.
4. Higher rate of reduction in reinforcement elongation than those of its strength is caused by the effective area of the residual sections of corroded reinforcement varying substantially along its length, as measured using penetration ratio and localization length.

References

1. Maslehuddin M, Allam I M, Al-Sulaimani G, Al-Mana A I and Abduljauward S N (1990), Effect of Rusting of Reinforcing Steel on Its Mechanical Properties and Bond with Concrete, ACI Materials Journal, Vol. 87, No.5, pp496-502.
2. Palssom R and Mirza M S (2002), Mechanical Response of Corroded Steel Reinforcement of Abandoned Concrete Bridge, ACI Structural Journal Vol.99, No.2 March-April 2002, pp157-162
3. Andrade C, Alonso C, Garcia D and Rodriguez J (1991), Remaining Lifetime of Reinforced Concrete Structures: Effect of Corrosion in the Mechanical Properties of the Steel, Life Predication of Corrodible Structures, NACE, Cambridge, UK, pp12/1-12/11.
4. Lee H S, Tomosawa F and Noguchi T (1996), Effect of Rebar Corrosion on the Structural Performance of Single Reinforced Beams, Durability of Building Materials and Components, Vol. 7, Edited by C Sjostrom, E & FN Spon, London, pp571-580
5. Morinaga S (1996), Remaining Life of Reinforced Concrete Structures after Corrosion Cracking, Durability of Building Materials and Components, Edited by Sjostrom C, Published by E & FN Spon, pp127-137

6. Zhang P S, Lu M and Li X Y (1995), The Mechanical Behaviour of Corroded Bar, Journal of Industrial Buildings, Vol. 25, No.257, pp41-44.

7. Almusallam, A A(2001), Effect of Degree of Corrosion on the Properties of Reinforcing Steel Bars. Journal of Construction and Building Materials, Vol. 15, No.8, pp361-368

8. Du Y G, Clark L A, and Chan A H C (2005), Residual Capacity of Corroded Reinforcing Bars. Magazine of Concrete Research, Vol.57, No.3, pp135-147.

9. Du Y G, Clark L A, and Chan A H C (2005), Effect of Corrosion on Ductility of Reinforcing Bars. Magazine of Concrete Research, Vol.57, No.7, pp407-419.

10. Du, Y G, Xu Y and Tang C W (2007) A Numerical Model To Assess Mechanical Properties of Corroded Reinforcement, Paper submitted to the Proceedings of the International RILEM Workshop on Integral Service Life Modelling of Concrete Structures, 5-6 November 2007, Guimarães, Portugal

Some Design Improvements for Integral Bridges

Prof. George L. England, Emeritus Professor, Imperial College London, UK.
Dr Neil C M Tsang, Lecturer, Imperial College London, UK.
Dr Pedro Ferreira, Lecturer, University College London, UK.
Mr Bruno S Teixeira, Research Assistant, University College London, UK.

Abstract

The design of integral bridges[1] is currently limited to lengths of no more than 60m in the UK because the behaviour of longer bridges is not yet fully understood. Longer bridges experience larger thermal length changes and abutment movements. These lead to: (i) a continuous flow behaviour in the backfill, leading to progressive settlements (close to the abutments) and, (ii) greater lateral soil/structure interface stressing. Current design methods provide only an empirical upper limit to the lateral stress escalation. They do not address the flow-induced settlement in the backfill at all.

Attention is focussed here on stabilizing the strain behaviour of the backfill while maintaining the integral form of construction, such that bridges of almost any length may be constructed. This is achieved by introducing into the beams a load bearing compression expansion joint - called a displacement compensation unit (DCU) – designed to operate at a load which maintains backfill stability and prevents the development of granular flow within. These units act as expansion joints operating under an essentially constant design force.

Model studies on a 1/12 scale stiff retaining wall have been used to simulate bridge lengths up to 120m. Recorded pressures on the face of the retaining wall and backfill soil movements over many cycles are presented. These studies are then compared with results for a model bridge of 60m length containing DCUs to accommodate the simulated thermal movements. The effectiveness of the DCUs is then quantified in relation to their effectiveness for longer bridges.

Advantages of the technique include: (i) definable maximum lateral stresses on the abutments, (ii) reduced road surface settlement close to the abutments, (iii) reduced stressing at the junctions of deck beams and abutments by eliminating hogging bending moments caused by repeated expansion/contraction of the deck, and (iv) an extension of integral form to long bridges.

---

[1] An integral bridge is defined here as one in which the deck beams are continuously connected to the abutments without mechanical joints and the abutments act additionally as earth retaining structures

Bridge design, construction and maintenance 2007, Thomas Telford, London

Introduction

Integral bridges with full height abutments, as currently in use in the UK, are restricted to lengths of no more than 60m [1]. This is because of the uncertain performance of longer bridges over their expected lifespan of 120 years. Figure 1(a) shows a typical integral bridge as currently constructed. For the longer bridges daily and seasonal temperature changes of the deck lead to repeated cyclical displacements being imposed on the backfill soil to the abutments [2] that are of such magnitude as to cause concern with regard to (a) the progressive cyclical build up of interface soil pressures, to limits which are not fully understood, and (b) to the progressive development of backfill settlements close to the abutments. In the longer term these result in depressions of the road surface and to ponding in wet weather. If these two features of adverse behaviour can be significantly reduced or eliminated, bridges of greater length than 60m can be constructed with confidence. This paper thus addresses these features through the introduction, into the span of the deck beams, of a compression/expansion joint called a displacement compensation unit (DCU) [3] that is designed to operate under a permanent compressive force so chosen as to: (c) limit the build up of lateral soil pressures on the abutment faces, and (d) stabilise the backfill soil such that the cyclically induced settlements adjacent to the abutments, common to most integral bridges, are eliminated.

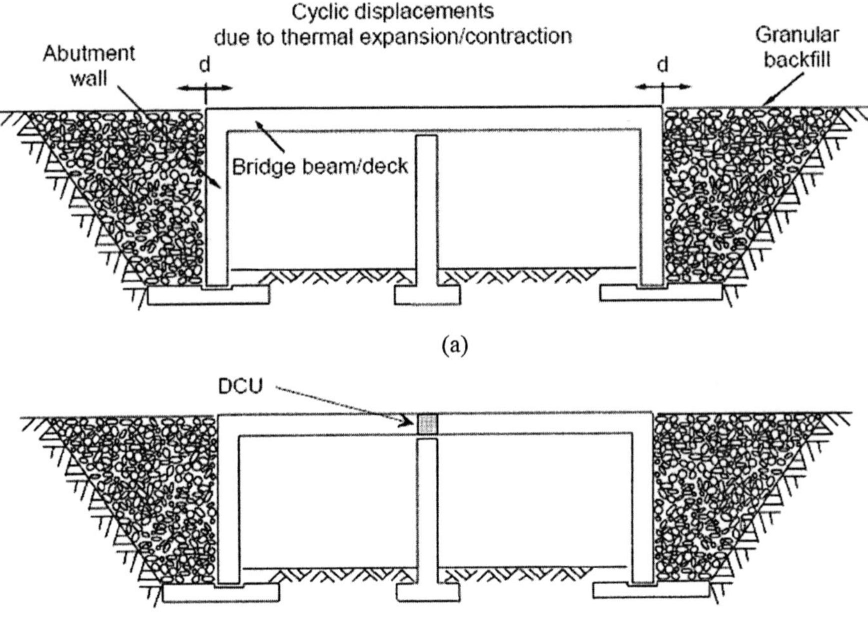

Figure 1   (a) Typical integral bridge with full height abutment,
(b) Integral bridge with DCU

Explanations describing the changes to backfill behaviour are given in the context of current bridge behaviour and the performance generally of granular backfill materials to cyclic strain and cyclic stress loading. Experimental results for the 1/12 scale model retaining wall are then presented for a range of bridge lengths up to 120m. Table 1 gives the details for a concrete bridge with 6m high abutments. Temperature data, (viz. effective bridge temperatures), have been taken from the work of Emerson [4] and relate to the Birmingham area of the UK for which the maximum annual variation is 42°C. A fuller report of daily and seasonal variations of temperature and their effect on bridges of concrete, steel and composite construction, is provided in [5]. Comparisons are then made with experimental results from two 60m model bridge simulations to demonstrate the effectiveness of the inclusion of DCUs on reducing lateral stressing of the abutments and the elimination of settlements close to the abutments.

Table 1. Tests for bridges of three different lengths

Experiment	Deck length, m	Seasonal d/H (%)*	No of seasonal cycles
Exp. 1	60	0.25	300
Exp. 2	120	0.50	180
Exp. 3	160	0.70	180

* d is the maximum overall movement of the top of the abutment resulting from the maximum annual change of effective bridge temperature, i.e. the bridge has been assumed to be symmetrical with each abutment being subjected to half the overall expansion and contraction, and thus to the same rotations, d/H, where H is the abutment height.

The ability of the DCUs to limit the soil-abutment interface stresses and settlements is clearly demonstrated from a comparison of the 'with' and 'without' results obtained from two tests using the same experimental set-up. Figure 2 shows the general form of the apparatus used to investigate repeated wall movements caused by the bridge deck expansion and contraction, the latter being applied by an eccentric cam as shown in inset (A). In this apparatus the DCUs are placed between the cam and the wall and may be seen in inset (B). When the DCUs are 'blocked' off as in inset (A) the cam displacement is transferred to the wall without reduction by compression of the DCUs.

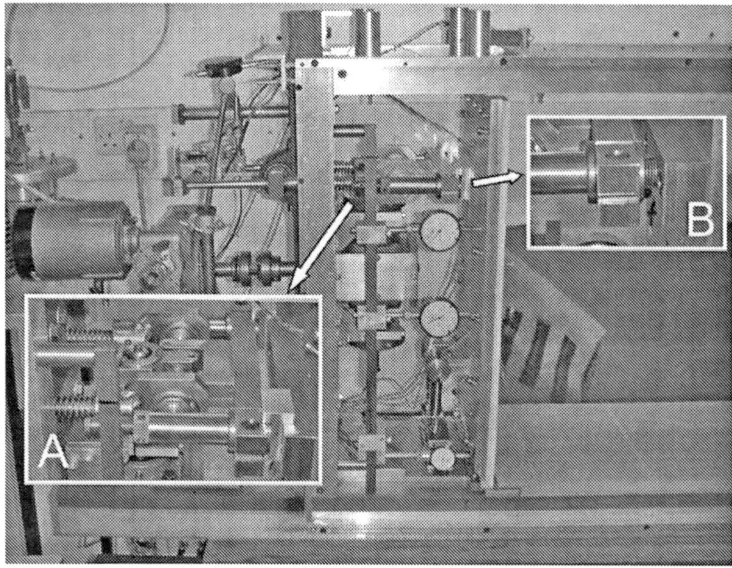

Figure 2   Photos of model wall apparatus (A) Displacement cam; (B) DCU springs

Here, an experiment containing DCUs (the 'with' test) was run for approximately 1500 cycles, the DCUs were then locked to produce a 'normal' stiff structure and a second test (the 'without' test) was run for comparison. By running the two experiments in series in this manner, variations of the backfill properties between tests, e.g. degree of compaction, were avoided. This feature can be readily appreciated because soil disturbances were absent during the running of the first test.

The experiments have simulated the magnitude of seasonal temperature cycles and have been extended well beyond the normal range for a 120 year lifespan (1500 cycles) for the purposes of defining limiting values of stresses and settlements and then determining the extent to which these limits are likely to be approached in practice. Daily temperature cycles are of smaller magnitude and when introduced in addition to the seasonal cycles they do not substantially change the overall stress performance, but can lead to increased settlements [2] in integral bridges as currently constructed.

Behaviour of Granular Backfill and Previous Results

Drained granular materials respond differently to cyclical stress and cyclical strain imposed loading and this is most marked when reversals of principal stress directions occur [6]. Cyclical stress changes lead to a progressive build up of 'structure' and to a stiffening of the soil due to reducing strains after each cycle. Little volume change generally occurs. Stress cycling without changes of principal stress direction and with a stress ratio $\sigma_1/\sigma_3<3$ leads to stable soil behaviour and with only little densification. Repeated cyclical straining usually leads to reversals of principal stress directions at each half cycle, to progressive densification, increased stiffening and to increasing principal stress ratios. The peak stresses are not easily determined and depend to some extent on the reasons for the repeating imposed deformations. When they relate to the interaction between the granular soil and an elastic structure, as in the case of a circular biological ring-tension filter bed [7], for example, the peak interface stresses are limited by the flexibility of the ring wall to the bed. The greater the flexibility, the lower are the interface stresses. For the abutment of an integral bridge the stiff nature of the structure encourages lateral interface stresses to increase cycle by cycle without restriction by the structure itself. The peak stresses will stabilise eventually by the soil reaching one of two possible states. Either to a shakedown state[2] [6,8] or to a steady-state flow [9] condition[3].

The former condition, if it genuinely exists for the integral bridge, can only occur in the backfill close to the abutment and for small wall rotations, i.e. d/H<0.0025. (*Here, d is the overall horizontal cyclical displacement of the abutment at bridge deck level (i.e. backfill soil level) and H is the height of the abutment above its centre of rotation at foundation level*). In this condition, repeating soil displacements are confined to a wedge of soil close to the wall. No active failure planes have been observed below this value of d/H.

For larger d/H values the soil behaviour during passive and active wall movements[4] is very dis-similar. Close to the abutment, rapid lateral expansion of the soil occurs during the active phase, accompanied by vertical contraction and settlement. This is followed, in the passive

---

[2] Defined as cyclically repeating stresses and strains, with no overall change of volume or shape during any complete cycle.
[3] Cyclically repeating stresses and strains with no overall change of volume, but with a ratcheting change of shape.
[4] *Passive* indicates abutment movement towards the backfill while *active* refers to movement away from the backfill

phase, by lateral compression which leaves a net downward deformation close to the wall and an approximately horizontal movement from the wall. A similar cyclical flow behaviour [9] occurs at points distant from the wall where the net movements over each cycle are horizontal compression and vertical expansion, thus leading to heave at the soil surface. These movements have been observed to take place at constant volume. Active failure planes have also been observed for those cases of d/H>0.0025. Although the limiting surface profile for this steady-state flow condition will not be reached during the normal lifespan of an integral bridge the large settlements close to the abutment face are of importance. So also, are the larger interface stresses which approach steady-state (i.e. repeating) values well within the lifetime of the bridge.

Figures 3 and 4 show the results of some earlier model studies for the three d/H values listed in Table 1. They compare the soil/structure interface stress distributions (close to steady state) and the surface settlements in the backfill soil behind the abutments.

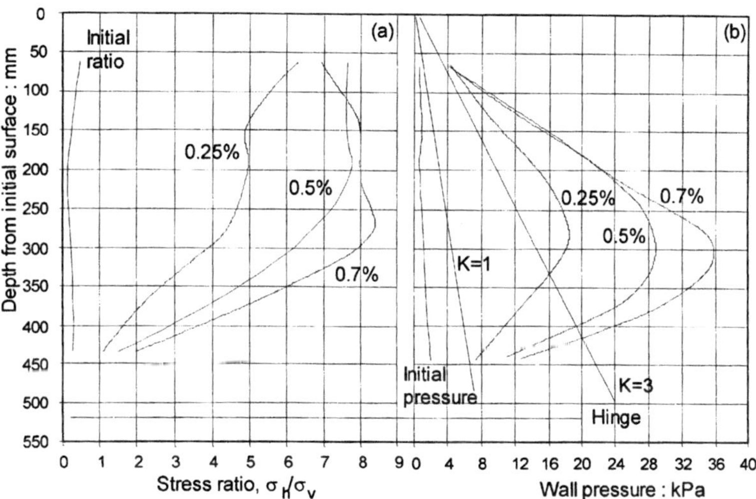

Figure 3   Variation of (a) stress ratio and (b) wall pressure at the wall face with depth below the soil surface; for three values of wall displacement, d/H = 0.25, 0.5 and 0.7%

Requirements of the DCUs

The two requirements of the DCUs are: (a) to prevent the cyclical escalation of lateral interface stresses on the contained side of the abutment faces and (b) to reduce or eliminate settlements behind the abutments.

These objectives are achieved by appropriate design of the DCUs such that the nature of the cyclic loading problem is transformed from one of displacement imposition at the abutment face (with associated 90° reversals of principal stress directions) to one of stress imposition (without changes of principal stress directions) during the cyclic process.

Ideally the requirement is to absorb the thermal expansion and contraction of the bridge deck under no change of stresses on the backfill. This could be achieved by interposing within the deck a hydraulic system operating at constant pressure with fluid displacement allowed during temperature length changes of the deck. This would require either an active feedback control system continually monitoring the hydraulic pressure, or a passive dead-weight control system on fluid flow. Other alternatives are elastic materials or springs provided they can be pre-compressed to specific design loads sufficient to support the backfill pressures during thermal cycling. Springs with non-linear load/displacements characteristics are best suited because they allow variations of backfill pressures to be maintained between close limits which themselves maintain good soil stability, i.e. little or no densification and therefore little or no settlement. The most appropriate limits within which to control the backfill behaviour are for, $3 > \sigma_1/\sigma_3 > 1$.

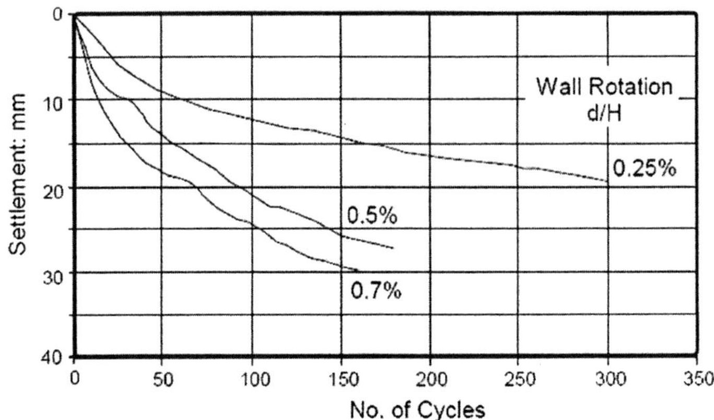

Figure 4   Progressive settlement of soil behind the model wall with wall rotational cycles for three values of wall displacement, d/H = 0.25, 0.5 and 0.7%

Figure 5 shows diagrammatically the three types of DCU described and Figure 6 gives the load/displacement characteristic for the DCUs used in the model experiments. Here the DCUs consist of two parallel banks of 10 disc springs each, so arranged to give an overall stiffness equal to one fifth of a single spring.

Results from new tests, with and without DCUs

The details of two comparative experiments , with and without the inclusion of mDCUs, are given in Table 2.

Table 2. Tests with and without DCUs

Test No.	d/H (%)	Effective bridge length (m)	With DCUs
4	#	*	Yes
5	0.25	60	No

  * This test is for the same deck expansion/contraction as Test 5, but reflects equally the behaviour for greater d/H values also, since the stiffness of the DCUs can be reduced proportionately for the increased deck expansion cases, i.e. longer bridges.

  # This has no meaning for the DCU test because the cyclic wall displacement is reduced to zero by the presence of the DCU. However longer bridges require DCUs of lower stiffness to achieve this result. This can be easily arranged by, for example in the disc spring DCU case, increasing the number of disc spring elements in the spring stack.

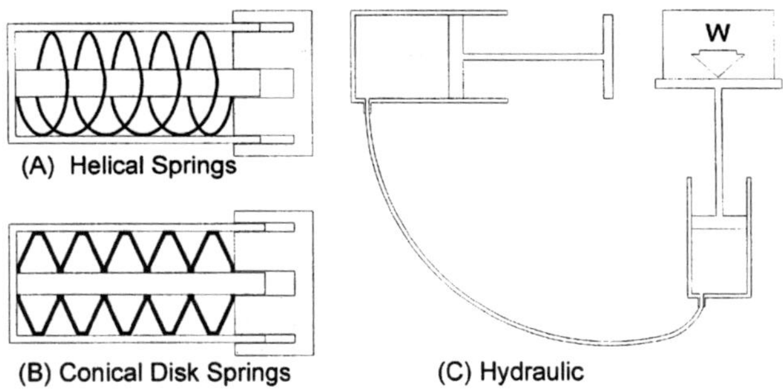

Figure 5. Diagrammatical representation of the three types of DCU

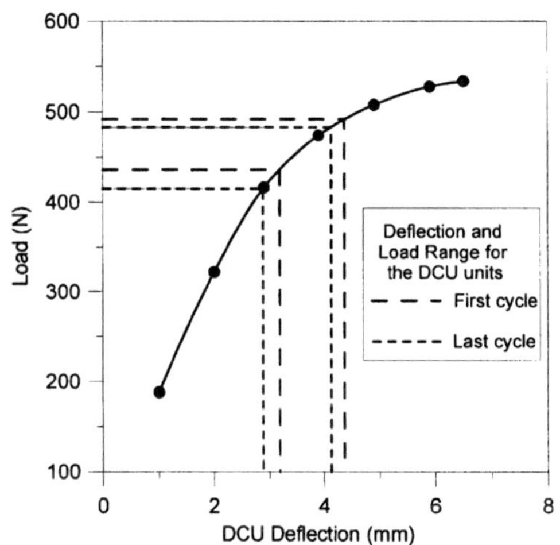

Figure 6.   Load/displacement characteristic for the DCUs used in the model experiments

Figure 7 illustrates the influence of the DCUs by comparing two experiments with the same simulated bridge deck thermal length changes. The 'without' results show the characteristic results for the passive (bridge deck expanded) pressures, with a peak stress at approximately half depth of the backfill amounting to approximately 16kPa. During the unloading phase of the displacement cycle stresses fell everywhere to lower than 2kPa. The result of including the DCUs shows a reduction of peak stress to around 10kPa, falling by an insignificant amount during the unloading phase. In other words, the cyclic stressing of the soil had become of very small amplitude and without change of principal stress directions during the cyclic process. The influence of this changed behaviour may be seen in Figure 8, where during release of the pre-compression in the DCUs, the wall moved into the backfill by approximately 1.6mm and thereafter showed little extra movement for the next 100 cycles (equivalent real time of 100 years), and eventually to around 1300 cycles with little change.

The reduction of backfill settlements are shown in Figure 9. Progressive settlements occurred in Test 5 (the 'without' test, Figure 10) and these are shown for comparison in Figure 9 with zero settlement for Test 4 (the 'with' test).

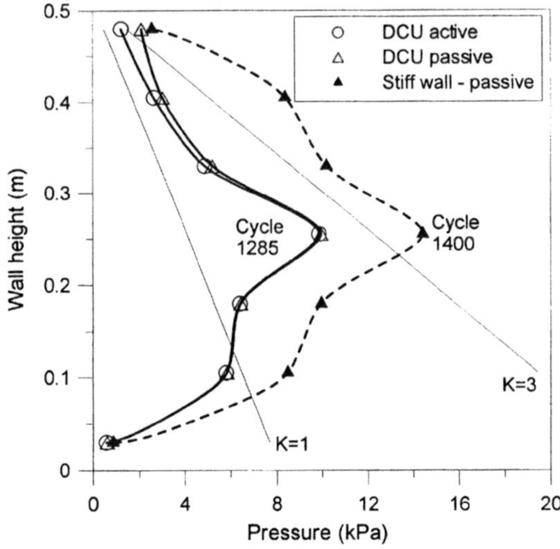

Figure 7.   The influence of the DCUs on soil stresses with the same simulated bridge deck thermal length changes.

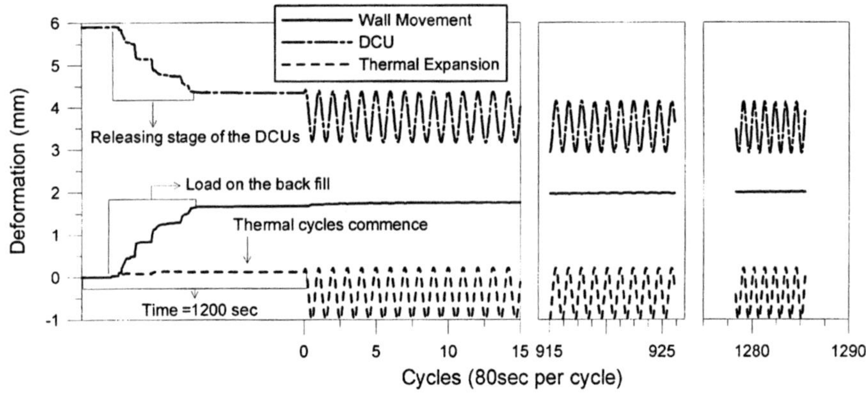

Figure 8.   Relative movements of bridge deck, DCU and the top of the model wall.

Discussion and Conclusions

Figure 11 shows schematically the behaviour of a soil element close to the wall at approximately half depth of the backfill, as influenced by the DCUs. The primary effect of the DCUs is to reduce the magnitude of the cyclic displacements imposed on the backfill, even though in the test reported here there was an initial lateral soil displacement on release of the pre-compression in the DCUs; Figure 8. This can be attributed to the loose packing density of the backfill sand used (voids ratio approximately 0.65). With appropriate compaction of the backfill it will be possible to significantly reduce or eliminate this initial displacement. Tests are currently being prepared to confirm this behaviour.

The 'with' test has demonstrated that cyclic displacements from the expanding and contracting bridge deck no longer produce reversals of principal stress directions at each half cycle. The displacement imposed problem has thus been transformed into a cyclic stress imposed problem for soil elements close to the wall; Figure 11. The fluctuating stresses have a small amplitude and over the upper portion of the wall these variations bring about soil stability at stress ratios in the range $3>\sigma_h/\sigma_v>1$. The stability to cyclic stressing is confirmed by the small amount of ratcheting horizontal displacement created, Figure 8, and lack of settlement, Figure 9.   In contrast, the 'without' test revealed progressively increasing settlements cycle by cycle with no indication of a limit being approached.

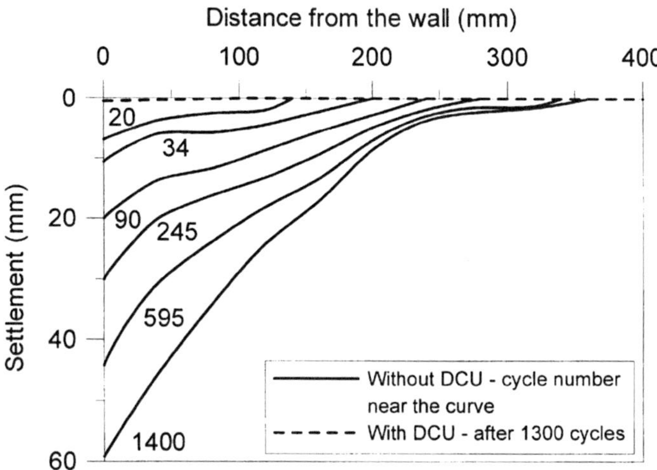

Figure 9 Soil settlement profile behind the model wall with number of wall rotation cycles indicated, showing substantial reduction of settlement by the DCU,

Figure 10.   Photo records of soil settlement for Test 5, without DCU (the spacing of the gridline on the glass wall is 40mm).

Although the example test quoted here refers to a deck displacement for a bridge of 60m length, the same behaviour can be expected for larger displacements. This is because the stiffness characteristic of the DCUs can be modified (by the addition of more series mounted discs in the disc spring case) to accommodate the larger displacements for the same force variation.

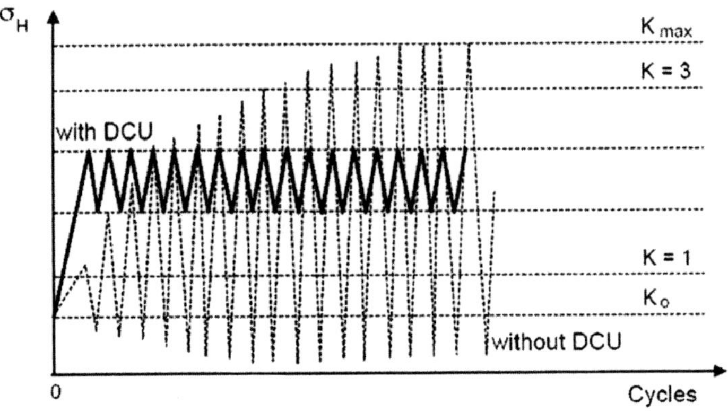

Figure 11. Schematic representation of soil stresses in an element at half-depth next to wall with and without DCU installation.

Good compaction of the backfill, prior to release of the DCUs will prevent any significant horizontal compression of the backfill during this phase. Good soil stability will be maintained thereafter.

Finally there are two practical issues that need to be addressed for prototype designs using DCUs. Firstly, they need to be designed such that they can be installed in a pre-compressed state and released after the backfill has been placed and compacted. Secondly they need to be protected (i.e. surrounded or encased) in a low modulus rubber so that the joint remains water-tight at all times.

Acknowledgement

The authors are grateful to Mr Paul K Mennell of the Geotechnic Engineering Laboratories, University College London for the high quality manufacturing of equipment for the upgrade of the model wall.

Reference

[1]   Highways Agency (2003), BA42/96 Amendment 1, The Design of Integral Bridges, Design Manual for Roads and Bridges, The Stationary Office, London, Vol.1.2.12. May 2003.

[2]   England, G.L., C.M. Tsang and D. Bush, (2000), Integral Bridge – a fundamental approach to the time-temperature dependence problem, Thomas Telford, UK. ISBN 0-7277-2845-8.

[3]   International (PCT) Patent Application No,. PCT/GB/2006/001383

[4]   Emerson, M.(1973), The Calculation of the Distribution of Temperature in Bridges, TRRL Report 561, Department of Environment, UK, 1973

[5]   EMERSON, M., Extreme values of bridge temperatures for design purposes, TRRL Report LR744, Transport and Road Research Laboratory, Department of Environment, UK, 1976

[6]   England, G. L., C.M. Tsang, T. Dunstan and R.G. Wan (1997), "Drained granular material under cyclic loading with temperature-induced soil/structure interaction", Appl Mech Rev, 50(10), 553-579.

[7]   England G.L. (1994), The performance and behaviour of biological filter walls as affected by cyclical temperature changes. Serviceability of Earth Retaining Structures, ASCE Geotechnical Special Publication No. 42, 1994, pp57-76.

[8]   England, G.L. and T. Dunstan, (1994), Shakedown solutions for soil containing structures as influenced by cyclic temperatures. Pro. of the 3rd Kerensky Conf.; Global Trends in Structural Engineering, Singapore, pp. 159-170.

[9]   England G L, Dunstan T, Tsang C M, Milhajlovic N, Bazaz J B (1995), Ratcheting Flow of granular materials, Static and Dynamic Properties of Gravelly Soils, Geotechnical Special Publication No. 56, M Evans and R Fragaszy (ed), ASCE, 1995, pp64-76

Influence of locked-in girder stress on the design of suspension bridges

Jian Ren Tao, Ph.D., P.E., Principal Engineer, SC Solutions, Sunnyvale, CA, USA
Tie Zong, MS, P.E., Senior Project Engineer, HNTB, Bellevue, WA, USA

Abstract
A fundamental assumption used in designing suspension bridges is: under the total dead load on the bridge the cable is parabolic and the stiffening girder is unstressed at mean temperature. Since the total dead load includes self-weight (DL) and superimposed dead load (SDL), the assumption implies that the SDL should be applied before stiffening girder has become integrated with the main cable, otherwise the stiffening girder will be stressed. The stress in stiffening girder due to SDL is thereafter referred to as locked-in girder stress. How to calculate the locked-in girder stress? Should it be considered in design? To answer these questions is the objective of this paper. First, an iteration algorithm is developed to determine the locked-in girder stress, and then results of two case studies are presented. The first case is Tsing Lung Bridge which has been recently designed, and the second one is The Second Tacoma Narrows Bridge which is currently under rehabilitation construction. The conclusion drawn from this study is: for either ne w design or rehabilitation evaluation, the locked-in girder stress is too significant to be ignored.

Introduction
Over the past one and half century, remarkable progress has been made in design and construction of suspension bridges [1, 2]. The theory of suspension bridges has evolved from Rankine theory, to elastic theory, three dimensional deflection theory [3] and finally today's nonlinear finite deformation theory [4]. Despite its high sophistication and superior accuracy, the modern theory of suspension bridges has kept one of the fundamental assumptions rooted in its origin: under the total dead load on the bridge the cable is parabolic and the stiffening girder is unstressed at mean temperature. The assumption clearly sets an initial condition for a suspension bridge to receive various loads subsequently applied, and its responses under the live load, temperature and wind has thus become the focus of various theories, whereas the fundamental assumption itself has left unquestioned. The assumption also implies that zero stress in the stiffening girder can be achieved by taking a specific erection sequence: lift stiffening girder in segments and hang them freely from the main cables, and then connect all joints after SDL is applied. In reality this ideal erection sequence is never strictly followed. For stability and practical considerations the stiffening girder is connected before SDL is applied. This is particularly true for modern suspension bridges where the streamlined steel box girder segments are jointed by butt weld prior to applying the SDL [5]. For truss type stiffening girders corrective actions may be used to release the locked-in girder stress due to erection process [6]. Even with the best attention and adjustment, it is unavoidable that there will be a stress residing in the stiffening girder due to erection process. Although the need to modify the fundamental assumption is so apparent and logical, it is often ignored by

Bridge design, construction and maintenance 2007, Thomas Telford, London

practicing engineers. This is mainly due to: 1. the difference between originally assumed and actually applied is easily overlooked; 2. the locked-in girder stress is considered negligibly small; 3. the correction is simple in concept but difficult to implement, because when cable is coupled with the stiffening girder under total or partial SDL, its initial shape becomes an unknown and it is not easy to find the solution.

Is the locked-in girder stress due to total or partial SDL really too small to be considered? To answer the question is the main purpose of this paper. In the paper, the SDL applied after the joints of the stiffening girder are connected is treated using a different approach from that of for live loading, and an iteration algorithm is developed to determine the cable geometry under total dead load and girder stress under SDL. The analysis is carried out using nonlinear finite element methods. Results of two case studies are presented. The first case is Tsing Lung Bridge which has been recently designed, and the second one is The Second Tacoma Narrows Bridge which is currently under rehabilitation construction. In conclusion, presence of locked-in girder stress will generally lead to increase in total girder stress under service load conditions. For the two cases presented in the paper, the increase in total girder stress is significant and locked-in girder stress should not be ignored.

Numerical Solution

Our objective is to define the cable shape and the girder stress under the total dead load. With presence of geometry nonlinearity of cables and locked-in girder stress, cable curve can not be solved explicitly and numerical solution has to be sought. Since the locked-in girder stress is introduced during construction, its distribution pattern and magnitude will vary depending on particular construction method and sequence used. It is impractical to consider all the possible erection schemes. For simplicity, we will restrict attention in this paper to the case where (1) the total dead load is applied sequentially as DL followed by SDL; and (2) SDL is applied after the stiffening girder is connected and become integrated with the main cables. To calculate the locked-in girder stress, we proceed with the following assumptions:

1. Cable is perfectly flexible.
2. Dead weight of cable is uniformly distributed along its horizontal projection.
3. Under the total dead load (DL+SDL) on the bridge the towers are vertical and moment free at mean temperature.

To seek numerical solution, the first Rankine assumption is no longer necessary. Main cables are subjected to uniform load of its own dead weight and concentrated loads from suspenders. Once dead weight of cable and suspender forces are known, cable shape can be calculated based on the above first two assumptions and using equilibrium equations. The difficulty is, however, when the stiffening girder is coupled with the main cables, suspender forces are not readily calculable. An initial distribution of suspender forces must be first assumed and then revised by iteration process. The convergent solution should satisfy the additional condition as stipulated in the third assumption for most practical purposes.

For a suspension bridge with general three-span arrangement and under the total dead load, the proposed numerical algorithm comprises the following four calculation steps.

Calculation Step 1

To obtain the first approximation, let's consider a virtual bridge as shown in Figure 1. The substitution is identical to the real case in which we are interested except that the stiffening girder is unstressed under DL + SDL. With absence of locked-in girder stress, the suspender forces can be calculated based on the given load distribution. According to the first two assumptions, the funicular cable curve can be expressed in the given coordinate system as [2]:

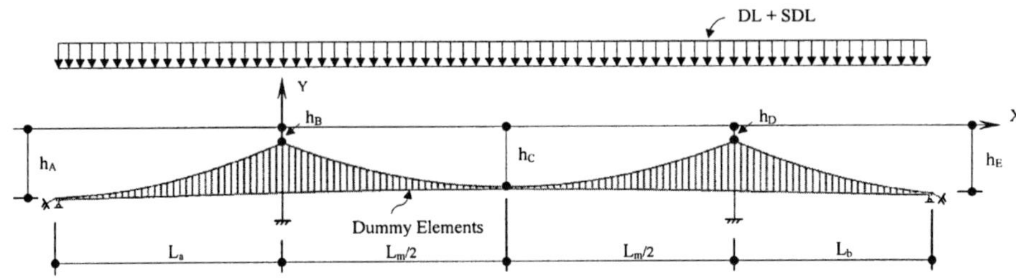

Figure 1: Apply DL+SDL to Virtual Bridge

$$y = \begin{cases} -\dfrac{M_a(x)}{H} + \dfrac{(h_A - h_B)x}{l_a} - h_B & -l_a < x < 0 \\[2mm] -\dfrac{M_m(x)}{H} - \dfrac{(h_D - h_B)x}{l_m} - h_B & 0 < x < l_m \\[2mm] -\dfrac{M_b(x)}{H} - \dfrac{(h_E - h_D)}{l_b}(x - l_m) - h_D & l_m < x < l_m + l_b \end{cases} \tag{1}$$

Where $M_a(x)$, $M_m(x)$ and $M_b(x)$ are the moments at section x of simply supported beams with lengths l_a, l_m l_b subjected to DL+SDL. DL should include the self weight of main cables, suspenders, and stiffening girder. The horizontal component of the cable force can be found from:

$$H = \frac{M_m(x = l_m/2)}{h_c - (h_B + h_D)/2} \tag{2}$$

where $M_m(x = l_m/2)$ is the simple moment at the main span center and $h_c - (h_B + h_D)/2$ is the cable sag at midspan. The span lengths, tower heights and cable sag at midspan are all predetermined. The bridge can be modeled using any finite element program taking into account of large deflection effects. Elements representing cables, suspenders and towers shall be given proper initial strains such that under total dead load their position will coincide with the curves defined by Equation (1). In order for the stiffening girder to be unstressed, it should be modeled using dummy elements with zero bending stiffness. Practically, it may be necessary to assign the dummy elements with negligible bending stiffness that is needed to maintain numerical stability. To be correct, the dead load of stiffening girder and suspenders including SDL is applied as concentrated forces acting at each hanger deck connection.

Calculation Step 2

After the imaginary status as described above is obtained, we go one step backward by removing the SDL from the bridge. Again the reversed load is applied as concentrated forces acting at the lower ends of the suspenders. During the process, Towers will deflect towards the side span side, the suspenders will shorten and the main cables move upwards. Possessing no bending stiffness, the stiffening girder will experience the total associated displacements. The deformed shape of the bridge is as shown by the dashed line in Figure 2.

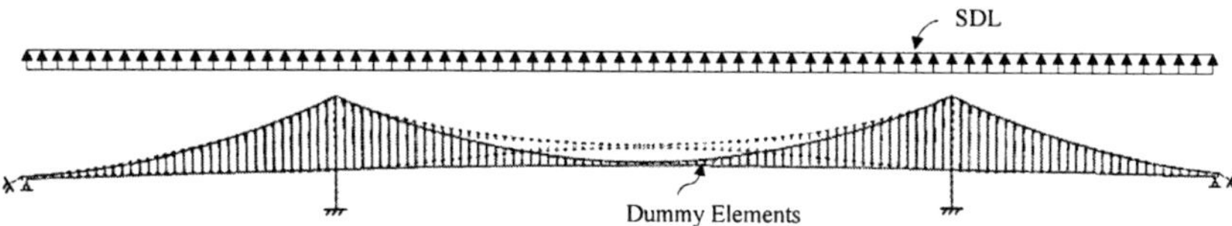

Figure 2: Remove SDL from Virtual Bridge

Calculation Step 3

The purpose of this step is to create a load case where the cable shape can be adjusted. Let's first present here as how to simulate the cable adjustment numerically, and explain why and when this adjustment becomes necessary in the next section. As illustrated in Figure 3, the cables are subjected to uniform temperature change of T_m in the main span and T_a and T_b in left and right sides spans, respectively. The change in temperature can either be rise or fall or both. Under the imposed thermal load, cable lengths, cable sags and cable span lengths will change, and towers will displace in the longitudinal direction accordingly. The values of T_m, T_a and T_b are determined depending on the results obtained from the next step. For the first iteration, however, they are taken as zero. Since the virtual temperature change does not alter the equilibrium equations of the bridge under externally applied forces, it only causes redistribution of internal forces among towers and main cables.

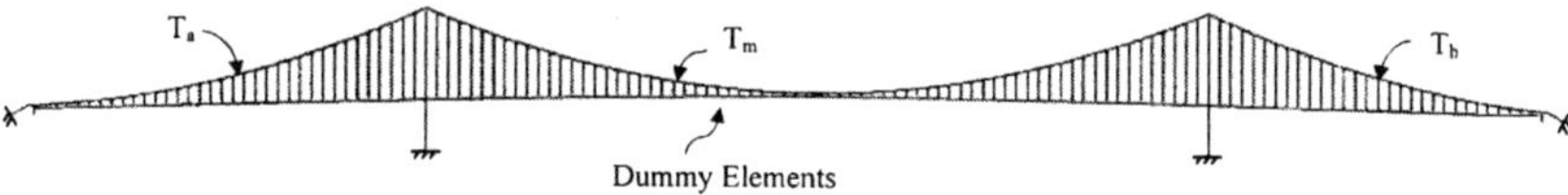

Figure 3: Adjust Cables Using Virtual Temperature

Calculation Step 4

At the end of Step 3 we have reached an initial condition for the stiffening girder to receive SDL together with the main cables. We proceed with replacing the dummy elements by beam elements or any types of appropriate elements possessing actual stiffness of the stiffening girder. This can be done by using death and birth options available in many finite element programs. The newly born stiffening girder elements will take the geometrical location coincide with that of dummy elements just removed. The analysis is continued by applying the same SDL back to the bridge. The situation is depicted in Figure 4.

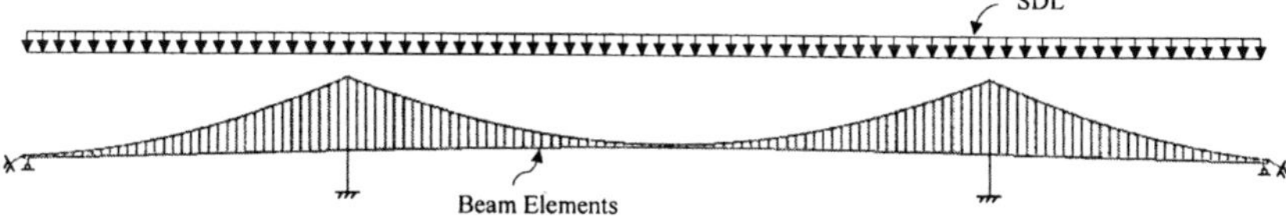

Figure 4: Apply SDL to Composite Structure

The final cable shape obtained in this step may not be exactly the same as that defined in the calculation step 1, because SDL is unloaded from and reloaded onto the different structural systems. If the horizontal components of cable force differ between the main span and the two side spans, towers would not stand plumb and are thus subjected to undesirable bending. In order to satisfy the third assumption, cable adjustment as described in Step 3 is required.

The particular values of T_m, T_a and T_b are chosen such that the obtained solution satisfies the following conditions simultaneously: the cable sag at midspan equals to what is used in the first calculation step; and the horizontal components of cable force in three spans are equal to each other. Practically, this is done through trail and error methods. Calculation Step 3 and Step 4 are repeated on each new set of temperature change until the above conditions are reached within acceptable tolerance. Theoretically, three conditions for three variables, the sole solution is ensured. As long as all equilibrium equations and boundary conditions are satisfied, the obtained results, which include suspender forces, the cable shape and the stiffening girder profile with presence of locked-in girder stress, are valid and final.

Case Studies

The methodology of numerical approach described in the preceding section is now applied to two actual cases and the analyses results are presented below.

Case 1: Tsing Lung Bridge

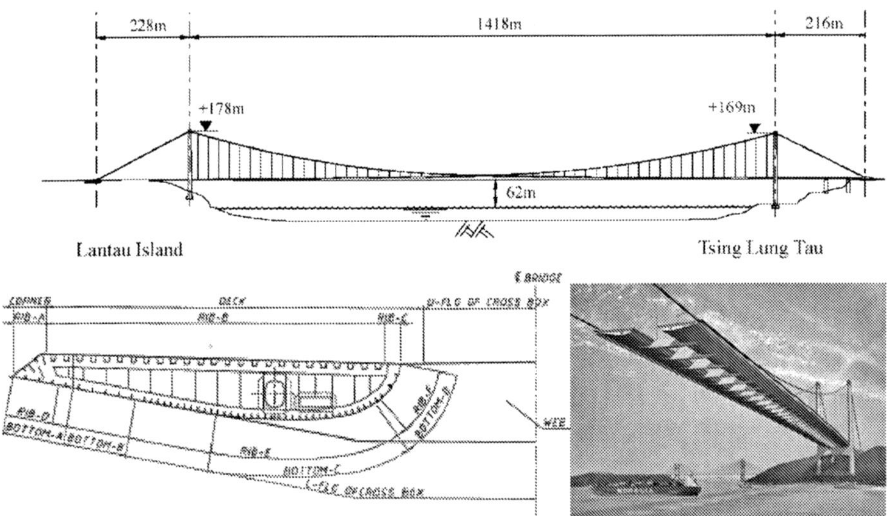

Figure 5: Tsing Lung Bridge – Elevation, Half Cross Section and Rendering

This bridge has been designed jointly by Maunsell Group, Brown Beech Associate and Chodai Co. Ltd. According to what has been adopted in the final design, Tsing Lung Bridge will be a single span suspension bridge. Representing much modernized design trend, the suspended span has a streamlined vented twin steel box girder spanning 1418m, as shown in Figures 5. The two bridge towers are made of concrete and have different heights, which are restricted by aviation requirements. The side spans are noticeably short. Due to increased inclination, the main cables in side spans are under higher tension comparing to that of in the main span, and thus have a slightly larger size. In the main span, the cable is composed of 61 strands each contains 504 high strength wires measured 5.3mm in diameter; whereas in the side spans, it has four extra strands each contains 432 wires of same size. Suspender ropes are spaced at 25.5m. The dead weight of cable is computed as 57.2kN/m in both side spans and 54.4kN/m in the main span including wrapping wires and cable bands. The stiffening girder has an average moment of inertia of 1.39 m^4 and weights about 144kN/m including cross beams. The SDL is approximately 56kN/m. The bridge is modeled using ADINA program.

For simplicity, the concrete approaches in the side spans are not included. In the computer model, cables and suspenders are modeled using truss elements, whereas towers and stiffening girder are represented by 3-D beam elements.

Case 2: The Second Tacoma Narrows Bridge

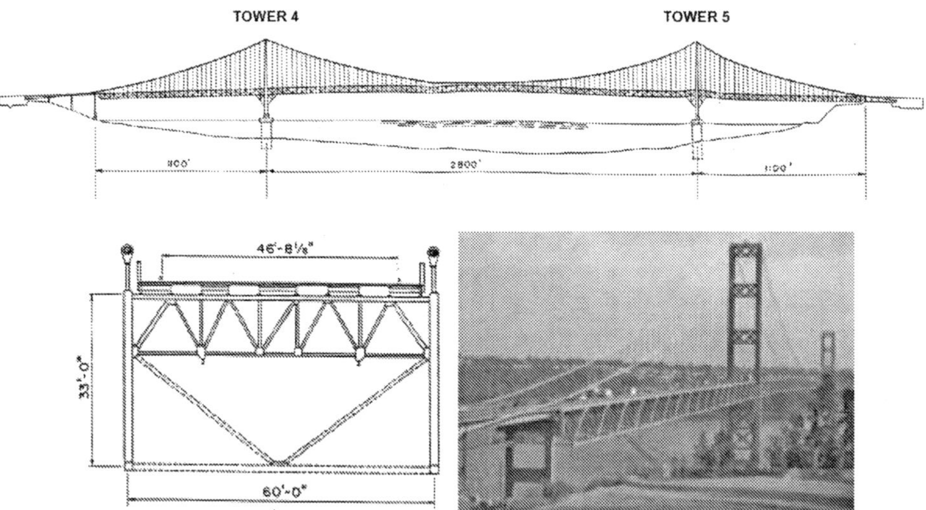

Figure 6: The 2nd Tacoma Narrows Bridge – Elevation, Cross Section and Photo

Completed in 1950, the Second Tacoma Narrows Bridge, as shown in Figure 6, replaced an earlier bridge which collapsed in 1940, four months after it was opened to traffic, due to lack of aerodynamic stability. The bridge has a conventional truss type stiffening girder and cast-in-situ light weight concrete deck. Each main cable is composed of 8702 No. 6 wires measured 0.192 inch in diameter and weights about 900 Lbs/ft including wrapping wires and cable bands. Suspender ropes are spaced at 30.5ft approximately. The stiffening truss has an average moment of inertia of 332 ft^4 and weights about 6.6 kips/ft including 3.3 kips/ft for assumed SDL. The bridge is modeled using HNTB in-house program T187. The vertical, lateral and diagonal members as well as truss chords are explicitly modeled using 2-D truss elements.

An on-going rehabilitation construction of the subjected bridge started in 2003. Study of historical documents reveals that the contractor when building the bridge made great effort to minimize the locked-in girder stress due to erection process [6]. The top and bottom chords of lifted truss were connected using temporary connections, and the final full strength connections were not made until after complete erection of the stiffening truss and the deck concrete placed. It was believed that by taking these measures very little stress would exist in the stiffening truss due to full dead load. However, a residual bending stress range of 1 to 5 ksi was found in the truss chords by filed measurements conducted in 1991 by Avid Grant [6]. And it was concluded by the investigator that this stress was locked-in during erection process. In the following analysis concrete deck is treated as part of SDL and is applied after the stiffening truss is presumably fully connected. By doing this, we are not suggesting that temporary connections used in the erection did not work as planned and the weight of the concrete slab was the source of the measured stress. Our focus here is just to study how a suspension bridge with conventional heavy stiffening truss would react in this somewhat speculated load case. To determine the true cause, probably multiple causes, of this residual stress is outside the scope of this paper.

Results and Discussions

The calculations are carried out following the numerical procedure described in the preceding section, and for both cases the geometric non-linearity due to large deflection of main cables is taken into account. In the results presented below, the legend 'Ideal' stands for conditions where the stiffening girders are perfectly unstressed under total dead load; whereas 'Actual' represents cases where the stiffening girders are initially unstressed under their own dead weight and then become integrated with the main cables before SDL is applied.

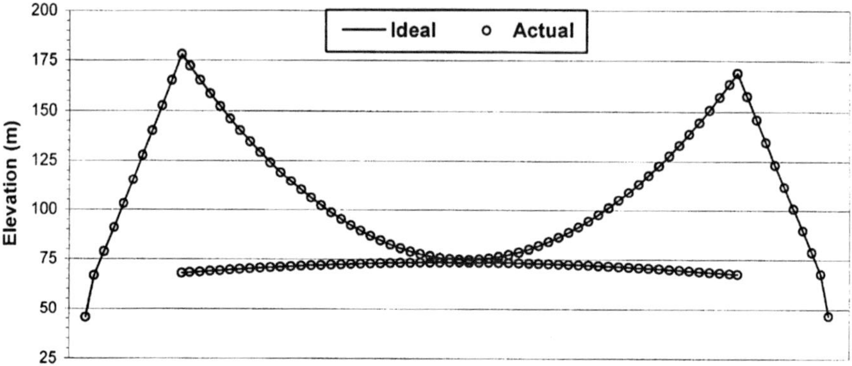

Figure 7: Comparison of Cable and Girder Geometry (Tsing Lung Bridge)

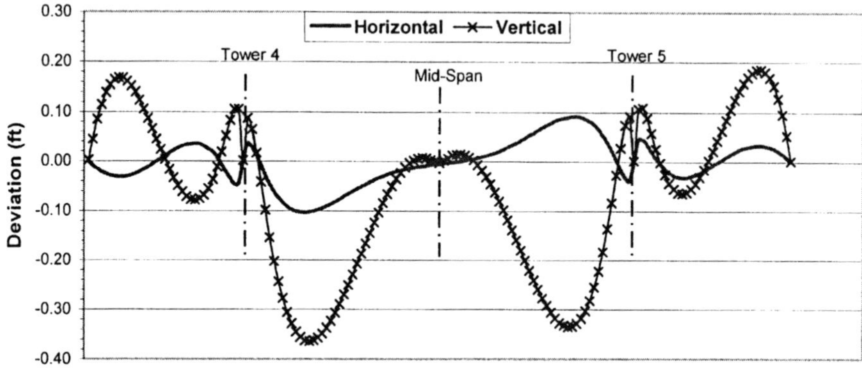

Figure 8: Change of Cable Curve (The 2nd Tacoma Narrows Bridge)

Figure 7 and Figure 8 show the shape change of main cables due to participation of stiffening girder during application of SDL. For Tsing Lung Bridge, the actual cable curve nearly coincides with the Ideal cable curve; the largest deviation is less than 10 millimeter. Similar results are observed in case of The Second Tacoma Narrows Bridge; where the maximum deviation of Actual cable from the Ideal one is 0.36 foot horizontally and 0.18 foot vertically. Compare to the span length of cables, these deviations can be totally ignored for most of the practical purposes. Since the difference in cable shape is so negligibly small, the calculation

Step 3 as proposed in this paper can be omitted for both bridges without causing noticeable changes in the final results. It can be concluded that for the two bridges selected for this study, cable geometry is insensitive to particular construction procedure adopted, and whether or not there is locked-in girder stress can not be determined by examining the cable curve alone.

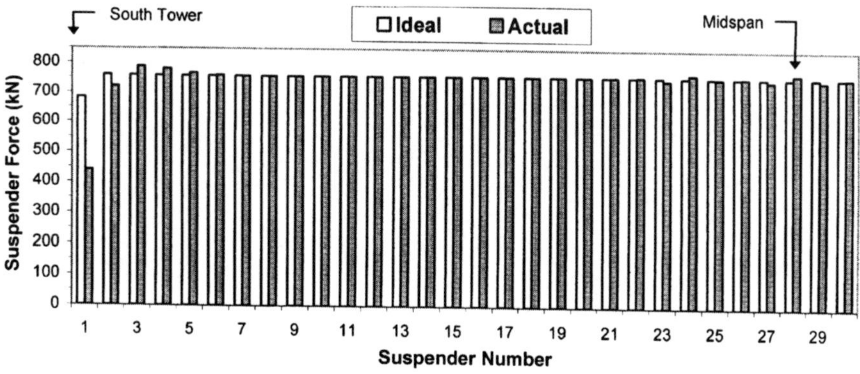

Figure 9: Suspender Forces Due to SDL (Tsing Lung Bridge)

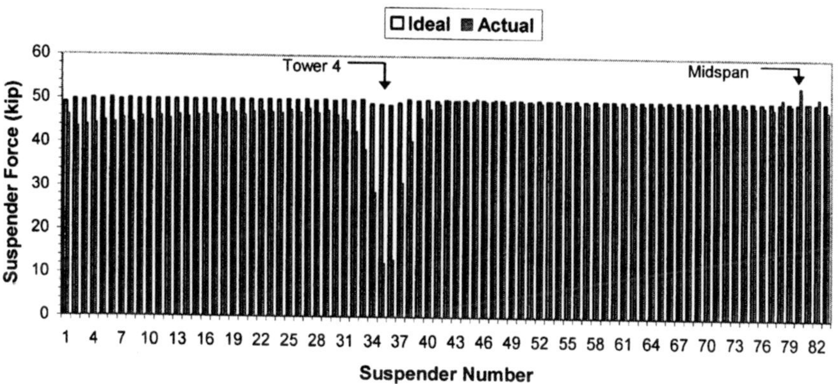

Figure 10: Suspender Forces Due to SDL (The 2nd Tacoma Narrows Bridge)

Another fundamental assumptions adopted in designing the suspension bridges is that both the dead weight of the stiffening girder and the weight of the SDL are completely transferred into main cables through suspenders. This is no longer true for Actual conditions. Figure 9 and Figure 10 show the suspender force due to SDL. The difference between the Actual suspender forces and the Ideal ones is the portion taken by the stiffening girders. Although the two bridges are not exactly symmetric about the midspan, the suspender forces are only shown for half of the bridge for clarity. It can be seen that for Tsing Lung Bridge the Actual suspender forces differ from that of the Ideal ones (Figure 9) in locations adjacent to towers and near the midspan, and the difference is more pronounced at the two ends of the suspended girder. The remaining 65% of the suspenders are left unaffected. Compare to the Ideal condition, the total suspender force in Actual case only reduced about 1%. Being extremely slender and flexible, the twin steel box girder shares negligible amount of the load, and the main cables carry 99% of the SDL. Figure 10 shows more noticeable change in suspender forces for the case of The Second Tacoma Narrows Bridge. In two side spans, the Actual suspender forces are generally smaller than that of the Ideal ones. Suspenders adjacent to two bridge towers show more significant force decrease, whereas in the main span suspender forces are only slightly fluctuated. The total force in Actual suspenders reduced about 10% comparing to that of the Ideal ones. This indicates a force distribution ratio of 1:9 between the stiffening truss and the main cables. Compare to the twin steel box girder, the stiffening truss is considerably stiffer in flexural and thus shares more load.

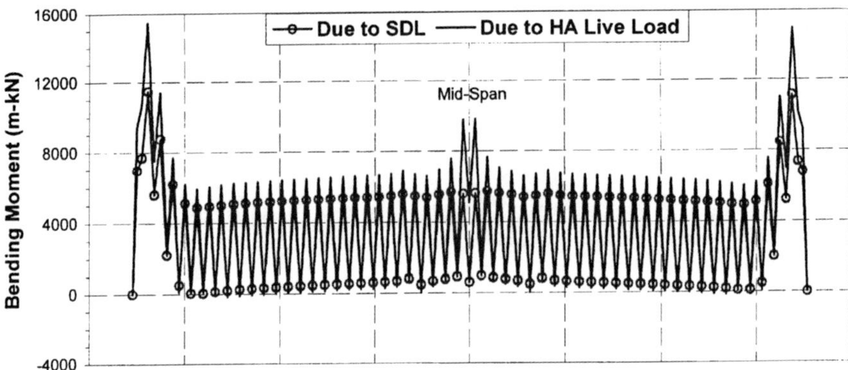

Figure 11: Deck Bending Moment (Tsing Lung Bridge)

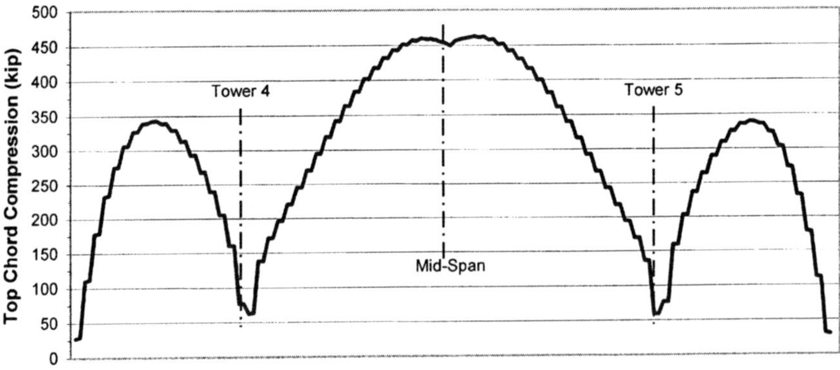

Figure 12: Stiffening Truss Top Chord Force (The 2nd Tacoma Narrows Bridge)

Let's now answer the key question posed in the beginning of this paper, how would locked-in girder stress due to SDL influence the design of suspension bridges? Figure 11 shows the SDL induced bending moment in twin steel box girder adopted in Tsing Lung Bridge. To appreciate the magnitude of the locked-in forces, Figure 11 also presents the bending moment in the box girder due to HA Live Load. Although HA Live Load may not be the governing load case for designing the stiffening girder, it is no doubtly one of the most important load cases the bridge is designed for. It can be seen that the bending moment due to SDL constitutes more than 70% near towers and over 50% near mid-span of that caused by HA Live Load. The resulted peak tensile stresses at the bottom of the steel box are calculated as 37.5Mpa near towers and 20.2Mpa in midspan; and the maximum compressive stress at the top of the steel box are computed as 19.0Mpa adjacent to the ends of the suspended girder and 12.9Mpa near mid-span. The locked-in girder tensile stress due to SDL is roughly in between

10% to 20% of what is allowed for Grade BE S355 steel and must be considered in the design. Since the allowable compressive stress is controlled by the local stability criterion, inclusion of permanent compressive stresses of magnitude 12Mpa ~ 19.0Mpa at the top of the steel box girder have significant influence on choosing the plate thickness and design of the local stiffeners. Despite the fact that the stiffening girder only shares 1% of SDL, the locked-in girder stress is significant and it is mainly resulted from large displacement imposed by cables, because the main cables sag 2.76 meter at the midspan under the influence of SDL.

Figure 12 shows SDL induced compression force in top chords of the stiffening truss of The Second Tacoma Narrows Bridge. The corresponding peak compressive stresses are computed as 3.1 ksi at side spans and 4.8 ksi at mid-span, respectively. The tensile stresses in bottom chords have similar magnitudes. Had this stress really existed, it should be included in future assessment of the stiffening truss to carry additional loading, and special consideration should be given to the side spans where locations of peak stress caused by locked-in SDL coincide with that of due to live loading. Since the stiffening truss takes up to 10% of SDL, the locked-in stress is caused by both the direct loading and the large displacement imposed by cables, which sag about 8.5 foot at the midspan due to SDL.

Conclusions

The locked-in girder stresses due to applied SDL in suspension bridges are too significant to be ignored. For designing new bridges, under what condition will the SDL be applied onto the bridge has to be clearly defined prior to carrying out the detailed designs. If the intended erection process will cause the stiffening girder to be stressed under SDL, the locked-in stress has to be quantified and included in the design. For rehabilitating an exist suspension bridge, it is essential to first determine whether there is locked-in stress in the stiffening girder due to SDL. If present, this stress must be included in developing new retrofit schemes, especially when the stiffening girder is expected to carry additional loading. For the particular erection process where the SDL is applied after the stiffening girder being completely integrated with the main cables, the numerical method presented in this paper can be utilized to computer the resulted stress. For the two bridges studied in this paper, we have learned the followings: (1) The main cables are insensitive to presence of locked-in girder stress due to SDL; (2) The shallow streamlined steel box girder adopted in modern long span suspension bridges shares negligible forces with the main cables and is stressed due to imposed large displacement; (3) The traditional stiffening truss used in early suspension bridges shares about 10% of the SDL with the main cables and the locked-in stress is induced by both the direct loading and the large displacement imposed by cables; (4) The peak value of the locked-in girder stress due to SDL is roughly in the range of 40% to 70% of that caused by un-factored live loading.

References

[1]   ASCE (1979), *Long Span Suspension Bridges: History and Performance*, Proc. ASCE National Convention, Boston.

[2]   Gimsing, N. J. (1998). *Cable Supported Bridges: Concept and Design*, Wiley, New York.

[3]   Pugsley, A. (1968). *Theory of Suspension Bridges*, 2nd Ed., Edward Arnold, London.

[4]   Brotton, D. M. (1966). "A General Computer Programme for the Solution of Suspension Bridge Problems." *Structural Engineering*, Vol. 44, No. 5.

[5]   *East Bridge.* (1998). A/S Stobæltsbindelsen, Storebælt Publications, Copenhagen.

[6]   *Tacoma Narrows Bridge Structural Response Investigation.* (1991). Arvid Grant Associates Consulting Engineers, State of Washington Department of Transportation.

Influence of analysis models and damping method to nonlinear earthquake response of complex bridges

Song Bo, Professor, Department of Civil and Environment Engineering, University of Science and Technology Beijing, Beijing 100083, China
Wang Xiaoyue, Postgraduate, Department of Civil and Environment Engineering, University of Science and Technology Beijing, Beijing 100083, China
Li Yue, Doctor, Department of Civil and Environment Engineering, University of Science and Technology Beijing, Beijing 100083, China

Abstract
From previous studies of damaged bridges, it is known that the use of non-linear dynamic analysis is important for estimating the safety of bridges. Already much effort has been put into studying the earthquake response of bridges not only of their non-linear behaviour but also on the selection of the analysis models. This paper describes and discusses a study of the influence of the analysis models and damping method used to assess the non-linear earthquake response of complex bridges. A comparative analysis of three different analysis models and damping methods is carried out on a complex continuous bridge, and the paper provides an example of the earthquake analysis of such a bridge.
Keywords: complex bridges, non-linear earthquake response analysis, analysis model, damping

Introduction

A numerical analysis is the main method of analysing the dynamic behaviour of bridge structures, and the experimental results show that the model parameters, load intensity and damping ratio are important factors when assessing the dynamic response[1]. Of these the most important factor that influences the analytical results is the dynamic analysis model itself, and in particular, whether it is necessary to ensure that it properly reflects the shape, mode of action of the load and the vibration characteristics of the actual structure. Some scholars have built finite element bridge models using 2D and 3D representations of the bridge structure, and then researched the bridge performance compared to experimental data[2-3].

Long span continuous steel bridges with high piers are the main type of bridge that have been studied to date. Some researchers have studied the high pier stability and stability factor calculation formula using the energy method for such structures[4]. The moment curvature

Bridge design, construction and maintenance 2007, Thomas Telford, London

analysis of a pier's critical section targets a particular section, but under earthquake loading a pier failure actually occurs over a region. A seismic analysis must also reflect the effect of other factors such as the pier's failure being cumulative due to the cyclic loading of an earthquake, so agreement between the section analysis results and the experimental results is essential to be confident of the reliability of the section analysis. For the non-linear seismic response of bridge structures a beam element moment curvature model is now generally used. Liu presents a three-dimensional non-linear dynamic analysis of a steel arch bridge under strong seismic motion[5-6]. Both the geometrical and material non-linearities were considered in the analysis.

When analysing the earthquake response of bridges, it is necessary to represent the hysteretic behaviour of the structure in the basic segments. Based on an experimental force-displacement curve, an analytical hysteretic model for stiffness and strength degradation has been developed[7]. In this study the Takeda model was used in which the framework curve considers not only the rupture of the concrete and the yielding of steel, but also the loss of the unloading rigidity after yielding and the effects of reverse loading[8].

A time-history analysis is an important method of checking the earthquake performance of a structure. Several piers of different height have been analysed using a non-linear time-history analysis method and the effects of the displacement of the top of high piers and the response of the rubber plinth to the piers displacement were researched during high earthquake intensity motion[9]. A direct numerical integration method, such as the Newmark-β method, is the main method for analysing the earthquake response of structures, and is not only used for single-freedom systems but also for the linear and non-linear vibration response of multiple-degree-of-freedom systems[2].

In this paper, the influence of the analysis model and damping method on the non-linear earthquake response of high pier, continuous bridges were studied and the finite element method used to analyse the earthquake response of a complex bridge is presented.

Analysis models

This paper describes three non-linear dynamic analysis models; of which the hybrid model is found to be more precise than the others. Moreover, the calculations and the study was carried out on a real, complex bridge structure and it was found that the damping method also has an influence on the non-linear earthquake response analysis results. The three analysis models used in these non-linear dynamic analyses are shown in Fig.1.

In the M-Φ model, non-linear beam elements are used for the main body of the piers. For areas of concentrated damage such as the bottom 3 meters of the piers, the elements are further subdivided. The relationship between the bending moment M and the curvature Φ was determined from a three gradient, linear model which consists of three components modelling cracking, initial yielding and final fracture.

In the M-θ model, the elements in the main body of the piers are considered as rigid bodies, the crack at the bottom of piers cannot be ignored and is simulated by an elastic-plastic rotation spring model. This uses a bilinear model which considers the yielding point. As load is imposed at the position of the inertia force, when the piers reach yield the bending moment at the bottom of piers is M, which at the same time generates a horizontal displacement δy at the inertia force position. The distance from the bottom of the pier to the inertia force location is H, thus the yielding rotation angle is θy (θy＝δy/H).

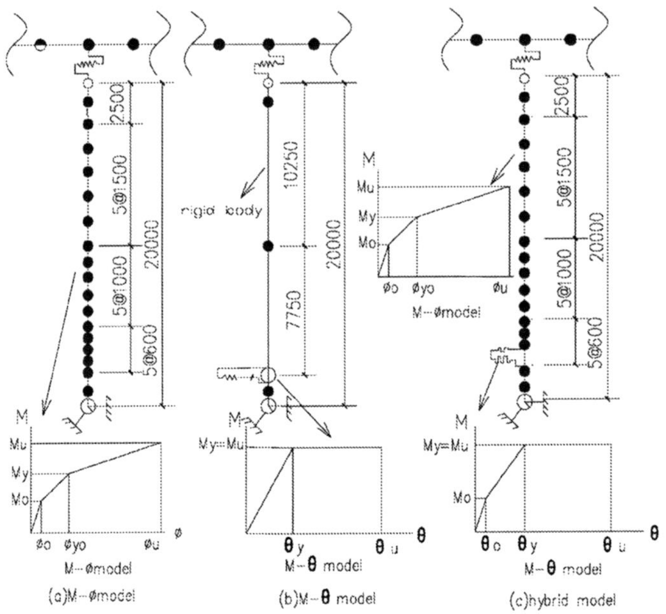

Fig.1 Three analysis models figure

In the hybrid model, an elastic-plastic rotational spring is used to simulate the plastic area and non-linear beam elements are used to simulate the main bodies of the piers outside the range of the plastic area. Furthermore, the relationship between the bending moment M and rotation angle θ and again consists of three components modelling cracking, initial yielding and final fracture.

Comparative analysis of the three models

Through calculating and comparison of the three analysis models, it is found that the M-Φ model, which considers the full body deformation of the bridge piers outside the plastic area, has similar analysis results to the hybrid model. However, the M-θ model concentrates the damage in the plastic area and ignores the body part deformation. The bending moment, bending deformation and support deformation have larger results than the other two models.

Comparing the M-θ model with the other models, the maximum bending deformation of M-θ model is more than three times that of the hybrid model. Most of the results from the three models are similar, in as much as the difference is not very significant. The results of the M-θ model are conservative.

By studying a continuous bridge with different height of piers, this paper will discuss the influence of the three analysis models on the non-linear dynamic analysis of the bridge and from this the following conclusions were derived:

(1) All three of the analysis models has its own characteristics. As the M-Φ model and hybrid model have many more elements, the amount of work required is much greater and there is a higher dependency on appropriate calculation tools. If the bridge is a large complex structure, it needs much more calculation time. The calculation elements in the bilinear M-θ model are fewer and the M-θ model is simpler and faster than other models.

(2) In relation to the appearance of the model, the M-θ model is closest to reality and the action of an earthquake. The bottom of piers are more likely to become plastic before the main body of the pier reaches yield, and causes damage to the piers and collapse of bridge etc.

(3) The results of the three analysis models are similar. The hybrid model is identified as being more precise, and the results of bending deformation in M-θ model are a little conservative, which means the analysis results are on the safe side.

(4) For the continuous bridge with different pier heights, the degree of plasticity is different according to the height of the piers. High-order modes of vibration have greater influence on an earthquake response analysis. This complex condition it still requires further study of the vibration characteristics.

Numerical analysis of complex bridge

Analysis object

The complex type of bridge with 14 spans of continuous steel girders is shown in Fig.2. The bridge has different pier heights and includes a 3 span continuous box girder between P15 and A1. The piers are reinforced concrete structures.

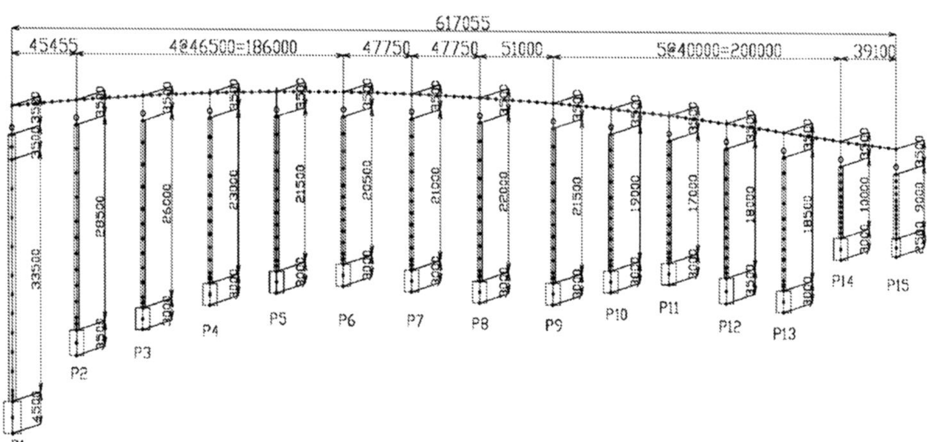

Fig.2 The 14 Span bridge model with different Pier Heights

A hybrid model was used in the analysis of the complex bridge shown in Fig.2. Non-linear beam elements and the elastic-plastic string of concrete elements used to model the piers used the Takeda model. In this paper, a time-history analysis was used to calculate the earthquake response of the bridge, with the Newmark-β (β=1/4) method being applied in the analysis for t=0.002sec.

A strain-energy proportional damping method and Rayleigh's damping method were carried out during the analysis of the bridge. The damping of every part was: 1% in the main beams, 2% in the piers, and 10% in the foundations. The earthquake wave used in the calculations is an actual earthquake wave, which is shown in Fig.3.

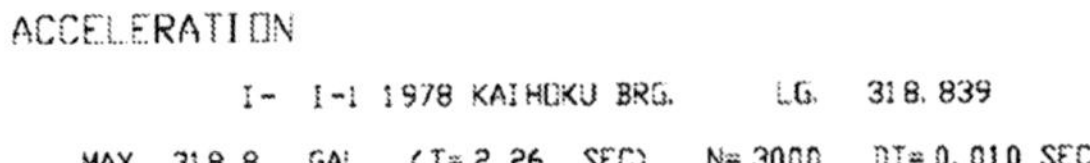
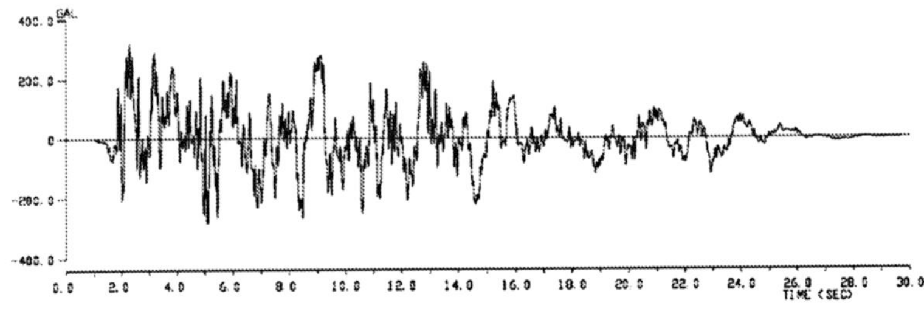

Fig.3 Earthquake wave used in the analysis

Analysis results

By calculation, the analysis model was found to have 796 vibration modes in all. The characteristic values of vibration modes from 1 to 10 are shown in Table 1.

Table 1 Characteristic values of vibration mode

Mode No.	Frequency (1/s)	periods (s)	effective mass ration			Strain-energy damping
			X	Y	Z	
1	0.75	1.33	0.141	0.000	0.000	0.0407
2	0.82	1.23	0.443	0.000	0.000	0.0198
3	1.06	0.94	0.017	0.000	0.000	0.1024
4	2.21	0.45	0.013	0.000	0.000	0.0343
5	2.66	0.38	0.011	0.000	0.000	0.0298
6	2.75	0.36	0.013	0.000	0.000	0.1237
7	2.84	0.35	0.000	0.001	0.000	0.0238
8	3.17	0.32	0.012	0.000	0.000	0.0359
9	3.26	0.31	0.014	0.000	0.000	0.1310
10	3.39	0.29	0.014	0.000	0.000	0.0598

Fig.4 shows the 4 different cases for the Rayleigh's damping parameters for this complex bridge. The results of the strain-energy proportional damping method and the Rayleigh's damping method are compared in Fig.5.

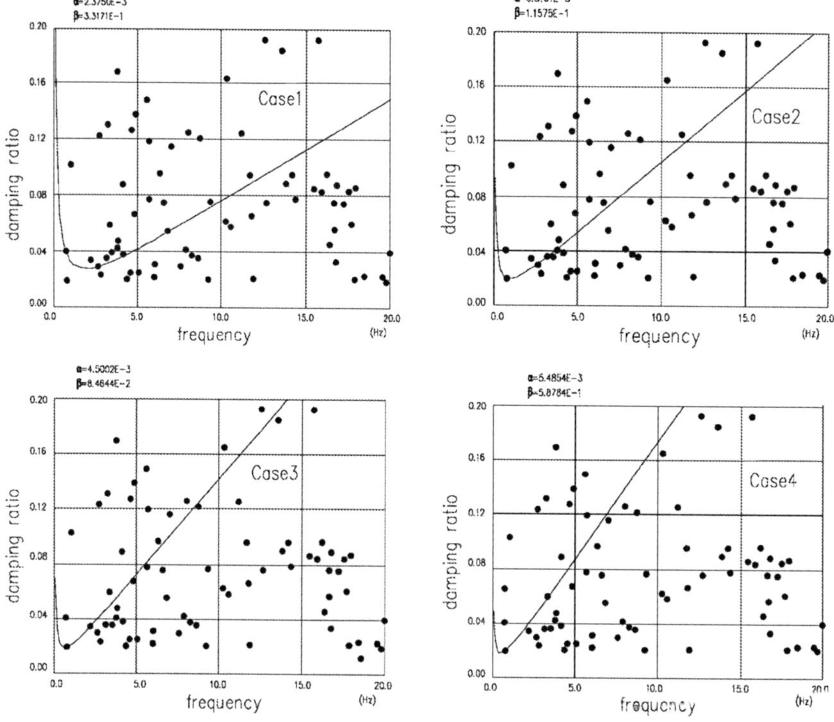

Case1: mode 1, 5; Case2: mode2, 8; Case3: mode2, 4; Case4: mode2, 10

Fig.4 The determination of Rayleigh's damping parameters

A comparative study of the damping method

Using a time-history analysis the hybrid model gave a good estimate of the behaviour of the complex bridge studied here. For damping of the structure, it is very important to determine the two parameters required by Rayleigh's damping method. Usually, the two main modes are used to determine these Rayleigh damping ratios, but, for the complex bridge in this paper, there are many methods for estimating the damping parameters from the different vibration modes, and the difference in the results can be large. As a result Rayleigh's damping method for such a complex bridge is comparatively less precise and less useful. What needs to be considered is an energy attenuation method.

By studying the damping method used in analysing this complex bridge, a verification of the safety of the piers can be carried. Fig.5 shows a comparison of the maximum bending curvature from the strain-energy proportional damping method and Rayleigh's damping method. It is found that the maximum bending curvature from the strain-energy proportional damping method is larger than that from Rayleigh's damping method except for pier 1. In this case all of the results are all very similar but for piers P4-P6 the error can be as much as 52%

in the most severe cases.

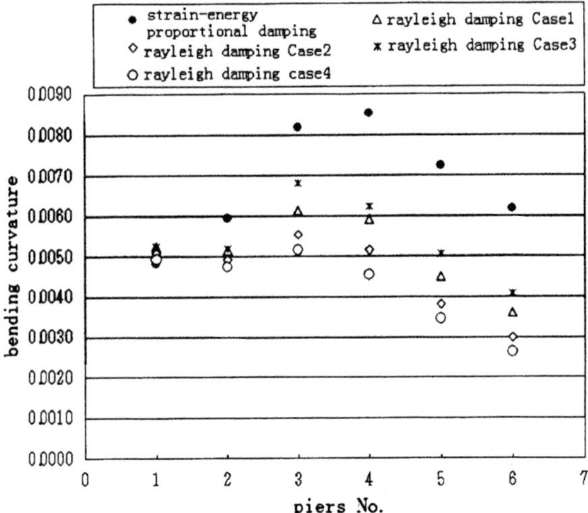

Fig.5 Comparison of maximum bending curvature by strain-energy proportional damping method and Rayleigh's damping method

For the nonlinear dynamic analysis results of complex structures, just as the bridge in this paper, Rayleigh's damping method is improper and the using of strain-energy proportional damping method is desirable.

Conclusions

In this paper, three analysis models were studied and it was identified that each method had its own merits. In the actual process of design and analysis, it is necessary to choose an analysis model according to the requirements of the actual situation for a non-linear earthquake response analysis. For the complex bridge structure described in this paper, the hybrid method was preferred as it had a higher precision.

From a non-linear earthquake response analysis of the hybrid model, the results accord with the actual situation. At the same time, damping is shown to influence to the non-linear dynamic response analysis results of complex structures. Damping evaluation was made using both Rayleigh's damping method and a strain-energy proportional damping method.

By comparing the results, considering the maximum bending curvature at the bottom of the piers, it was found that the Rayleigh's damping method lead to a comparatively lower evaluation of the bending curvature than the strain-energy proportional damping method. For such a complex structure, Rayleigh's damping method is conservative and therefore inappropriate. Attention should be paid to the selection of the parameters α and β if Rayleigh's damping method is used, but using the strain-energy proportional damping method is better.

References

[1] Bryant G. Nielson and Reginald DesRoches. Influence of modeling assumptions on the seismic response of multi-span simply supported steel girder bridges in moderate seismic zones. Engineering Structures, 2006, 28(8): 1083-1092

[2] Xie Xu. Seismic response and earthquake resistant design of bridges [M]. Bei Jing: China communications press, 2005

[3] Zongfen Zhang and A. E. Aktan. Different levels of modeling for the purpose of bridge evaluation. Applied Acoustics, 1997, 50(3):189-204

[4] Lv Yigang. Stability analysis of the continuous rigid frame bridge with large span and high piervia energy methods[J]. Journal of Changsha university of science and technology(natural science), 2005, 2(4):22-26

[5] Niu Songshan. Experimental research on moment-curvature relation of pier's plastic hinge under cyclic loading[J]. Technology of highway and transport, 2005, 10(5):91-95

[6] Liu Yuqing, Guo Yanlin. Nonlinear dynamic analysis of steel pipe arch bridge under strong seismic motion[J]. Earthquake engineering and engineering vibration, 2003, 23(1):44-49

[7] Li Yongzhe. Determination of the degradation coefficient of stiffness and strength for circular piers of reinforced concrete bridge[J]. China safety science journal, 2003, 13(8):53-55

[8] Xiong Zhongming. Theoretical study of nonlinear earthquake response by energy analysis of structure[J]. Journal Xi'an University of architecture and technology (natural science edition). 2005, 37(2): 204-209

[9] Luo Guoji. Nonlinear time-history analysis of simple supported bridge with high piers[J]. Technological development of enterprise, 2007, 26(1): 36-38

Long-term earth pressure measurements of two large-span flexible culverts in Norway

B. Kunecki, The Norwegian University of Science and Technology, Trondheim, Norway
J. Vaslestad, The Norwegian Public Roads Administration, Oslo, Norway
A. Emdal, The Norwegian University of Science and Technology, Trondheim, Norway

Abstract

This paper shows the first results of a research project concerning the full-scale tests of two flexible culverts in Norway. The project is called: "Large-Span Flexible Corrugated Steel Culverts" (CULVERTS) and is realized at The Norwegian University of Science and Technology in Norway, thanks to financial support provided by The European Community under a Marie Curie Intra-European Fellowship.

Several full-scale tests have been performed in the field to validate the long-term performance and load bearing capacity of these structures. The primary objective of the CULVERTS project is connected with realization of long-term field earth pressure measurements of two existing large-span flexible culverts in full-scale under live loads and various climate conditions in Norway. The first culvert is a horizontal ellipse located at Dovre, about 350 km north of Oslo and serving as a road tunnel for Euroroad no. 6. The second structure Tolpinrud is a pipe-arch culvert located at Hønefoss, about 60 km North of Oslo and serves as a rail-road tunnel under a main road.

In the CULVERTS project experimental results obtained in the years 1982-1988 by dr Jan Vaslestad are compared with experimental results obtained in the years 2005-2006. The originality of the project consist in the fact that such long observation period of changes in this kind of structures has never been done so far. The results are presented on figures and charts

Introduction

Corrugated steel culverts are increasingly used in road and railway projects as alternative solutions to bridges and tunnels.

The long-term observation of the behavior of flexible structures is essential to understand the behavior of such structures. It is well recognized that the flexible structures undergo changes in earth pressure and structural response as time progresses after installation. Peck and Peck (1948) reported the behavior of several flexible steel culverts, with emphasis on the long-term deflection. They concluded that if the soil is adequately compacted, a moderate deformation of the culvert will establish a state of nearly uniform all-around pressure. They also concluded that field supervision of backfilling operations is of outstanding importance. The present paper shows the results from field tests conducted during a period of 21 years on two flexible culverts. Selig et al. (1979) reported results from the instrumentation of a 7.9 m span high profile arch with 7.0 m of cover. Unfortunately the long-term results were collected only for 4 months after construction. Long-term field observations on the Dovre and the Tolpinrud structures were reported by Vaslestad (1989). Other instrumented buried steel structures with long-term measurements are reported by Vaslestad, Madaj and Janusz (2002) and Vaslestad, Madaj, Janusz and Bednarek (2004). The fundamental behavior of the buried steel structures is analyzed by Kunecki (2006).

Description of the instrumented structures

Dovre structure

The culvert is located at Dovre, about 350 km north of Oslo. The structure is a horizontal ellipse with a span of 10.78 m, a rise of 7.13 m and a total length of 35 m. Culvert serves as a road tunnel for Euroroad no. E6. The height of cover over the crown is 4.2 m. Built in 1985 in cut-and-cover operation through a soil ridge, this was the largest long-span flexible steel culvert in Scandinavia in 1985. The cross section of the structure is shown in figure 1.

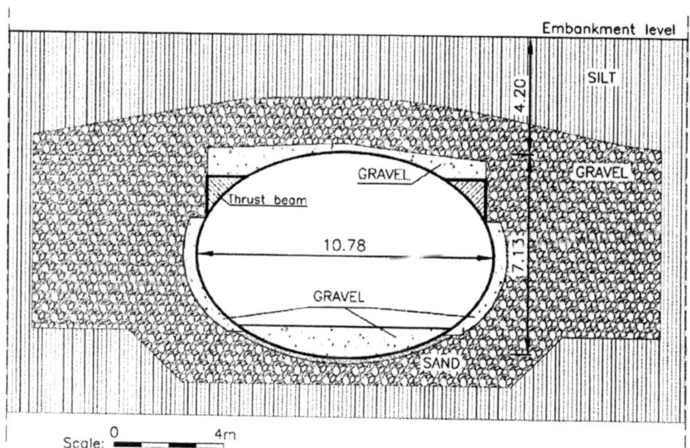

Figure 1. Geometry and soil layers for Dovre structure

The culvert was built using 7 mm thick steel plate with corrugations of 200 mm pitch and 55 mm depth. The plates were assembled in the field, using 20 mm diameter high-strength bolts in 25 mm diameter holes. There are 15 bolts/m of longitudinal seam. The structure has longitudinal stiffeners of reinforced concrete beams.

High-quality well-graded gravel was used for backfilling in a zone extending 6 m out from the springline and 2 m above the crown. The remaining backfill consisted of sandy silt. The in situ soil consisted of relatively dense sandy silt. The well-graded gravel was placed in layers of maximum 30 cm and compacted to minimum 97 % Standard Proctor.

Tolpinrud structure

The Tolpinrud structure, located at Hønefoss, about 60 km north of Oslo, is a pipe arch with span of 7.81 m and a rise of 6.92 m. A cross section is shown in figure 2. The structure is constructed of steel plates with corrugation 200 by 55 mm and thickness 6.8 mm. The structure, which serves as railroad tunnel under road has 106 m long. It was the first super-span structure in Norway and construction was completed in 1982.

The backfill consists of gravel and sand compacted to minimum 97 % of Standard Proctor. The backfill was placed in 20 cm thick layers and compacted with a 1100 km vibrator roller and 12 tons bulldozer. The depth of cover over the crown varied from 1.1 m to 1.6 m. The in situ soil is a medium stiff clay with undrained shear strength 40 to 80 kPa.

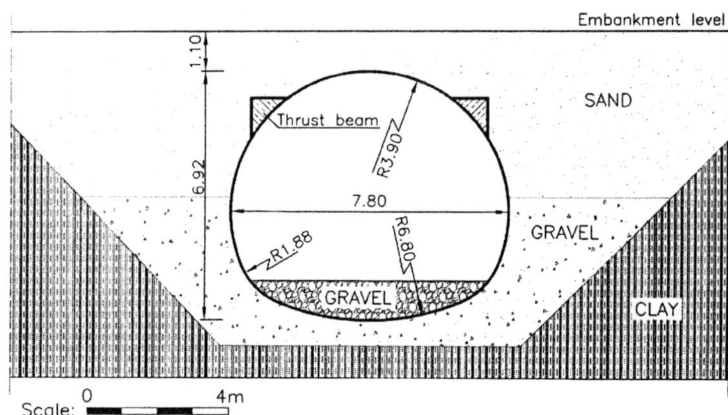

Figure 2. Geometry and soil layers for Tolpinrud structure

Instrumentation and measurements

Earth pressure cells for Dovre structure

Hydraulic earth pressure cells of the Glötzl type were installed at one cross section near the middle of the structure. The size of cells is 30 by 40 by 0.5 cm. The cross section with location of earth pressure cells for Dovre structure is shown in figure 4.

The working principle of the Glötzl earth pressure cell is shown in figure 3. Calibration of cells was performed by the manufacturer and was found to agree well with calibration checks in laboratory. Calibration of the cells with temperature variations from +20 °C to -30 °C was also performed in laboratory.

With this cell the pressure shown on the gauge during circulation of oil is related to the earth pressure, σ, by the following expression:

$$P_A + P_B = P_o + A\,\sigma + P_c$$

where:
P_A – the pressure given by the cell pressure gauge
P_B – the difference between the level of the cell land pump
P_o – the pressure required to cause circulation of oil
A – the cell action factor
P_c – a pressure term due to temperature change

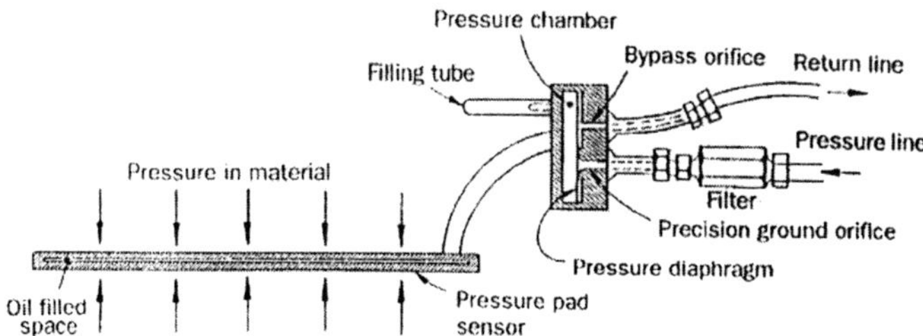

Figure 3. The principle of Glötzl earth pressure cell

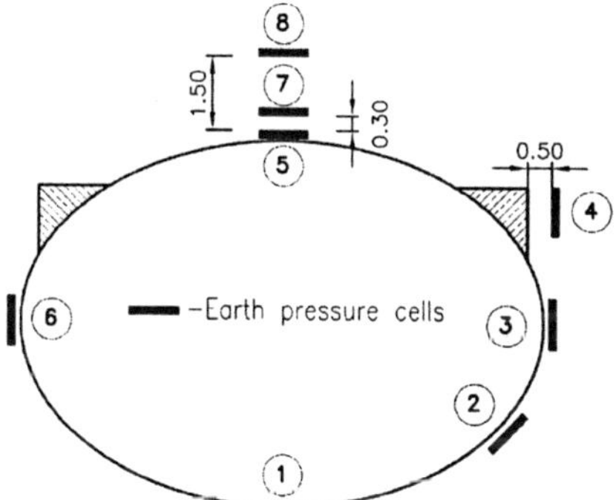

Figure 4. Location of earth pressure cells for Dovre structure

Eight earth pressure cells were used. Five of them were bolted directly on the steel structure using specially constructed brackets. The brackets have the same corrugation and radius as the culvert structure and they have 60 by 80 cm. One cell was placed horizontally in the sand under the structure, two cells were placed vertically on the sides of structure to measure the horizontal earth pressure. Two cells were placed horizontally 0.3 m and 1.5 m over the crown to monitor arching effects. One cell was cast vertically in the concrete on the thrust beam to measure the lateral earth pressure on the beam. The instrumentation is described by Johansen (1987).

Earth pressure cells for Tolpinrud structure

Earth pressure cells of the Glötzl type were mounted on the structure at two sections, one 25 m from the northern end and the other 50 m from the northern end and. Six earth pressure cells were in each section. The location of the gauges is shown in figure 5. Four of the earth pressure cells were bolted on the structure using the specially constructed brackets the same type as on the Dovre structure. One cell was placed horizontally in the sand under the culvert, and one cell was placed vertically in the backfill 0.5 m from the thrust beam to measure the horizontal earth pressure.

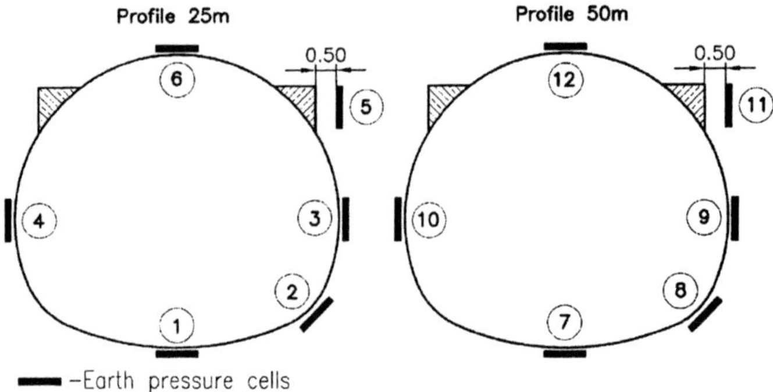

Figure 5. Location of earth pressure cells for Tolpinrud structure

Results of long-term earth pressure measurements

Dovre structure

The measurements show that the earth pressure below the structure and in the haunch area is quite low. The earth pressure at the haunch has even decreased with time. In Vaslestad (1989) it is also shown that the earth pressure in this area is not varying much with temperature.

Figure 6 shows the measured earth pressure around the structure. The measurements from gauges no. 1, 2, 3, 5, 6 taken in September 1989 and November 2006 are shown below. The figure shows that the earth pressure has decreased mostly in the haunch area, there is also a smaller decrease on the top and on the sides.

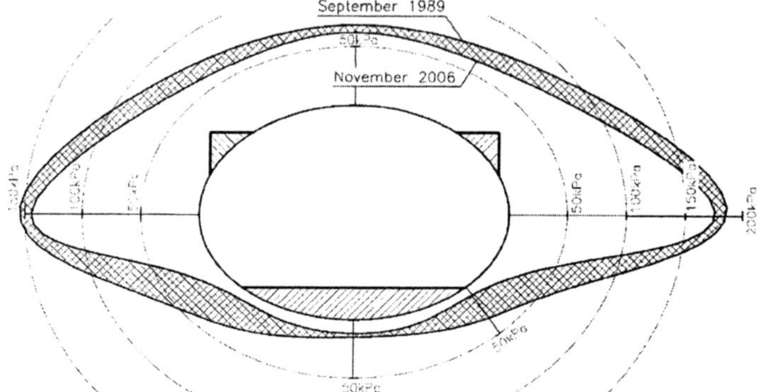

Figure 6. Earth pressure measured around the Dovre structure

Figure 7 shows the lateral earth pressure against the thrust beam recorded in 1989, 1997 and 2006. The earth pressure is up to 1.89 times larger than the overburden.

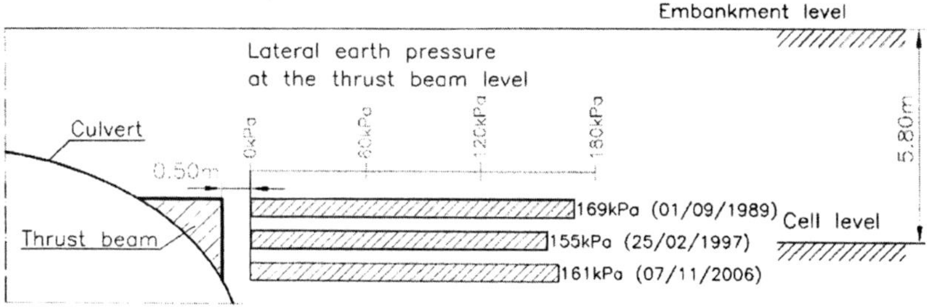

Figure 7. Earth pressure measured against the thrust beam for Dovre structure

Figure 8 shows the positive arching above the structure recorded in gauges no. 7 and 8. The arching is still working after 21 years. The arching curve is very similar to that reported by Terzaghi (1936) from the trap door test. Terzaghi pointed out the following statement: „Conclusive information on the importance of the percentage reduction (arching) can only be obtained by direct measurement under field conditions" (Vaslestad et al. 2007).

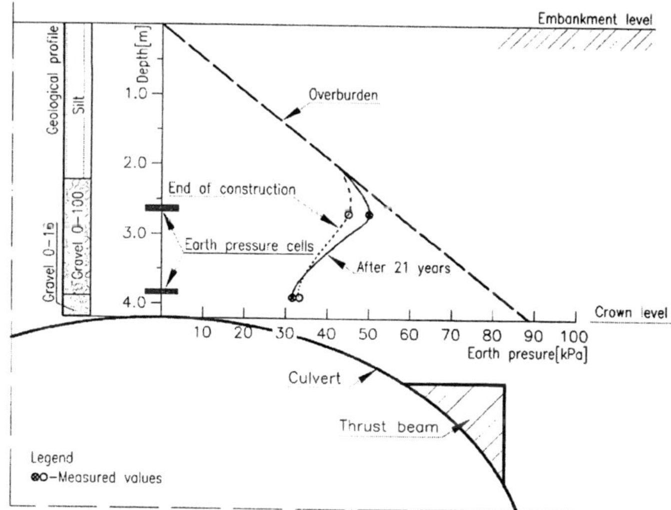

Figure 8. Earth pressure measured above the structure (arching) for Dovre structure

Tolpinrud structure

The comparison of earth pressure measurements performed for both profiles (25 m and 50 m) in the same period of time is shown in figure 9. There was a significant difference in the earth pressure changes between the two profiles. The maximal decrease of earth pressure was observed at the bottom and on the right side of the 50 m section. Only a slight increase of the earth pressure was detected in the crown of the structure in both profiles during analyzed period of time.

The seasonal measurements show that the earth pressure depends on the temperature.

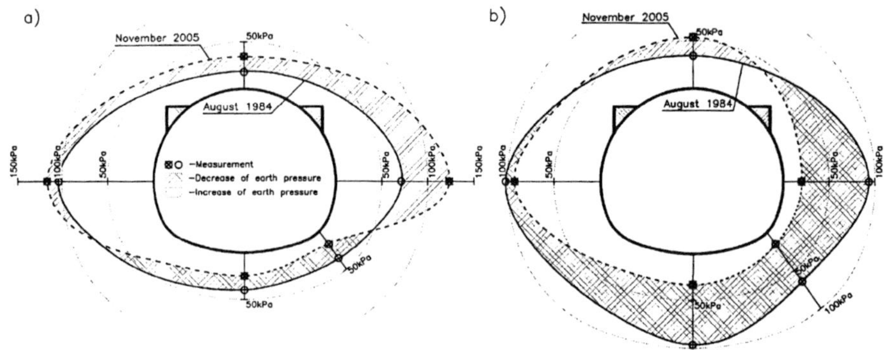

Figure 9. Earth pressure measured around the Tolpinrud structure for profile (a) 25 m and (b) 50 m

The changes of the lateral earth pressure from cells no. 3 and 4 is shown in figures 10 and 11. The measured lateral earth pressure at the springline was about 50 % of the vertical overburden pressure at this level at the end of construction. After about one year the lateral stresses increase to about 100% (Vaslestad, 1989).

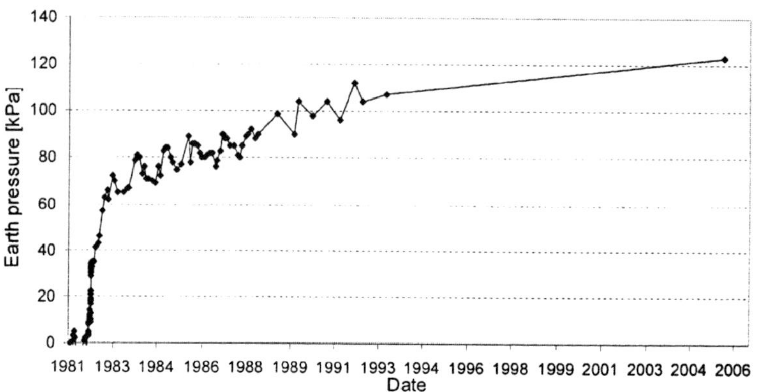

Figure 10. Earth pressure measured on the right side (cell no. 3) for the Tolpinrud structure

The regular increase of the lateral stresses has been observed during 23 years of service at the springline of the structure. The last measurement shows nearly symmetrical lateral pressure approximately 120 kPa (that is about 133% of the vertical overburden pressure at this level). The gauge on the left side of the structure was more sensitive to the changes of the temperature.

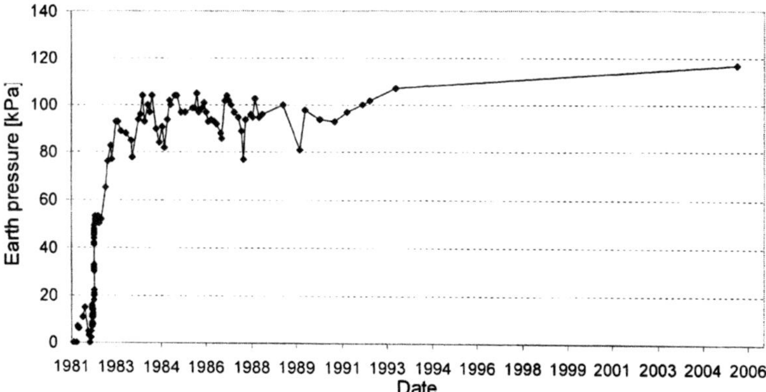

Figure 11. Earth pressure measured on the left side (cell no. 4) for the Tolpinrud structure

The average lateral earth pressure against the thrust beam measured 6 years after the assembling of the structure was about 2.8 times lower in comparison to the maximal pressure observed just after installing. The seasonal differences in earth pressure were larger for the 50 m profile. The final earth pressure ranged between 8 and 12 kPa.

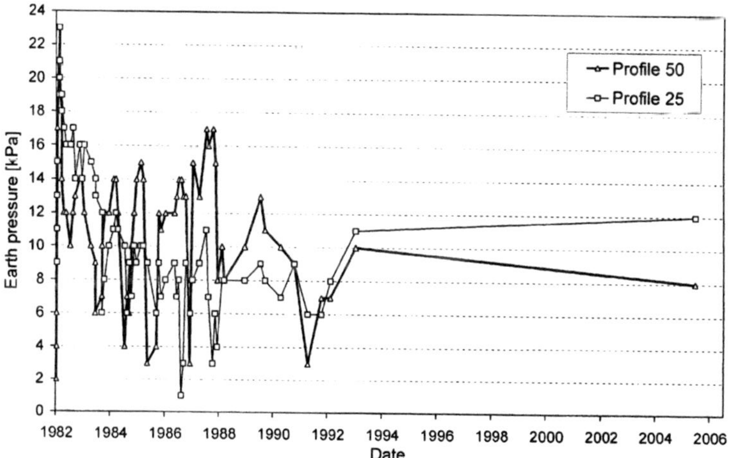

Figure 12. Earth pressure measured against the thrust beam for both profiles for the Tolpinrud structure

Conclusions

The results from long-term field measurements on two large-span structures have been presented. Measurements were taken to assess structure, soil behavior and culvert-soil interaction during long-term service.

The long-term behavior of the flexible steel structure presented in this study shows that buried steel structures undergo changes in earth pressure distribution and structural response as time progresses after construction.

The Glötzl hydraulic earth pressure cells have been in service and are still working 23 years after installation. The strain gages used to measure steel stress stopped working about 5 years after installation for Dover structure.

The regular increase of the lateral stresses has been observed during 23 years of service at the springline of the Tolpinrud structure for profile 25 m. On the contrary, for profile 50 m the decrease of the lateral stresses has been recorded. The location of profiles under traffic loads can be the cause of the difference in the lateral stresses between profiles.

The changes of the earth pressure around the structure depend on the seasonal changes of the temperature.

The arching effect above the structure is quite stable and is still working after 21 years for Dovre culvert. The calculation method that is now used to design steel structures in Sweden, Norway, Poland and also other countries, Pettersson and Sundquist (2006) depends on the arching factor Sar. The arching factor Sar is developed by Vaslestad (1990). These measurements show that the arching factor is stable with time.

Acknowledgements

The present paper was prepared thanks to financial support provided by The European Community under a Marie Curie Intra-European Fellowship: Large-Span Flexible Corrugated Steel Culverts (EC Contract no: MEIF-CT-2006-024775). The authors also wish to thank the Norwegian Public Roads Administration for supporting this research and Tor Helge Johansen for his valuable work on the instrumentation and field operation.

References

Terzaghi,K., (1936): *Stress distribution in dry and saturated sand above a yielding trap-door*, Proc. 1st International conference on soil mechanics and foundation engineering, Cambridge, Massachusetts, Vol. 1, pp 307-311.

Peck O.K., Peck R.B., (1948): *Earth pressure against underground constructions. Experience with flexible culverts through railroad embankments*. Proc. 2nd International conference on soil mechanics and foundation engineering, Rotterdam. Vol. 2, pp. 95-98.

Selig E.T. et. al., (1979): *Measured performance of Newtown Creek culvert*. Journal of the Geotechnical Engineering Division, ASCE, Vol. 105, No GT9, pp 1067-1087.

Johansen T.H., (1987): *Long-span steel structure Dovre. Instrumentation.* [in Norwegian], Oslo, Norwegian Road Research Laboratory, Internal Report no. 1318.

Vaslestad J., (1989): *Long-term behavior of flexible large-span culverts*, Transportation Research Record 1231, Transportation Research Board, National Research Council, Washington DC, pp. 14-24.

Vaslestad J., (1987): *Long-span steel structure Dovre. Field measurements of deformation, steel stress, earth pressure and temperature.* [in Norwegian], Oslo, Norwegian Road Research Laboratory, Internal Report no. 1334.

Vaslestad J., (1990): *Soil structure interaction of buried culverts.* Doctor of Engineering thesis 1990:7. Norwegian Institute of Technology, Trondheim.

Vaslestad J., Madaj A., Janusz L., (2002): *Field measurements of long-span corrugated steel culvert replacing corroded concrete bridge,* Transportation Research Record 1814, Transportation Research Board, National Research Council, Washington DC, pp. 164-170.

Vaslestad J., Madaj A., Janusz L., Bednarek B.,(2004): *Field measurements of old brick culvert slip lined with corrugated steel culvert,* Transportation Research Record 1892, Transportation Research Board, National Research Council, Washington DC, pp. 227-234.

Kunecki B., (2006): *Behaviour of orthotropic buried arch shells under static and dynamic live load.* [in Polish] – PhD thesis, Wroclaw, Wroclaw University of Technology, Report no. 14/2006.

Pettersson L., Sundquist H., (2006): *Design of buried flexible steel structures*, Stockholm, Royal Institute of Technology, Report 58, Structural Engineering - 3rd edition.

Vaslestad J., Kunecki B., Johansen T.H., (2007): *Twenty one years of earth pressure measurements on buried flexible steel structure*, Archives of Institute of Civil Engineering 1/2007, Buried flexible steel structures, Rydzyna, pp. 233-244.

Three-dimensional Seismic Response Analysis of Self-anchored Cable-stayed Suspension Bridge with Response Spectrum Method

Zhu-Weizhi, PhD student of Civil Engineering, Bridge Science Research Institute, Dalian University of Technology, Dalian 116023, China
Zhang- Zhe, Professor of Civil Engineering, Bridge Science Research Institute, Dalian University of Technology, Dalian 116023, China.
Yu-Baochu, Associate Professor of Civil Engineering, Dalian Fisheries University, Dalian 116023, China.

Abstract: The Dalian Gulf Bridge to be constructed according to schedule is a kind of self-anchored cable-stayed suspension bridge with 800m main span. This type of bridge cancels the huge anchorage so that it can reduce the construction difficulties and costs. Furthermore, this type of bridge can provide a better example to other similar construction of huge span bridges under deep-sea soft foundation conditions and it will have bright application prospects. At present, there are fewer researches on this type of bridge so it is necessary to do some researches on the seismic response analysis of this bridge. In this paper, the seismic response is analyzed with response spectrum method. The CQC method is applied to the response spectrum computation. Finally, some conclusions are drawn which will be helpful to the preliminary design and construction of the similar bridge.

Introduction

The response spectrum method (RSM) is widely used in the analysis and design of earthquake-resistant structures, especially at the preliminary stage of the structure design. The earthquake ground motion can be affected by some factors, such as the characteristic of earthquake center, epicentral distance and propagation medium, so it has strong randomness. In order to remove the effects caused by the randomness of earthquake, the design codes of different countries consider the smooth curve coming from the statistical average of many response spectrum curves as the design response spectrum. Because the mode shapes cannot occur simultaneously and their influences on the response of structures are also different, the response of structures cannot be computed through simple addition of each mode shape. In the view of the present situation, it is the various combined projects based on the random vibration theory that are widely used, such as the method of CQC and the method of SRSS. For the long span cable-stayed bridge, the base period is very long. In order to use the RSM in the earthquake-resistant design of long span bridges, it is necessary to deal with the long-period response spectrum of the earthquake ground motion firstly. Professor Xiang Haifan put forward pieces of advice and made some revisions on the specification of earthquake resistant design for highway engineering of 1977, which has proved to be effective. In addition, the structures of long span bridges all have a kind of nonlinear characteristics, while the RSM only applies to the linear structures. And the seismic excitation made by each supporting point of the long span bridges differs. In order to overcome the limitations of the RSM, the RSM for multisupport seismic excitation is proposed by many researchers, such as the method of MSRS by Der Kiureghian and Neuenhofer. Because of its simple computation, high computing efficiency, and general acceptance by civil engineers, nowadays the RSM is still the main tool for preliminary design. In a word, the RSM is the most widely adopted method in the analysis of earthquake-resistant structures the entire world.

The basic theory of the response spectrum method

The equation governing the relative displacement of a s ingle-degree-of-freedom system subjected to uniform ground acceleration can be expressed as:

$$M\ddot{y} + C\dot{y} + Ky = -M\ddot{x}_g(t) \tag{1}$$

can also be written as:

$$\ddot{y} + 2\varsigma\omega_0\dot{y} + \omega_0^2 y = -\ddot{x}_g(t) \tag{2}$$

where M, C, K are mass, damping coefficient, and stiffness, respectively, y is the relative displacement between the structure and its base, ω_0 is the natural frequency, ς is the fraction of critical damping. Because the ground acceleration is unknown and uncertain, the solution y to equation (1) and (2) is uncertain. According to the RSM, based on enveloping curve control, the solution is:

$$y = \alpha g / \omega_0^2 \tag{3}$$

where α is seismic affected coefficient from code, g is acceleration of gravity.

The equation governing the deformation (relative display cement) of a 3D system subjected to arbitrary direction earthquake action can be expressed as:

$$[M]\{\ddot{y}\} + [C]\{\dot{y}\} + [K]\{y\} = [M]\{E\}\ddot{x}_g(t) \tag{4}$$

Here, [.] designates a square $n \times n$ matrix, while {.} represents a column vector with n components. In general, the mass matrix $[M]$, the damping matrix $[C]$, and the stiffness matrix $[K]$ may have all elements different from zero. For most structural models, however, these matrices have only the diagonal and few off-diagonal terms that differ from zero. $\{E\}$ is indication vector of inertial force. In current codes, it is assumed that the span of structure is not so long and all the ground nodes of structure are considered to move with same acceleration and same-phase. Equation (4) is always solved with the mode superposition method. The first q order self vibration circular frequency $\omega_i ,(j=1,2,\cdots\cdots,q)$ for the structural and its corresponding a set of uncoupled coordinates $[\phi]$ are computed. Then according to the mode shapes, Equation (4) reduces to

$$\{y(t)\} = [\phi]\{u(t)\} = \sum_{j=1}^{q} u_j \{\phi\}_j$$

(5)

Equation (4) is left multiplied by $[\phi]^T$ and substitute Equation (5) into it, under the assumption of cross damping, q single-degree-of –freedom equations are obtained as

$$\ddot{u}_j + 2\varsigma_j\omega_j\dot{u}_j + \omega_j^2 u_j = -\gamma_j\ddot{x}_g(t)$$

(6)

Here, ς_j is the damping ration of the j order mode shape while γ_j is the coefficient participation of the j order mode shape

$$\gamma_j = \{\phi\}_j^T [M]\{E\}$$

(7)

By multiplying Equation (3) with γ_j, the solution to Equation (6) is obtained as

$$u_j = \gamma_j \alpha_j g / \omega_j^2$$

(8)

For each u_j, its corresponding $\{y\}_j=u_j\{\phi\}_j$ is computed. If the No. k element y_k wants to be solved, all the No. k element for $\{y\}_j$ should be taken out, forming a Vector $\{y\}_k$ with q element. Then substitute this vector into the following expression and the y_k becomes

$$y_k = \sqrt{\{y\}_k^T [\rho]\{y\}_k}$$

(9)

Here, $[\rho]$ stands for the correlation matrix of correlativity between every order mode component. According to the theory of random vibration, Wilson and Kiureghian deduced the following expressions of non-diagonal elements for $[\rho]$

$$\rho_{ij} = \frac{8\sqrt{\varsigma_i\varsigma_j\omega_i\omega_j}\left(\varsigma_i\omega_i +\varsigma_j\omega_j\right)\omega_i\omega_j}{\left(\omega_i^2 +\omega_j^2\right)^2 + 4\varsigma_i\varsigma_j\omega_i\omega_j\left(\omega_i^2 +\omega_j^2\right)+ 4\left(\varsigma_i^2 +\varsigma_j^2\right)\omega_i^2\omega_j^2}$$

(10)

This is the CQC (Complete Quadratic Combination) method of combining modes in the response spectrum analysis that is widely used the entire world. If $\rho_{ij}=0,(i\neq j)$, the method in Equation (9) becomes the SRSS (Square Root of Sum of the Squares) method. Because of its convenience in computation, the above response spectrum methods are the most widely used method in earthquake resistant design.

Engineering Background of Dalian Gulf Bridge

The Dalian Gulf Bridge to be constructed according to schedule is the main bridge of the large project that connects the main city of Dalian with the development zone of Dalian. According to the design, it is suggested to be the self-anchored cable-stayed suspension bridge with the 800m main span (the following refers to it as "Dalian Gulf Bridge"). With the overall length of 1326m, the span of the bridge is 263+800+263m and the ratio between the side span and the medium span is 0.33. Each side span is designed with an auxiliary pier .The pylon is designed as the twin columnar tower of "H" type with the radial-type cable plane (the general arrangement refers to Figure 1). The overall width of the bridge is 34.0m. The main beam is the box girder with the flat streamline shape. In the middle point, the height of beam is 3.5m. The main beam of the bridge can be divided into two parts, the steel girder part and the prestressed concrete girder part. The steel box girder is applied to the suspension part of the span, while the prestressed concrete box girder is applied to the cable-stayed part (its cross-section refers to Figure 2 and Figure 3). There are altogether 288 cable stays and 31pairs of slings for the whole bridge. As a new structural system bridge with large span, the self-anchored cable-stayed suspension bridge combines both merits of self-anchored suspension bridge and cable-stayed bridge. Because of the adoption of self-anchored system, the huge anchorages are not necessary and the main cable can be anchored on the back of the main beam, so that it can reduce the construction difficulties and costs. This type of bridge can provide a better example to other similar construction of huge span bridges under deep-sea soft foundation conditions so it will have bright application prospects. At present, there are fewer researches on this type of bridge. Based on the 3D FEM, the RSM is applied to analyze the space earthquake responses for the Dalian Gulf Bridge in this paper. In addition, other 3D finite element models for cable-stayed bridges and suspension bridges with the same span and materials to the Dalian Gulf Bridge have been established. The computing analyses have been done to the models and the results have been compared with that of Dalian Gulf Bridge. Finally, some valuable conclusions have been drawn which can provide reference for the preliminary design and construction of the similar type of bridges.

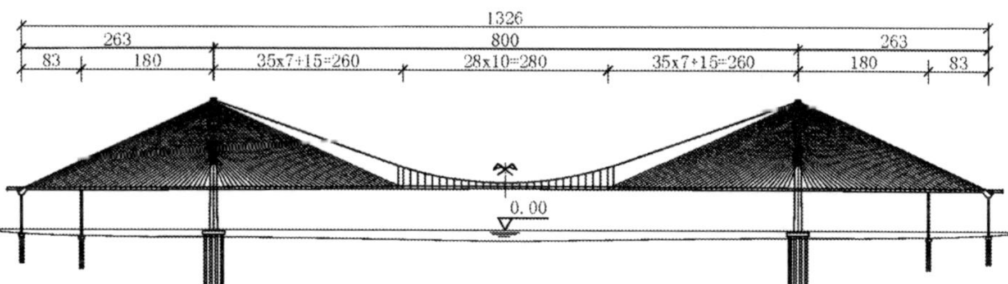

Fig. 1 Arrangement diagram of The Dalian Gulf Bridge   (unit: m)

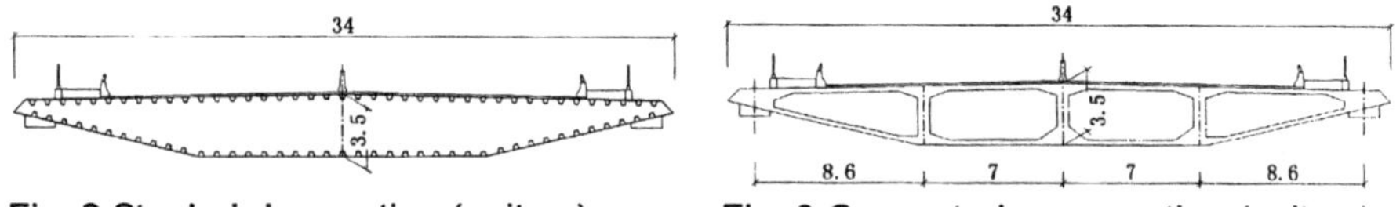

Fig. 2 Steel-girder section (unit: m)   Fig. 3 Concrete-beam section (unit: m)

3D Finite Element Model

Being a kind of computing program for 3D FEM, Midas/Civil is applied to the structure computation. The main forcing components of the whole bridge are the following, cable, beam, tower and pier, etc. The analogy for the computing model is focused on keeping coherence between the structural stiffness, the quality, and the boundary condition of the model and that of the actual structure.

Analogy for Bridge Structure

The single-beam type finite element model is applied to this bridge. The analogy for the three-dimensional beam element is applied to the main beam, the tower, the pier and the pile, while the analogy for the plate element is applied to the bearing platform. The elastic support is used to simulate the influences of the elastic resistance of soil on the pile. In order to actually simulate the mass distribution, mode shape and the seismic force distribution, the element division for the pylon should be much finer compared with other elements. For the huge span type of bridges, the nonlinear of the main cable, the sling and the cable stay appears especially obvious, so the contribution from the axially tension stiffness should be considered. That is to say the linear second-order influences of dead load cable force should be taken into account through geometric stiffness matrix. Otherwise much unreal local vibration frequency would be set at the first of the frequency sequence, covering the actual overall vibration frequency of the suspension bridge. The cable elements analogue is applied to the cable stay, the sling and the main cable of this bridge. The initial stress stiffness matrix is introduced to the computation. There are altogether 3376 nodal elastic supports on the bridge. The bearing support is simulated with the elastic fixing.

Mass Distribution and Consideration of Initial Tension

Because it is to prevent the wheels from abrading the bridge deck carriageway and to distribute the concentrated load of wheel weight, the bridge deck pavement shouldn't be designed as forcing components. That is to say the bridge deck pavement can't be together with the main beam to bear structural internal force. While building models, it is the mass of the bridge deck pavement that should be simulated, not its stiffness. For the cable of high stress, the structural out-of-plane stiffness is greatly affected by the in-place stress of cable in structure. This coupling between in-place stress and out-of-plane stiffness is called stress stiffening. In Midas/Civil, The function of initial forces for geometric stiffness can be used to add cable tension to its relative cable element, and then make it turn into initial stress matrix and add to principal stiffness matrix.

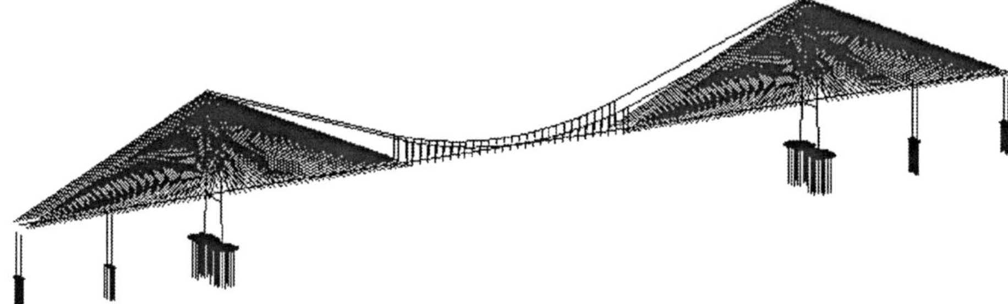

Fig. 4 3D FEM of The Dalian Gulf Bridge

There are altogether 7597 nodes and 7364 elements in the structural computing model, including 4164 beam elements, 2782 plate elements and 418 cable elements (the computing model refers to Figure 4). From the results of mode shape analysis, it is known that that the

Gulf Bridge if of long basic period and intensive mode shape distribution. Being a kind of flexible structure, the basic period of the bridge is 10.1s; belong to the long period structure. Compared with normal bridges of small span, the distribution of mode shape in the Dalian Gulf Bridge is more intense. There are more mode shapes in narrow frequency scope, such as 30 order mode shapes in the frequency range from 0.0989HZ to 0.9468HZ. In order to ensure computing accuracy, the earthquake response spectrum of self-anchored cable-stayed suspension should be analyzed with the first 120-order mode shape with the CQC method, not the SRSS method.

Analysis for Earthquake Ground Response

Earthquake Ground Motion Input

Earthquake ground motion input is the base of earthquake response analysis. Therefore, choosing the proper earthquake ground motion input is the first and important step in analyzing the structural earthquake response. The transmission of seismic wave in strata can lead to the ground motion, while ground motion can also cause the vibration of bridges and other buildings to bear the earthquake effect. Among the characteristics of ground motion, the most important factors on destroying structures are earthquake intensity, frequency characteristics and endurance period of strong earthquake. In order to make these three characteristics meet the demands when choosing the earthquake ground motion, the objective and the level for earthquake resistant requirement need to be determined. Then according to the design data of Dalian Gulf Bridge and the earthquake effect preliminary evaluation report for bridge location, the response spectrum curve of earthquake ground motion can be determined in the end.

The Dalian Gulf Bridge is located in a moderately active earthquake region, with an intensity degree of 7 (which corresponds to maximum acceleration of 0.1g). However, because of its importance, according to the code requirements, the project should in crease one degree, considering 8 degree (which corresponds to maximum acceleration of 0.2g) as protected earthquake intensity. The design follows a two-level seismic criteria and two-stage design requirement. The first seismic criteria level is 10% probability of exceedance in 50 years, referred to as "P1 probability". Correspondingly, the first design stage is to ensure the strength of the bridge's main components from going into the inelastic range under the first seismic criteria level; the second design stage is to ensure the structure's ductility, avoiding unexpected collapse of the deck from the side spans under the second seismic design criteria level. The second criteria level is 2% probability of exceedance in 50 years, referred to as "P2 probability". In this paper, the peak acceleration for P1 probability is 0.2g, which corresponds to an intensity degree of 8, and for P2 probability is 0.3g, which corresponds to an intensity degree of 9. According to the specification of earthquake resistant design for highway engineering (JTJ 004-89), the soil condition of Dalian Gulf Bridge site belongs to category III. The standard response spectrum curve for the category III soil condition refers to the Fig.5.

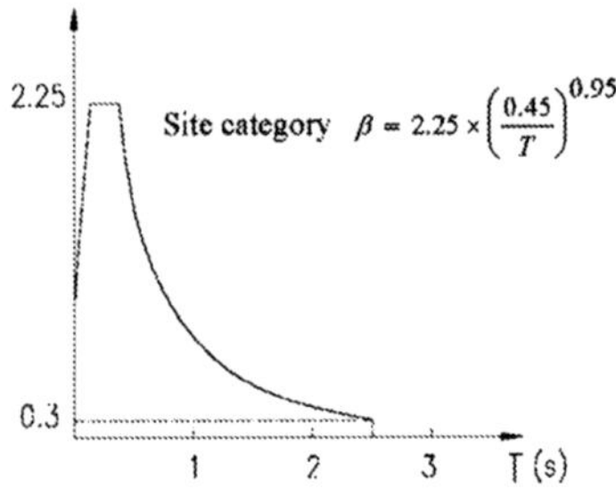

Fig.5 The standard response spectrum curve for site category III

Structural Response under Action of Different Directions Vibration Component

Under the action of longitudinal earthquake wave, the response of Dalian Gulf Bridge is mainly about the longitudinal and vertical vibration of main beam and pylon. The effect of transverse vibration is very slight and the longitudinal vibration can't be coupled with the transverse vibration. The maximum bending moment and the maximum shear of main beam are both located in the section near auxiliary pier, while the maximum blending moment and the axial force of bridge tower is located at the bottom of bridge tower.

Under the action of transverse earthquake wave, the response mainly appears to be the transverse vibration of main beam and pylon. The displacement of longitudinal and vertical vibration for main beam and pylon is zero. Moreover, the axial force of main beam is very small and the longitudinal vibration can't be coupled with the transverse vibration.
Under the action of vertical earthquake wave, the response is mainly about the longitudinal and vertical vibration of main beam and pylon. The displacement of transverse vibration for main beam and pylon is zero and the transverse internal force is very small.

Combined Result of Three-dimensional Orthogonal Vectors

The tested acceleration peak value of vertical earthquake component from many earthquakes in recent years is very large, for example, the largest horizontal acceleration peak value of the earthquake in Hanshin Japan is 833gal and the largest vertical acceleration peak value is 332gal. Another example is in America. From the Newhall's record of the earthquake called "Northridge" happened in Los Angels, USA, the vertical component and that of the north-south and the west-east all get to 0.6g, with the same intensity. So when analyzing the earthquake response in recent years, the combination of transverse earthquake effect and vertical earthquake effect is considered as ground motion inputs. According to the specification of earthquake resistant design for highway engineering (JTJ 004-89), the ground motion inputs are: (1) 100% longitudinal+ 100% vertical; and (2) 100% transverse + 100% vertical, with the vertical input equals to 67% of the horizontal value. The response peak value of main beam bending moment under the effect of the combination is shown in Fig. 6and Fig. 7. Table 1 and Table 2 list the response peak value of the main beam and the pylon control section.

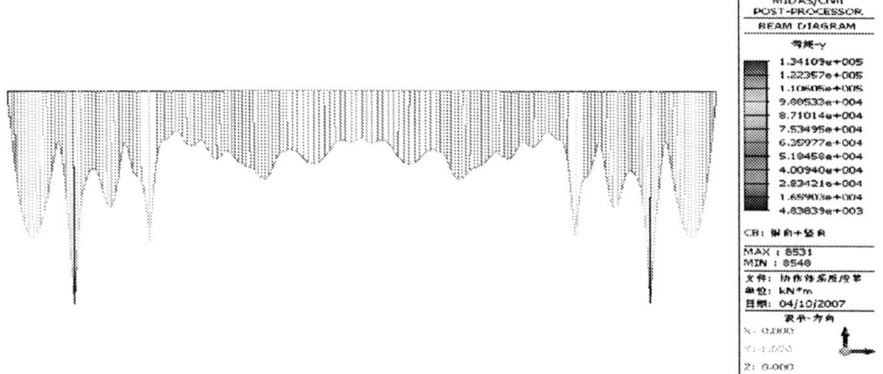

Fig.6 The response peak value of main beam bending moment M_3 under the ground motion input longitudinal+ vertical

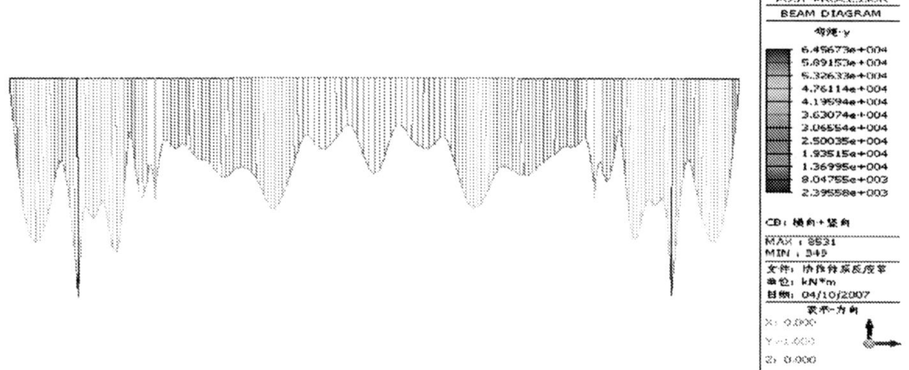

Fig.7 The response peak value of main beam bending moment M3 under the ground motion input transverse+ vertical

Table 1. The seismic response peak value of main beam

Level	Input	Internal force (KN • m)				Displacement (m)		
		N	Q_3	M_2	M_3	u	v	w
P1	(1)	35319	7111	23887	134109	0.148	0.001	0.330
	(2)	11122	4280	1393034	64567	0.006	0.413	0.207
P2	(1)	70638	14222	47774	268218	0.296	0.002	0.660
	(2)	22244	8560	2786068	129134	0.012	0.826	0.414

Remark: N stands for axial force, Q_3 stands for vertical shear force, M_2 stands for transverse bending moment, M_3 stands for longitudinal bending moment, u stands for longitudinal displacement, v stands for transverse displacement, w stands for vertical displacement

Table 2. The seismic response peak value of pylon control section (P1 probability)

Input	Section position	N (KN)	Q_2 (KN)	Q_3 (KN)	M_2 (KN·m)	M_3 (KN·m)	u (m)	v (m)
(1)	A	5027	1340	2927	16547	49692	0.196	0.001
	B	13328	1722	3593	13157	211924	0.171	0.001
	C	15613	1109	2232	8768	239666	0.107	0.001
	D	18710	1064	13897	13837	482205	0.007	0.001
(2)	A	2833	2960	1442	34558	24299	0.035	0.130
	B	9362	4557	976	58050	51465	0.028	0.116
	C	19455	4316	1376	63091	34981	0.012	0.083
	D	39326	11270	1996	282866	63893	0.001	0.006

Remark: A stands for the pylon section around upper cross beam, B stands for the pylon section around upper cross beam, C stands for the pylon section around lower cross beam, D stands for the pylon bottom section, N stands for axial force, Q stands for shear force, M stannds for bending moment, 2 stands for transvers, 3 stands for longitudinal, u stands for longitudinal displacement, v stands for transverse displacement

Conclusions

According to the geologic condition of Dalian Gulf Bridge, the earthquake response, which is under the action of the transverse, vertical and longitudinal earthquake wave and also under the combined action of three orthogonal vectors, is computed with the response spectrum method under the earthquake influence of two probability levels (P1, P2). The response peak value for internal force and displacement of main beam control section and pylon control section is got through the computation. Then, the earthquake response analytical result is compared with that of the cable-stayed bridge and suspension bridge having the same span and structural material. The results of the study lead to the following conclusions:

(1) For the self-anchored cable-stayed suspension bridge, because the main cable is anchored at the two ends of stiffened girder, the stiffened girder bear the horizontal force from main cable so as to reduce structural stiffness. The vibration modes of both transverse bending and vertical bending appear earlier, so the problem for structural buckling stability is comparatively prominent. For the self-anchored cable-stayed suspension bridge, the vibration modes of both transverse bending and longitudinal float are similar with that of the suspension bridge, while the torsional vibration mode is similar with that of the cable-stayed bridge. The torsional vibration mode of the self-anchored cable-stayed suspension bridge appears to be the vibration together with cable, girder and tower. According to the frequency value, the value of the first order transverse bending frequency lies between that of the suspension bridge and cable-stayed bridge. The value of the first order torsional vibration frequency is close to that of the suspension bridge. The value of the first order longitudinal float frequency also equals to that of the cable-stayed bridge. The value of vertical bending frequency is much more than that of suspension bridge and cable-stayed bridge.

(2) Under the action of longitudinal earthquake wave, the response of Dalian Gulf Bridge is mainly about the longitudinal and vertical vibration of main beam and pylon. The effect of transverse vibration is very slight. Under the action of transverse earthquake wave, the response mainly appears to be the transverse vibration of main beam and pylon. The displacement of longitudinal vibration for main beam and pylon is zero. Moreover, the axial force of main beam is very small. Under the action of vertical earthquake wave, the response is mainly about the longitudinal and vertical vibration of main beam and pylon. The displacement of transverse vibration for main beam and pylon is zero and the transverse internal force is very small. Under the effect of both longitudinal earthquake wave and vertical earthquake wave, the response of Dalian Gulf Bridge is mainly about the longitudinal and vertical vibration of main beam and pylon with small displacement of transverse vibration and small transverse internal force. Under the joint action of both transverse earthquake wave and vertical earthquake wave, the response of Dalian Gulf Bridge is mainly about the transverse and vertical vibration of main beam and pylon with small displacement of longitudinal vibration and small longitudinal internal force. The vertical earthquake component takes up a large portion in the combined action of three orthogonal vectors to the bridge, so it cannot be neglected.

(3) Compared with that of the cable-stayed bridge and suspension bridge having the same span and structural material, basically the earthquake response result to the pylon of the self-anchored cable-stayed suspension bridge lies between that of the cable-stayed bridge and the suspension bridge, much smaller than that of the cable-stayed bridge. The initial force of pylon and main beam for cable-stayed bridge is over three times larger than that of the self-anchored cable-stayed suspension bridge. While the initial force of the suspension bridge is near to the self-anchored cable-stayed suspension bridge, the different value between them is not more than 30%.

References:

[1] Yamamura N, Tanaka H. Response analysis of flexible MDF system for multiple-support seismic excitations. Earthquake Engineering and Structure Dynamics. 1990; 19

[2] Berrah M K, Eduardo Kausel. A modal combination rule for spatially varying seismic motions. EESD, 1993; 22

[3] Kiureghian A D, Neuenhofer A. A discussion on seismic random vibration analysis of multi-support seismic excitaions. Journal of Engineering Mechanics, 1995

[4] Trifunac MD. 70anniversary of Boit spectrum. 23 rd ISET annual lecture. ISET Journal of Earthquake Technology, 2003; 40(1)

[5] Ministry of Communication of China (1999). Specification of earthquake resistant design for highway engineering, JTJ 004-89, People's communications press, Beijing, China.

[6] Said M·Allam a, T·K· Datta. Analysis of cable-stayed bridges under multi-componen random ground motion by response spectrum method. Engineering Structures, 2000; 22

[7] Todorovska, M.I. Full-scale experimental studies of soil-Structure Interaction. ISET Journal of Earthquake Technology, 2002; 39

Theme three:

Durability, structural reliability, risk management and health monitoring

ASSESSMENT OF THE DYNAMIC BEHAVI THE JAMUNA MULTIPURPOSE BRIDGE TH ____GH NOISE STUDY

R. Ahsan, Associate Professor, Department of Civil Engineering, Bangladesh University of Engineering and Technology, Dhaka, Bangladesh
M. A. Ansary, Professor, Department of Civil Engineering, Bangladesh University of Engineering and Technology, Dhaka, Bangladesh
T. M. Al-Hussaini, Associate Professor, Department of Civil Engineering, Bangladesh University of Engineering and Technology, Dhaka, Bangladesh
S. Z. Rahman, Graduate Student, Department of Civil Engineering, Bangladesh University of Engineering and Technology, Dhaka, Bangladesh

Abstract

Jamuna Multipurpose Bridge, located in a seismically active region, is as yet the longest and most important bridge in Bangladesh. For seismic protection of the bridge, devices consisting of steel pin dissipating elements and shock transmitter units were placed in between girders and piers. The bridge has later been instrumented with accelerometers in order to monitor the performance of the seismic devices as well as its over-all dynamic behaviour. A noise study has been conducted by analysing records of both traffic and ambient vibration. The objective of the noise study was to properly filter earthquake data off the frequency contents of traffic and ambient vibrations. However, the present paper shows the usefulness of the traffic data in monitoring different traffic conditions on the bridge. The study also shows that ambient vibration records and records of very mild tremors are also useful in understanding the dynamic behaviour of the bridge.

Introduction

In the service life and safety of bridge structures, dynamic response of bridge has long been recognized as one of the significant factors. Dynamic response is a most significant factor but it has not been clearly identified. A major requirement of any research programme designed to address these issues is to develop a convenient, accurate, and reliable analysis methodology so that the bridge engineer can easily construct a computer model of a bridge structure to predict the dynamic response. Vibration monitoring of civil engineering structures (e.g. bridges, buildings, dams) has gained a lot of interest over the past few years, due to the relative ease of instrumentation and the development of new powerful system identification techniques. Sometimes it is doubted whether the measured deviations of dynamic properties (eigen frequencies, mode shapes) are significant enough to be a good indicator of damage or deterioration. The comparison of original and new dynamic properties can also be hampered by natural changes caused by environmental influences (e.g. temperature changes). Another issue of continuing research is the localization of damage starting from any observed difference in dynamic properties. To convince the engineering ᴊmmunity that vibration monitoring is a valuable technique for structural assessment a proof of feasibility is essential. The ambient vibration properties of the main bridge are determined through field-testing under traffic-induced excitation. The purpose of measuring the ambient vibration properties is to determine the mode shapes and the associated natural frequencies.

The 4.8 km Jamuna Bridge is located in a seismically active region that can be subjected to moderately strong earthquakes. Special earthquake protection devices have been used in bridge. This bridge connects Bhuapur in Tangail (East Bank) with Sirajganj (West Bank). The bridge is located in a seismically active region and has been designed to resist dynamic forces due to earthquakes with peak ground acceleration as high as 0.2g. JMB is the first bridge in the country where seismic pintles have been used. The pintles act as an isolation device for protection against earthquakes. Jamuna Multipurpose Bridge Authority (JMBA) took necessary steps for the installation of seismic instruments on and around the Jamuna Bridge. Accelerometers and displacement sensors have been installed at various locations of a seven-span module of the bridge close to the west end of the bridge and also on the west abutment. In addition, triaxial accelerometers have been installed to record the free field station at seven sites. Also one borehole accelerometer has been placed near the west end.

Bridge information

The main bridge is slightly curved, about 4.8 km long, prestressed concrete box-girder type, and consists of 47 nearly equal spans of 99.375 m and 2 end spans of 64.6875 m. The total width of the bridge deck is 18.5 m. The main bridge is supported by twenty-one 3-pile piers and twenty-nine 2-pile piers. There are 128 m long road approach viaducts at both the ends of the main bridge. There are six hinges (expansion joints) that separate the main bridge structure into seven modules; two end modules, four 7-span modules and a 6-span module in the middle. The bridge consists of four lane roads with a single-track metre gauge railway and a footpath. The crossing has been designed to carry a dual two-lane carriageway, a dual gauge (broad and metre) railway, a high voltage (230 kilo volts) electrical inter-connector, a 750 mm diameter high-pressure natural gas pipeline and telecommunication cables. The carriageways are 6.315 m wide separated by a 0.57 m width central barrier; the rail track is located along the north side of the deck. On the main bridge, electrical interconnector pylons are positioned on brackets cantilevered from the north side of the deck. Telecommunication ducts run through the box girder deck and the gas pipeline is located under the south cantilever of the box section. The main bridge deck is a multi-span precast prestressed post-tensioned concrete segmental structure, constructed by the balanced cantilever method. Each cantilever has 12 segments (each 4 m long), joined to a pier head unit of 2 m long at each pier and by an in-situ stitch at mid span. The deck is internally prestressed and of single box section. The width of the box section is 18.5 m but the depth of the box varies between 6.5 m at the piers and 3.25 m at mid-span. An expansion joint is provided every 7 spans by means of a hinge segment at approximately quarter span. The segments were precast and erected using a two-span erection gantry.

The substructure of each module consists of three 3-pile piers and three or four 2-pile piers for the six and seven span module. The bridge is supported on tubular steel piles of 80-85 m long, filled with concrete, driven into the riverbed. Sand was removed from within the piles by airlifting and replaced with mass concrete. The bridge is supported on 2.5 m diameter 2-pile piers and 3.15 m diameter 3-pile piers, which were driven by powerful (240 mt) hydraulic hammer to withstand predicted scourge and possible earthquakes. Pile toe levels vary from −70.0 m PWD (Public Works Datum) to −82.0 m PWD, with a head level of +11 m

PWD. The thickness of the steel tube varies along the length of the pile. Pile caps are of precast reinforced concrete shell with in-situ reinforced concrete infill construction. They have a base level of +11.0 m PWD, and so the piles are embedded some 7 m within the caps. The pile caps carry pier stem, which in turn support the bearing

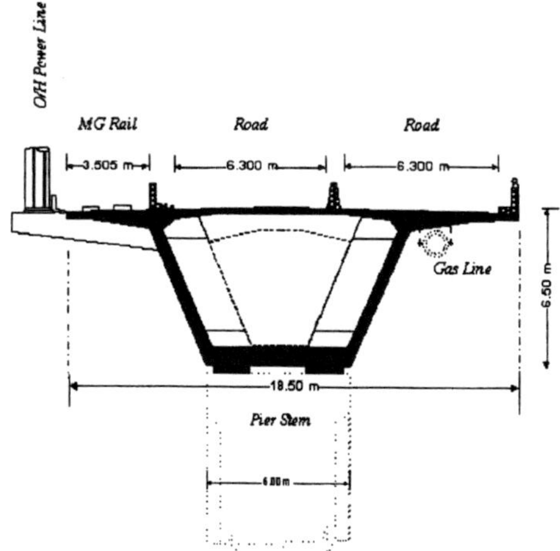

Figure 1. Deck cross-section

The height of the pier stem varies from 2.72 to 13.05 m. They are founded on concrete pile caps, whose shells were precast and filled with *in-situ* reinforced concrete. The cross-section of the pier stem are shown in Figure 1. The reinforced concrete pier stems support pier heads which contain bearings and seismic devices which allow movement of the deck under normal loading conditions but lock in the event of an earthquake to limit overall seismic loads through the structure and minimize damage. The hollow section of the pier stem is filled with concrete up to 3 m of its height. The isolation system consisting of flat sliding bearings, pin dissipating elements and shock transmitter units have been placed in between girders and the piers. Pin dissipating device has been used as base-isolator for protection against earthquakes. Two types of dissipating device are used depending on the mechanism of operation.

- The centre of portion of each module has the fixed-type, i.e., in these locations (at the 3-pile piers), there are multipin elastoplastic devices in which all horizontal movements other than those occurring in very short durations are accommodated by the elastic deformation of the pin elements.
- At the 2-pilepiers, mobile-type devices are used that include shock transmitters with the multipin elements to allow slowly-occurring movements (like the thermal expansions and contractions of the bridge superstructure) through adjustments by the shock transmitters.

At sudden onset of loads, such as during an earthquake, the horizontal forces are resisted by pin elements in both the fixed and mobile-type devices. The sharing of the loads is achieved because the shock transmitter in the mobile locations locks up and transmits the loads to the pin elements. The plan comprises of the following major components (FIP Industriale, 1996):

- A central body with pin dissipating elements.

- An upper and a lower plate, between which the pins are affixed. The top plate transmits the horizontal seismic loads from the superstructure to the pier through the pins.
- A frame with two tapered faces is affixed to the superstructure. Two frames, one on either side of the central body, with an outer surfaces tapered to match the taper of the inner frame is attached to the top of the pier maintaining the required clearances for the deformation of the pins. The inner and outer frames together form a fail-safe mechanism.

The dissipating device for the mobile locations includes the same components as described for the fixed devices plus a shock transmitter unit made of a one-piece hydraulic cylinder that has both the end closed a double-headed piston rod that creates two chambers within the cylinder. Bolt (1987), in his report on the seismicity studies for the Jamuna Bridge, mentioned that the adopted site of the bridge (24.42°N, 89.75°E) could experience shaking from both great and moderate-sized distant earthquakes and from moderate near-site earthquakes during the lifetime (considered as 100 years) of the bridge structure. According to Bolt only one seismic source needs to be considered in postulating strong ground shaking at the Jamuna Bridge site: Zone D at a distance of 25 to 50 km. The design peak ground acceleration is 0.2 g.

Noise study

It is essential to perform necessary corrections to the data collected from the accelerometers for any analysis. Sensors collect input data (earthquake excitation or noise) and record them Etna. There is a chance of acquiring erroneous data from sensors in the process. This error may occur for sensor offsets, instrumental setup, baseline error etc. It is essential to perform noise study to identify sensor offsets, instrumental errors and any other problems that may interfere with real earthquake data. So noise study is essential for data analysis. Noise data is generated by giving impulse in the sensors. After collecting data, it is necessary to process this data by software. For this study traffic data and ambient data were collected in different times. It was seen that the truck and bus passed through the bridge is in the weight range 10 tons to 22 tons. The speed is also very important for this purpose. The speed range is 35 mph to 85 mph. the bus have relatively higher speed than the truck. It is seen that as the weight of traffic increases the acceleration, velocity and displacement of bridge structure increases. It is also seen that as the speed of vehicles increases the response of the bridge increases. From Figure 2 to Figure 7 describes the above findings. Acceleration, velocity and deformation can be observed from these findings. Transverse and longitudinal directional response can be described from these plots. From the maximum limit of weight and speed, maximum limit of vibration data such as acceleration, velocity and deformation can be determined. After that if any vehicles cross this maximum limit it can be identified. Figure 8 shows the FFT of ambient vibration data. From this Figure it is observed that predominant frequency of bridge in transverse direction is 1 Hz. Figures 9-11 show the FFT of train and truck movement data. These also show the predominant frequency of bridge is about 1 Hz.

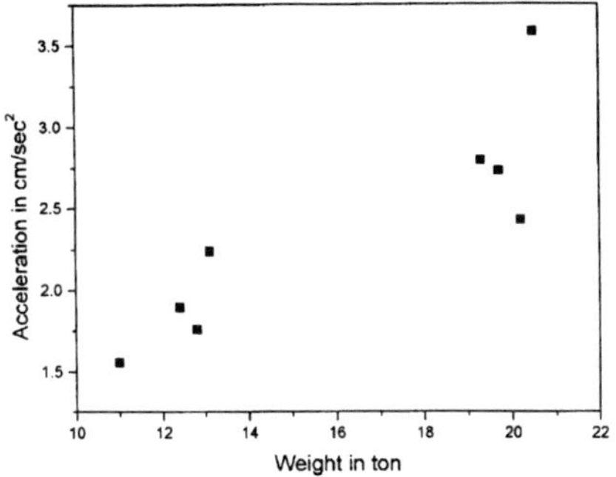

Figure 2. Acceleration of bridge deck in transverse direction vehicles from east

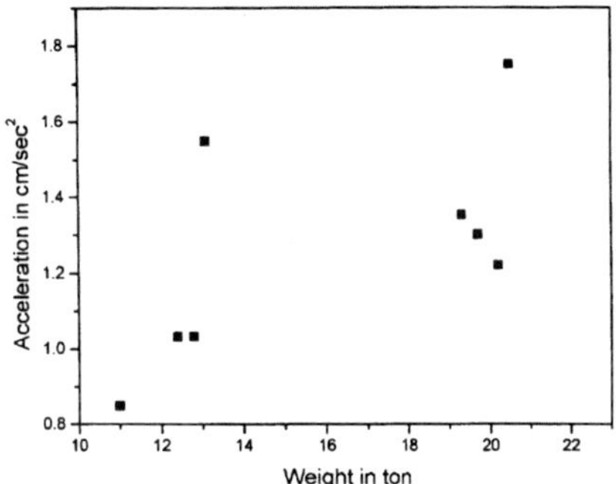

Figure 3. Acceleration of bridge deck in longitudinal direction vehicles from east

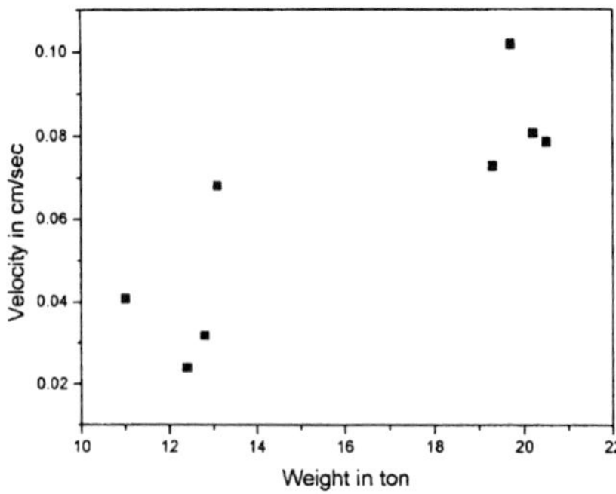

FIgure 4. Velocity of bridge deck in transverse direction vehicles from east

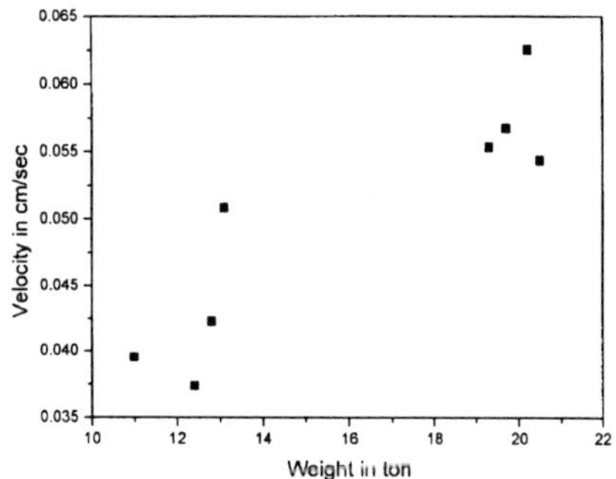

Figure 5. Velocity of bridge deck in longitudinal direction vehicles from east

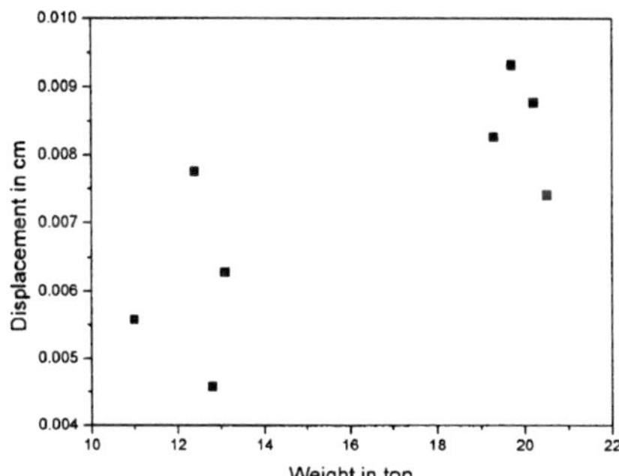

Figure 6. Displacement of bridge deck in transverse direction vehicles from east

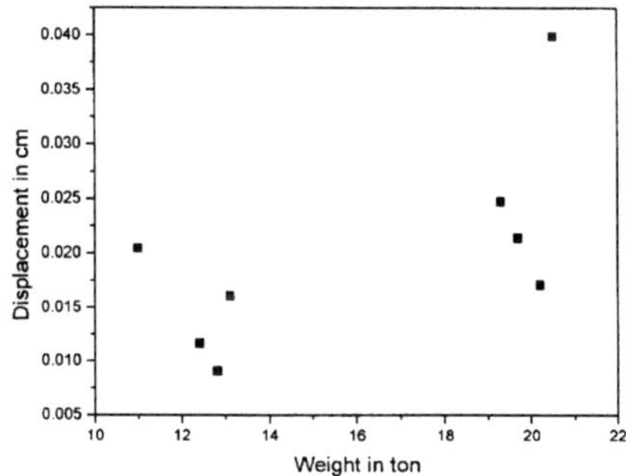

Figure 7. Displacement of bridge deck in longitudinal direction vehicles from east

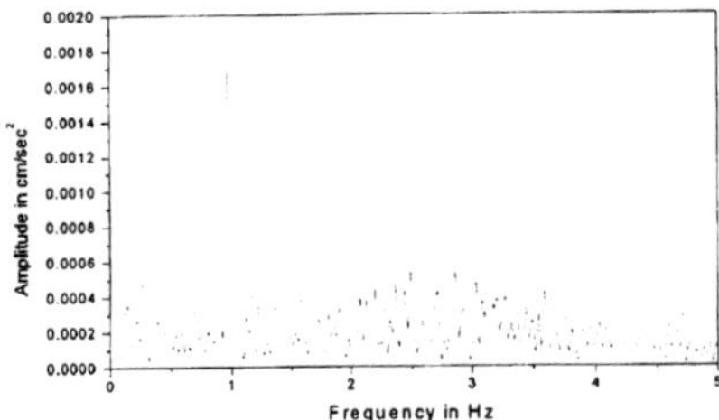

Figure 8. FFT in transverse direction at Bridge deck for ambient vibration data

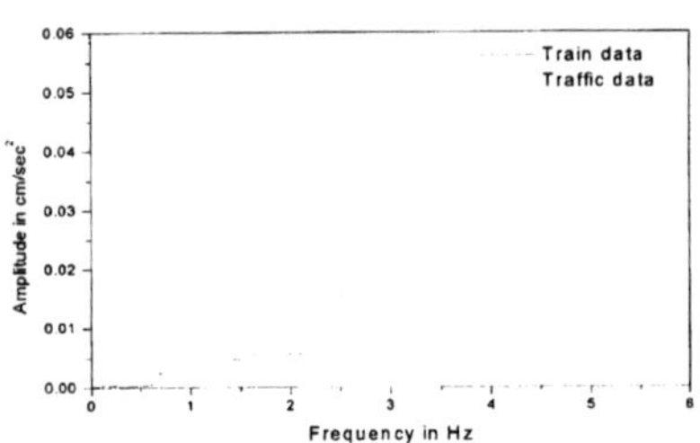

Figure 9. FFT in transverse direction at bridge deck for train and traffic data

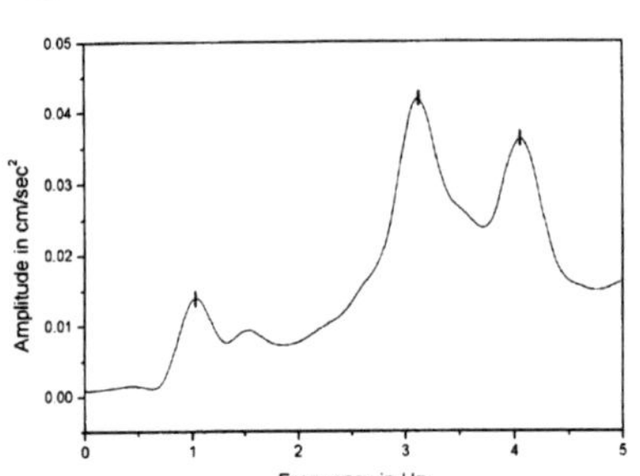

Figure 10. FFT of bridge deck transverse vibration due to truck movement from east

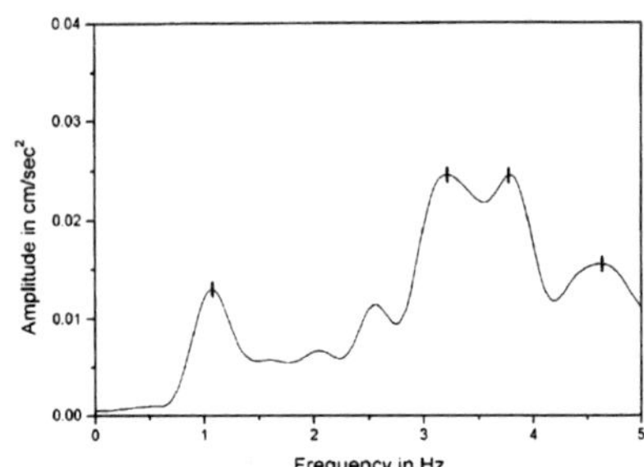

Figure 11. FFT of bridge deck transverse vibration due to truck movement from west

Effect on bridge due to train movement

Train data were collected from bridge at different times from the 13 sensors. This data is filtered and processed with software SMA. This corrected data was taken as input of Origin5.0. Then find the maximum displacement for different dates at different locations. Table 1 gives the maximum displacement of train data. This data is compared with the earthquake data of 16/06/04. Displacement due to earthquake is always higher than that of train. But at the mid span of deck the displacement in transverse direction due to train movement is higher than the displacement due to the small earthquake of 16/06/04. Due to decrease of train speed in recent times the response of bridge decreases. Table 1 also gives the maximum acceleration of train data. These data are compared with the earthquake data of 16/06/04. Due to train movement the acceleration of the deck at mid span is higher than the acceleration due to small earthquake. Due to earthquake the response is much higher at the pile cap than the bridge deck.

Table 1. Comparison of Displacement due to train and Earthquake

Channel ID	Sensor Location	Maximum Displacement of Train Data (cm)		Earthquake data (cm)	Maximum Acceleration of Train Data (cm/sec^2)		Earthquake data (cm/sec^2)
		Speed 40 mph	Speed 20 mph		Speed 40 mph	Speed 20 mph	
BR-1X	Pile Cap at Pier P10	0.0052	0.0045	0.0300	5.09	2.97	13.89
BR-2Y		0.0086	0.0041	0.0188	1.48	1.03	5.98
BR-3Z		0.0030	0.0018	0.0040	3.63	3.02	2.83
BR-5X	Deck at Pier P10	0.0088	0.0084	0.0261	7.67	4.54	7.73
BR-6Y		0.0112	0.0061	0.0334	2.53	1.43	4.61
BR-7Z		0.0031	0.0051	0.0043	8.05	3.94	3.39
BR-12X	Deck at Mid Span	0.0344	0.0190	0.0317	17.01	11.88	7.74
BR-13X		0.0420	0.0161	0.0279	17.76	11.49	10.54

Note: X means orientation across the bridge (transverse direction)
         Y means orientation parallel to the bridge (longitudinal direction)
         Z means vertical direction

Earthquake records

In the second half of the first year of the monitoring phase one earthquake was detected on June 17, 2004 at 05:36:53 hrs BST, (23:36:53 hrs GMT, June 16, 2004) at the Bridge West End free-field station which was the first earthquake recorded by the stations after installation. All accelerometers inside the bridge and bridge west-end, east-end and Mymensingh free-field stations were activated at this time. Other four free-field stations were not activated which may be due to the fact that the intensity of ground motion at those sites being lower than the trigger level set for the starting devices. Epicenter of this earthquake lies close to the bridge site. Fifteen out of sixteen sensors on the bridge recorded the motion of the bridge. The West-End free-field station is nearest to the instrumented bridge module (7-span module next to West-end module). The ground acceleration at the West-End free-field station had a maximum value of 42.2 cm/sec^2 in the NS direction, and the maximum acceleration in the bridge occurred at the pile cap with a value of 13.9 cm/sec^2 also in nearly NS direction. The maximum acceleration in the deck (10.5 cm/sec^2), which occurs in nearly NS direction, is smaller than the acceleration at the pile cap. The earthquake data is recorded through the sensors installed in the bridge. A typical example of earthquake response of pile-cap in transverse direction and corresponding earthquake response in the deck in transverse direction are shown in Figures 12 and 13. The Fast Fourier Transforms (FFT) of these earthquake data are shown in Figures 14 and 15. From Figure 14 the predominant frequency is found to be 1 Hz. The deck vibration also shows a frequency 1 Hz as can be observed in Figure 15. The predominant frequency of ground motion of west-end free-field station near west-end of the bridge is found 8 Hz.

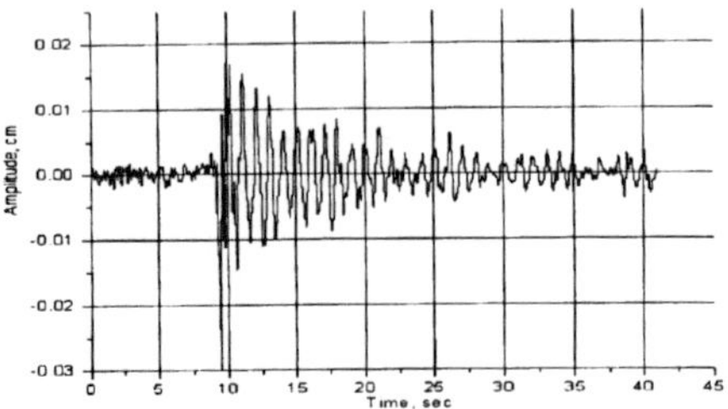

Figure 12. Amplitude vs. time plot of the earthquake data at pile cap of bridge in transverse direction

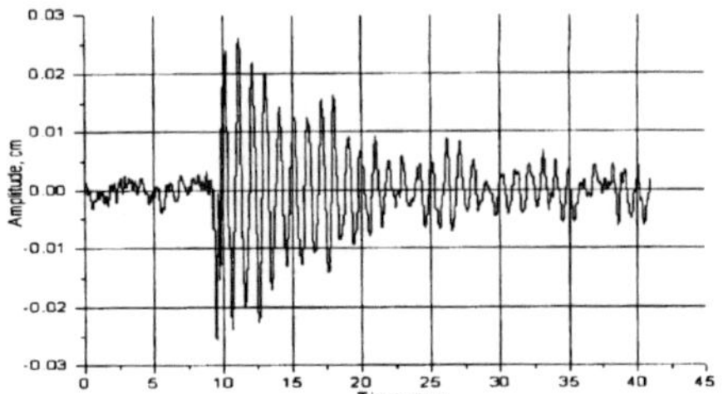

Figure 13. Amplitude vs. time plot of the earthquake data of bridge deck in transverse direction

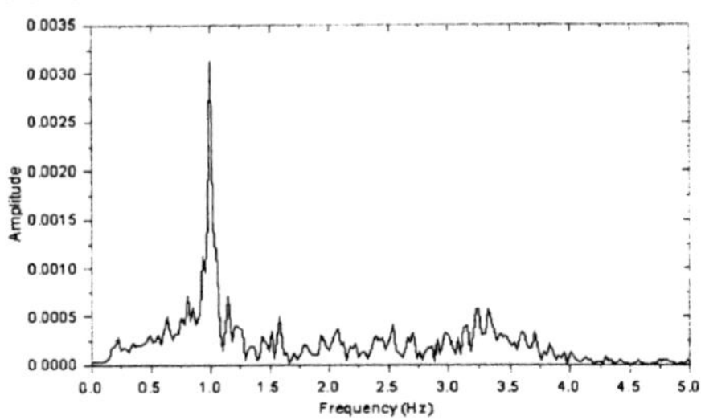

Figure 14. FFT of earthquake data at pile cap of bridge in transverse direction

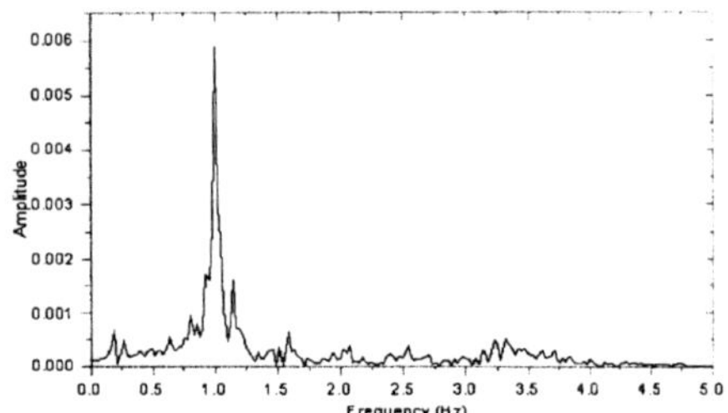

Figure 15. FFT of earthquake data of bridge deck in transverse direction

Transfer ratio

Transfer ratio is the ratio of the FFT of the amplitude of the bridge deck to the FFT of the ground motion of the earthquake. TR is calculated using the FFT of the amplitude recorded at transverse direction at deck and FFT of the source motion recorded at the west-end free-field station in the north-south direction. The calculated TR is plotted against frequency. Using Origin 5 peak points of TR were evaluated. Those TR data were then plotted against the corresponding frequency which shows that TR is highest at frequency 0.24 Hz. The behavior of the phase angle plot is similar to noise vibration motion. So the average value of θ is assumed to be zero, which may be limitation of the analysis. The graphical representation of the derived TR function is shown in Figure 16.

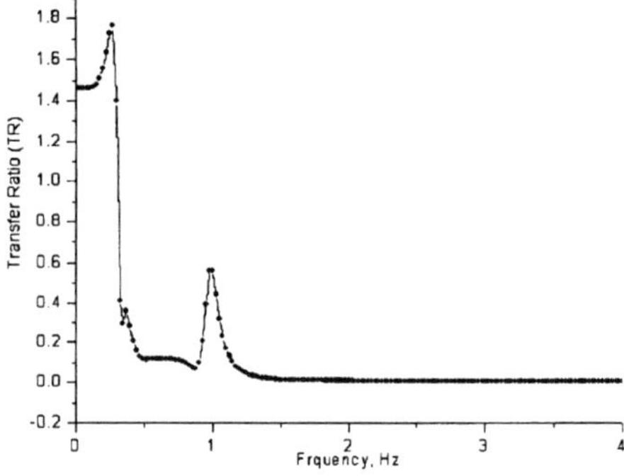

Figure 16 Plot of TR function

The plot represents that first two peaks were found at frequency below 1 HZ and the third peak was found at frequency 0.98 Hz. After 2 Hz the TR values approaches towards zero. So earthquake records having high magnitudes before 2 Hz will have impact on the TR plot. This graphical representation is further used to observe the responses of the Jamuna Multipurpose Bridge for earthquake of different magnitudes. It could be mentioned here that amplitude of input excitations corresponding to the frequency of the first two peaks of TR was very small compared to the response for 1 Hz.

The prepared TR function plot is useful in many aspects. The TR function plot can be used to observe the response of the bridge and to take necessary remedial measures to predict any possible damage of the bridge. The TR values found from the TR function multiplied with earthquake ground motion of different magnitudes to find the amplitude of that earthquake. The obtained amplitude represents the response of the bridge due to the applied earthquake force and helps to take necessary action in this regard. The TR function plot for the recorded earthquake of the Jamuna Bridge is used here to observe the response of the bridge. The application of the earthquake on the TR function was done through following steps.

$$TR = \frac{\hat{u}_1}{\hat{u}_g} = \mathbf{T}\,e^{\,i\theta} = \mathbf{T} = \sqrt{T_R^2 + T_I^2}$$

Let, $\hat{\ddot{u}}'_g$ = FFT of ground motion of the applied earthquake = $u_p \times e^{iu_\theta}$

u_p = absolute value of $\hat{\ddot{u}}'_g = \sqrt{d^2 + e^2}$

d = real value of $\hat{\ddot{u}}'_g$

e = imaginary value of $\hat{\ddot{u}}'_g$

u_θ = phase of $\hat{\ddot{u}}'_g = \tan^{-1}\dfrac{e}{d}$

$\hat{u}'_1$ = FFT of the displacement of Jamuna Bridge due to the applied earthquake

$= \mathbf{T} \times \hat{\ddot{u}}'_g = \mathbf{T} \times u_p \times e^{iu_\theta} = \left|\hat{u}'_1\right| \times e^{iu_\theta} = f + ig$

$\left|\hat{u}'_1\right|$ = absolute value of FFT of the displacement of the Jamuna Bridge due to the applied earthquake

$= \sqrt{f^2 + g^2}$

f = real value of $\hat{u}'_1 = \left|\hat{u}'_1\right| \cos u_\theta$

g = imaginary value of $\hat{u}'_1 = \left|\hat{u}'_1\right| \sin u_\theta$

u'_1 = displacement of Jamuna Bridge due to the applied earthquake = $\left|\hat{u}'_1\right| \times e^{i\theta_u}$

θ_u = phase of $\hat{u}'_1 = \tan^{-1}\dfrac{g}{f}$

When the values of f and g were found, Backward Fourier Transformation (BFT) has been done to get the displacement (u'_1) of the Jamuna Bridge due to the applied earthquake motion. The response of the bridge due to the applied earthquake motion has been studied through the displacement found by Backward Fourier Transformation. If the Elcentro earthquake occurs near the Jamuna Bridge site and the time history of the acceleration near the bridge west-end is like Figure 17 in NS direction, the response of the bridge deck above the pier in transverse direction will be like the time history shown in Figure 18. The maximum deformation in transverse direction will be 20.3 cm. So, response of the bridge can be predicted by TR for different earthquakes.

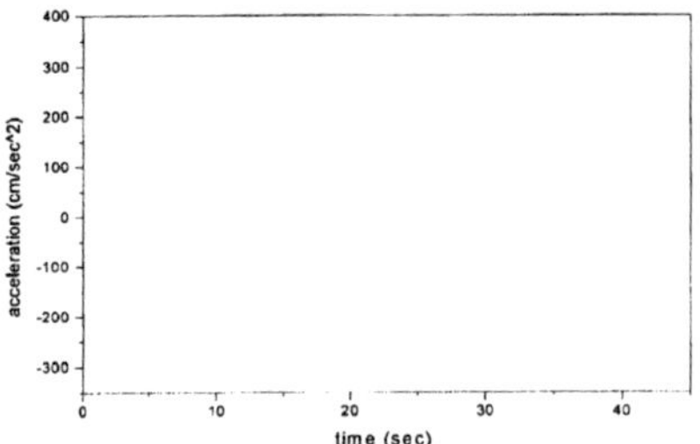

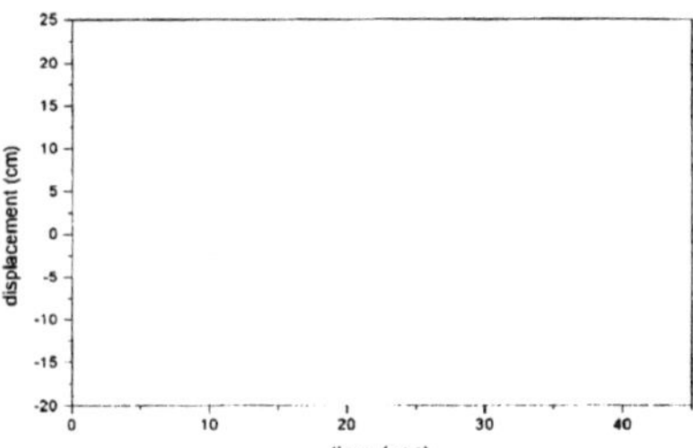

Figure 17. Time history of Elcentro earthquake

Figure 18. Response of Jamuna bridge deck in transverse direction due to Elcentro earthquake.

Conclusions

A noise study has been conducted by analysing records of both traffic and ambient vibration. The objective of the noise study was to properly filter earthquake data off the frequency contents of traffic and ambient vibrations. However, the present paper shows the usefulness of the traffic data in monitoring different traffic conditions on the bridge. The study also shows that ambient vibration records and records of very mild tremors are also useful in understanding the dynamic behaviour of the bridge. It is seen that as the weight of traffic increases the acceleration, velocity and displacement of bridge structure increases. It is also observed that as the speed of vehicles increases the response of the bridge increases. From the results of the analysis it can be concluded that the TR function derived gives ample opportunity to study the response of the bridge and provides an excellent opportunity to examine the performance of various components of the bridge systems. Performance evaluation of the bridge system will suggest necessary action to be taken to protect the bridge against earthquakes.

References

Ahsan, R., Al-Hussaini, T.M., Ansary, M.A and Rahman, M.M. (2005). Identification of dynamic parameters of the Jamuna Multipurpose Bridge in ambient transverse vibration. Japan-Bangladesh Joint Seminar on Advances in Bridge Engg.

Ali, M.H. and Chaudhury, J.R. (1994). Assessment of Seismic Hazards in Bangladesh. Proc. of workshop on implementation of global seismic hazard assessment program in Central and Southern Asia. Beijing, China, pp.43-58.

Bolt, B.A. (1987). Site Specific Seismicity Study of Seismic Intensity and Ground Motion Parameters for Proposed Jamuna Bridge, Bangladesh. Report prepared for seismic design of Jamuna bridge.

BUET (2003), "Final Report: Installation Phase, Jamuna Multipurpose Bridge Seismic Instrumentation Project", report submitted to JMBA.

FIP Industrial (1995), "Bearings and Seismic Devices for the Jamuna Multipurpose Bridge", Test report on pin dissipating device, Padova, Italy.

Reliability Analysis of Prestressed Bridge Girders: Comparison of Chinese Codes (1989), (2004), and AASHTO LRFD Specifications (2005)

Genmiao Chen, Research Assistant, Univ. of Nebraska-Lincoln, Omaha, USA
Jingjuan Li, Research Assistant, University of Washington, Seattle, USA
George Morcous, Assistant professor, Univ. of Nebraska-Lincoln, Omaha, USA
Andrzej S Nowak, Professor, University of Nebraska-Lincoln, Lincoln, USA

Abstract

The objective of this paper is to evaluate the flexure capacity reliability level for prestressed concrete bridge girders designed using old Chinese Bridge Code (CBC 1989), new Chinese Bridge Code (CBC 2004), and USA AASHTO LRFD Bridge Specifications (AASHTO LRFD 2005). Typical I-girders with variable section-dimensions, span lengths and spacing are considered. Rachwitz-Fiessler method is applied to analyze the reliability indices, β, defined as a function of probability P_F: $\beta=-\Phi^{-1}(P_F)$. Considerable differences existing in the three codes unavoidably lead to different reliability levels for designing a bridge. The conclusions drawn from the reliability analysis and comparisons may provide helpful guidance for the evaluation of bridges under old Chinese code (CBC 1989) and new one (CBC 2004); also provide a reference between Chinese code and USA code.

Introduction

The Chinese Bridge Code 2004 (CBC 2004) is a reliability-based code. It provides guidance on determining bridge design loads and load-carrying capacities of structural members. Safety is ensured by application of load partial factors for load effects of actions and resistance partial factors for structural load-carrying capacity. The Chinese Bridge Code 1989 (CBC 1989) specified the bridge design using load safety factors and material safety factors. Significant changes exist between the two Chinese codes. During the fifteen years period before 2004, thousands of bridges were designed and constructed against the CBC 1989. Most of them are reinforced or prestressed concrete bridges. This paper will compare the reliability levels for prestressed concrete bridge girders designed using CBC 1989 and CBC 2004.

The AASHTO LRFD bridge specification (AASHTO LRFD 2005) is also a reliability-based code. Bridge design is specified by load factors and resistance factors. Considerable differences exist between Chinese bridge codes and AASHTO LRFD bridge code. These differences include design life-time, dead load model, live load model, and limit state functions, et al, which unavoidably lead to the different reliability levels. This paper will also compare the reliability level of prestressed concrete bridge girders designed against the AASHTO LRFD bridge code and Chinese bridge codes.

The reliability analysis is based on the typical prestressed concrete bridges. Rachwitz-Fiessler method is applied. Six types of AASHTO LRFD standard I-girders with variable section-dimensions, span lengths, and spacing are considered. The reliability levels of flexure capacity defined by the limit state using three selected codes for these bridges are analyzed. The load effects, resistances, reliability indices of different bridge situations are calculated. The necessary statistical parameters of variables, distribution type (CDF), coefficient of variation (V), and bias factor (λ) are gathered from the available literatures. The conclusions are drawn from the results of reliability analysis for different girder bridges specified by three codes.

Scope of bridge

The reliability analysis is performed for typical prestressed concrete bridge girders with different span lengths and spacing. All six sections of type I through type VI of AASHTO standard girders are included, the dimensions of which are presented in Fig.1 below.

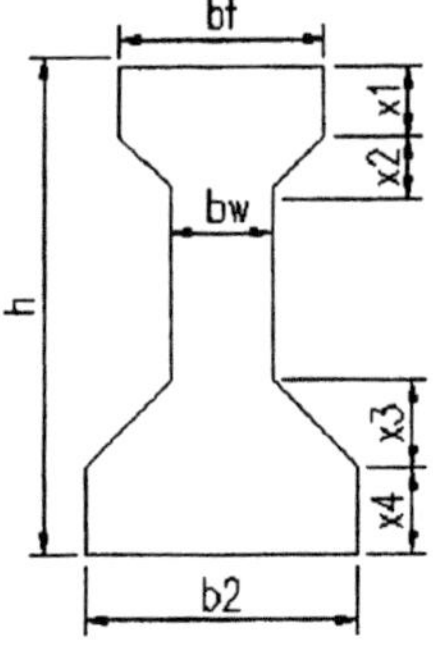

I-Girder	bf	x1	x2	b2	x3	x4	b_w	h
	mm	mm	mm	mm	mm	mm	mm	m
Type 1	305	102	76	406	127	127	152	0.711
Type 2	305	152	76	457	152	152	152	0.914
Type 3	406	178	114	559	191	178	178	1.143
Type 4	508	203	152	660	229	203	203	1.372
Type 5	1067	127	178	711	254	203	203	1.600
Type 6	1067	127	178	711	254	203	203	1.829

Fig.1 Dimensions of AASHTO prestressed concrete I-girders

The bridge girders are simply supported. The spans vary from 14 m to 40 m. The total bridge width for all bridges is 15.55 m with 4 lanes; bridge section configuration is shown in Fig.2. The girder spacing varies from 1.74 m to 2.73 m, and the number of girders changes with the spacing. The girder type, span length, spacing, and number of girders are presented in Table 1 below.

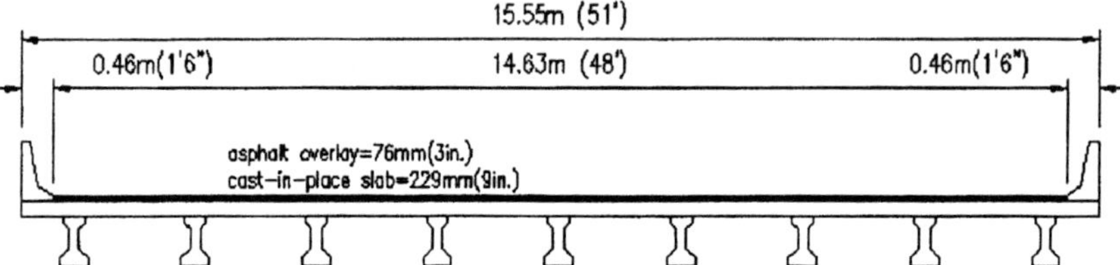

Fig.2 Cross-section of bridges, 1feet=0.3048m

Table 1 Prestressed concrete girder bridges

Girder type	Span m(ft)	Girder spacing m(ft)	Number of girder
Type I	14 (46)	1.74 (5.75)	9
Type II	20 (66)	1.97 (6.5)	8
Type III	25 (82)	2.28 (7.5)	7
Type IV	30 (99)	2.28 (7.5)	7
Type V	35 (115)	2.73 (9.0)	6
Type VI	40 (132)	2.73 (9.0)	6

Limit state function

The reliability analysis is based on the limit state function in this paper, which is expressed in the following equation *(1)*,

$$g = R\text{-}Q \tag{1}$$

Where, R=resistance, Q=total load effect. The load and resistance are treated as random variables. The total load is a sum of several components, as expressed in the following equation (2),

$$Q = (D1 + D2 + D3) + (L + I) \tag{2}$$

Where, $D1+D2+D3$=dead load; $D1$=self-weight of precast girders; $D2$=cast-in-place concrete, including slab, barrier and parapets; $D3$=Asphalt overlay; $L+I$=Vehicular live load; L=vehicular load; I= Dynamic load caused by vehicle (impact).

Flexure design function

The flexure capacity (bending moment) is considered. For the three codes, the ultimate design flexure capacity of a girder, $\emptyset M_n$, should greater than or equal to the factored load effect, M_u, i.e. $\emptyset Mn \geq Mu;$ $\emptyset$ is resistance or safety factor. Large differences exist in between the AASHTO LRFD 2005 and the CBC 1989/2004. For the Strength-I limit state, the AASHTO LRFD 2005 define the ultimate moment M_u as expressed in the following equation (3),

$$Mu = 1.25M_{D1+D2} + 1.50M_{D3} + 1.75M_{L+I} \tag{3}$$

For the two Chinese codes, 1989 and 2004, M_u is defined by the following equation (4),

$$Mu = 1.20M_{D1+D2} + 1.40M_{I,+I} \tag{4}$$

Load model

Dead load model

The three major load components for highway bridges are considered in this paper, dead load, live load (vehicular load) and dynamic load. Dead load is the gravity load due to the self-weight of structural and non structural elements permanently connected to the bridge. Three components of the dead load are considered, D1=dead load due to factory made elements (precast concrete); D2=dead load due to cast-in-place materials (concrete slab or deck); and D3=dead load due to asphalt overlay, but this load is not considered in two Chinese codes. Based on the studies by Dr. Nowak, *et al [1]*, all components of dead load may be treated as normal random variables. The statistical parameters, distribution type (CDF), coefficient of variation (V), and bias factor (λ) are available in literature *[2]*, as presented in Table 2 below.

Table 2 Dead loads statistical parameters

Dead load	λ	V	CDF
D1	1.03	0.08	Normal
D2	1.05	0.10	Normal
D3	76mm (mean)	0.25	Normal

The unfactored dead load moment are calculated and presented in Table 3 for three codes. It is assumed that the concrete materials and unit weights for the structural components are same for three codes.

Table 3 Unfactored dead load moment (kN-m) per girder

Bridge No.	Span m (ft)	AASHTO LRFD 2005			CBC 1989, 2004		
		M_{D1}	M_{D2}	M_{D3}	M_{D1}	M_{D2}	M_{D3}
Type I	14 (46)	104	263	62	104	263	0
Type II	20 (66)	283	605	144	283	605	0
Type III	25 (82)	671	1090	259	671	1090	0
Type IV	30 (99)	1362	1570	373	1362	1570	0
Type V	35 (115)	2380	2556	609	2380	2556	0
Type VI	40 (132)	3329	3339	795	3329	3339	0

Live load model

Live load produced by vehicles moving on the bridge includes the static and dynamic effect. Load effect depends on many parameters including span length, truck weight, axle loads and configuration, position and number of vehicles on the bridge, girder spacing, and stiffness of united structures. Based on the studies by Dr. Nowak, *et al [1]*, live load may be treated as normal random variables. The statistical parameters for moment are available in literatures. In this paper, the coefficient of variation, V=0.12, and bias factor, λ=1.0.

Considerable differences for live load exist in three codes, including (vehicular load model, multi-lane factor, and dynamic load model. For sake of comparison, the live load level is considered for the same highway level in three codes. The vehicular load for CBC1989 is Super-20, Highway-I for CBC 2004, and HL-93 for AASHTO LRFD 2005.

Vehicular load

In Chinese Bridge Code 1989, the Super-20 vehicular load is specified, which consists of a single heavy truck (550kN) and series of standard trucks (20kN) before or after at specified intervals, as shown in Fig. 3. The trailer loading is required to check the truck loading. In this paper, the trailing loading does not control the design and therefore not addressed.

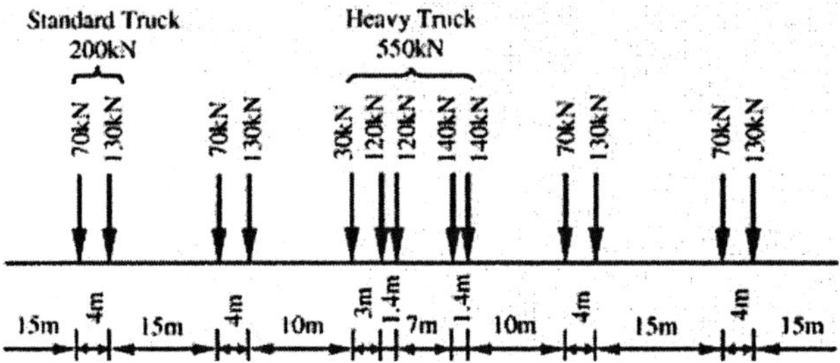

Fig.3 Chinese Bridge Code 1989 Super-20 loading

In Chinese Bridge Code 2004, the Highway-I vehicular load consists of a concentrated force, P_k and a uniform lane load, q_k The concentrated force, P_k is a function of span length; which is presented in the following equation (5),

$$P_k= \begin{array}{ll} 180 \ kN \ (42 \ kips), & when \ L \le 5 \ m \ (16.5 \ ft) \\ 360 \ kN \ (84 \ kips), & when \ L \ge 50m \ (165 \ ft) \\ 180 \sim 360 \ kN \ (42 \sim 84 \ kips), & when \ 5 \ m \ (16.5ft) < L < 50 \ m \ (165 \ ft), \ linearly. \end{array} \quad (5)$$

$$q_k = 10.5 \ kN/m \ (0.72 \ kips/ft)$$

The vehicular load model is shown in the following Fig.4,

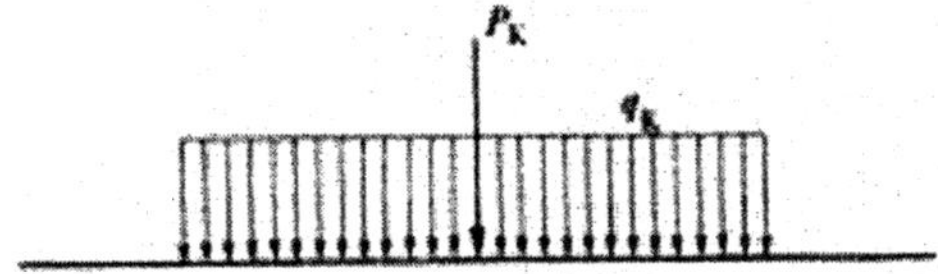

Fig.4 Chinese Bridge Code 2004 Highway-I loading

The AASHTO LRFD specifies HL-93 load which consists of a truck load with three axle concentrated forces, 35, 145, 145 kN (8, 32, 32, kips), and a uniform lane load 9.3 kN/m (0.64, kip/ft), as shown in the Fig. 5. Dynamic load is specified as 0.33 for the truck load only. The tandem load is required to check the truck loading, but not addressed in this paper.

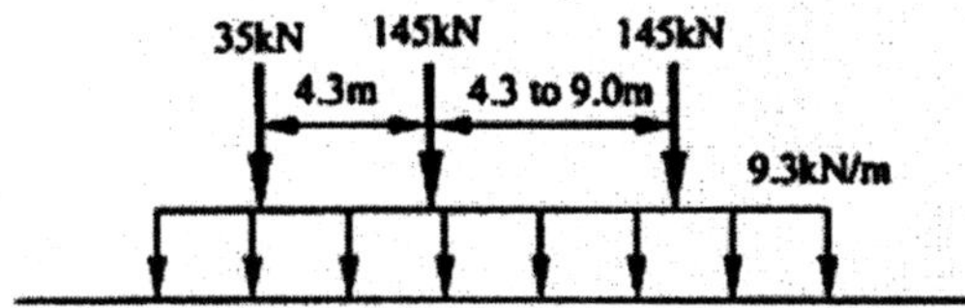

Fig.5 AASHTO LRFD 2005 HL-93 loading

The unfactored live load moments per lane are presented in Table 4 below.

Table 4 Unfactored live load moment per lane, kN-m, impact not including

Bridge No.	AASHTO LRFD		CBC 2004		CBC 1989
	M_{truck}	M_{lane}	M_{force}	M_{lane}	$M_{super-20}$
Type I	758	229	760	260	982
Type II	1237	470	1208	530	1442
Type III	1637	735	1636	829	2429
Type IV	2038	1059	2114	1193	3092
Type V	2440	1442	2643	1624	4248
Type VI	2841	1884	3223	2121	5308

Multi-lane reduction factor

The multi-lane factors for live load are different for three codes. These factors are presented in Table 5 below. In this study, the number of lanes is 4 for three codes; correspondingly, the multi-lane reduction factors are 0.7, 0.67 and 0.65, separately for CBC 1989, CBC 2004 and AASHTO LRFD 2005.

Table 5 Multi-lane reduction factors specified by three bridge codes

Lanes	2	3	4	5	6	7	8
CBC 1989	1.00	0.80	0.70	0.70	0.70	0.70	0.70
CBC 2004	1.00	0.78	0.67	0.60	0.55	0.52	0.50
AASHTO LRFD	1.0	0.85	0.65	0.65	0.65	0.65	0.65

Impact factor

The dynamic load allowances for the vehicular load are also different for three codes. The AASHTO LRFD 2005 specified the dynamic load, IM, as 0.33 for the HL-93 truck only.

The impact factor specified in CBC 1989 is determined by span length, L (m), as presented in the following equation (6),

$$v= 0.3 \qquad\qquad when\ L \leq 5m$$
$$0.0075(45-L) \qquad when\ 5 \leq L \leq 45\ m \qquad (6)$$
$$0 \qquad\qquad when\ L \geq 45\ m$$

In Chinese new code.CBC 2004, the impact load factor is determined by the structural basic frequency. For the simple, typical bridge structure, the basic frequency, f, can be approximately calculated using equation specified in code; the linear relationship between the impact factor and the natural logarithmic scale of frequency is specified as equation (7) below. For other bridges, the finite element analysis is required to obtain the basic frequency of bridge.

$$v= 0.05, \qquad\qquad when\ f < 1.5\ Hz$$
$$0.45, \qquad\qquad when\ f > 14\ Hz \qquad (7)$$
$$0.1767 \ln f - 0.0157,\ (i.e.0.05 \sim 0.45), \quad when\ 1.5\ Hz \leq f \leq 14\ Hz,\ nonlinearly$$

The dynamic factors for three codes are presented in Table 6 below.

Table 6 Impact factor (IM) for three codes

Bridge No.	Span m(ft)	AASHTO LRFD IM	CBC 2004 f (Hz)	CBC 2004 IM	CBC 1989 IM
Type I	14 (46)	0.33	3.85	0.223	0.233
Type II	20 (66)	0.33	2.59	0.153	0.188
Type III	25 (82)	0.33	2.31	0.132	0.150
Type IV	30 (99)	0.33	2.15	0.12	0.113
Type V	35 (115)	0.33	2.01	0.108	0.075
Type VI	40 (132)	0.33	1.79	0.088	0.038

Note: *0.33 for HL-93 truck only*

Lateral load distribution factor

Live load effect caused by vehicles is distributed to each girder by multiplying live load distribution factor (LLDF). For the girder bridge, girder distribution factor (GDF) for moment is specified as the following equation (8) below by AASHTO LRFD 2005,

$$GDF = 0.075 + (\frac{S}{2900})^{0.6}(\frac{S}{L})^{0.2}(\frac{K_g}{Lt_s^3})^{0.1} \qquad (8)$$

Where, S=girder spacing (mm), L=span length (mm), t_s=thickness of slab (mm), and K_g=stiffness parameter.

The girder distribution factor may be calculated by eccentrically-loading method in the Chinese Bridge codes. According to study *[4]*, the GDFs are very close to each other using AASHTO LRFD specifications and Chinese Bridge codes. The girder distribution factors for three codes are shown in the Table 7.

Table 7 Girder distribution factors (GDF)

Bridge No.	Span m(ft)	Spacing m(ft)	AASHTO LRFD	CBC 2004	CBC 1989
Type I	14 (46)	1.74 (5.75)	0.524	0.540	0.564
Type II	20 (66)	1.97 (6.5)	0.554	0.571	0.596

Type III	25 (82)	2.28 (7.5)	0.618	0.637	0.665
Type IV	30 (99)	2.28 (7.5)	0.624	0.643	0.672
Type V	35 (115)	2.73 (9.0)	0.708	0.730	0.763
Type VI	40 (132)	2.73 (9.0)	0.703	0.725	0.757

Live load moment per girder

Including the dynamic factors and girder distribution factors, the unfactored live load moments per girder for three codes are shown in Table 8.

Table 8 Unfactored live load moment per girder, (kN-m)

Bridge No.	Span m(ft)	AASHTO M_L	CBC 2004 M_L	CBC 1989 M_L
Type I	14 (46)	648	673	683
Type II	20 (66)	1171	1143	1021
Type III	25 (82)	1800	1776	1857
Type IV	30 (99)	2351	2380	2311
Type V	35 (115)	3320	3450	3485
Type VI	40 (132)	3983	4212	4169

Factored load

The factored loads including dead and lived load are presented in Table 9 below.

Table 9 Factored moment per girder, (kN-m)

Bridge No.	Span m(ft)	AASHTO M_{factor}	CBC 2004 M_{factor}	CBC 1989 M_{factor}
Type I	14 (46)	1258	1035	1045
Type II	20 (66)	2525	2005	1879
Type III	25 (82)	4302	3467	3551
Type IV	30 (99)	6261	5169	5098
Type V	35 (115)	9694	8126	8161
Type VI	40 (132)	9979	10511	10466

Resistance model

Resistance is a variable representing the load carrying capacity. It can be affected by uncertainties in materials, dimensions and analysis. It is treated as the lognormal distribution, and the statistical parameters were derived by Dr. Nowak *et al* [1]. for prestressed concrete girders in this paper, coefficient of variation, V=0.075, and bias factor, λ=1.05.

The resistance of the prestressed concrete girder may be obtained by two ways. It can be determined when the number of strands is known. Another way is that the minimum required resistance, μ_R, may be calculated by design limit functions. For three codes, the following equations 9), (10) and (11) are specified. AASHTO LRFD 2005,

$$\mu_{R-LRFD}=1.25\mu_{D1+D2} +1.50\mu_{D3} +1.75\mu_{L+1}/\varnothing \tag{9}$$

$$\mu_{R-1989}=(1.20\mu_{D1+D2} +1.40\mu_{L+1})/K \tag{10}$$

$$\mu_{R-2004}=(1.20\mu_{D1+D2} +1.40\mu_{L+1})/\gamma \tag{11}$$

The minimum required moments are obtained by these equations for three codes in this paper. It should be noted that, the resistance factor Ø is specified as 1.0 for prestressed flexure structure in AASHTO LRFD 2005; in CBC 1989, the resistance factor is presented by a safety factor, 1/1.25; and in CBC 2004, the γ is divided into two partial resistance factors, concrete (1.45) and prestressed strand (1.47), in this paper the combined factor γ is assumed at 1.45. The mean values for minimum moment are presented in the Table 10,

Table 10 Minimum required resistance, (i.e. mean moments, kN-m)

Bridge No.	Span m(ft)	AASHTO $\mu_{R\text{-}LRFD}$	CBC 2004 $\mu_{R\text{-}2004}$	CBC 1989 $\mu_{R\text{-}1989}$
Type I	14 (46)	1705	1755	1771
Type II	20 (66)	3423	3397	3184
Type III	25 (82)	5833	5876	6017
Type IV	30 (99)	8488	8760	8640
Type V	35 (115)	13143	13771	13832
Type VI	40 (132)	16832	17813	17737

Reliability Analysis

Reliability analysis is performed for typical prestressed concrete girders. The reliability index, β, defined as a function of probability of failure, P_F, is presented in the following equation (12),

$$\beta = -\Phi^{-1}(P_F) \tag{12}$$

Where, Φ^{-1} is inverse standard normal distribution function. The Rackwitz-Fiessler method is applied to calculate the reliability index by an iteration procedure. The limit state function is formulated as the following equation (13), where L is for live load. The R is lognormal probability distribution, and the others are normal,

$$g = R - D1 - D2 - D3 - L \tag{13}$$

The reliability indices for six considered girder bridges are presented in the following Table 11 and Fig.6. It shows that for the considered design bridges, the reliability indices β=3.93-4.31 for AASHTO LRFD specifications; for CBC 1989, β=4.83-4.94; and β=6.31-6.52 for CBC 2004.

Table 11 Reliability indices, β, for three codes

Bridge	AASHTO	CBC1989	CBC2004
Type 1	4.31	4.83	6.31
Type 2	4.22	4.89	6.40
Type 3	4.13	4.92	6.46
Type 4	4.03	4.94	6.51
Type 5	3.93	4.94	6.51
Type 6	3.93	4.94	6.52

It can be seen that the Chinese new code (CBC 2004) has a higher reliability level than the old code (CBC 1989). The major objective of Chinese new code is to increase the safety; this is confirmed for prestressed concrete bridge by this study. It also can be seen that the two Chinese codes have higher reliability for the same prestressed concrete girder than USA LRFD code; this can be partly explained that the resistance factor does reduced ($\varnothing$=1.0) for prestressed concrete flexural capacity in LRFD code. The design life time, 100 years for CBC 2004 and 75 years for AASHTO LRFD code, contributes the higher safety requirements for Chinese code. The AASHTO requires the reliability level is 3.5; and CBC 2004 requires 4.7-5.2 for class-I bridge structures.

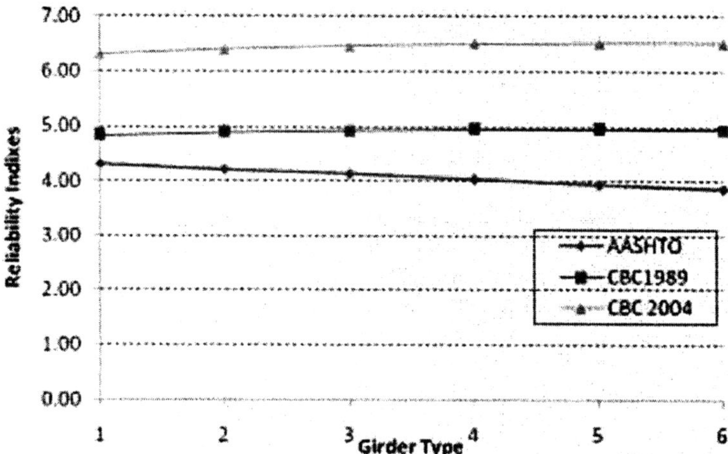

Fig.6 Reliability indices for three codes

From the Fig.6, it can be seen that, the reliability indices changes in different ways for Chinese codes and USA code while span length changes. When the span length increases, the reliability indices increase for both Chinese codes; but it decreases for AASHTO code. This may be verified by a preliminary parameter study for Type-VI girder by changing its span length, which is not included in this paper. Furthermore, it can be seen that the reliability level for prestressed girder bridge relatively has the smaller derivation for Chinese codes when span length or girder spacing changes.

Conclusion

The reliability analysis is carried out for flexure capacity of typical prestressed concrete girder bridges designed using the USA AASHTO LRFD bridge specification (2005) and the Chinese bridge codes (1989, 2004). The loads and resistance are treated as random variables. The dead load and live load are calculated corresponding to the three codes, and the minimum required resistance is calculated by the strength-I limit state function. An iteration procedure is performed to calculate reliability indices by applying the Rackwitz-Fiessler method. The conclusions are drawn from this study:

(1) The factored loads specified by both Chinese codes are very close although the vehicular load models are quite different.

(2) The reliability level for the new Chinese bridge code (CBC 2004) is higher than the old Chinese bridge code (CBC1989) due to higher resistance partial factor (1.45) for former than the later (1.25 safety factor).

(3) When span length increases, the reliability indices increase for both Chinese codes; but it decreases for AASHTO code.

(4) It confirmed that the Chinese codes have a higher safety requirement than USA code: CBC 2004 requires the reliability level of 4.7-5.2 for class-I bridge structures and the AASHTO LRFD requires 3.5.

References

[1] Nowak, A. J., & Collins, K. R. (2000), Reliability of Structures, New York: McGraw-Hill.

[2] Nowak, A. S., Park, C., & Casas, J. R. (2001), Reliability analysis of prestressed concrete bridge girders: comparison of Europe, Spanish Norma IAP and AASHTO LRFD. *Structural Safety, 23,* 331-344

[3] Nowak, A. S. (1999). Calibration of LRFD Bridge Design Code, Washington, D.C.: National Academy Press.

[4] Du, J. S., & Au, F.T.K (2005), Deterministic and Reliability analysis of prestressed concrete bridge girders: comparison of Chinese, Hong Kong and AASHTO LRFD codes. *Structural Safety, 27,* 230-245

[5] AASHTO LRFD Bridge Design Specifications, Washington, D.C: 2005 interim.

[6] General Code for Design of Highway Bridges and Culverts, China: the People's Transportation Press, 1989 (JTJ 021-89)

[7] General Code for Design of Highway Bridges and Culverts, China: the People's Transportation Press, 2004 (JTG D60-2004)

[8] Code for Design of Highway Reinforced Concrete and Prestressed Concrete Bridges and culverts, 2004 (JTG D62-2004)

Experiences and some results in experimental testing of bridges in Croatia

Prof. dr. sc. **Zdravko Kapovic**, University of Zagreb, Faculty of Geodesy
Prof. dr. sc. **Mladenko Rak**, University of Zagreb, Faculty of Civil Engineering
Domagoj Damjanovic, Bsc, University of Zagreb, Faculty of Civil Engineering

Abstract

The paper presents the importance and procedures of displacement and strain measurements in experimental testing of bridges. During these procedures structure is loaded according to certain phases, while displacements, strain, rotation angles, dynamic characteristics, temperature etc. are measured by means of geodetic and physical methods. The analysis of the results should help in defining whether the discrepancies in measured and theoretical displacement are significantly different for various types of structures. Detailed testing results will be indicated for 4 important bridges built lately in Croatia. These bridges are different types of structures and are built from different materials. Results and comparison of theoretical and experimental displacements values for these four bridges are presented in this paper, as well as comparison of maximal values of displacements for about twenty other bridges.

The importance of experimental testing of bridges

It is legally regulated[1] in Croatia that all road bridges with spans larger than 15 m and all railway bridges with spans larger than 10 m should be tested before their opening to traffic. The purpose of these testings is to check whether the behaviour of some bridge in reality is the same as that foreseen by project, whether it has been built in accordance with regulations, and finally, to prove that it is capable of accepting the traffic load planned by the project.

At present time, when more and more attention is paid to bridge management that should make it possible to achieve as convenient ratio of investments and construction durability, it is very important to define some initial construction parameters. The results collected during test loading can be of great importance in diagnosing the structure state during its exploitation. One can say that test loading is the first step in high quality bridge management, and the beginning of monitoring the behaviour of bridges during their useful life.

Apart from that, the importance of these testings is also in verification and improvement of various numerical models of theoretical static and dynamic structure analysis.

Testing of the Dubrovnik Bridge
Description of the bridge structure
The bearing structure of the Dubrovnik Bridge consists of a 147.4 m long prestressed concrete access viaduct, and a 324.7 m long main nonsymmetrical cable-stayed bridge which is composite structure, steel-concrete. The spans of the bridge are 87.35+304.05+80.7=472.1 m. The prestressed access viaduct has the span of 87.35 m and it is console-extended toward the

large span for 60.05 m. The main bridge with slant braces is nonsymmetrical with spans of 244+80.7=324.7 m. The pylon is made of concrete, it is A-shaped with the vertical peak, and its total height is 141.5 m. The sides of the pylon have a box cross section with external dimensions 4.0 x 5.0 m, being at distance of 32.0 m at the foundation. (Figure 1)

Figure 1. The Dubrovnik Bridge

Testing of the Dubrovnik Bridge under static load

Static testing of the bridge was carried out through 8 phases of loading[2] in order to achieve extreme magnitudes of internal forces and displacement of the main bearing elements of the structure. During these load phases displacements and strain were measured.

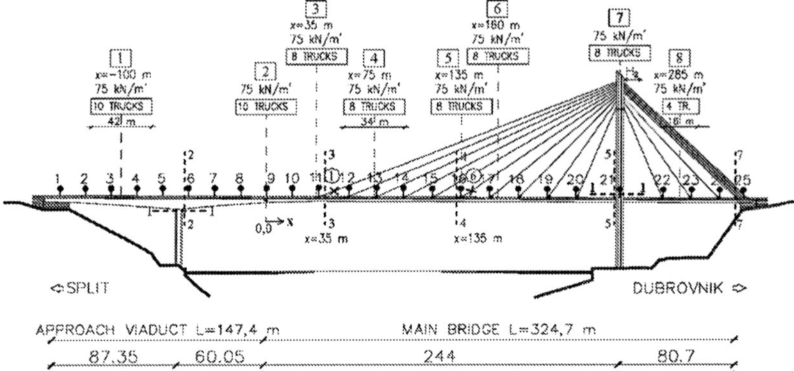

Figure 2. – Overview of load schemes and measuring points

Vertical displacements were measured along the two lines of the bridge, in quarter and half points of the superstructure of the approaching viaduct and on locations of the cable hinges of the main bridge[3]. Vertical displacement measurement was performed using the modified method of geometric levelling, where two precise levelers of type WILD-NA2 with coplanar slabs were used. For horizontal displacements measurement on the top of the pylon, as well as for some controls of superstructure vertical displacements, GPS system was used.

Results of deflection measurement[4]

Values of measured deflections are directly compared to the theoretical values determined from the FEM model of the bridge. Maximal values of measured deflection for each load phase, together with corresponding theoretical value and residual deflections after unloading are shown in Table 1. Deflections of the bridge during load phase nr. 4 which induced maximal overall deflections are shown graphically in figure 3.

Table 1. Experimental, theoretical and remaining deflections

Load phase	Measuring point	Maximal deflections		Residual deflections (mm)
		Experimental (mm)	Theoretical (mm)	
1st	3	11,2	14,8	0,0
2nd	9	75,7	98,0	2,2
3rd	12	250,0	259,0	2,0
4th	13	289,0	298,0	4,0
5th	15	168,0	169,0	2,0
6th	17	118,0	124,0	2,0
7th	20	9,0	-	0,0
8th	22	12,0	21,58	1,0

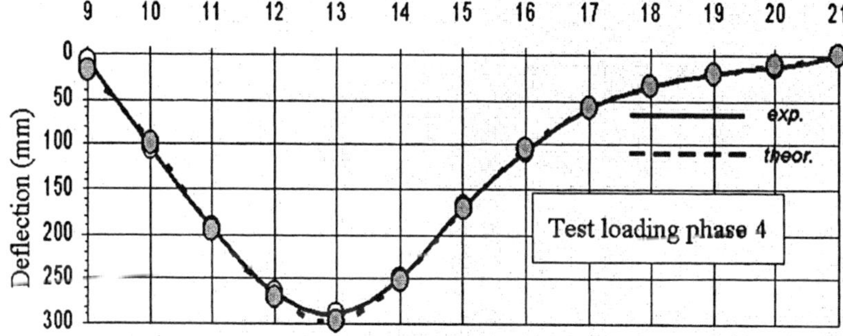

Figure 3. – Deflections for the 4th phase of loading

Horizontal displacements of the pylon top

Horizontal displacements of the pylon top were measured with a GPS device which was placed at the top of the pylon. The results of the horizontal displacements of the pylon top in the direction of the roadway, during load schemes 3 and 4 are shown in Table 2.

Table 2. Horizontal displacements of the pylon top

Load phase	Horizontal movement of the pylon top
3rd	36,0
4th	53,0

Testing of the Maslenica Bridge
Description of the bridge structure[5]

The main bearing structure of the bridge consists of two «two-hinged arches» (Figure 4). The arch span is L =155,0 m, arrow f = 41,45 m, relation f/L = 1/3,75. The arch axis, intrados and extrados are formed as parabolas of second degree through three given points. Cross-section of the bridge is constructed as a two-walled tin-slab girder. Height of the arch ridge at the vertex is 2,80 m, and at the supports it is 2,40 m. The supports of arches are hidden, so visually, the arch seems as if it is fixed. Two arches are mutually stiffened with a wind couple of crossover diagonals.

Figure 4. The Maslenica Bridge

The roadway structure of the bridge is 315,30 m long and it is a continuous assembly over 17 spans. It is a composite structure composed of 25 cm thick reinforced concrete roadway slab, transverse and longitudinal steel girders. The roadway slab is a stabilizing element of the bridge structure for the forces in the horizontal plane (wind couple).

Longitudinal girders of the roadway are welded tin-slab girders with the constant height of 1200 mm, and reinforcement is done "towards inside". Lower band of transverse girders is horizontal, while the upper band follows the transverse inclination of the lower edge of roadway slab. That is why the height of the ridge varies from 730 mm along the longitudinal girder to 810 mm in the middle of the bridge.

All columns are square, made from the slabs. «Small» (slender) columns are constructed as «pendulum-columns» with sides of size 650, 750 and 850 mm. The "large" (portal) columns are fixed into foundations of arches, and their sides are 2300 mm.

Testing of the Maslenica Bridge under static load[6]

Static loading of the bridge was inflicted using 8 heavy trucks. The trucks were positioned to achieve maximal inner forces and deflections. Static load testing was carried out through total of 18 phases of loading, 4 arch loading phases and 14 roadway loading phases. This paper presents the results of deflection and strain measurement only for phases of arch loading.

Displacements were measured on the main bearing elements of the structure along the two lines (A and B), over the supports and in the middle of each span. On the arch, horizontal and vertical displacements were measured in the quarter of the span and in the vertex.

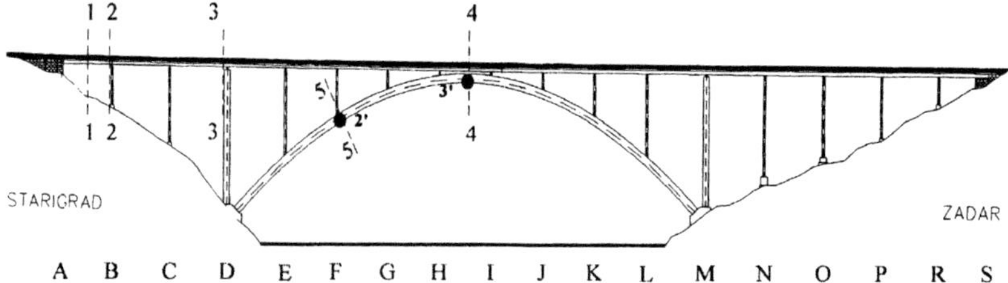

Figure 5. Cross-sections of strain measurement and points for movement measurement

Results of deflection measurement

Maximal deflections measured during the phases of arch loading are shown in Table 3, where they are compared to the corresponding theoretical deflection values. Graphical presentation of experimental and theoretical deflections for arch loading phases is shown in figure 6.

Table 3. Comparison of experimental and theoretical deflection of the arch

Measuring point	Load phase	Vertical (mm)		Horizontal (mm)	
		Experimental	Theoretical	Experimental	Theoretical
Arch quarter	14th	116	122,7	96	99,5
Arch vertex	16th	52,0	54,6	0,0	0,0
Arch quarter	18th	-69,0	-82,2	-69,0	-73,1
Arch vertex	21st	-15,0	2,0	-23,0	0,0

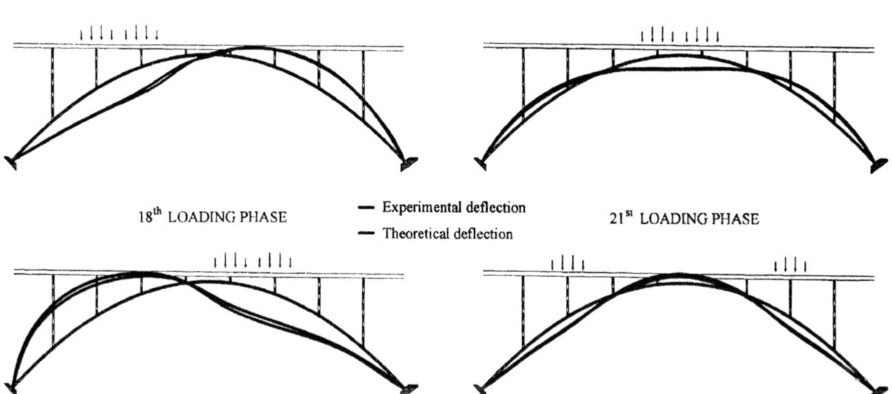

Figure 6. Experimental and theoretical deflections for arch load phases

Testing of the Krka Bridge
Description of the bridge structure[7]

Main bearing structure of the bridge is hinge less reinforced concrete arch over a 204,0 m span with arch rise of 52,0 m (f/L = 1/3,92) (Figure 6).

Figure 6. The Krka Bridge

Arch has a box cross section with two chambers. It is 10,0 m wide and 3,0 m high, horizontal walls are 40 cm and vertical walls are 50 cm thick. In the vicinity of the abutment horizontal walls are strengthened to 60 cm. Columns S7 and S8 have solid rectangular cross section (180x220 cm) while all the rest have box cross section of various dimensions. Bridge deck is a lightweight composite structure made of two main steel girders, steel cross girders at every 4 m and reinforced concrete deck slab (25 cm). The total length of the bridge deck is 360,0 m, it is supported by abutments and columns continuously over 12 spans (4x32,0 + 3x28,0 + 3x32,0 + 28,0 + 24,0 = 360,0 m). Transversally bridge deck is a cantilever beam.

Testing of the Krka Bridge under static load[8]

16 heavy trucks were used as static load for phases of arch loading and 8 for single span loading phases. Static testing was carried out through 20 loading phases and 13 phases of load releases. Most interesting phases are the first four loading phases in which the arch was loaded. During these 33 phases of testing deflections and strains were measured.

Deflections were measured by surveying method of levelling above all columns and in the middle of spans along 4 lines (A, B, C and D) and at the spans where nonsymetrical load was applied there were 6 lines (A, B, C, D, E and F). At the arch there were 6 points for displacement measurement during arch loading phases[9].

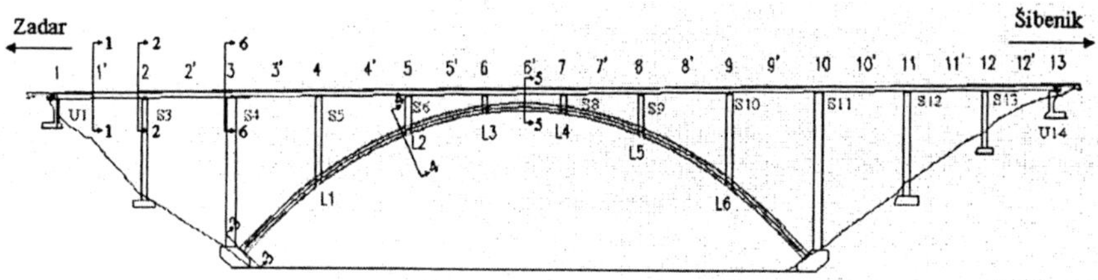

Figure 7 : Movement measuring points

Results of deflection measurement

Maximal deflections measured on the arch for symetrical load consisting of 16 heavy trucks are shown in Table 4, these experimental values are compared to corresponding theoretical deflection values. Experimental and theoretical deflections for arch loading phases are shown graphically in figure 8.

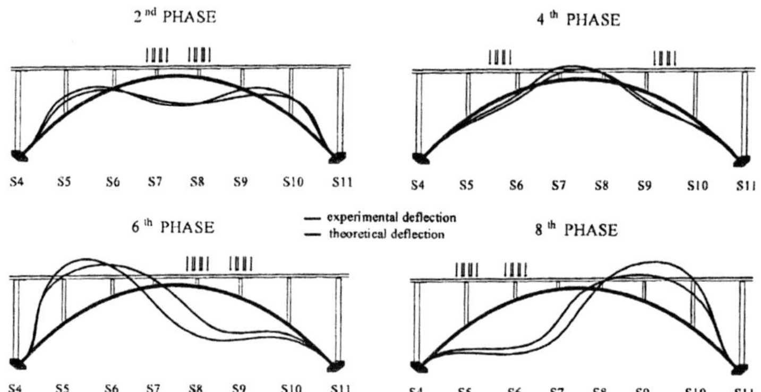

Figure 8. Experimental and theoretical deflections for arch load phases

Table 4. Comparison of experimental and theoretical deflections of the arch

Measuring point	Load phase	Vertical (mm)		Horizontal (mm)	
		Experimental	Theoretical	Experimental	Theoretical
Arch vertex (S8)	2nd	17	16,9	2	0,4
Arch vertex (S7)	4th	-8	-5,0	-2	0,1
Arch quarter (S6)	6th	25	29,1	18	24,2
Arch quarter	8th	22	29,8	-13	-19,3

Testing of the Mirna Viaduct
Description of the structure[10]

Span assembly consists of two longitudinal steel girders with constant height (2750 mm) at the distance of 550 cm that are composite with concrete roadway slab, and of transverse girders. The static system of the assembly is continuous over 22 spans with total length of 1354,86 m (51,07 + 15x66,49 + 70,10 + 2x50,07 + 63,09 + 42,56 + 30,54 = 1354,86 m), in horizontal and vertical curve.

The main girders are solid welded I-type girders. Upper flange is 700 mm wide, and lower 800 mm along the whole bridge length. The basic thickness of the upper flange is 30 mm (35 mm in the span of 70 m) in the middle of the span, and it runs up to 90 mm (100) over the columns, at the length of 2 m and it is reduced towards the field with the declination of 0,55%. The basic thickness of the lower flange is 35 mm (40 mm), and it is 90 mm (100 mm) over the pylons. The web thickness is 16 mm along the whole bridge length.

Transverse girders in the spans are welded with the web of 760x12 mm and flanges of 300x20 mm, and they are placed at 700 mm below the upper edge of the longitudinal girders. Near the bearings they are at the distance of 4,8 m, and in the middle part at the distance of 7,54 m. Transverse girders over the bearings are strenghtened to web dimensions of 1500x25 mm and flange dimensions of 500x30 mm. The steel quality is St 52-3 according to DIN.

Figure 9. Mirna Viaduct

Reinforced concrete roadway slab has got basic thickness of 25 cm, being thickened to 35 cm over the main girders. External edges are thickened to 54 cm, and the traffic barrier is anchored in them. In transverse direction, the slab is a cantilever beam (230 + 550 + 230 = 1010 cm). The span assembly is supported by abutments, and by altogether 21 pier. The abutments and ending piers have got shallow foundation in the rock while other piers are founded on steel piles. The abutments and piers are made of concrete B 35.

Testing of the Mirna Viaduct under static load[11]

Static loading was   performed with eight (8) heavy trucks with average total mass of about 40,23 t. The trucks were placed in such a way that they provoke maximal internal forces and displacements. Testing of viaduct with static load was carried out in altogether 45 phases, 28 loading phases and 17 phases of load releasing when residual displacements were measured. Out of 28 loading phases, during 22 phases each span was symmetrically loaded using 8 trucks, and on initial spans there have been also 4 phases of unsymmetrical load consisting of 4 trucks. In one phase the trucks were positioned to achieve negative bending moment over the first pier (P1), and in one phase the trucks were positioned to cause the maximal reaction over the pier P16.

During static loading displacements of the main bearing elements of the structure were measured along two parallel lines (A and B), over the supports and in the middle of each span (45 cross sections). The displacements have been measured by means of trigonometric levelling method.

Results of deflection measurement

Values of measured deflections were directly compared to the theoretical values determined from the FEM model of the structure. Maximum values of measured deflections for some characteristic symmetrical load phases, together with corresponding theoretical values are presented in Table 5. Graphical presentation of experimental and theoretical deflections during loading of span L13, which gave maximal deflections is shown in figure 10.

Table 5. Comparison of some experimental and theoretical deflections

Measuring point (span length)	Load at span	Experimental (mm)	Theoretical (mm)
Middle of span L1 (51,07 m)	L1	39,5	41,5
Middle of span L4 (66,49 m)	L4	66,0	71,7
Middle of span L13 (66,49 m)	L13	66,5	72,1
Middle of span L17 (70,09 m)	L17	63,0	74,9
Middle of span L18 (50,07 m)	L18	31,0	33,7
Middle of span L20 (63,09 m)	L20	53,0	56,9
Middle of span L21 (42,56 m)	L21	15,5	19,0
Middle of span L22 (30,54 m)	L22	5,5	7,2

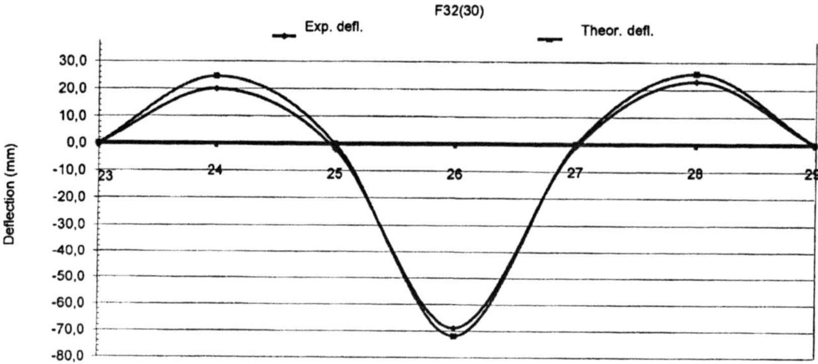

Figure 8. Experimental and theoretical deflections at span L13

Comparison of experimental and theoretical displacement for additional 21 structures

Along with the detailed presentation of displacement measurement at previous 4 objects, some results of experimental testing for additional 21 object are processed. Table 6 shows maximal experimental and theoretical values of deflections for given objects.

Table 6. Experimental and theoretical values of deflections for additional 21 object

Bridge	Material	Measured deflection	Theoretical deflection	Difference (mm)	Congruity (%)
Krčki bridge	concrete	6,4	7,2	0,80	88
The bridge over Pazinska pit	concrete	30,4	36,8	6,40	79
The bridge over river Glina	steel	10,0	10,1	0,10	99
Viaduct Zalesina	concrete	8,7	11,1	2,40	72
Viaduct Mosor	concrete	4,0	4,2	0,20	95
Viaduct Majdan	concrete	3,5	4,0	0,50	86
Viaduct Novo Mjesto	concrete	10,9	11,3	0,40	96
Viaduct Polonje	concrete	10,2	11,3	1,10	89
Viaduct Komin-junction	concrete	9,8	9,9	0,10	99
The Sava bridge by Orašje	steel	292	341	49	83
Flyover Vranjići	concrete	1,65	1,95	0,30	82
Subway Smiljike	concrete	1,3	1,47	0,17	87
Viaduct Raštević	concrete	5,0	5,3	0,30	94
The bridge Kličevica	concrete	7,3	9,0	1,70	77
Flyover Grabar	concrete	1,5	1,47	0,03	98
Subway Paljevine (right)	concrete	0,5	0,5	0,00	100
Subway Paljevine (left)	concrete	0,45	0,5	0,05	89
Railway bridge "Strug"	steel	12,0	12,0	0,00	100
Viaduct Arambašići	concrete	2,75	2,85	0,10	96
Viaduct Blaževci	concrete	6,75	7,25	0,50	93
Viaduct Tukcetin	concrete	3,7	4,0	0,30	92

A superficial look at the results of the previous table shows very good congruity of theoretical and experimental deflection values for various types of structures. Table 7 presents the research results for theoretical and experimental values of all deflections measured on the objects given in the table 6 (181 sample).

Table 7. Statistical indicators of congruity between theoretical and experimental deflections

Construction	Sample (number of deflections)	Average value	Standard deviation	Median	Mode
Steel constructions	66	89.1	11.6	91	100
Concrete constructions	115	84.1	15.5	89	100

Conclusions

The results of the analysis carried out offer a few conclusions:

Measured maximal deflection values during the experimental testing of all tested objects were within the expected limits and were in accordance with the calculated values. Residual values of deflections and strain after unloading were small so the behaviour of the structures can be observed as elastic.

The analysis of the results obtained during the test loading was meant to show whether the discrepancies between experimental and theoretical deflections differ significantly for various types of structures. Experimantally obtained deflections of steel structures have indicated better congruity with theoretical values as related to the concrete constructions, which is in accordance with expectations.

The stated conclusion proves that during the testing the optimum accuracy of displacement measurement has been achieved.

In spite of some deficiencies in the application, the indicated results show that GPS method can also be used for experimental testing of the structures or for monitoring over a longer period of time.

References:

[1] Regulation HRN U.M1.046: Bridge testing with test load, Official Journal No. 60/1984.
[2] M. Rak, Z. Šavor, M- Friedl: Testing program for the 'Bridge over Dubrovnik River', Faculty of Civil Engineering, Zagreb, Department of Engineering Mechanics, No. 180-54/02, April 2002
[3] M. Rak, J. Krolo, V. Čalogović: Report on experimental testing of the 'Bridge over Dubrovnik River', Faculty of Civil Engineering, Zagreb, Department of Engineering Mechanics, No. 180-84/02, April 2002.
[4] Kapović, Zdravko; Herceg, Ljudevit; Krolo, Joško. Test loading of the Dubrovnik Bridge. //AVN Allgemeine vermessungs-nachrichten. 7/2005 (2005) , D1103; 258-262.
[5] Štorga, S. 2004. Project of renewal of Maslenica Bridge. Roads and briodges (Ceste i mostovi) 10-12, Year (God.) 50 (2004)10-12: 56-66. Zagreb.
[6] M. Rak, J. Krolo, V. Čalogović & D. Damjanović, 2005. Report on experimental testing of the Maslenica Bridge. Faculty of Civil Engineering, Zagreb, Department of Engineering Mechanics, No. 180-138/05, June 2005.
[7] Savor, Z. 2003. Design of the Krka bridge, Parts 1-4, Structural department, Faculty of Civil Engineering, University of Zagreb.
[8] M. Rak, J. Krolo, V. Čalogović, D. Damjanović 2005. Report on test loading of "The Krka Bridge", Department of Technical Mechanics, Faculty of Civil Engineering, University of Zagreb. No. 180-149/05, June 2005.
[9] Rak, Mladenko; Bjegović, Dubravka; Kapović, Zdravko; Stipanović, Irina; Damjanović, Domagoj. Durability Monitoring System on the Bridge over Krka River // Bridges / Radić, Jure (ur.). Zagreb : SECON HDGK, 2006. 1137-1146.
[10]   Savor, Z. 2003. Design of the Mirna Viaduct, Structural department, Faculty of Civil Engineering, University of Zagreb.
[11]   M. Rak, D. Damjanović, 2005. Report on test loading of "Mirna Viaduct", Department of Technical Mechanics, Faculty of Civil Engineering, University of Zagreb. No. 180-102/05, April 2005.

Numerical analysis of carrying capacity deterioration and repair demand of existing reinforced concrete bridge

Dr ZHAO Dong-bing, Tongji University, Shanghai, P.R. China

Professor FAN Li-chu, Tongji University, Shanghai, P.R. China

Abstract

By combining concrete carbonation model and rebar corrosion model, using the stochastic finite element method to analyze the time-dependent failure probability and reliability of reinforced concrete bridge within life cycle, new repair demand prediction model based on target reliability index and repair cost benefit is proposed. Applied in one tied arch bridge and compared with different repair plans, this model predicts repair demand during life cycle and verifies its feasibility.

Introduction

When it comes to the assessment and management of existing reinforced concrete bridges, such factors as material property deterioration, structure reliability and repair demand prediction have to be considered in a comprehensive way. Among various factors leading to the declining of bridge's load carrying capacity, rebar corrosion resulting from environmental corrosive medium is the primary cause that makes structural performance degrade, even be destroyed. Under normal atmospheric conditions, the prerequisite of rebar corrosion is the loss of inert environment on the surface of rebar caused by concrete carbonation.

Stochastic finite element method, a new reliability computation method gradually developed in recent years, is a combination of finite element method and stochastic analysis which relies on rapidly-developed computer technology to realize the reliability computing of large-scale and complex structure. According to concrete carbonation model and rebar corrosion model, this paper aims to use the stochastic finite element method to analyze failure probability of structure and reliability of existing reinforced concrete bridges during life cycle, proposes the bridge repair demand prediction model on target reliability index and repair cost-benefit ration and take example for predicting bridge repair demand.

Concrete Carbonation Model

Theoretical concrete carbonation model, put forward by Papadakis, Greek Scholar (Papadakis, Vayenas and Fardis, 1989), describes physical diffusion and chemical reaction in the process of carbonation by establishing mass equilibrium equation of gaseous CO_2, solid and dissolved $Ca(OH)_2$, calcium silicate hydrate (CSH) and non-hydrated silicate compounds. The model can be further simplified as follows (ZHANG and JIANG, 2003), on that condition the active ingredient in cement is completely hydrated and humidity is relatively stable.

The paper is supported by China Scholarship Council (CSC).

$$\frac{\partial}{\partial x}(D_e \frac{\partial[CO_2]}{\partial x}) = [CO_2]\{K_{CH}[Ca(OH)_2] + 3K_{CSH}[CSH]\} \qquad (1)$$

$$\frac{\partial}{\partial t}[Ca(OH)_2] = -K_{CH}[Ca(OH)_2][CO_2] \qquad (2)$$

$$\frac{\partial}{\partial t}[CSH] = -K_{CSH}[CSH][CO_2] \qquad (3)$$

among them, D_e represents diffusion coefficient of CO_2 in concrete (square meter per second). [j] denotes the molar concentration of constituent j in concrete (moles per cubic meter), j= CO_2, $Ca(OH)_2$, CSH. K_{CH}, K_{CSH} is respective carbonation reaction constant of $Ca(OH)_2$ and CSH. x (in meter), t (in second) indicates the distance from one certain position in concrete to surface and the reaction time respectively. De can be expressed as

$$D_e = 8 \times 10^{-7}(w/c - 0.34)(1 - RH)^{2.2} \qquad (4)$$

in which, w/c is concrete water-cement ratio; RH is atmospheric relative humidity (in percent).

Rebar Corrosion Model

After the concrete cover is completely carbonated, rebar begins to corrode. The process of corroding belongs to electrochemical reaction. The corroded production will hinder the spread of Fe ion, and corrosive velocity will be nonlinearly slowed down with the increasing of corrosion time. Under normal atmospheric conditions, the relation between rusted rebar diameter and corrosion time is outlined as (Yalcyn and Ergun, 1996)

$$D(t) = D_0 - 0.0232 \int_0^t i_{corr}(t)dt \qquad (5)$$

$$i_{corr}(t) = i_{corr}(i) \cdot 0.85t^{-0.29} \qquad (6)$$

$$i_{corr}(i) = 37.8(1 - w/c)^{-1.64}/C_b \qquad (7)$$

among them, D(t) refers to rusted rebar diameter after t years' rebar corrosion (in millimeter); D_0 is rebar diameter before corrosion (in millimeter); $i_{corr}(t)$ represents corrosion current density at time t (in micro ampere per square centimeter); $i_{corr}(i)$ denotes initial corrosion rate (in millimeter per annum); w/c is concrete water-cement ratio; C_b is thickness of concrete cover (in millimeter).

The Analysis of Stochastic finite element reliability

Based on structural finite element model, stochastic finite element method, citing stochastic analysis and making the use of computing ability of finite element method to form response surface between uncertain variables and structural response, simulates structure failure probability and reliability with the help of Monte Carlo method. Its analyzing steps can be divided into: (1) citing finite element model; (2) defining random variables; (3) describing ultimate state equations; (4) building response surface; (5) running analysis; (6) Monte Carlo simulation.

Bridge repair demand prediction

The purpose of bridge repair demand prediction is to determine when and how to repair within life cycle. The model aims at predicting repair demand by analyzing cost benefit to meet the requirement of minimum target reliability index. The existing models (Frangopol, Lin and Estes, 1997) ignore the reduction of structural failure probability and existence of cumulative failure probability due to repair measures, which makes the repair demand prediction inapplicable.

The new model, which integrates target reliability index and repair cost-benefit analysis, is suggested as

$$B_{\text{fail}}(T_{\text{int}}) = C'_{\text{fail}}(X, T_{\text{ref}}) - C''_{\text{fail}}(X, T_{\text{ref}}) \tag{8}$$

$$C(T_{\text{int}}) = \int_0^{T_{\text{int}}} [C_{\text{PM}}(t) + C_{\text{INS}}(t) + C_{\text{REP}}(t)] \cdot (1+r)^{-t} \, dt \tag{9}$$

$$\max[B(T_{\text{int}}) - C(T_{\text{int}})] \in (\beta_L > \beta_L^*), B(T_{\text{int}}) > C(T_{\text{int}}) \tag{10}$$

$$C_{\text{fail}}(X, T_i) = P_f^{\text{cum}}(X, t_i) \cdot C_f \cdot \int_0^{t_i} (1+r)^{-t} \, dt \tag{11}$$

among them, $B_{\text{fail}}(T_{\text{int}})$ denotes repair benefit; $C'_{\text{fail}}(X, T_{\text{ref}})$ is structural failure cost without repair measure; $C''_{\text{fail}}(X, T_{\text{ref}})$ refers to structural failure cost with repair measure, T_{ref} is life cycle; T_{int} indicates repair time; $C(T_{\text{int}})$ is generalized repair cost concerning capital interest r (including bridge inspection cost C_{PM}, preventive maintenance cost C_{INS} and repair cost C_{REP}); β_L^* is target reliabilty index; $C_{\text{fail}}(X, T_i)$ reprensents structural failure cost at time t_i; $P_f^{\text{cum}}(X, t_i)$ is structure cumulative failure probability at time t_i, C_f is structural construction cost.

The principle of repair plans selection is to meet the requirement of target reliability index, maximize the difference between repair benefit and cost, and lower the risk of structural failure.

Example

Take one tied arch Bridge located in Shanghai for example, the net bridge width is 36.5m with design load vehiclec-20 level, trailer-120. The standard cross section of tied box beam is 1.4m wide, 1.6m high with hollow part 0.6m wide, 1.1m high. Life cycle is assumed to be 50 years.

Analysis of concrete carbonation

The tied beam uses concrete grade C50, 525# ordinary Portland cement with its content $526kg/m^3$ and water/cement ratio 0.39. The net concrete cover is 30mm in thickness; According to meteorological information, the average relative humidity is 80%; Atmospheric CO_2 concentration is assumed to be 0.045% in accordance with its industrial environment. (Saetta and Vitaliani, 2004); Based on equation (1) - (4), the computing results indicate that the concrete cover will be completely carbonated in 25 years and then rebar will start to rust.

The Analysis of stochastic finite element reliability

Based on the "Unified standard for reliability design of highway engineering structures" (GB/T50283-1999), the target reliability index of tied beam is 4.2 (2nd level of ductile failure). In finite element model, the concrete adopts 8-node hexahedral 'SOLID65' brick element and rebar 2-node 'LINK8' element. The stress-strain relation between concrete and rebar uses data of "Code for design of concrete structures" (GB50010-2002). Concrete failure criterion uses Willam-Warnke five parameter model. Rebar area is set as normal random variable, mean value of which changes according to equations (5) - (7). The coefficient of variation is 3% (CHANG and JIANG, 1995). The results show that structural reliability index of tied beam will be below the target value after the use of 29.4 years.

Comparison of repair plans and repair demand prediction

Repair plans

Anti-carbonation coating on the surface of concrete has recently been used as preventive repair measures, which can prevent moisture and gas from penetrating into beam in five years. There are three repair plans for you to choose from:

(1) Plan 1: repair one time every 5 years after the service of tied beam.

(2) Plan 2: the first repair is done when the concrete cover is completely carbonated (at the end of 25 years since its service), and then repair one time every 5 years.

(3) Plan 3: the first repair is done when the reliability of beam is lowered to near 4.2-the taget reliability index (at the end of 29 years since its service), and then repair one time every 5 years.

In accordance with the result of time-dependent failure probability analysis, cumulative failure probability of three repair plans is showed in Figure 1.

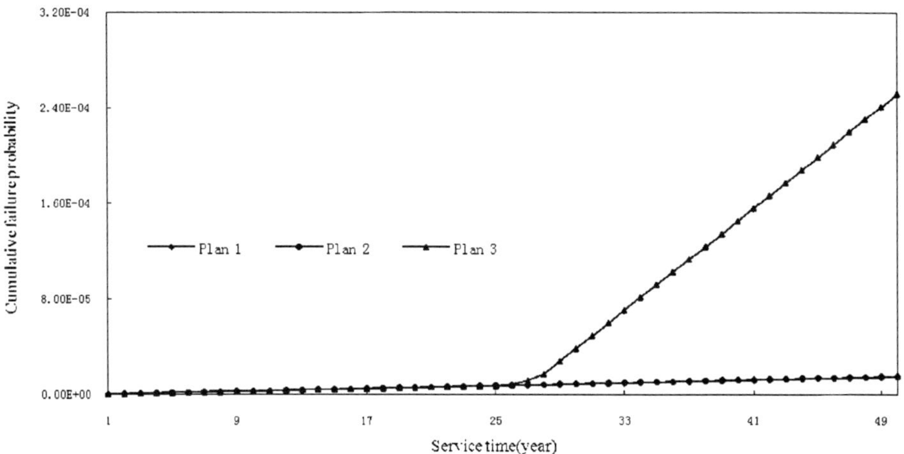

Figure 1. Cumulative failure probability of three repair plans

Comparison of repair plans and demand prediction

➤ Comparison of repair plans

The reduction between repair benefit and cost is calculated by equation (8) - (11). The results of three repair plans are showed in Figure 2. If the capital interest r ranges from 1% to 10%, the reduction between repair benefit and cost of plan 1 is the lowest, and plan 1 has little feasibility. If the capital interest r is less than 4%, plan 2 is the best plan; however, if the capital interest r is more than 4%, plan 3 is the optimum plan. The higher the capital rate is, the more favorable it is to put off repair time.

➤ Repair demand prediction

When capital interest is more than 4%, the first repair is done at the end of 29 years since its service and then repair one time every 5 years, 5 times in total in life cycle. When capital interest is less than 4%, the first repair is done at the end of 25 years since its service and then repair one time every 5 years, 6 times in total in life cycle.

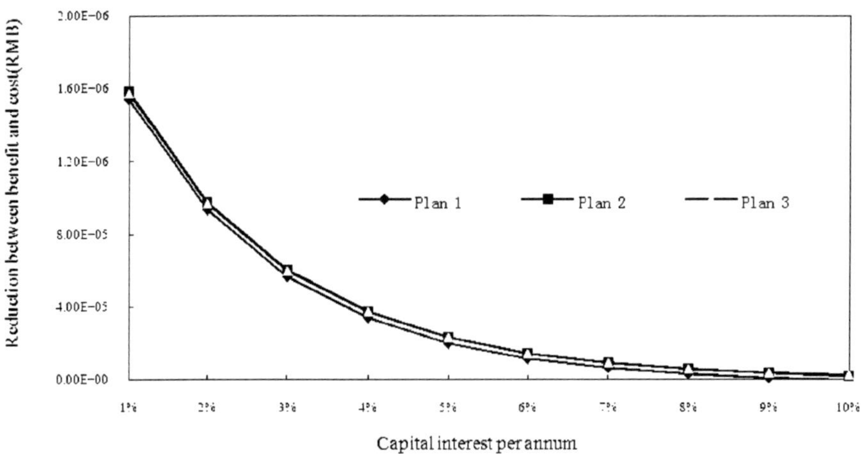

Figure 2. Three repiar plans' reduction between repair benefit and cost

Conclusion

By synthesizing concrete carbonation model and rebar corrosion model, using the stochastic finite element method to analyze the time-dependent failure probability and reliability of existing reinforced concrete bridge within life cycle, and building repair demand model, this paper predicts bridge repair demand. The suggested analyzing method concerns performance assessment of existing reinforced concrete bridge and key process of repair management. The example shows that the method can meet the practical requirements of bridge assessment and management.

References

CHANG Da-min, JIANG Ke-bin: Reliability analysis and design of bridge structure, China Railway Publishing House, 1995.

Frangopol DM, Lin KY, Estes AC.: Life-cycle cost design of deteriorating structures. Journal of Structural Engineering, ASCE, 1997, Vol. 123, pp. 1390-1401.

Papadakis VG, Vayenas CG, Fardis MN.: A reaction engineering approach to the problem of concrete carbonation. AICHE Journal, 1989,35(10):1639-1650.

Saetta AV, Vitaliani RV.: Experimental investigation and numerical modeling of carbonation process in reinforced concrete structures Part I: Theoretical formulation. Cement and Concrete Research, 2004, Vol. 34, pp. 571-579.

Yalcyn H, Ergun M.: The prediction of corrosion rates of reinforcing Steels in concrete. Cement and Concrete Research, 1996, Vol. 26, pp. 1593-1599.
ZHANG Yu, JIANG Li-xue, et al: Durability of concrete structures, Shanghai scientific & Technical Publishers, 2003.

Theme four:

Bridge management and its relationship with infrastructure management

Accelerating the renewal of America's highway bridges

Dr. Z. John Ma, Associate Professor, University of Tennessee, Knoxville, TN, USA

Abstract

The American Association of State Highway and Transportation Officials' (AASHTO) Highway Subcommittee on Bridges and Structures (HSCOBS) has identified a total of seven grand challenges. The top two grand challenges are Optimizing Structural Systems and Accelerating Bridge Construction. The availability of powerful means of lifting and transporting, high-performance materials, like high performance concrete (HPC) and fiber reinforced polymer (FRP) composites and refined structural analysis methods, along with combined use of prefabrication, pretensioning, post-tensioning, segmentation and splicing, and in-situ construction techniques, have led to considerable changes in the design and construction of concrete bridges in favor of precast construction. Taking advantages of their expertise and interests in these areas, members at the Tennessee Bridge Research Laboratory (TBRL) at the University of Tennessee Knoxville (UTK) are actively involved in the research area of accelerating renewal of highway bridges (web.utk.edu/~zma2). Three active example projects, sponsored by National Cooperative Highway Research Program (NCHRP) and National Science Foundation (NSF), at UTK are given in the paper.

Introduction

The highway network is the backbone of America's transportation system, making it possible to meet the mobility and economic needs of communities, regions, and the nation as a whole. It carries more than 90% of passenger trips and 69% of freight value (BTS, 1999). However, after decades of constant use, often exceeding facilities' original design life, much of the highway system is in need of extensive renewal. Renewal work, in contrast to the construction of new highways, must be performed while the facilities remain in service. This practice has introduced significant safety, mobility, and economic concerns: tens of thousands of lives lost and millions of injuries each year on America's highways; hours of congestion and delay due to highway construction work zones, crashes, and other incidents. A recent Texas Transportation Institute study (Schrank, et al., 2001) estimates the cost of congestion in just 68 urban areas has grown from $21 billion in 1982 to $78 billion in 1999 (36 hours per driver a year and 6.8 billion gallons of wasted fuel). The study also estimated that congestion results in 4.4 billion person hours of delay annually in the 68 urban areas studied. The problems have wide-ranging impacts on the nation's economy and quality of life.

The first Strategic Highway Research Program (SHRP), conducted in the late 1980s and early 1990s, focused on a few critical infrastructure and operations problems faced by state transportation agencies. Given the success of SHRP and the pressing need to find solutions for the problems challenging the highway system today, the Congress requested in 1998 that the Transportation Research Board (TRB) "conduct a study to determine the goals, purposes, research agenda and projects, administrative structure, and fiscal needs for a new strategic highway research program." In December 2001, the AASHTO Board of Directors passed a resolution supporting a Future SHRP (F-SHRP) funded at the $450 million level over six years through a Trust Fund takedown and

Bridge design, construction and maintenance 2007, Thomas Telford, London

authorizing a NCHRP project to develop these plans (TRB, 2001). Federal Highway Administration (FHWA) matched the NCHRP funds. Work began on the interim planning phase of F-SHRP in January 2002 (NCHRP, 2003). The overarching theme for F-SHRP is to provide outstanding customer service for the 21st century. In keeping with the overarching theme of providing outstanding customer service for the 21st century, the committee decided to conduct an extensive outreach process to identify highway needs and research opportunities. The outreach process identified hundreds of highway needs and research opportunities. From this vast array of possibilities, the committee selected four strategic focus areas (TRB, 2001). The top focus area is to accelerate the renewal of America's highways that is rapid, causes minimum disruption, and produces long-lived facilities.

A quarter of America's 590,000 bridges are currently classified as structurally deficient or functionally obsolete. Recognizing the importance of accelerating renewal of highway bridges, the AASHTO's Highway Subcommittee on Bridges and Structures (HSCOBS) has recognized a more structured procedure for prioritizing research was needed. While the F-SHRP committee was working on interim planning phase of F-SHRP, HSCOBS conducted two workshops, in February 2000 and April 2005 respectively, to develop a strategic plan for bridge engineering. The products of these workshops are a focused set of critical problems called "grand challenges" that, if solved, would lead to significant advances in bridge engineering (HSCOBS, 2005). A total of seven grand challenges have been identified by HSCOBS. The top two grand challenges are Optimizing Structural Systems and Accelerating Bridge Construction. Conventional bridge reconstruction is typically on the critical path because of the sequential, labor-intensive processes of completing the foundation, the substructure, the superstructure components (girders and decks), the railings, and other accessories. New bridge systems are needed that will allow components to be fabricated off-site and moved into place for quick assembly while maintaining traffic flow. The availability of powerful means of lifting and transporting, high-performance materials, like high performance concrete (HPC) and fiber reinforced polymer (FRP) composites and refined structural analysis methods, along with combined use of prefabrication, pretensioning, post-tensioning, segmentation and splicing, and in-situ construction techniques, have led to considerable changes in the design and construction of concrete bridges in favor of precast construction. Such advances allow engineers to design and build, precast, prestressed concrete girder bridges that are comparable to those cast in place, due to their structural efficiency, appearance and economic competitiveness. Longer spans are now within reach. This technology is applicable and needed for both existing and new bridge construction.

Taking advantages of their expertise and interests in these areas, members at the Tennessee Bridge Research Laboratory (TBRL) at the University of Tennessee Knoxville (UTK) are actively involved in the research area of accelerating renewal of highway bridges (web.utk.edu/~zma2). Researchers at UTK understand that in order to use all of these materials and systems in a structurally efficient, durable and cost effective manner, research is needed to better characterize their properties and optimize their use, and develop efficient design and construction systems, standards and details. Example activities in this research area at UTK are discussed below.

Optimized bridge systems with accelerating renewal features

In order to meet the "grand challenges" of optimizing structural systems and accelerating bridge construction, TBRL at UTK is currently involved with three active research projects: NCHRP 10 – 71 *Cast-in-Place Reinforced Connections for Precast Deck Systems*, NCHRP 12 – 69 *Design and Construction Guidelines for Long-Span Decked Precast, Prestressed Concrete Girder Bridges*, and National Science Foundation CAREER *Behavior and Design of FRP-Decked Concrete Bridges*.

Cast-in-place reinforced connections for precast deck systems

This project was funded by National Research Council (NRC) through TRB's NCHRP program. This 39-month project, NCHRP 10 – 71, started in July 2006. The objectives of the project are to (1) develop detailed design, fabrication, construction, and performance criteria that must be satisfied to provide durability, strength, fatigue resistance, seismic resistance, and rapid construction; (2) develop conceptual designs for cast-in-place (CIP) reinforced concrete connections: at least two for longitudinal connections for precast slab superstructures, and at least three for longitudinal or transverse connections between full-depth deck panels or deck flanges. Connections should not require overlays or post-tensioning. Emphasis should be placed on increasing construction speed while achieving durability and ride quality; and (3) develop specification language and commentary for recommended changes to the *AASHTO Load and Resistance Factor Design (LRFD) Bridge Design Specifications* and the *AASHTO LRFD Bridge Construction Specifications* as necessary to implement the recommended connection details. The typical sequence of erecting bridge superstructures in the United States is to erect the precast prestressed concrete or steel beams, place either temporary formwork or stay-in-place formwork such as steel or concrete panels, place deck reinforcement, cast deck concrete, and remove formwork if necessary. This project focuses on systems that eliminate the need to place and remove formwork thus accelerating on-site construction and improving safety.

Long-span decked precast, prestressed concrete girder (DPPCG) bridges

This 4-year project, NCHRP 12 – 69, started in June 1, 2004. This project addresses I-beam, bulb-tee, or multi-stemmed girders with integral decks cast and prestressed with the girder. The girders are erected abutting flanges of adjacent units (Figure 1a), and load is transferred between the adjacent units using shear-connector connections (Figure 1b). Figure 2 shows the computational model of the DPPCG bridge. Despite the major benefit of eliminating CIP concrete decks, the construction of this type of bridge has not shown the growth it deserves and has been mostly limited to the Pacific Northwest states of Alaska, Idaho, Oregon, and Washington. The reason is two-fold: the concerns and limitations in design and construction using DPPC girders, and the lack of understanding due to limited research in this area. These issues include connections between adjacent units, live load distribution (Ma et al., 2007a; Millam and Ma 2005), and other factors. During Phase I study of the project, the performance of longitudinal joints and connections was identified as an issue with high priority. Because of the concern and interest in the durability of the type of longitudinal joints currently being used, the research team concluded that the study should include an investigation of a potential improved moment and shear connection between decked bulb-Tees. A study to develop improved longitudinal joint details over the conventional shear-connectors for DPPCG bridges is being conducted as part of the ongoing NCHRP 12-69 project.

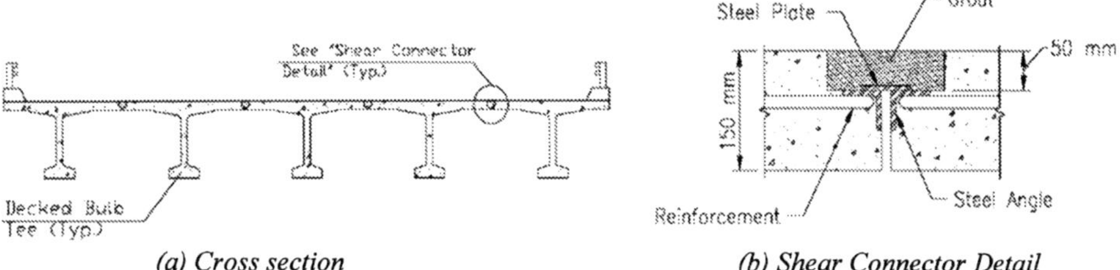

(a) Cross section   (b) Shear Connector Detail

Figure 1: DPPCG bridge system

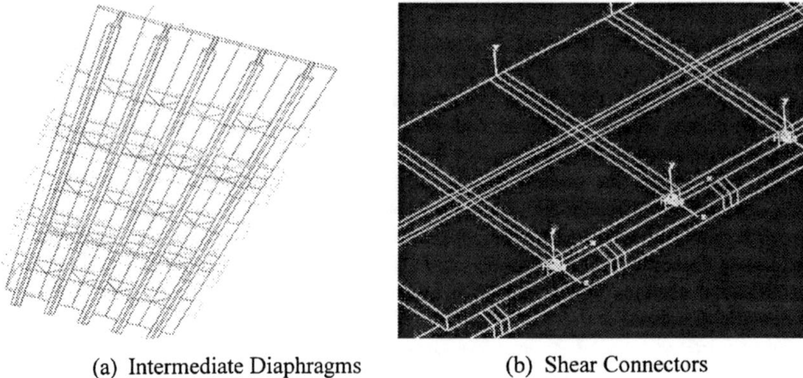

(a) Intermediate Diaphragms   (b) Shear Connectors

Figure 2: Modeling transverse connections

Because the systems do not incorporate transverse post-tensioning to protect against the development of longitudinal and/or transverse cracking between sections, the placement of the mild reinforcement across transverse/longitudinal joints and the detail of the concrete at that joint need to be considered carefully. An example of a joint detail for this type of connection is the loop bar detail shown in Figure 3 (Ralls et al., 2005). Other connection concepts to be considered for either longitudinal or transverse joints in the project include straight reinforcing bars with spiral confinement to minimize splice lengths, headed reinforcing bars, and welded wire reinforcement. Computational models of prototype bridges will be developed using commercial finite element analysis tools to determine the service load demands on the various connections selected for study. Critical combinations of moment and shear will be obtained considering camber leveling forces and live load forces. Test procedures will be developed and refined for static and fatigue loading of the laboratory test specimens representing the selected connections. A series of parametric studies will be performed for live load forces considering girder depth, girder span, girder spacing, single- and multi-lane loading, angle skew, diaphragm spacing and stiffness, and location of wheels relative to the joints in the precast deck. The prototype bridge models developed will be modified as needed for seismic analysis to determine force (stress) and displacement (strain) demands.

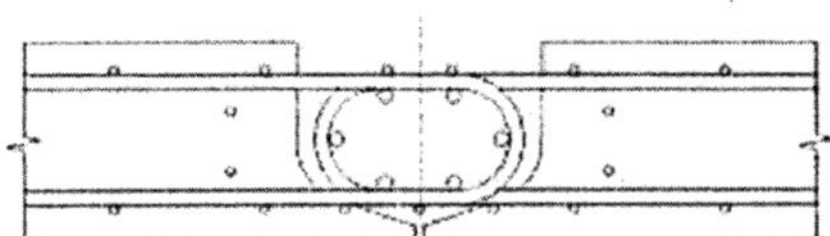

Figure 3: Example of concept loop bar detail

Behavior and design of FRP-decked concrete bridges

This 5-year project was funded by the National Science Foundation (NSF) in July 2004. The research deals with an integrated research and education plan to evaluate and develop a durable bridge system with accelerating renewal features for use in various applications and environments including very cold temperatures. The bridge system consists of glass fiber reinforced polymer (FRP) composite deck and precast/prestressed concrete I-girders. The primary objectives are: (1) understanding the synergistic effects of low temperature and low-temperature thermal cycling combined with fatigue loading on the performance of FRP composite bridge decks, and (2) quantifying the horizontal shear stiffness of typical deck-to-girder connection systems and developing the connection design criteria based on the shear stiffness rating of the connection. FRP composites are often chosen over conventional materials for their durability. However, there remains a "durability gap" for FRP composites (Karbhari, et al., 2003). Based on a recent study performed by Dutta et al. (2002), it was observed that the low-temperature (-30^0C) stiffness in the load-deflection curves was higher than at the high temperature (50^0C). In cold regions like Alaska, the temperature during the winter season may go well below -50°C. It is difficult to do cast-in-place concrete for bridge decks in winter. As a result, the use of the FRP composites to replace existing, deteriorated bridge deck systems offers a very attractive alternative. The 2-year study of the 5-year project was focused mainly on the study of the behavior of the FRP deck panel at very cold temperature (below -50^0C) cycling combined with a service load cycling.

The FRP deck panel specimen was obtained from Kansas Structural Composites Inc. (KSCI), Kansas. It had an overall geometry of 2.134m in length, 0.340m in width, and 0.178m in thickness. The specimen was of sandwich type construction with a vertical corrugated core and was manufactured using the wet lay-up method, as discussed by Plunket (1997). Figure 4 shows the specimen and the testing setup. The sensor instrumentation can also be seen in the figure. The testing was conducted using a self-equilibrating structural steel frame in a cold room. The cold room is 4.3m by 4.3m in size and was designed to provide a low temperature of -70^0C. Figure 5 shows the control panel of the cold room while one of tests was in progress. It was decide to use the service load level of 35.6 KN to test the specimen at various temperatures. At each temperature, the specimen was loaded up to 35.6KN at a loading rate of 2.2 to 4.4 KN per minute. Figure 6 shows the loading and temperature cycling history of the experiment. The detailed testing program was reported elsewhere (Ma et al., 2007b).

Figure 4: Three-Point Bending Test Setup

Figure 5: Control Panel of the Cold Room

After the first stage room temperature testing, the temperature of the cold room was lowered from 26^0C to 0^0C as shown in Figure 6. The panel was tested at 24 hours and 48 hours from the time of lowering the temperature. The data points in Figure 7 shows the relationship between the applied load and the deflection of the panel at mid-span. Figure 7 also shows that the panel stiffness, as represented as the slope of the trend line, is 10.82 KN/mm. The load vs. deflection curve for tests at other temperatures has the same pattern as shown in Figure 7, but with different stiffness values. Figure 8 shows a comparison of the panel stiffness at different temperatures. Please note that the testing conditions remained the same throughout the experiment. The only variable in each data set of the experiment was the temperature. Referring to Figure 8, at the end of the 4th cycle loading test, the panel stiffness increased the maximum of about 3.8% as the temperature decreased from 26^0C to 0^0C and further to -20^0C. However, when the panel was kept in the -20^0C for another 24 hours the stiffness did not further increase. Instead, it decreased 5.0% comparing with the stiffness at room temperature. When subjecting the panel to further colder temperature as cold as -55^0C, the panel stiffness showed a gradual decrease, and stabilized around the value of 9.22 KN/mm. When comparing with the stiffness at original room temperature, it decreased 11.5%. Most importantly, this degradation in stiffness could not be recovered by raising the panel temperature, as shown in Figure 8. Figure 9 shows an example comparison of the AE testing data at different temperatures. On comparing the two figures [(a) and (b)], it is seen that more activity happened at -31^{0}C. There was also more energy at -31^{0}C. A higher amplitude range (50-60 dB) was observed at -31^{0}C when compared to a temperature of 26^{0}C. Please note that in Figures 9 (a) and (b) the longer the time of observation the greater is the load applied on the specimen.

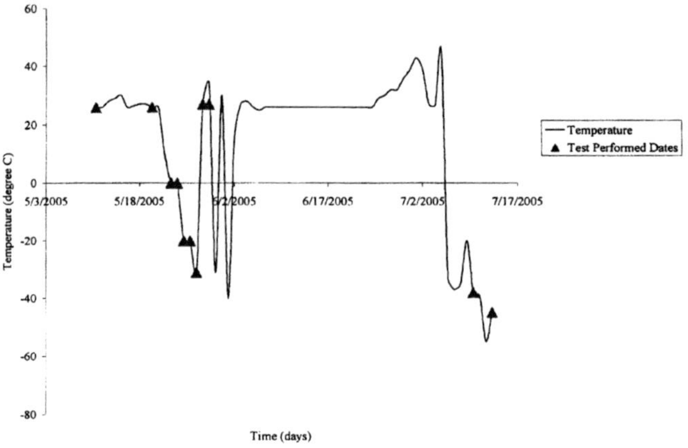

Figure 6: Loading and Temperature Cycling History

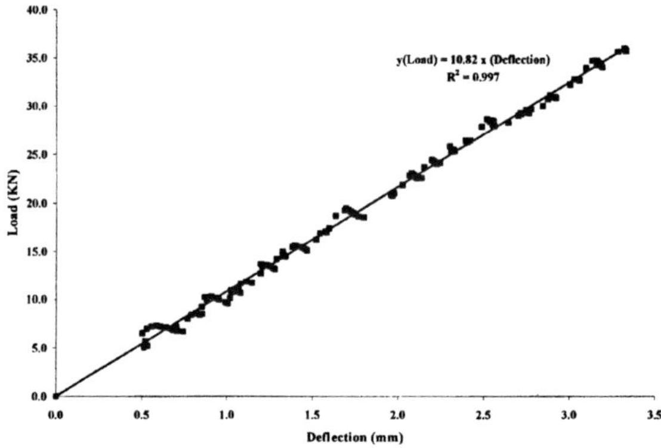

Figure 7: Load vs Deflection at 0^{0}C (24 hours test)

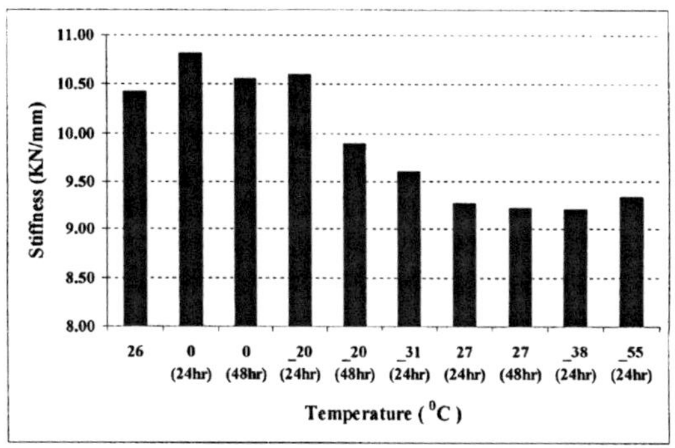

Figure 8: Stiffness vs Temperature and Time

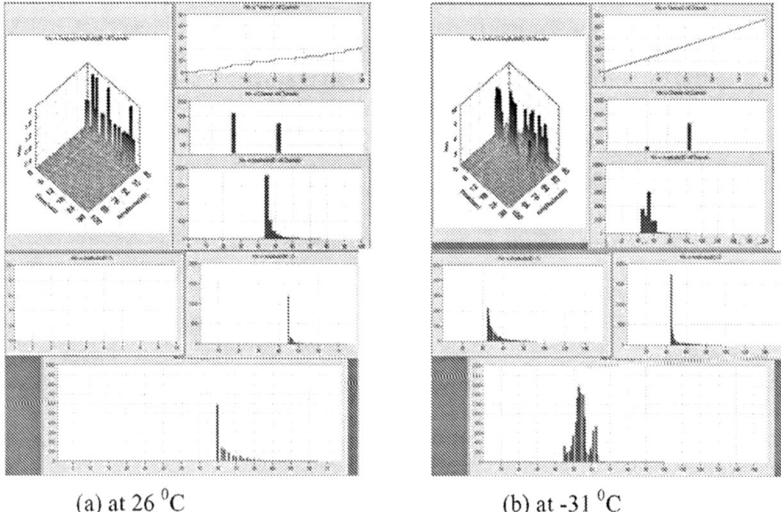

(a) at 26 ^{0}C   (b) at -31 ^{0}C

Figure 9: Overview of the AE Data at Different Temperatures

CONCLUSIONS

The top focus area selected by the Future Strategic Highway Research Program committee is to accelerate the renewal of America's highways that is rapid, causes minimum disruption, and produces long-lived facilities. The AASHTO's Highway Subcommittee on Bridges and Structures (HSCOBS) conducted two workshops to develop a strategic plan for bridge engineering. The products of these workshops are a focused set of critical problems called "grand challenges" that, if solved, would lead to significant advances in bridge engineering. To meet these challenges and take advantages of their expertise and interests in these areas, researchers at the Tennessee Bridge Research Laboratory (TBRL) at the University of Tennessee Knoxville (UTK) are actively involved in the three funded on-going projects. Based on the preliminary research efforts carried out so far, it has been found that precast/prestressed concrete bridge systems with the improved connection details have the potential to achieve the goal of accelerating renewal of highway bridges. Based on the first phase cold-room testing of FRP deck panels, the following conclusions have been made: (1) The low-temperature stiffness was about 3.8% higher than at the high temperature as long as the low temperature was above a certain degree (about -20^0C under the testing condition presented in the paper); (2) When the temperature was further lowered from -20^0C to -55^0C, the low-temperature stiffness did not further increase. Instead, it decreased 11.5% when comparing with the stiffness at the original room temperature. Most importantly, this degradation in stiffness could not be recovered by raising the panel temperature; and (3) Very cold temperature exposure resulted in matrix hardening, matrix microcracking and fiber-matrix debonding as demonstrated by the AE testing data at different temperatures.

REFERENCES

BTS (1999). *Transportation Statistics Annual Report 1999*. BTS99-03. Bureau of Transportation Statistics, U.S. Department of Transportation.

Dutta, P.K., S.C.Kwon, G.Dullel, and R. Lopez-Anido (2002). Fatigue Evaluation of Composite Bridge Decks. *Proceedings of 10th U.S. Japan Conference on Composite Materials*. September 16-18, Stanford University, Stanford, C.A., DEStech Publications, 336-344.

HSCOBS (2005). *Grand Challenges: A Strategic Plan for Bridge Engineering*. AASHTO Highway Subcommittee on Bridges and Structures, Woods Hole, Massachusetts, April 18 – 20.

Karbhari, V.M., et al. (2003). Durability Gap Analysis for Fiber-Reinforced Polymer Composites in Civil Infrastructure. *Journal of Composites for Construction*, Vol. 7, No. 3, August, 238 – 247.

Ma, Z. J., Chaudhury, S., Millam, J., and Hulsey, J. L. (2007a). Field Test and 3D FE Modeling of Decked Bulb-Tee Bridges. ASCE Journal of Bridge Engineering, Vol.12, No.3.

Ma, Z. J., Choppali, U., and Li, L., (2007b). Cycling Tests of a Fiber-Reinforced Polymer Honeycomb Sandwich Deck Panel at Very Cold Temperatures. *International Journal of Materials and Product Technology (IJMPT)*, Vol. 28, No. 1/2, pp. 178 – 197.

Millam, J. and Ma, Z. J. (2005). Single Lane Live Load Distribution Factor for Decked Precast/Prestressed Concrete Girder Bridges. *Journal of Transportation Research Board*, TRR No. 1928, 142 – 152.

NCHRP (2003). *NCHRP Report 510: Interim Planning for a Future Strategic Highway Research Program*, TRB, National Research Council, Washington, D.C.

Plunket, J. D. (1997). *Fiber-reinforced Polymer Honeycomb Short Span Bridge for Rapid Installation*. IDEA Project Report, November.

Ralls, M. L., et al (2005). *Prefabricated Bridge Elements and Systems in Japan and Europe*. FHWA-PL-05-003, March, 48pp.

Schrank, D., and Lomax, T. (2001). *The 2001 Urban Mobility Report* at mobility.tamu.edu. Texas Transportation Institute, College Station, TX.

TRB (2001). *Special Report 260: Strategic Highway Research: Saving Lives, Reducing Congestion, Improving Quality of Life*, TRB, National Research Council, Washington, D.C.

Achieving best value and minimising disruption through bridge management plans

Dr Donald Pearson-Kirk, Structures Technical Director, AccordMP/Mouchel Parkman
Exeter, United Kingdom

Martin Lynch, Highways Agency, Department for Transport, United Kingdom

Abstract

Many European Government Agencies have now generally accepted that their concrete structures – for example, buildings and bridges - will be managed more cost-effectively if the condition of the structures is accurately known. For agencies with a large number of structures, there is often a problem in prioritising the structures for investigation as the only records available generally consist of visual inspection reports. However, visual inspections alone are not always sufficient. The condition of structures and the causes of any deterioration must be accurately determined by testing and monitoring, if the structures are to be managed cost-effectively and if safety and operational requirements are to be achieved. The ideology currently used by maintenance engineers has to change.

The paper outlines how structures may be prioritised for condition monitoring by developing three-phase management plans. The first phase is a desk study where previous inspection reports are reviewed and relevant data inputted into a rating equation that is used to prioritise structures for investigation. The second phase comprises preliminary site inspections of selected structures, the results of the inspections being used to check the validity of the rating equation and its amendment if considered necessary. The third phase consists of site investigations to the higher priority structures, where the causes, extent and severity of deterioration are determined. A case study is presented to demonstrate the very significant benefits of this three-phase management process.

Keywords: Best value; management plans; condition monitoring

Bridge design, construction and maintenance 2007, Thomas Telford, London

Introduction

Highway structures are a key component of highway infrastructures and their long-term performance in service is clearly of great importance socially, environmentally and economically. Traffic volumes and loadings are increasing, ageing structures are deteriorating and there are often significant constraints on the funding available for rehabilitation or replacement. In Europe it is estimated that repairs will account for 40% of total construction contract costs [1] and, in the United States, the Federal Highways Administration has stated that 10,000 bridges are to be rebuilt or rehabilitated each year at a cost of $7 billion per annum [2]. These are challenging statistics in the context of sustainable development and the social and economic costs of undertaking the necessary maintenance.

Parsons Brinckerhoff, with the Highways Agency, has been developing techniques for early identification of deterioration and prioritisation of maintenance and repair of highway structures. These techniques enable enhanced bridge management decision-making and help expedite the maintenance and repair of deficient structures at a significantly lower cost, while minimising disruption.

Performance of Bridges

The unexpected early deterioration of many concrete and of some steel structures has been well documented and is of increasing concern in many countries in all climatic zones. In the United States, the magnitude of the problem of corrosion related deterioration of concrete bridge decks was first recognised in the early 1960's, resulting in co-operative research programmes funded by the Federal Highway Administration (FHWA) to determine how the durability of concrete bridges could be improved. Early projects involved the investigation of the benefits of using surface impregnation and the provision of adequate concrete cover to reinforcement in aggressive environments [3].

Since the mid 1980's, agencies around the world have been reporting problems with their structures. In 1986 the Commonwealth Bureau of Roads, Australia [4] reported that 39% of the bridges on interstate highways were deficient; in 1988 the Organisation for Economic Co-operation and Development (OECD) reported that 40% of 800,000 structures inspected in 18 developed countries were structurally deficient [5]; in 1989 a report for the UK Department of Transport [6] stated that only 12% of concrete bridges studied were in good condition, and in 1992 the FHWA in the United States estimated that it would cost $11 billion to rehabilitate corrosion induced deterioration of structures on the Interstate Highway System [7].

Some post-tensioned bridges with grouted ducts were found to have problems in Europe, and in 1992 the UK Department of Transport placed a moratorium on this form of construction. The moratorium was only partly lifted in 1996. In France during the 1990's, some 22% of pre-stressed concrete bridges were found to have serious defects [8] and in 2001, the FHWA issued a memorandum to all States requiring that higher risk post-tensioned bridges be checked.

The FHWA, in consultation with the States, produce an annual bridge condition survey inventory. The 2003 bridge inventory stated that 26% of highway bridges – a total of 153,000 – were structurally deficient or functionally obsolete. The United States is building, replacing or rehabilitating 10,000 of these structures each year, at an annual cost of $7 billion.

One question that has been asked is "are the bridges that bad?" The inventory states that the condition of most bridges is not accurately known as the data in the inventory is substantially based on visual inspections. Many forms of deterioration do not result in external visual evidence until the deterioration process is well advanced, and is then costly to repair with significant disruption to road users.

Early detection of deterioration is essential. The condition of bridges must be accurately known to enable the bridge portfolio on the highway network to be managed cost-effectively and meet safety and operational performance requirements. The management plan approach outlined in this paper enables this improvement to be achieved.

Time-Dependant Deterioration of Structures

Many of the deterioration processes identified on highways structures, and especially concrete structures, are time-dependant [9] with an increasing effect of the deterioration mechanism with time.Visual detection of the processes is often difficult at early stages without intrusive investigation.   This demonstrates the importance and essential requirement for early identification of deterioration processes on structures and highlights the necessity for management studies in order to appropriately focus limited funds.

There are many mechanisms of deterioration that can affect the performance of structures in service. Management studies provide bridge managers with the tools to identify the extent of the problem on the networks using cost-effective techniques and thereby achieve best value of the limited resources.

Some structural features warrant particular attention, for example: in-span joints supporting suspended spans, failed corrosion protection, hinge joints, welding defects etc.

Changes in Operation & Standards

Changes in operational requirements, such as loading and the development of standards and specifications with improving knowledge can also initiate programmes of structures maintenance and upgrading. Such programmers could also be implemented more effectively through the development of Management Plans.

Typical change programmes can be to be initiated by:

- Assessment and strengthening for increased normal and abnormal traffic loadings
- Assessment and strengthening for vehicular impact on piers
- Assessment and strengthening of bridge parapets

The Bridge Management Role

Following the substantial completion of the core highways infrastructure network in the United Kingdom the Bridge Management Role is changing substantially, from one of improving the asset to striving to maintain the asset value at a steady state. With ever increasing traffic growth this maintenance must be carried out with minimum impact on network availability. The competing demands for Government funding between schools, hospitals and railways etc has seen the budgets available for bridge maintenance decline in recent years. As a result the bridge managers are under increasing pressure to minimise disruption on the network and also to ensure that best value is obtained from the limited resources available. One way of addressing this problem is to use whole life costings to compare different maintenance regimes. Lowest initial cost does not necessarily equate to best value. Accurate forecasting of cost and network disruption also plays a key part in the bridge management role. Overspend on one structure means fewer funds for other structures and network programmes.

Value of Management Studies

Management studies provide invaluable knowledge for the bridge managers into the integrity and condition of the asset. This information is stored as a reference manual that can also contain information on best practice and current applicable standards, which enables deficiencies or shortcomings to be easily identified. From this information future works programmes can be developed and prioritised. By providing the bridge manager with the full extent of the defect or shortcoming, predictability of cost to rectify the problem is maximised. This information can be developed into a steady state maintenance programme to maintain the best value of the asset.

Bridge items that have been the subject of management plans include:
- Post tensioned bridges
- Sulfate attack of above ground components of structures
- Bearings
- Waterproofing
- Corrosion Protection – painting
- Weld defects and imperfections
- In-span joints
- Parapets
- Pier Impact

It is considered worthwhile producing discrete management plans given the limitations of the UK national Structure Maintenance Information System (SMIS). The SMIS database is unable to concentrate on particular generic deterioration problems.

Case study: Alkali-Silica reaction in Southwest England

ASR Management Plan

Concrete is one of the most versatile and widely used construction materials, due largely to its strength and durability. However, many concrete bridges are now showing signs of deterioration including chloride and sulfate ingress, cracking, staining and spalling and in a number of cases in the UK, bridges have suffered serious damage due to alkali-silica reaction (ASR). A report on the performance of 200 highways bridges for the UK Department of Transport [6] indicated that over three-quarters of the Department's concrete bridges contained potentially reactive aggregates.

For several years, there have been numerous Principal Inspections carried out on the bridges in the Southwest of England (Highways Agency Area 1) and the reports contained references to a visual deterioration of the concrete structures, which was often attributed to ASR. Parsons Brinckerhoff acting as the Area 1 Bridge Managers for the Highways Agency since 2002 needed to determine the extent and severity of this problem in order to assess the remaining service life of each structure affected by damaging ASR and hence the maintenance requirement on the Network.

The approach adopted was to implement a Management Plan applicable to all structures susceptible to ASR. This type of plan had been proven in the past by Parsons Brinckerhoff to be a cost effective tool in the identification of deficiencies or shortcomings on transportation networks.

The Management Plan initially focused on the 225 structures located on the A38 trunk route in Area 1. The Plan proposed to adopt the proven three-phase approach to the problem to accurately and cost-effectively assess the likelihood that structures are suffering from ASR. The work began by reviewing all available information from the Principal Inspections (PI), Post-Tensioning Special Inspections (PTSI) and Special Inspections (SI); maintenance records; Roads 277 Forms and BE11/94 Forms (both forms contain visual condition assessments of the structures' components). The review showed that approximately 40% of structures were showing signs of potentially damaging ASR.

ASR Management Plan Phase 1 – Desk Study

This phase of the Management Plan was a desktop study aimed at identifying all structures that are susceptible to damaging ASR and to place them into a prioritised programme for future preliminary site investigations in Phase 2. Phase 1 desktop study comprised of developing a Total Assessment Rating (TAR) equation based on known and available information that could potentially indicate ASR. A deterioration specific assessment using the TAR was then carried out on each of the structures to determine its susceptibility to ASR. A TAR value was then determined for each structure and a priority factor was assigned, using equation 1.

Each individual rating factor used as part of the TAR equation was assigned a numerical grading defining the potential to identify ASR and also given a certain weighting to reinforce its significance in the overall process of ASR identification.

$$TAR = \sum \left(4R_a + R_{st} + R_{to} + R_{tu} + R_{sf} + 2R_{od} + 3R_{asr} \right) \qquad (1)$$

The main visual features indicative of damaging ASR in concrete are considered to be map-type cracking, pop-outs, efflorescence, exudations and crazing. The map-type cracking has been identified with particular reference to Technical Guide 2 'Guide to Testing and Monitoring the Durability of Concrete Structures [10]. The Guide shows Cracking Reference numbers 4, 7 or 9 to follow the map-type pattern. Table 1 below shows the individual rating factors and their numerical grading to define their potential to identify ASR deterioration.

Table 1 Rating factors and their numerical grading for assessment

ASSESSMENT FACTORS	SYMBOL	RANGE OF ASSESSMENT RATING
Age	R_a	1-5
Structure Type	R_{st}	1-5
Traffic Over	R_{to}	1-5
Traffic Under	R_{tu}	1-5
Structure Formation	R_{sf}	1-5
Observed Defects	R_{od}	1-5
ASR Specific Defects	R_{asr}	1-5

The results of Phase 1 of the Management Plan are shown Table 2 below. This identified 42 structures as Priority 1, this represents only 18% of the structures and not 40% as originally thought were likely to be suffering from damaging ASR.

Table 2 Results of the Phase 1 TAR assessment

PRIORITY OF STRUCTURES	CLASSIFIED STRUCTURES		POTENTIAL ASR DETERIORATION
	NUMBER	PER CENT	
Priority 1 Structures	42	21	High
Priority 2 Structures	60	30	Medium
Priority 3 Structures	41	20	Low
Priority 4 Structures	24	11	Very Low
Priority 5 Structures	36	18	None

From these results, a programme of preliminary site inspections (Phase 2) was completed including all of the Priority 1 structures, a representative proportion of Priority 2-5 structures and all of the 22 structures where there was no information available in Phase 1.

ASR Management Plan Phase 2 – Preliminary Site Inspections

The Phase 2 investigations were then used to update and confirm the rating and priority factors determined in Phase 1 and plan for Phase 3 site investigation works. The results of Phase 2 preliminary site investigations highlighted the need to make modifications to the original assessment factors and hence re-calculate the TAR values and priority factors. For example, the "structure type", where previously all pipes scored relatively low. On inspection, however, it was discovered that it was the headwall and not the pipe that showed signs of ASR. Consequently, the headwall, allowing for availability of water, now scores relatively higher in the defects matrix than the pipe.

The investigations were all carried out by a single team of inspectors that produced uniformity in the visual assessment of any deterioration and the extent and severity of that deterioration. The modified priority ratings are shown in Figure 1.

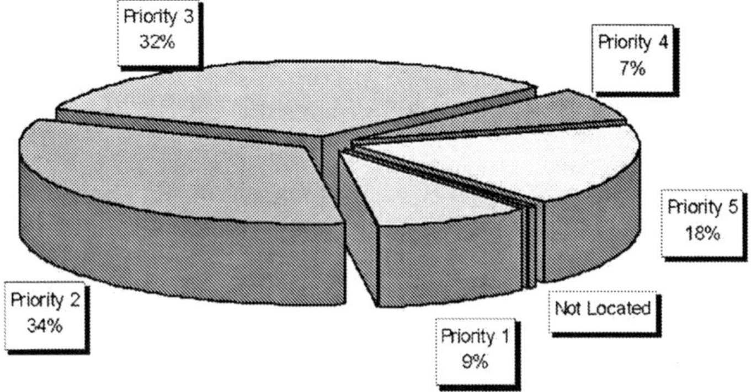

Figure 1 Graph to show the amended Priority after Phase 2 Visual Inspections

The amended ranking factors of Phase 2 reduced the number of Priority 1 structures that exhibit signs of ASR to 9%, which was a large reduction on the Phase 1 results (18%) and a significant reduction on the original estimate of 40%.

ASR Management Plan Phase 3 – Special Inspections

The results of the Phase 2 preliminary site inspection works determined that 20 structures were categorised as Priority 1. PB has proposed to investigate the Priority 1 structures during the Phase 3 Special Inspections, which is on going at present.

The Phase 3 Special Inspections will positively identify the structures that are suffering from damaging ASR by intrusive investigation and laboratory examination and determine the accuracy of the TAR equation to identify structures with potential ASR based on normal Inspections. Phase 3 has the potential to expand the plan to Phase 4 if required – 'Monitoring and Repair of Structures'.

Conclusions

The observed early deterioration of concrete bridges is of increasing concern in many countries throughout the world because of the clear social, environmental and economical consequences. Many agencies around the world have been reporting problems with their bridges since the mid 1980's, but it has been found that there is often a problem in prioritising these structures for investigation, as only limited or incomplete records are available to bridge managers.

The Bridge Management role is generally one of improving the asset value and striving to maintain that value at a steady state. However, the budgets available for bridge management and maintenance have declined in real terms in recent years, and as a result, has placed increased pressure on managers to ensure that disruption on the network is minimized and best value is obtained from the limited resources available. Management Plans have been shown to provide invaluable knowledge for bridge managers into the extent and condition of each asset.

The ASR Management Plan described focused on 225 structures (Figures 2 and 3)on one route within the Area 1 Network, in the Southwest of England. Ultimately the management strategy for Area 1 is to address the most severe cases of damaging ASR on the Network and prioritise the future maintenance work within the constraints of the future budget / bid allocation. The results of Phase 1 of the Plan prioritised the structures in terms of suspected damaging ASR and indicated that only 18% of the structures were, in fact, suspected to be at highest risk from damaging ASR. The results of Phase 2 of the plan further reduced the number of suspected highest risk structures to just 9%, thus reducing the number of Priority 1 structures by 50%. The estimated cost of repair / replacement was reduced by over 75% from £120 million (US $216 million) to less than £30 million (US $54 million).

The benefit of the Management Plan focusing on a single form of deterioration was clearly demonstrated to the client with a significant reduction in the estimated cost of repair/replacement and it also allowed PB to develop a maintenance specific programme for affected structures at low costs.

Figures 2 & 3 Damaging ASR in the wingwalls of Area 1 bridges.

References

1. NEWMAN, J., Concrete Repair Strategies, Essentials of Concrete Investigations, Institution of Engineers of Ireland, 2003.

2. FEDERAL HIGHWAY ADMINISTRATION, Delivering Infrastructure for America's Future, US Department of Transportation, June, 2002.

3. CLEAR, K., Improving the Performance of Concrete Bridge Decks, Federal Highway Administration, US Department of Transportation, March 1973.

4. COMMONWEALTH BUREAU OF ROADS, The Condition of Bridges on Interstate Highways, Canberra, May. 1986.

5. ORGANISATION FOR ECONOMIC COOPERATION AND DEVELOPMENT, The Durability of Concrete Road Bridges, Road Transport Research Program, Paris. 1988.

6. WALLBANK, E.J., The Performance of Concrete in Bridges: A Survey of 200 Highway Bridges, Her Majesty's Stationary Office, April 1989.

7. FEDERAL HIGHWAY ADMINISTRATION, Rehabilitating Corrosion Damaged Concrete Bridges', US Department of Transportation, 1992.

8. GODART, B., 'Status of the Durability of Post-tensioned Tendons in France, FIB Durability of Post-tensioning Tendons Workshop, Ghent, November. 2001.

9. RICHARDSON, M. G., Fundamental of Durable Reinforced Concrete Spon Press, 2002.

10. CONCRETE SOCIETY, 'Guide to Testing and Monitoring the Durability of Concrete Structures, Technical Guide 2', Concrete Bridge Development Group, 2002.

Effective Governance of Bridge Management: The UK Example

Richard Fish BSc CEng FICE FIStructE FIHT
Director of Planning, Transportation and Estates, Cornwall County Council, UK
Chairman of the UK Bridges Board and
County Surveyors' Society (CSS) Bridges Group

Abstract

The paper describes the various levels of national and local government interest in the management of the United Kingdom's bridge stock and the governance structures that are in place to support bridge owners in achieving a consistent and cohesive approach to bridge management. The respective roles of the three key bodies, CSS Bridges Group, the UK Bridges Board and the Bridge Owners' Forum, are reviewed before identifying the basis in which they complement each other and the net benefit gained from the existing arrangement. Special attention is given to the identification and management of bridge related research to illustrate the advantages of a coordinated approach. Finally, suggestions are given for possible future improvements to ensure that the management of the UK's bridges remains as effective as possible.

Introduction

In line with most countries, the United Kingdom has many interested parties when it comes to the management of the nation's bridge stock. Owners include national government for the motorway and trunk road network, a large number of local authorities (large and small) for other primary routes and local roads and many other public and private owners. Furthermore, the day to day practical elements of bridge management are often in the hands of agents who could be either public or private sector bodies.

There is also a wide range of stakeholders with a keen interest in ensuring that the highway network remains safe, serviceable and sustainable, including the travelling public, public transport operators and freight organisations.

With the emphasis shifting from new or replacement bridges, and strengthening only through significant intervention, to more demanding but robust principles of management, monitoring and maintenance, most private sector consultants and contractors are interested in providing an holistic service as managing agents.

There is further interest from the academic sector with regard to research and development initiatives.

How, then, can all these interests be co-ordinated? How can strategy and policy be successfully promulgated to all parties? How can local issues be tested and channelled to the

Bridge design, construction and maintenance 2007, Thomas Telford, London

appropriate body? And, how can research into theoretical analysis, materials behaviour and management systems be targeted to the requirements of bridge owners?

In the UK, this has been achieved by establishing a number of groups: the Bridges Board to deal with high level policy, strategy and funding issues; the CSS Bridges Group and a network of Area Bridge Conferences providing a forum for local authorities; and the Bridge Owners' Forum focussed on developments at practitioner level and dealing with research initiatives.

This paper will explain the background and composition of the various groups and how they communicate and interact to give a cohesive structure in which all aspects of bridge management can be optimised and effectively addressed to the net benefit of all interested parties. The views herein are those of the author and not necessarily those shared by his employer, the members of the CSS Bridges Group or the members of the UK Bridges Board.

Geography and Administration

The United Kingdom (of Great Britain and Northern Ireland) is actually four countries: England, Scotland, Wales and Northern Ireland. Occasionally we might be known as *The British Isles* but this refers to the same four nations plus an independent state, Eire. For the purposes of this paper, Eire has not been included in the areas that are being described.

The UK has a population of some 60 million people and an area of about 94,000 square miles (or 243,500 square kilometres). This gives a density of some 630 people per square mile of the land mass.

In terms of Government, all four countries are part of the monarchy of Queen Elizabeth II and are governed to a greater or lesser extent by the Parliament and the British Government based in London. For England, this is the only central administration but the other three countries have had powers devolved in recent years. Wales has a Welsh Assembly, Scotland and Northern Ireland have respective Executives although each has a slightly different legal framework and range of powers.

Each Country has effectively both a national road network and a local road network. The national network comprises all motorway and trunk roads which are predominantly strategic and high speed. These roads are owned and managed by the national governments, including the formulation of policy and the setting of budgets for both capital improvements (which includes major maintenance) and routine maintenance and management of the infrastructure. A Government Minister, or Secretary of State, manages a Department (or Ministry) with these responsibilities. For example, in England, the Secretary of State for Transport, together with a number of junior ministers runs the Department for Transport. Most of the network management responsibility, however, is discharged through the Highways Agency, a distinct body with a reasonable degree of autonomy.

There are two principal exceptions as far as the strategic network is concerned: Firstly, there are 8 DBFO Companies (Design, Build, Finance and Operate) which have long term (up to 30 years) contracts with Government for full responsibility for all management, maintenance and improvement of a specific route.

The second exception is major estuarial long span crossings which are either effectively established as Highway Authorities in their own right or, again, are privately owned and operated. The management of such crossings is generally funded from toll revenue. These facilities include tunnels as well as bridges and an *ad hoc* support group has recently been established to cover areas of mutual interest, mostly associated with operational matters such as traffic control and toll collection systems. This group is known as the Big Bridges Group.

Local roads are owned and managed by Local Highway Authorities (also known as Transport Authorities since legislation introduced in 2000) which are the various types of unitary or first tier Councils, all of which are democratically elected bodies.

England

England has 24 County Councils, 47 Unitary Councils and 36 Metropolitan Councils. Counties retain a two tier system of Local Government, with the lower tier being District Councils which have no highway powers. Unitary Councils have all the powers and duties shared by Counties and Districts in two tier areas. Metropolitan Councils have similar powers and duties but cover densely populated cities or urban conurbations. The exception to these definitions is London, the capital city, where strategic roads are managed by the Mayor through an organisation known as Transport for London (TfL) and local roads are the responsibility of 33 London Boroughs who are also Highway Authorities.

As well as these organisations the country is split into 8 regions (excluding London) and within each region, there are 3 distinct governmental bodies:

Government Offices:	Responsible for managing central government policy and strategy at a regional level.
Regional Assemblies:	Responsible for regional planning strategy and for coordinating the network of local authorities within the region.
Regional Development Agencies:	Responsible for leading and implementing regional economic development initiatives.

Scotland

Scotland has a devolved government, the Scottish Executive, which discharges its highway functions through an agency, Transport Scotland. This body manages all of the trunk road and motorway network with local roads, including some of the Primary Route Network (PRN), being the responsibility of 33 Unitary Councils of varying size and urban/rural mix. Incidentally, Scotland has its own legal framework and, under Scottish law, there are no "Highway" Authorities but the Councils are technically "Road" Authorities.

Wales

The devolved administration in Wales is the directly elected Welsh Assembly which has powers for the strategic motorway and trunk road network. There is no "governing agency" but various lengths of road are managed through Contracts. Local Government in Wales is single tier with 22 Unitary Councils of varying size and rural/urban split.

Northern Ireland

All roads in Northern Ireland are managed by the Northern Ireland Roads Service, an Agency under the Department of Regional Development within the new devolved administration. Although none are highway or road authorities, there are presently 26 Councils in the province but plans are currently being considered to reorganise into a smaller number of between 7 and 15 Councils. A decision on the powers and duties of the new bodies is also under review and it is possible that a smaller number could take on Highway Authority responsibilities.

Europe

The United Kingdom is a full member of the European Union and, as such, has to comply with all directives, usually leading to harmonisation across member states. Whilst it is accepted that this is a worthy intention, there is an argument that the UK tends to remain somewhat isolated in terms of benefits to be gained by further engagement in significant research programmes. This is an area in which opportunities for improved working still remain to be grasped.

Traditional Support Groups

In the decades that followed the Second World War, the United Kingdom entered into the planning and construction of major infrastructure improvements including the motorway programme and bypasses on the trunk and primary road network. This was coordinated by central government's various Transport Ministries or Departments in which a strong team was established with a high level of expertise. Detailed guidance in the form of Technical Memoranda was made available to all engineers, in both public and private sectors, engaged in the design programme. At the same time, a number of software packages (which required the use of main frame computers, usually on overnight runs) for design and analysis were developed and managed through this central team.

As the road building programme developed in the early 1970s, a number of regional Road Construction Units (RCUs) were established, normally linked to the traditional County Council Highway Departments where RCU Sub-Units were accommodated alongside Local Authority staff. This enabled the cross fertilisation of ideas and best practice all of which emanated initially from the centre.

The early 1990s saw the Government establish the Highways Agency as a delivery body for the management of the motorway and trunk road network, leaving the Department for Transport (or its equivalents) to concentrate on higher level policy and budgetary considerations. Technical advice was transferred to the Agency.

County Councils themselves had been set up as early as 1889 and the post of County Surveyor became established. Little changed in the various reorganisations of local government until 1994 when a number of unitary authorities came into being. It was at this time that financial and political pressures led to some County Councils either restructuring to combine highways, planning and other related functions or to externalise large elements of their design capacity by means of new contracts with the private sector and transferring staff. Both of these events meant that there was a tendency towards reduced or limited capacity either in smaller Unitary Councils or in those in which the majority of staff had been externalised.

CSS Bridges Group

From the outset of the introduction of the post of County Surveyor, a support group had been established: The County Surveyors' Society. This remains in place today although is now formally known by the initials, CSS. Within CSS, there are a number of committees and working groups, with one of the former being the Engineering Committee below which sits one of the latter, the CSS Bridges Group. Below Bridges Group sit 8 Area Bridge Conferences (ABCs) conterminous with the English regional bodies. ABCs are open to all local authority bridge managers in that particular region. There are also equivalent conferences in Wales (CSS Wales) and Scotland (SCOTS), and the London Boroughs and TfL also enjoy a working group, LoBEG (or London Bridge Engineers' Group). The inter-relationship between Bridges Group and the ABCs is both "top down" with the capability to cascade information to every local authority in the Country, and "bottom up" giving the opportunity for any bridge manager to bring issues to a higher level and ultimately to challenge nationally agreed policy.

Bridges Group meets 3 times a year and arrangements are normally made for meetings to follow soon after the CSS Engineering Committee and for Area Bridge Conference meetings to be held a few weeks later to facilitate the cascading of information. The chairman of Bridges Group is a full member of CSS and a member of its Engineering Committee.

Recent Developments

For most of its early life, CSS Bridges Group was concerned with issues around bridge design and construction and meeting agendas consisted mainly of technical matters.

Things started to change in 1991 when the UK Government, in line with a European Union (EU) directive but derogated until 1999, embarked on a major programme of bridge assessment and strengthening in order to bring the bridge stock up to a European standard of 40 tonne vehicles and 11.5 tonne maximum axle load. Initially for the Primary Route Network, but later extended to local roads, the programme was funded centrally through grant allocations to local Highway Authorities against annual bids made through an individual authority's Transport Policy and Programmes (TPP) submission. It was at this time that Bridges Group opened an invitation for central government staff to attend meetings as well as developing much increased liaison through a number of specific meetings on the subject of the work in hand.

Up to the early 1990s, Bridges Group had also extended membership to central government and/or the Highways Agency and also to the largest of the private bridge owners, British Rail, and its privatised successor organisation, Railtrack (now renamed Network Rail). Bridges on disused railway lines were initially the responsibility of another body, British Rail Property Board who also attended meetings.

As far as road-over-rail bridges are concerned, the ownership is not always obvious. A rule of thumb says whichever piece of infrastructure was there first, the other party built the bridge and is therefore the owner. For bridges carrying highway loading, Railtrack also embarked on an assessment programme known as Bridgeguard 3 which was to be funded, in part, by the local highway authority but using government grant. This led to problems of prioritisation of both projects and funding.

All of above (with changes within the Devolved Administrations, the Highways Agency, Local Government, externalisation of services, privatisation of rail companies and a major investment programme) presents a picture of a professional discipline starting to fragment at a time when there was a huge need for coordination and cooperation. Rather than all parties meeting together, there was a very large number of separate meetings and information and/or decisions had to be communicated to other groups, often at second or even third hand.

As well as this predicament specifically with bridges, there was a parallel level of concern over road maintenance funding and management in the round.

The Solution – The UK Roads Liaison Group

Like solutions to many problems, the answer here was obvious: simply arrange for all appropriate parties to get together to share their perspective of the various issues and agree a plan of action. But, as in most cases, the difficult part is not the "what should happen" but the "how do we get there".

In this case, the initiative was seized by civil servants in the Department for Transport who drew up the concept of an over arching liaison group with a number of boards concentrating on specific professional disciplines. This was the genesis of the UK Roads Liaison Group and its four Boards: Roads, Bridges, Lighting and Traffic Management. (NB It is accepted that the last of these is somewhat anomalous in that the first three are dealing with the hard infrastructure of the network whereas the fourth concerns itself predominantly with the end user, the traffic using the network. This debate is ongoing and there has been a recent move to recast this Board into one of Network Management, thereby embracing the emerging technology around intelligent transport systems etc.)

The Roads Liaison Group is chaired by a senior civil servant in the Department for Transport. The precedent has been established for specialist sub-groups to sit below Boards but, to date, this has only happened with the Roads Board. The simple structure is shown below (Figure 1):

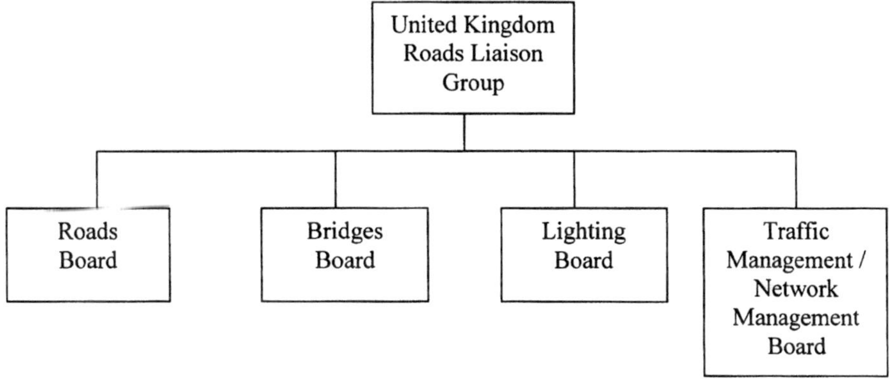

Figure 1. The UK Roads Liaison Group and Boards

The function of the secretariat, and that of administration of meetings is currently also held by the DfT.

The full membership of Bridges Board is as follows:

Department for Transport	Senior civil servants with responsibility for Road Maintenance policy and funding – the Department presently also acts as the secretariat.
Highways Agency	Senior representative
CSS	Members of the CSS Bridges Group, including a representative of the Metropolitan Authorities.

CSS Wales	Senior Bridge Engineer and Chair of Conference
SCOTS	Senior Bridge Engineer and Chair of Conference
Transport Scotland	Senior Bridge Engineer
Welsh Assembly	Senior Bridge Engineer
Northern Ireland Roads Service	Senior Bridge Engineer
Transport for London	Senior Network Manager
LoBEG	A London Borough Bridge Engineer
London Underground	Senior Bridge Manager
Network Rail	Senior Bridge Manager
British Waterways Board	Senior Bridge Manager

The Board meets 3 times a year. All representatives are empowered to act with the authority of their employer and/or parent organisation. The chairman, along with counterparts from other Boards, is also a member of the UK Roads Liaison Group thereby enabling a communication link to and from Boards to the RLG.

The Bridge Owners' Forum

Along with the CSS Bridges Group and the UK Bridges Board, the third body in the triumvirate of bridge groups is the Bridge Owners' Forum. Established in 2000 through the initiative of Professor Campbell Middleton of Cambridge University, the Forum includes most UK bridge owners, (often the same individuals who also represent their organisation on the Bridges Board), including a representative of the UK Big Bridges Group, and with Professor Middleton in the Chair, the academic sector is well represented. The Board meets two or three times a year and has 3 key strands of work not routinely covered by either CSS or UK Bridges Board:

1. To identify areas of the industry that would benefit from additional research, and to undertake preliminary appraisals of proposals.
2. To take an objective view of similar proprietary methods and/or techniques in order to arrive at impartial advice to bridge owners.
3. To act as the "radar" of the bridge profession whether in terms of newly identified problems, new initiatives, new sources of funding, or connection with international bodies.

The forum presents an annual report to the UK Bridges Board on its work and activities.

Complementary Functions

At first sight, and this is a criticism that has been made in the past, there is a risk of duplication or even triplication of roles and responsibilities between the three groups. There has also been a perception (and even an allegation) that little of value is achieved and that each meeting could end up with the same people talking about the same subject matter.

Whilst acknowledging that there is indeed a risk that this could be the outcome, the diagram below (Figure 2) demonstrates the inter-relationship between the groups.

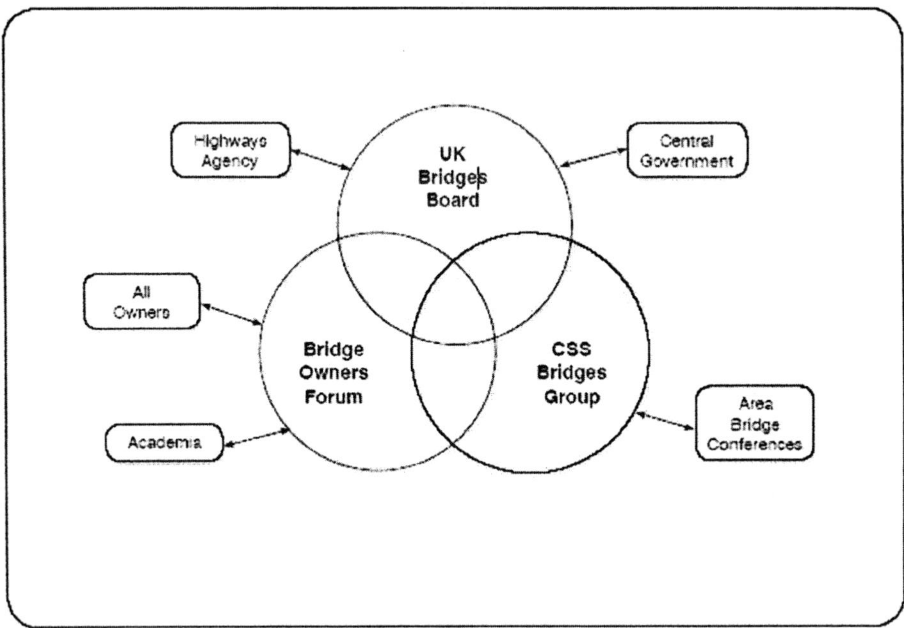

Figure 2. Inter-relationship of Bridge Groups

The UK Bridges Board concentrates on high level strategy, policy and funding issues whereas CSS Bridges Group covers the practical implementation of bridge management issues. The Bridge Owners' Forum offers a degree of independence to judge objectively emerging developments in research.

As well as the above, there are wider and softer benefits of this increased level of liaison between all relevant parties. To fully understand this position, one needs to be reminded of the degree of fragmentation and the opening schisms that were starting to take a hold in the bridge management profession in the middle of the last decade. These points, and the reasons behind them, have been described above and the net effect was one of perception of suspicion concerning the motives of the various parties at a time when closer working should have been a priority. Since 2000, that suspicion has started to be replaced by its dictionary opposite, trust. That starts with improved understanding of others' positions, the requirements of their senior managers and stakeholders, and the cultures of the respective organisations. The mechanism through which this trust is achieved and delivered is part of the role that each group should have as part of its *modus operandi.*

It is accepted that this outcome is only partly delivered at this stage and remains an aspiration of the author. It is clear, however, that this is "work in progress" and the direction of travel is positive.

Research

Bridge related research projects can be instigated by a number of different bodies and be based on many different needs. Examples might be the use of a new material or construction techniques having been developed by a private company, the development of analytical tools to enhance empirical methods of establishing structural behaviour or changes to standards of "finishings" such as joints, parapets or waterproofing.

Similarly, there are many funding streams which can be tapped into to pay for research. There are research budgets within central government through the Department for Transport and the Highways Agency in England and the devolved administrations in Scotland, Wales and Northern Ireland. The CSS has a "Research Club" funded by contributions from individual members and from this, budgets are identified for specific projects. Private bridge owners also have research funds. Maximum benefit can be gained from pooling funds and forming a partnership to promote and administer the project.

Although many research projects have been successfully delivered in this way, there is still room for improvement by structuring the process on project management principles, clearly identifying key roles and responsibilities. The various strands of work in this context are shown in the diagram below (Figure 3):

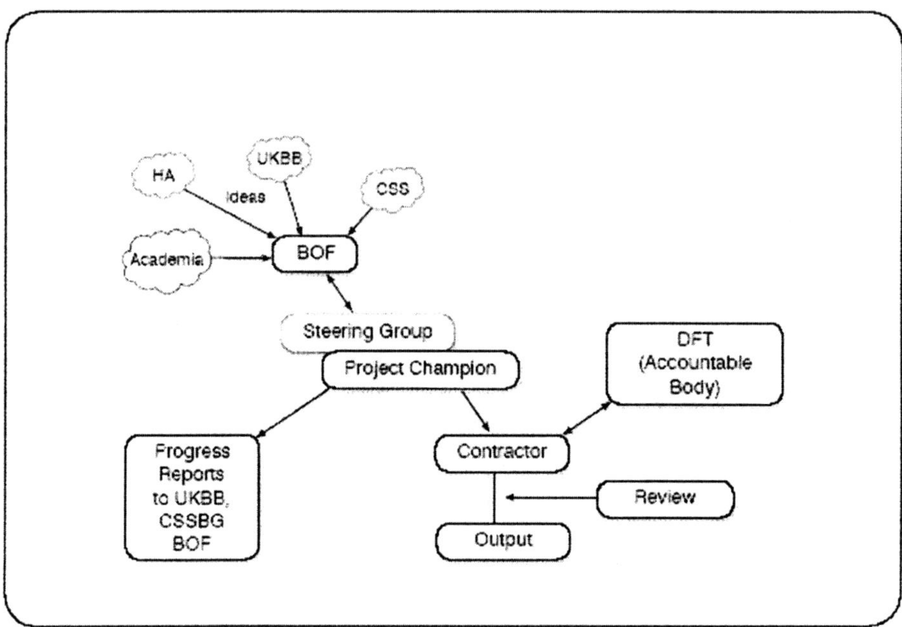

Figure 3. Model for managing bridge related research

The key to a successful research project is the definition of roles and the identification of key responsibilities. Project management principles should be employed with a project manager working for the "contractor" and a project "champion" representing the client interests. A steering group made up of all interested parties can be used to make decisions or changes or modifications to the original specification or to review early results or outputs. Other roles are the Accountable Body which must administer funding and, in this case, the role of the Bridge Owners' Forum which should act as a filter and conduit for research proposals.

Possible Further Improvements

Whilst excellent progress has been made over the last few years, it would be wrong to assume that the present position is as good as it gets. Indeed, the DfT instigated a formal review in 2006 to see if the principles behind the establishment of the UKRLG and its Boards were being applied as intended and to insure that the output from the various bodies remains both effective and adding value to the business of infrastructure management. The review concluded that there was a benefit in continuing with the present arrangements although recognised that there was room for improvement in the methods of working, the interaction between Boards and the possible conflict of interest in having one of the key partners also responsible for the secretariat and administrative functions. Plans are being prepared to see if this function can be passed to a professional body which does not have a role in the existing arrangement.

Other areas for consideration include further work on developing an improved mutual understanding of the issues facing various bridge owners and the need to more clearly identify areas of possible duplication between groups. The fact that a number of individuals sit on, or at least regularly attend, all three groups should help this aspiration.

Perhaps due to the nature of its geography, the United Kingdom has a tendency towards insularity and not looking at opportunities at a European and international level. With some notable exceptions, this is generally true within the field of bridge management and this is another area where further opportunities need to be explored. European research funding is a source which seems to be only occasionally accessed but there are also lessons to be learned from international colleagues who are inevitably to be faced with the same sets of problems and challenges as we have to deal with in the UK.

Conclusion

Although the numbers of bridge owners and other related stakeholders in the UK may seem excessive, it is clear that some degree of liaison and cooperation is essential if the bridge stock is to be maintained and managed as effectively as possible. The arrangement of groups and supporting networks, as described in this paper, ensure that this objective is achieved, at least in part. As with most other initiatives of this type, there will always be areas in which improvements can continue to be made, particularly in the research field, and all parties engaged in bridge governance are committed to this ethos of continuous improvement.

Design of rehabilitation and refurbishment of existing r.c. bridges

Prof. Claudio Modena, Department of Construction and Transportation Engineering, University of Padova, Padova, Italy
Dr. Paolo Franchetti, Department of Construction and Transportation Engineering, University of Padova, Padova, Italy
Dr. Manuel Grendene, Department of Construction and Transportation Engineering, University of Padova, Padova, Italy

Abstract

The design of rehabilitation and refurbishment interventions of existing r.c. bridges requires a very complex and comprehensive approach, deeply involving the use of conventional and unconventional structural analyses and technologies. The assessment of structural performances requires an appropriate campaign of experimental and theoretical investigations and consequently the use of tools like long term monitoring and dynamic identification.

The subsequent choice of the proper intervention (in terms of both material and application technique) is then dependent on several factors, the main ones being: the type of structural or non structural components (e.g. deck, piers, abutments, bearings etc.), the structural solutions (e.g. simply supported, continuous deck, arch bridges, etc.) and the type of action (static, cyclic, dynamic).

The paper describes the practical application of basic criteria for rehabilitation and refurbishment to three cast in-situ r.c. bridges crossing the river Adige, all part of the massive works of post-war rebuilding in northern Italy.

Introduction

The r.c. bridges built in Italy in the post-WWII period show currently dimensional, structural and functional deficiencies caused by different factors. Deterioration processes reduce the strength of structural components and the comfort level to road users. The codes, in the course of the bridges service life, have been updated due to changes to the vehicular traffic (dimensions, typologies and design speed of vehicles, etc.), thus imposing their static retrofit and widening. In particular, the updating of the site seismic classification can make necessary seismic retrofitting.

The assessment of the current and the prognosis of future structural performances, first of all, require an appropriate campaign of experimental and theoretical investigations. In addition to the more commonly addressed issues of strength and stability, consideration might also be given to a broad range of factors including stiffness, dynamic behavior, long-term deformations, durability and possibly even aspects.

The structural - static and dynamic - characterization of a bridge cannot come from standard procedures (in situ and laboratory tests by witch the characterization of material properties and of the principal effects of deterioration phenomena can be obtained). On the contrary it requires the use of such tools as long term monitoring and dynamic identification, which can, when appropriately applied, give substantial information on the overall structural behavior of the bridge, allowing, via model updating procedures, for appropriately selecting the most efficient intervention strategies and for controlling the efficiency of the applied interventions. Rehabilitation is usually coupled with refurbishment interventions, in order to improve the safety and comfort of road users. This is obtained, in particular, by widening the deck, eliminating the expansion joints (which are also one of the most critical elements for durability), modifying the approach spans and ancillary structures, laying life-lines on pavements, adapting road markings, safety barriers and parapets, and lighting.

This paper describes the practical application of basic criteria for rehabilitation and refurbishment to three of the most usual typologies of post-WWII r.c. bridges in northern Italy. In these cases, the decision to intervene was well justified by the numerous aspects of limited efficiency and performance of the bridges, with respect to current bridges codes.

General description of the three bridge studies

The three case studies represent the most usual typologies used in the post-WWII period. Their structural typologies make their rehabilitation and refurbishment interventions suitable for considerations of a general nature. In these cases, the decision to intervene was well justified by the numerous aspects of limited efficiency and performance of the bridges, with respect to current bridges codes.

- Albaredo bridge. The bridge was built on six spans with a total length of approx. 230 m entirely. Three spans, on the right bank of the Adige, have an effective span of approx. 23 m, and are composed of simply-supported beams, whereas the remaining three, with an effective span of approx. 53 m, are "Nielsen" type arches (a tied arch bridge with inclined hangers). The roadway was 6 meters wide, flanked by two footpaths of slightly more than 1 meter wide.
- Zevio bridge. The bridge consists of seven spans of approx. 32 m and two terminal spans of approx. 14.3 m, for an overall length of 252.7 m. The isostatic structure is formed of the "Gerber" type beams (statically determinate beam consisting of one or a greater number of beam members connected by hinged joints): the height of the transverse section varied from a minimum of 1.52 m on the abutments to a maximum of 2.67 m at the piers. The roadway was 6.00 m wide (inadequate for today's traffic) and two 1.05 m wide footpaths
- Sega bridge. The bridge had a deck with an overall width of 8.34 m, supported by a main arch with a span of 60 m and a secondary arch with a 25 m span. The section of the main arch is approximately $2.0 \times 1.8 \text{ m}^2$.

A general view of the bridges before and after the works is given in Figure 1.

Intervention for rehabilitation and refurbishment

Below are described all the interventions applied to the bridges just described. Particular attention was paid to the description of the "structural" - static and dynamic – part.

Treatments of the deteriorated parts of the structure

Deteriorated materials (both concrete and steel) were systematically removed and replaced. The concrete cover was completely hydro-demolished in the most serious cases, while more light treatment by blast sanding was used for the well preserved concrete (typically the tie rods of the Nielsen arches, where carbonation didn't penetrate more then few mm).

The entire surface area is then dressed by pressurised sanding, until clean degreased surfaces are obtained without any fine particles that could obstruct the adherence of the subsequent plastering. All the exposed rebars were sanded down to white metal, blown with pressurised air jets and treated with an anti-corrosive agent. New bars are positioned next to the oxidised bars, which are of the same diameter but with improved adhesion. Lastly, the new plastering was applied to the cover using ready-made thixotropic cement mortar, fibre-reinforced with polymers, which has strong adhesion.

Figure 1. View of the Albaredo bridge: (a) prior to the upgrading; (b) after the upgrading. View of the Zevio bridge: (c) prior to the upgrading; (d) after the upgrading. View of the Sega bridge: (e) prior to the upgrading; (f) after the upgrading.

Elimination of joints

An intervention of fundamental importance from both the static and functional points of view consists of eliminating the deck joints. This allows consolidation of the entire deck in terms of the transmission of horizontal forces. The system demonstrates high stiffness in the

horizontal plane, and at the same time assure the possibility of rotations, in correspondence to the saddles (in the case of Zevio bridge) or of the supports on the piers (in the case of Albaredo bridge), without altering the original static behaviour under vertical loads and therefore without significantly changing the distribution of the bending moments along the arches and reactions on the piers. Furthermore, the work allows to avoid the disposition of joints, which is critical in relation to the durability and viability conditions of the bridge.
In the cases of Albaredo bridge and Zevio bridge, the joints between adjacent spans were eliminated by means of laying a continuous supplementary concrete slab along all the spans. Expansion joints were only placed at the ends of the bridges, to allow temperature expansion.

Supports

In many existing bridges the supports are inefficient or sometimes partly lacking. It is therefore necessary to provide an efficient and adequate support system, which is easily accessible for inspection purposes.
For the Albaredo bridge, new multi-directional restraints beneath each span were positioned in such a way as to permit the natural expansions and contractions caused on the structure by temperature variations. The bearings are in steel with Teflon (PTFE). The size of the inserted bearings guarantees longitudinal creep proportional to their distance from the abutment on the left bank (fixed), to which the entire deck is fixed.
For the Zevio bridge, steel bearings with surfaces in Teflon (PTFE) were inserted in correspondence to each pier and each Gerber saddle, positioned in such a way as to permit the natural expansions and contractions caused by temperature variations on the structure.

Widening

The managing authorities of the examined bridges had expressly asked for the bridges to be widened, especially in order to obtain new cycle paths and footpaths raised above the roadway and separated from it by safety barriers.
The design choices were different in the three cases, taking into account both architectural and structural-functional considerations.
The more complex solution was adopted for Albaredo bridge (Figure 2). The widening of the deck to 14.60 m, was obtained by cantilevered reticular structures in steel that support the layers composed of corrugated metal sheet and a supplementary covering in light reinforced concrete. To achieve this it was necessary to arrange new steel tie-rods and cross-beams which allowed integration with the existing bridge and ease of installation of the metal structures used to build the new lanes reserved for pedestrians and cyclists.
Functional adaptation of the Zevio (Figure 3a) bridge was obtained, for both the main structures and the newly-built supported spans, using lateral cantilever concrete slabs supported by auxiliary metal structures in cor-ten steel. Deck width after the intervention is 14.00m.
In the case of Sega bridge (Figure 3b), taking into account the specific construction and geometric properties, the widening has been obtained with post-tensioned r.c. cantilevers. The prestressing was necessary in order to ensure continuity of the deck (along the joint between existing and new structure) subjected to heavy traffic loads, thus ensuring both traffic comfort condition and durability.

(a)   (b)

Figure 2. Albaredo bridge: (a) particular of new cross-beam – cantilever attachment; (b) completed bridge

(a)   (b)

Figure 3. (a) Zevio bridge, particular of new cross-beam – cantilever attachment; (b) Sega bridge, post-tensioned r.c.cantilevers

Static retrofit

The need to retrofit the bridge in order to bring the safety conditions of the main structures up to the current standards specifies the design of the interventions, that are defined by the type of structure, materials and state of preservation of both the substructure and superstructure: integration of the existing structure, its partial substitution or, sometimes, even complete demolition and rebuilding may be necessary. The solutions adopted for the examined bridges are described in detail below.

Albaredo bridge

The existing foundations of abutments and piers are adequate in terms of both load-bearing capacity and undermining phenomena.

The investigations and tests which had been carried out demonstrated the possibility of relying on the arch without any reinforcement work. The reinforcement of the hangers also resulted as being adequate, provided that the effective capacity was guaranteed of the concrete slab to absorb a significant ratio of the tensile stress inevitably transmitted by the arches and, possibly, that the bending effects would be limited. The tie-rods, lastly, were quite badly over-stressed and furthermore there were no sufficient guarantees about the durability of the reinforcements protected by the concrete of the covering.

The concrete slab instead showed marked deficiencies, due to inadequate thickness and the excessive deformability of the cross-beams, which triggered decidedly higher longitudinal bending stresses than those corresponding to the schematization adopted at the design stage, for which the cross-beams acted as restrainer.

This situation suggested a design option that could also preserve the most characteristic building components of the existing structure, i.e. the system of arches and longitudinal beams.

The problems described were resolved with:

- the addition, in parallel to the existing structures, of a new system of steel cross-beams and tie-rods, while still respecting the geometry of the existing structure;
- a supplementary covering on the existing concrete slab and the relative longitudinal pre-stressing;

The new tie-rods were positioned vertically at regular intervals, making use of the wheelbase imposed by the existing tie-rods, and applying an adequate layer of pre-stressing. They are composed of groups of four stainless-steel rods, thus avoiding any interference with the existing structure: anchorage to the arch was obtained with external plates, connected by a suspender that allows a self-equilibrated system of transmission of the stresses to be obtained.

The following effects were thus obtained:

- reduction of tension stresses in the existing rods;
- reduction of bending stress in the longitudinal beams;
- an efficient base for the new metal cross-beams, which are of the same shape as the existing ones in reinforced concrete;
- adequate safety conditions of the entire bridge, thanks to the size of the new suspenders, which can resist even if one of the existing suspenders should collapse.

The new cross-beams, which in themselves would have insufficient stiffness, like those existing, to contrast the above-mentioned behaviour, were positioned after imposing a pre-bending that allowed the desired reduction of bending in the existing concrete slab to be obtained thanks to the corresponding state of constraint introduced.

The intervention was completed by a supplementary covering on the concrete slab. This work was done using light concrete, except in brief terminal sections where high-resistance concrete was used to anchor the cables of longitudinal pre-stress that were simply laid in sheaths on the upper surface of the existing concrete slab and incorporated in the supplementary covering.

The longitudinal pre-stress can ensure the efficient transmission of a significant amount of the tensile stress transmitted by the arches, making the longitudinal beams adequate.

Regarding the three beamed spans external to the arches, the inadequacy of both the main beams and concrete slab-cross-beams system required the building of three new beams, formed by four beams in welded steel and a surface concrete slab in high-resistance concrete.

Lastly, to obtain the functional modification within the terms indicated for the arched spans, a cantilevered concrete slab was produced with the same typology as the metal structures used for the arched spans.

Reinforcement work was also carried out on the supports of the deck's restraint devices.

Zevio bridge

The foundations were given a radical intervention of strengthening, obtained by flanking the original caissons (stiffened by internal partitions) with a circle of micro-rods alternated with columns of compacted earth, inside which (and beneath the base of the caissons) high-pressure jet-grouting was carried out to ensure the necessary margins of safety against undermining phenomena. The abutments, given the static scheme adopted, had only the function of breast wall.

As regards the principal structural components making up the simple spans and cantilevers, the investigations and tests done demonstrated the possibility of relying on the existing sections, with interventions of reinforcement consisting of laying a suitably reinforced, supplementary concrete slab and the use of a pre-stress system that can achieve a state of constraint that contrasts the external stresses where the reinforcement deficiencies are greater. As regards the inadequate resistance and stiffness of the concrete slab and beams constituting the spans resting on the cantilevers, it was decided to substitute the secondary spans in reinforced concrete with new mixed-structure spans (steel cor-ten+cls) to avoid the need for supplementary concrete laying and onerous integrations of the existing reinforcement. This also permitted to reduce the loads on the supports. As mentioned above, the supplementary concrete slab was laid continuously on all the spans.

The interventions of consolidation have not altogether changed the stresses on piers and abutments. Reinforcement work was done on the pier heads and saddles by means of external hoops, especially in the areas where the transfer of loads from the supports to the substructure takes place, and localized pre-compressions.

Sega bridge

The bridge doesn't present major structural deficiencies of the main structural components: the robust concrete section of the arches, in fact, is fit for supporting the current traffic loads, so that the main structural component didn't require other interventions then those aimed to improve the lateral confinement of the longitudinal rebars. What was really lacking, in fact, was the transversal reinforcement: new stirrups were added (after removal of the concrete cover), and both the new and the exiting stirrups were securely anchored to concrete.

The deck, on the contrary, required severe interventions, both on extrados (deteriorated concrete removal, integration of rebars) and on the intrados. Particular weak points were the lack of transversal reinforcement the curved shape of the intrados, where the longitudinal (curved) bars were not at all adequately confined (against delaminating effects). For these reasons, transversal reinforcement was added on both sides of the deck, and the curvature eliminated by laying down new concrete, so making the intrados horizontal.

Seismic retrofit

Seismic protection of existing inadequately-designed bridge structures can be put into effect by adopting various techniques that contribute towards reducing the vibrations.

One of the most used is passive control of the vibrations, obtained through the seismic isolation of the structure (consisting in the disconnection of the structure from the foundations, so that the seismic action transmitted from the ground does not reach the upper parts of the bridge) or the use of systems that provide damping of the incoming energy (consisting in limiting the accelerations and the maximum transverse force transmitted to the structure). Alternatively, to prevent or limit the movements of the decks with respect to the supports, and to equally distribute the dynamic forces on the piers, it is possible to use deck connecting systems or consolidate the supplementary concrete slab between the different spans.

On the Albaredo and Zevio bridges, the seismic retrofit in a transverse direction was obtained through systems of passive control. In a longitudinal direction the horizontal actions are absorbed, in the former bridge, exclusively by a piling at the back of an abutment. The second bridge was instead adjusted longitudinally with two double-effect viscous dampers on each abutment.

Albaredo bridge

The interventions involved the realization of:
- a palisade, standing behind the abutment on the left bank of the Adige that can absorb all the horizontal forces that might develop longitudinally during an earthquake;
- a connecting system of the decks;
- elasto-plastic dampers on all the piers and on the abutment on the right bank of the Adige;

The abutment on the right bank of the Adige was also adapted through the insertion of tilted micro-rods to limit the horizontal push that are generated by an earthquake because of the presence of the raised banks.

Zevio bridge

The bridge was isolated seismically by means of an elasto-plastic damper on each pier to absorb transversal actions and two double-effect viscous dampers on each abutment to absorb the longitudinal seismic actions. The latter were installed on a connecting beam of vertical and tilted micro-rods, in correspondence to the abutments.

The transversal isolators were dimensioned to resist the action (the SLU) of wind in the elastic field. Using the same criterion, the longitudinal dampers were dimensioned to absorb braking actions, parasite forces and seismic forces.

Finishing work

The bridges, which had been upgraded as described above, were refurbished through interventions such as adaptation of the access ramps, laying of the pavements, adaptation of the road signs, safety barriers and parapets, the laying of pipelines for underground mains and cables; new lighting and new parapets, water-proofing of the deck by the laying of a bitumen-based polymer membrane on the roadway and a double bituminous sheath on the cycle paths, footpaths, kerbs and discontinuities.

Conclusions

Natural ageing, poor maintenance, severe environmental actions or changes in use of a structure, as well as increased safety requirements, necessitate repair and/or strengthening. Using three r.c. bridges built near Verona, Italy, as examples, the preparation and execution of repair and/or strengthening has been discussed. In any case, there is a fundamental decision to be made at the start of a bridge retrofit, particularly in seismic regions. The decision will be the level to which the bridge should be retrofitted. In the cases described, the structures were adapted to the new functional requirements dictated by the owners and the regulations in force, without conducting a detailed cost-benefit analysis.

However it is important to stress that regular inspection and maintenance is the most economic way of achieving an adequate service for bridges. A timely intervention is much cheaper, easier and more effective.

Acknowledgments

The offices of the Province of Verona, in specific Eng. E. Pellegrini e R. Castegini, and the technical staff of SM Ingegneria S.r.l. for the continuous and decisive support.

References

Modena, C. & Reginato, F., with the cooperation of Stoppa M., Mesaroli M., Morbin A., Corsini A., 2003, Interventi di consolidamento, adeguamento statico e sismico ed allargamento del ponte sul fiume Adige a Zevio lungo la S.P. n° 20 "dell'Adige e del Tartaro": progetto esecutivo – Relazione generale, *Verona, 16 December 2003.*

Modena, C., Castegini, R., De Zuccato, L. & Stoppa, M. 2004. Interventi di adeguamento e miglioramento sismico: l'esempio del ponte di Albaredo d'Adige in Provincia di Verona, *Proc. Giornate AICAP 2004, Verona, 26-29 May 2004.*
Strategies for Testing and Assessment of Concrete Structures, Guidance Report, Bulletin d'information n. 243, CEB-FIP, *Lausanne, May 1998.*

Theme five:

Innovation in bridge design, construction and repair

Strengthening of Bridges using CFRP composites

Engr. Najif Ismail, Catholic Relief Service, Islamabad, PAKISTAN.

Abstract
The use of advance composite materials in construction for repair and rehabilitation has become a frequent option in the last decade. FRP composites have many advantages over the traditional technique of steel bonding for a number of reasons:

1. Composites add little or no additional weight to a building, eliminating the need for costly foundation strengthening.
2. The application of FRP composites is faster than the traditional materials and requires no heavy and noisy machinery. This allows the structure to remain in use during the strengthening process.
3. FRP composites are very thin (1.2mm to 1.4mm). So there is no loss of floor space and negligible effect on the architectural aspect.
4. FRP composites do not corrode, which makes it long lasting.

However, the method has yet to become a mainstream application due to a number of economical and design related issues. Brittle de-bonding failure, the aging effect on bonding, a broad based awareness and proper design guidelines are the main concern for future research work. This paper is focused on the ultimate load carrying capacity of the CFRP-strengthened beams and their effect on the deflection and failures modes by varying the amount of CFRP content.
Keywords: Beams, CFRP, steel plating, strengthening, de bonding.

Introduction
There are considerable numbers of existing reinforced structures in the world that do not meet current design standards because of inadequate design and/or construction error or need structural upgrading to meet new seismic design requirements. Retrofitting of flexural concrete elements was traditionally accomplished by externally bonding steel plates to concrete. Although this technique has proved to be effective in increasing the strength and stiffness of reinforced concrete elements, it has the disadvantages of being susceptible to corrosion and being difficult to apply and install. Recent developments in the field of composite materials, together with their inherent properties, which include high specific tensile strength, good fatigue and corrosion resistance and ease of use, make them an attractive alternative to steel plates in the field of repair and strengthening of concrete elements.

The use of fiber-reinforced polymer (FRP) composites for strengthening reinforced concrete (RC) structures was first investigated as an alternative to steel plate bonding for beam strengthening at the Swiss Federal Laboratory for Materials Testing and Research (EMPA) (Meier et al. 1993) where tests on RC beams strengthened with CFRP plates started in 1984.

After this many research studies have been carried out and awareness, trust and confidence in FRP composites increased among the professionals particularly in USA, Japan and the countries of Europe. Today there are many case studies in these countries where this technique has achieved the desired results.

Objective

The objective of this paper is to compare the ultimate load carrying capacity of CFRP-strengthened beams and their effect on the mid span deflection and failure modes by varying the amount of CFRP content.

Experimental Work

Eight full-scale reinforced concrete beams with constant cross-section (7"x 12") and length (9'-0"), having the same reinforcement in all the beams. This was 2 #4 bars of grade 60 (60,000 Psi) on the tension side, 2 #3 of grade 40 (40,000 Psi) on the compression face and #3 stirrups @ 6" C/C as shear reinforcement throughout the length. Of the eight beams constructed, two were not retrofitted (B-1) and six beams were retrofitted by using different amounts of CFRP (B-2 to B-4) under the direct supervision of a Sika trained team. All beams were tested till failure and comparisons were made. Procedures for the construction and strengthening of the beams were as described below.

Retrofitting Technique
Substrate Repair and Surface Preparations
The behavior of concrete elements strengthened or retrofitted with FRP system is highly dependent on a sound concrete substrate and proper preparation and profiling of the concrete surface. Surfaces of the beams were ground and smoothed with the help of a hand grinder by the trained Sika team as per the manufacturer's requirements.

Characteristics as per manufacturer's Instructions
"Clean, free from grease and oil, dry, no loose particles or laitance. Concrete age, depending on climate, 3 to 6 weeks minimum. Blast-clean, scabble or grind. After grinding remove all dust from the surface with a vacuum cleaner. Planeness of substrate to be checked with a metal batten. Tolerance for 2 m length max. 10 mm"

Strengthening of Beams
Surface preparation of the beams, adhesive mixing, laying of the adhesive on the prepared surface of the beams and laminates was done by the trained team. Six beams were retrofitted as below:

B-2: Two beams - One S812 laminate was bonded in the center along the length of beams.
B-3: Two beams - Two S812 laminates were bonded to the prepared surface.
B-4: Two beams - One 4' 0" piece of S812 laminate was bonded at mid span in the center of the specimen.

These beams after retrofitting were left for 7 days to attain a satisfactory bond strength as per the manufacturer's instructions.

Test Specimens & Loading Arrangement
All eight beams had lengths of 9'-0" and cross-sectional dimensions of 7" x 12" as shown in **Fig. 1**. All the beams had the same flexural (longitudinal) reinforcement of two 4/8" diameter deformed steel reinforcing bars as bottom reinforcement and two 3/8" diameter deformed

steel reinforcing bars as top reinforcement. The flexural reinforcement was chosen to provide an under-reinforced section with a flexural-dominating behavior. The shear reinforcement consisted of 3/8" diameter plain mild steel reinforcing bars as closed stirrups spaced every 6" along the beam longitudinal axis as shown in **Fig. 1**. The shear reinforcement was designed to provide enough shear strength to exceed the shear forces associated with the flexural failure of all retrofitted beams.

Out of eight beams, two beams i.e, B-1 & B-1a served as the control beams (Concrete Reinforced beams without the addition of CFRP). The average of B-1 and B-1a, named as B-1 was used in the results. The other six beams were retrofitted using different amounts and configurations of CFRP as shown in **Fig. 2**.

The beams were tested in a staining frame. One dial gauge was installed at mid span to record the mid span deflection and two gauges were installed at the supports to measure settlement. Two point loads were applied at mid span and deflection and cracks were recorded for each increment of load.

CFRP Bonding and Loading Pattern

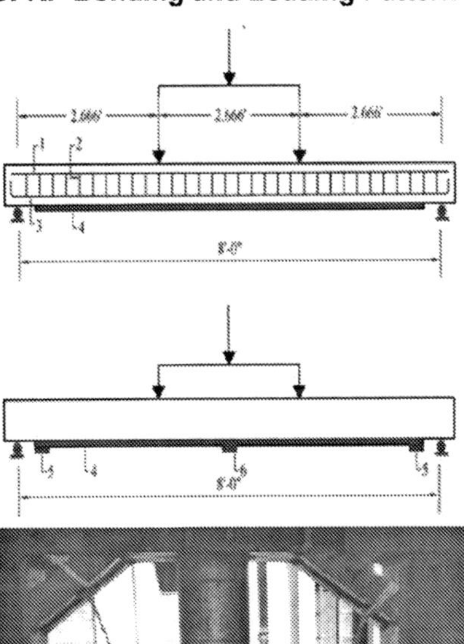

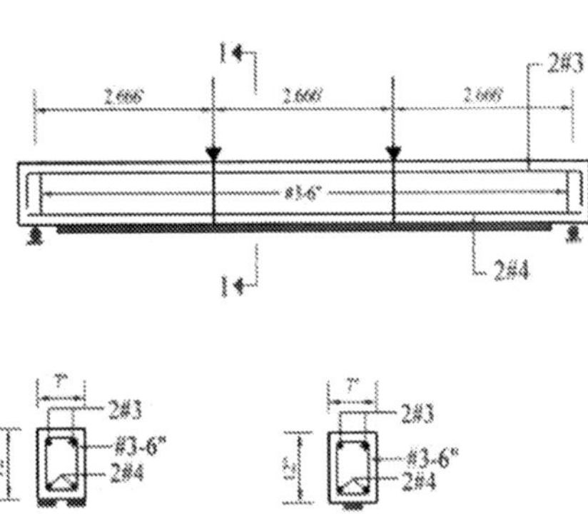

1. Top bars (2-#3)
2. Shear Stirrups (#3@6" C/C)
3. Bottom Bars (2-#4)
4. Carbon Fiber (S-812 Sika Carbodur)
5. Displacement Gauges at ends
6. Displacement Gauge at mid span

Figure 1

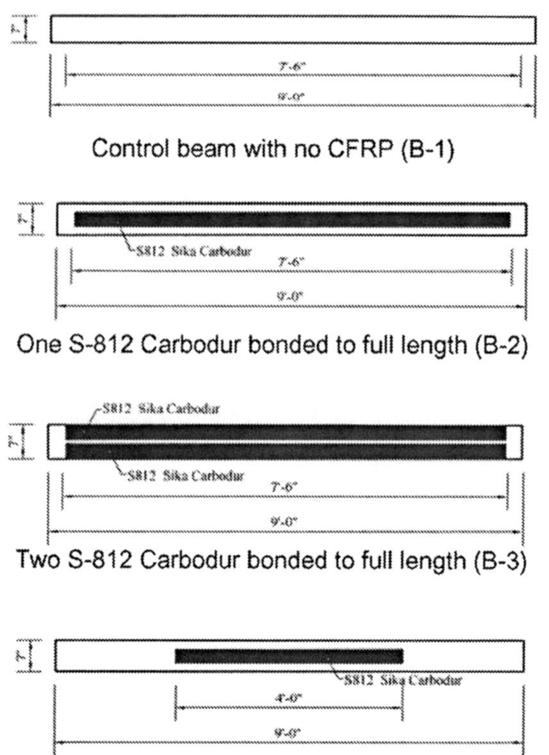

Control beam with no CFRP (B-1)

One S-812 Carbodur bonded to full length (B-2)

Two S-812 Carbodur bonded to full length (B-3)

One 4'-0" Long S-812 Carbodur bonded to beam (B-4)

Figure 2

Results and Discussions

Comparison of Experimental & Theoretical Nominal Moment Capacity

Type Of Beam	Experimental Mn (Ft-Kip)	Theoretical Mn (Ft-Kip)	Difference
B-1	18.752	18.820	(+) 0.40 %
B-2	36.93	39.83	(-) 7.280 %
B-3	49.311	51.544	(-) 4.332 %
B-4	25.385	28.85	(-) 12.65 %

1. Control Beam (B-1) shows negligible variation.

2. (B-2) and (B-4) exhibit an appreciable difference between the experimental and theoretical ultimate moment capacities. The strengthened beam (B-2) failed due plate end interfacial de bonding because of the high interfacial stresses near the ends of the bonded plates.

3. Strengthened beam (B-4) failed in tension without plate de bonding. The large variation between the experimental and theoretical moment capacities shows that the length of bonded plate was not sufficient.

4. Beam (B-3) failed due to concrete cover separation (spalling off) and shows less variation between the predicted and experimental capacity. This shows that the strengthened beam (B-3) was close to developing their full flexural capacity.

Graphical Representation of Experimental Results

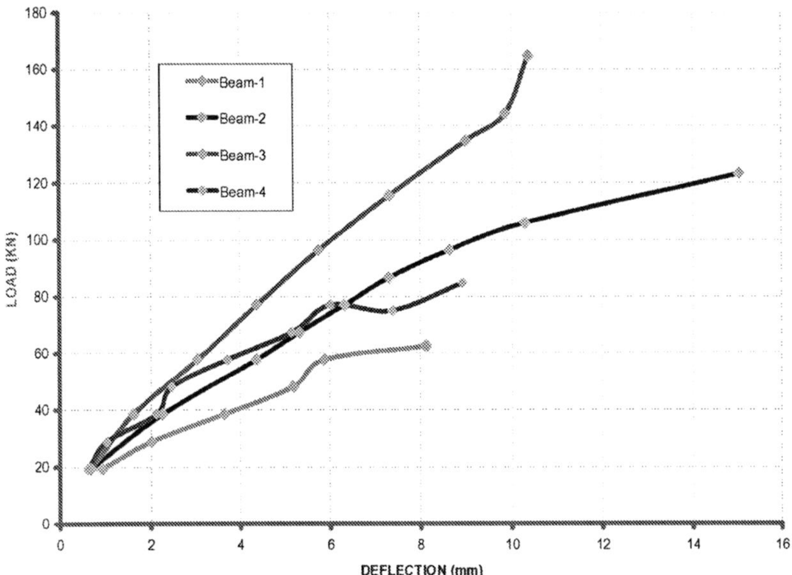

Load v Deflection Graph

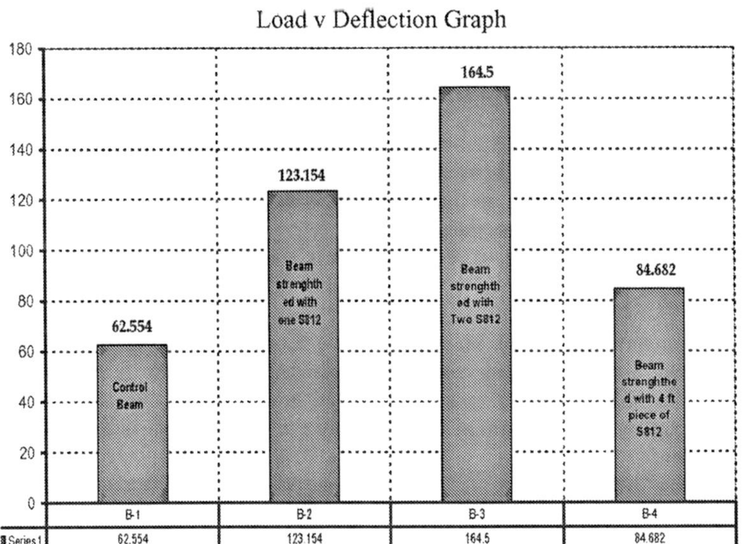

Ultimate Moment Capacity Comparison

De-Bonding

Cover Separation

Figure 3

Test Results & Conclusions

The following conclusions are drawn from the experimental program carried out in this investigation:

1. Strengthening of RC beams with externally bonded CFRP plates is effective and leads to an increase in load carrying capacity of between 96% to 162% over the control beams by doubling the amount of CFRP content.

2. Applying the CFRP along the full length, increased the bond of the CFRP to the concrete and enhanced the concrete ductility.

3. The retrofitted beams possessed less ductility compared to the un-retrofitted beams (control). This is a result of the brittleness of the failure modes of the retrofitted beams (concrete cover separation along with CFRP plates and de-bonding of CFRP plates from ends).

4. In order to achieve full capacity of the laminate and to avoid brittle failure by de-bonding, special attention should be paid to providing adequate anchorage length for the fibers either longitudinally or transversely.

5. The deflection values of all the CFRP-strengthened tested beams were reduced.

7. For shear, the FRP plate can be bonded to the web of the beam throughout its length, or it can be bonded to the areas where the highest shear is expected. Since fibre composites are anisotropic the effectiveness of the plate primarily depends on the orientation of the fibres, and the effectiveness will be different if the beam was cracked or not prior to strengthening. Also, the inclination of a crack will be different if it arises before or after the fibres are applied.

8. The case studies have shown the main flaw in the usage of CFRP strengthened beams is that the long term aging effects can cause de-bonding and ultimately failure if ignored. It requires a well-established monitoring system for the bridge before and after strengthening and the bonding agents to be modernized so that they can withstand damaging effects over a long period of time.

9. Even though the "transfer length" is small for composites bonded to concrete with epoxy, the anchoring is often the problem especially for T-beams. Strengthening of T-beam is portrayed in **Fig. 4.**

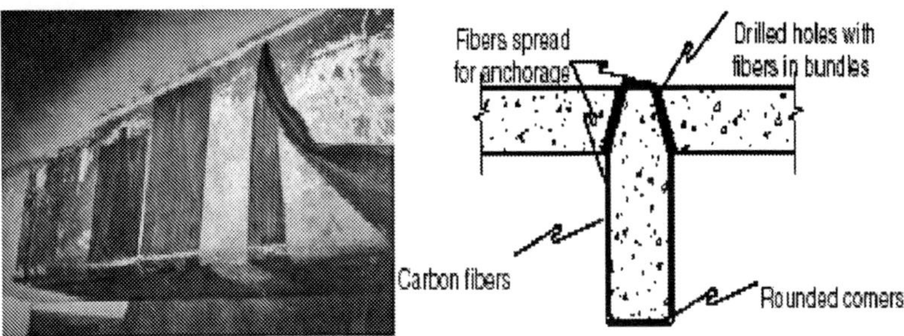

Figure 4.

Case Studies

In addition, advanced composites can be used for strengthening by confinement. Confining of columns can be done by use of pre-fabricated composite shells or by fiber winding. Fiber winding can be done with a dry or wet system. Column wrapping has been extensively used, especially in earthquake regions.

Bridge outside Sundsvall

This bridge was built in 1939 and the columns had a shortage of stirrups. The repair work started with rebuilding of the concrete cover of the column, which had spalled due to corrosion of the reinforcement. Almost 18 Columns were successfully strengthened.

Column strengthened with CFRP to the left and one column with uncovered corroded
steel reinforcement to the right
Figure 5.

Southbound Interstate 95 highway

An impact from a truck damaged the outer girder of a bridge on the Southbound Interstate 95 highway in West Palm Beach, Florida. The truck caused a longitudinal twist in the concrete beam. The owners considered that except for replacement, the only alternative was strengthening with carbon fibre due to the twisted shape of the beam. Replacement was not an acceptable alternative since it required the closure of one lane of the bridge for at least a month, Busel (1995). Therefore carbon fibre was used and work was undertaken in three, five hour-long night shifts.

Hiyosshikura Bridge

The deck of the Hiyosshikura Bridge in Japan was strengthened in 1994 because of demands for higher design loads. Measurements that were undertaken showed that the stresses in the old steel reinforcement decreased by 30-40% after strengthening, Aboudrar and Johansson (1998).

Pre-heater Tower of Fauji Cement Factory

In 2004-2005, the Pre-heater Tower of the Fauji Cement Factory in Jhang Battar, Pakistan was identified as being a deficient design in bending and torsion. It was strengthened with carbon fibre laminates and wrapping with carbon sheets.

References

Engr. Aftab Nadir, Ex- Student MSc, "MSC Thesis on modern retrofitting techniques" U.E.T Taxila.

Bazaa, I. M., M. Missihoun, and P. Labossiere (1996), "Strengthening of Reinforced Concrete Beams with CFRP Sheets," Proceedings of the First International Conference on Composites in Infrastructures (ICCI'96), Tucson, Arizona, USA, January 1996, pp 746-759.

Nanni, A., Di Tomasso, A., and Arduini, M., (1997), "Behaviour of RC Beams Strengthened with Carbon FRP Sheets," Research Report Part II: Continous Beams, University of Bologna, Italy, PP. 1-16.

Tumialan, G., Serra, P., Nanni, A. and Belarbi, A., "Concrete Cover Delamination in RC Beams Strengthened with FRP Sheets, SP-188, American Concrete Institute, Proc., 4rth International Symposium on FRP for Reinforcement of Concrete Structures (FRPRCS4), Baltimore, MD, Nov. 1999, PP.725-735.

Prof. J.G.Teng, Dr.S.T. Smith and Dr. L. Lam "Behavior and Strength of RC Structures Strengthened with FRP Composites" Structural Faults and repair 2001.

A. Aprile, Member, ASCE, S. Limkatanyu and E. Spacone, Engineering Dept., University of Ferrara, Via Saragat 1, 44100 Ferrara, Italy " Analysis of RC Beams Strengthened with FRP Plates" ASCE Structures Conference, Washimgton, DC, May 2001.

K.H.Tan "Durability of FRP System Under Tropical Climate" 28th Conference on Our World in Concrete & Structures: 28-29 August 2003, Singapore.

KL Muthuramu, G Jeyakumar, N Sadish-Kumar, M S Palanchamy " Strengthening of Reinforced Concrete Beam Using External Prestressing." 27[Th] Conference on Our World in Concrete & Structures: 29-30 August 2002, Singapore.

Externally Bonded FRP System for Strengthening Concrete Structures (ACI Committee 440).

"Externally Bonded FRP Reinforcement for RC Structures" Flib CEB-FIP
Technical report on the Design and use of externally bonded fiber reinforced polymer reinforcement (FRP EBR) for reinforced concrete structures. Bulletin – 14.

Design of two curved cable stayed bridges with overlapping girders supported by a single "X" shaped tower, Real Park Complex, São Paulo – Brazil.

Eng. Catão F. Ribeiro, ENESCIL Engenharia de Projetos, São Paulo, BR
Eng. Mauro Lemos de Faria, ENESCIL Engenharia de Projetos, São Paulo, BR
Eng. Heitor A. Nogueira Neto, ANTW Engenharia de Projetos, São Paulo, BR
Arq. João Valente Filho, Valente, Valente: Arquitetos, São Paulo, BR
Eng. Acir Mercio Loredo-Souza, Universidade Federal do Rio Grande do Sul, Porto Alegre, BR

Abstract

The objective of this paper is to describe the Real Park Complex bridge. This structure, conceived by the São Paulo Municipality City Hall and implemented by EMURB (Empresa Municipal de Urbanização – Municipality Urbanisation Company), is aimed at connecting the Jornalista Roberto Marinho Avenue with the Nações Unidas Avenue and the Marginal do rio Pinheiros Avenue, both of which are important city thoroughfares, and comprises 2400m of bridges and viaducts and a 580m cable stayed crossing of the Pinheiros River.

Introduction

The Real Park Complex, conceived by the São Paulo Municipality City Hall and implemented by EMURB (Empresa Municipal de Urbanização – Municipality Urbanisation Company), comprises 2400m of bridges and viaducts and a 580m cable stayed crossing of the Pinheiros River. The complex joins the Jornalista Roberto Marinho Avenue with the Nações Unidas Avenue and the Marginal do rio Pinheiros Avenue, both of which are important city thoroughfares.

The cable stayed bridges are composed of two curved decks with a constant radius of curvature of 275.10m (on the longitudinal deck axis), each of which is composed of two spans, one of 150m over the Pinheiros river and one of 140m over the Nações Unidas Avenue. Near the cable stay tower, the decks are superimposed one above the other with a 12m difference in height.

Bridge design, construction and maintenance 2007, Thomas Telford, London

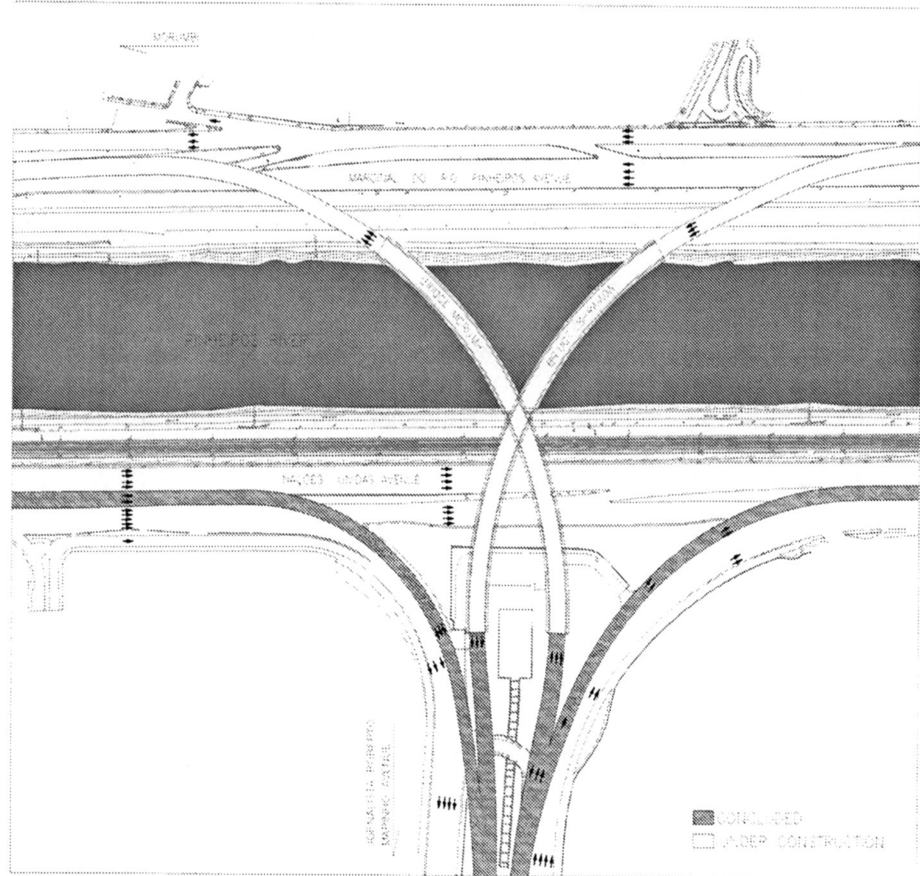

Foundations

The main pier foundation is composed of 112 excavated piles, 130cm in diameter and 40 "root piles" (piles excavated under pressure) 41cm in diameter, separated equally in 4 pile caps. This geometric configuration was necessary due to the many constraints at the work site. Of these the most important were the CTEEP transmission line, the river waterways and the CPTM train lines, the effect of all of which was amplified by the requirement that the main pier should not be built on the Pinheiros riverbed.

The two side piers, at the Jornalista Roberto Marinho Avenue, has its foundations built with 11 excavated piles of 110cm of diameter, and the side pier foundations built near the Pinheiros Marginal are composed of 50 root piles of 41cm of diameter. The option of root piles was selected due to the presence of a transmission line under the pier that did not allow the use of large equipment.

Meso-structure
Main Pier – Cable Stayed Tower

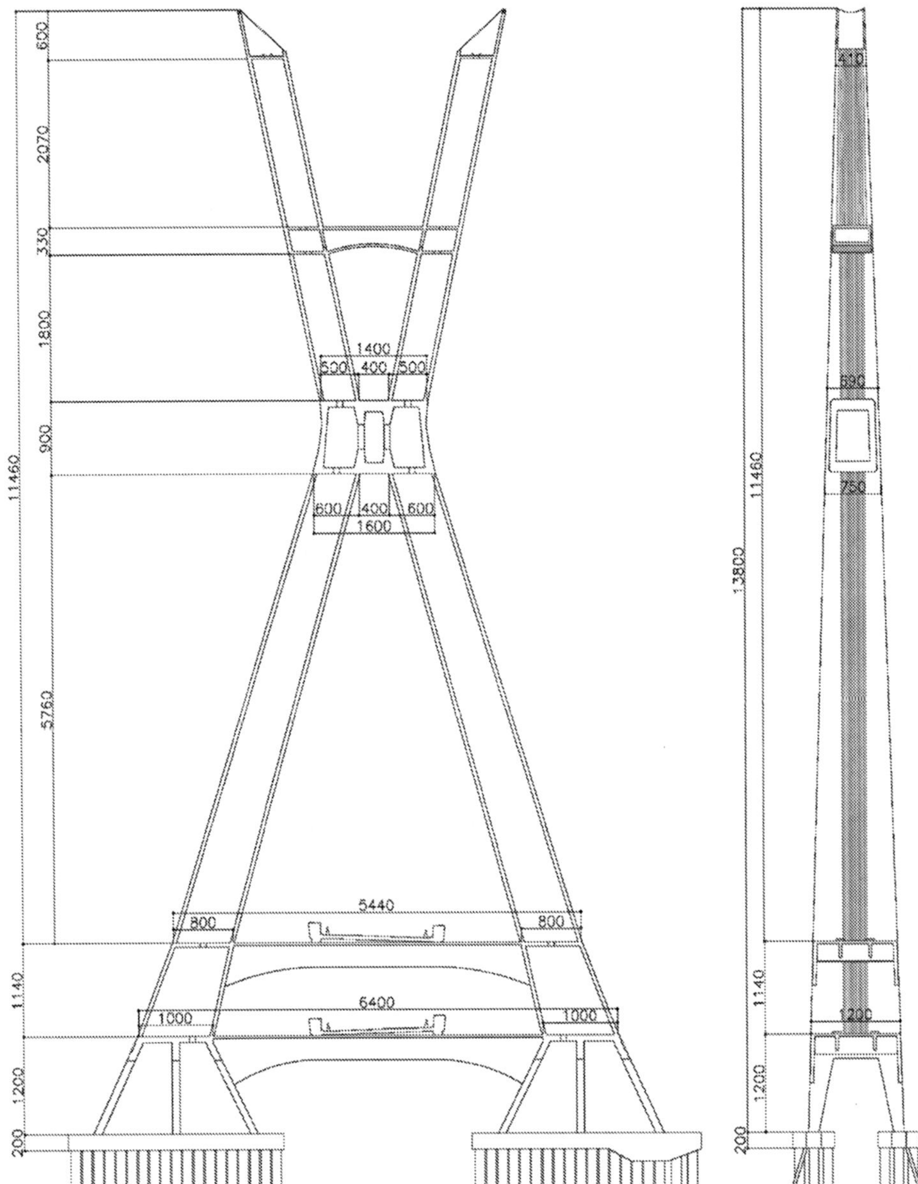

Figure 2. Tower front view.

The main pier is composed of 4 "wall columns" with variable cross section and 12m in height. These columns are transversally restrained two by two by tree beams and a slab, and longitudinally through two beams and a prestressed slab (that compose an inverted "U" cross-section). It is from this altitude that the inclined towers that compose the mast begin.

Figure 3. Tower under construction.

Figure 4. Virtual image isometric view.

These inclined towers, with variable hollow cross section, are 11.4m high where they are longitudinally restrained by a new group of two beams and a slab (again an inverted "U" cross-section). It is important to mention that over these longitudinal beam restraints, the cable-stayed decks are fixed to the main pier.

The two inclined towers start from this second beam restraint. Inclined at a gradient of 1:3 so that their tips converge at the same point, they are 57.6m high and have a variable rectangular hollow cross section. Where they converge a cellular box that is 9m high again restrains the tower tips.

The towers continue at an angle of 1:4.5 for another 42m. Throughout this length (where the stay cables will be anchored), the tower tips diverge, giving the mast an "X" shape. Each of these towers supports a deck and their inclinations were studied and proportioned so as to minimize the torsional effects through their height due to permanent loads. At the middle of these last segments another restraint beam joins the two towers.

Figure 5. Tower under construction.

Figure 6. Virtual maquette front view.

In summary, the main pier will reach 138m height and will consume 5600m^3 of concrete (f_{ck}=30MPa). Even though this height is greater than that of other cable-stayed bridges with the same span, it is necessary to assure the vertical clearance of 6.0m between the vehicles that are passing through the cable-stayed decks and the cables on the side piers.

Side Piers
The four side pier columns have a regular cross section that is 6m wide and varies in thickness from 1.2m (at the top of the pile cap) to 1.6m (at the top of the column). Throughout this height, the corners of this cross-section are rounded to a radius of 60cm.

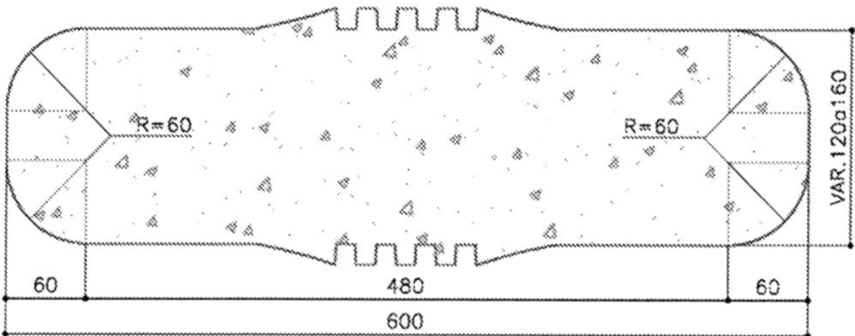

Figure 7. Side pier column cross section.

The transverse beams supported by these columns are composed of just one piece which is fixed to the girder. Because of this, after the deck has experienced all the shortening due to the concrete shrinkage it is estimated that the top of the columns will be displaced by 10cm.

These columns also support the spans that are constructed using pre-cast beams. These pre-cast beams are supported on movable metallic bearings.

The height of these columns varies between 6.5m and 17.2m, consuming almost 400m³ of concrete in their construction.

Cable Stays

The cable stays have a three-dimensional arrangement since the decks are curved on plan and the towers are inclined. There will be 18 pairs of cables supporting both spans of both decks, totalling 144 cables.

Figure 8. Virtual maquette top view.

The cables that support the span over the Pinheiros River are set at intervals of 7.0m along the girders, and the cables that support the span over the Nações Unidas Avenue are set at intervals of 6.5m.

Near the mast, on the riverside, the cables will entwine because they start at one of the inclined towers and end at the deck placed on the opposite side. Because of this, the intervals between the cable anchorages along the height of the tower vary from 4.5m to 1.3m, in such a way as to avoid any collision between them.

Each cable is composed of a group of parallel strands (composed of 9 to 22 strands), fixed to the girders by an adjustable anchorage and to the mast by a fixed anchorage. The strands are individually protected from corrosion by being galvanized and then covered in an HDPE (High-Density Polyethylene) sheath filled with wax.

The HDPE ducts cover the full length of the cables but in addition an anti-vandalism tube will be placed along bottom 3.0m length adjacent to the girder.

In summary, there will be used 374,350 m of strands, totalling 462 tons of steel (50kg/m² of deck).

Superstructure

The deck is built in concrete (f_{ck} = 40MPa) and comprises two inverted girders providing an overall width of 16m (Two side girders of 1.5m width, two 85cm wide pedestrian walkways, two 40cm wide concrete guard rails and a 10.5m carriageway). The girders, with an average height of 1.4m, support a 48cm of thick, transversally prestressed slab.

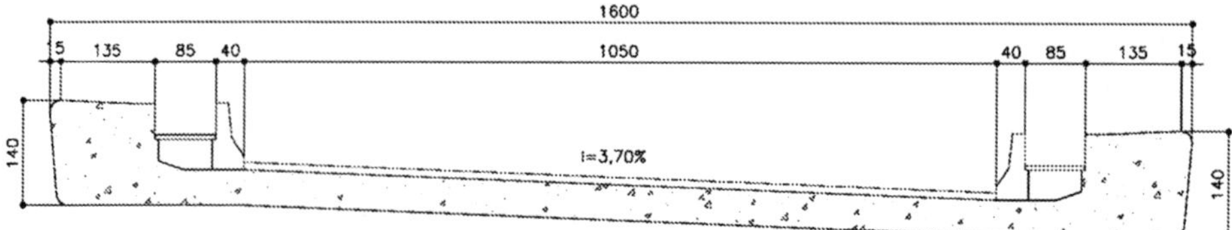

Figure 9. Bridge deck cross section.

The deck will be constructed using the free cantilever method with 7.0m long segments over the river, 6.5m long segments over the avenue and 2.5m long key segments, except for the first segment, over the transverse restraint which will be 29m long and will be built over a shoring system.

Due to the constant deck curvature, the segments are distinguished from each other only by the location of the housing tube and the outline geometry of the anchorage.

Once the steel temporary works have been positioned over the first segment, the formwork will be prepared, the steel bars will be positioned in the segment, together with the housing tubes and anchorage details After this, the concrete will be poured. When the concrete has reached the specified resistance, the segment will be prestressed longitudinally with tendons of 7, 9, 12 and 15 strands of Ø15.2mm, and the slab will be transversally prestressed with tendons at intervals of 40cm, each tendon being composed of 4 Ø15.2mm strands. After prestressing each tendon, the truss will be moved forward, while the cable tensioning and the preparation of the next segment will are completed. The HDPE duct will the be installed by threading the strands through them, after which they are tensioned.

After the key segments have been finalised and before the construction of the guard rails the pavement and the decks will be transversely prestressed with 19 strand tendons, Ø15.2mm. It is worth noting that this prestress was intended to counteract the flexure in both the vertical and horizontal directions.

Following the construction of the pavements, the cable stays will be re-tensioned, so that the bridge can be opened to the traffic.

Status of construction
At the present time the two viaducts connecting the Nações Unidas Avenue and the Jornalista Roberto Marinho Avenue have been completed and are in operation (See figure 1).

With respect to the bridges over the Pinheiros River, the foundations have been completed without problems except for the interferences on site. The tower has reached the segment where the cables are anchored and should be concluded in about three months time. The cable stayed decks have had their first segments erected over the restraints and one segment has been completed in each of the four side spans, the third segment of each span is under construction.

So far the worst problem that has been confronted is the vertical deformations of the temporary trusses under the wet concrete load. These deformations have proved to be larger then predicted by the truss supplier, and the deck has therefore dropped a few centimetres below that specified. Current studies aim to correct these deformations by adjustment of the next segments pulling the deck back into the original vertical layout.

Figure 9. Status of construction.

Conclusion

The need to establish a connection between three main thoroughfares within the São Paulo Municipality, and the complexity of the local interfaces, means that this project represents an opportunity to produce a structure of great architectural appeal.

Due to this visual appeal and the structural complexities inherent in both the management of such a project and the execution of the construction activities, this structure has all the qualities necessary for it to become on the great architectural monuments of the city, especially because of its location in one of the most valuable municipality regions.

New Developments in Underwater Pile Repair Using Fiber Reinforced Polymers

Dr. Rajan Sen, Dept. of Civil & Environmental Engineering, USF, Tampa USA
Dr. Gray Mullins, Dept. of Civil & Environmental Engineering, USF, Tampa USA

Abstract

The development of resins that can cure in water has led to interest in the possible application of fiber reinforced polymers (FRP) for the underwater repair of corrosion-damaged piles. Over the past four years, USF have completed four demonstration projects in which FRP was used to repair corroding piles. The FRP material was directly applied to the submerged, partially wet or dry concrete surface. Fiberglass and carbon were both used in conjunction with two types of resin - a water-activated urethane resin and an epoxy resin. In all the projects, selected piles were instrumented to allow evaluation of the post-wrap FRP corrosion mitigation performance. No particular problems were encountered in wrapping the partially submerged piles. Linear polarization measurements indicated lower corrosion rates in the FRP wrapped specimens in comparison to their unwrapped counterpart. This paper provides an overview of the studies and highlights some of the important lessons that were learned.

Introduction

Corrosion of reinforced or prestressed piles in tidal waters in sub-tropical environments is a world-wide problem. The combination of wet/dry cycles in a marine environment and high temperature/high humidity can be lethal since it allows chloride ions to readily penetrate the concrete cover and destroy the passive layer that normally protects steel in concrete's alkaline environment. Subsequent electrochemical reactions can cause steel to corrode rapidly, sometimes within eighteen months of service.

Traditionally, corroded piles are repaired using methods such as *'chip and patch'*, *shotcreting*, *epoxy injection* and *pile jacketing* (Fig. 1). Typically, chloride-free concrete is placed directly adjacent to chloride contaminated concrete. This sets up new corrosion cells that lead to more corrosion. The wisdom of following procedures that result in costly re-repair is increasingly under question.

Over the past decade, several highway authorities in Canada and US have explored the possibility of using FRP for corrosion repair[1]. In most applications no attempt was made to remove the chloride contaminated concrete and the wrapping was conducted under dry conditions. The availability of new resins that cure under wet conditions has spurred interest in extending FRP repair to partially submerged piles. Some full-scale FRP pile repairs were recently reported[2, 3]. However, as the piles were not instrumented the performance of the FRP in reducing corrosion cannot be quantitatively assessed from these studies.

Bridge design, construction and maintenance 2007, Thomas Telford, London

Figure 1. Jacketed pile repair (left) and view with jacket removed (right)

The University of South Florida (USF) have completed four demonstration projects in which corroding reinforced and prestressed piles were repaired using FRP. These projects were the culmination of several research studies that investigated the role of FRP in corrosion mitigation applications[4-6]. Several of the wrapped and unwrapped piles were instrumented to allow assessment of long term performance (these results are, however, not included in this paper). Moreover, as the sites for the repairs differed, alternative instrumentation approaches had to be explored. This paper provides a brief overview of the projects. Complete descriptions may be found elsewhere[7-9].

Background

The external bonding of FRP in underwater applications is more involved than similar repairs carried out under dry conditions. Changing tides, winds, wave action create unfavorable conditions that need careful attention. For this reason, existing application methods developed for repair under dry conditions have to be modified and new techniques developed to minimize the impact of the potentially adverse conditions.

As repair involves restoration of strength, bond is an important consideration. As with any bonded application, surface preparation is critical; good bond requires the substrate to have an open pore structure to ensure capillary suction of the epoxy[10]. In underwater application, however, pores will be saturated with water or small marine organisms or algae that are likely to adversely impact bond. New surface preparation techniques and new measures, e.g. bonding agents, may be required to ensure satisfactory performance.

FRP will only be used for repair if it leads to lower initial costs and better performance in comparison to traditional repairs. Thus, pile repairs need to be properly engineered so that FRP use is optimized. This requires reliable information not only on the capacity that must be restored but also on the relationship between corrosion and expansion. Available experimental data obtained from accelerated corrosion tests may not be applicable. Whereas laboratory studies result in symmetrical cracking, the combined action of wind and waves result in uneven chloride penetration and asymmetric cracking in piles. More importantly, the solubility of corrosion products from accelerated testing may differ from that under field conditions. Lack of information on expansion introduces uncertainty in design.

Finally, constructability and safety are of prime importance. The most careful attention has to be paid to all details so that the wrap is carried out safely and expeditiously.

Designing FRP Repairs

The design of the FRP wrap requires the same information needed for conventional repair, namely, an estimate of the capacity loss that has to be recovered. Additionally, information is required on corrosion expansion that has to be accommodated by the FRP; several researchers have independently demonstrated[1] that corrosion continues as long as the ingredients required for the electro-chemical reactions remain inside the wrap. Thus, the FRP material simply serves as a barrier element reducing future ingress of deleterious materials such as chlorides and moisture.

The role of FRP in pile repair is therefore twofold: first to restore lost flexural capacity due to corrosion of steel; second to provide resistance to withstand expansive forces caused by corrosion products. The former requires fibers to be oriented parallel to the direction of the reinforcing or prestressing steel, i.e. along the length, while the latter requires fibers in the transverse or hoop direction, i.e. perpendicular to the steel. This can be met either by using two different sets of uni-directional fibers - one for each direction or preferably, by using bi-directional FRP material.

Research has shown that there is no simple relationship between metal loss and strength. This uncertainty stems from its dependence on other factors such as the bond between steel and concrete, the confinement provided by the ties and ductility reduction in steel due to corrosion. Given this uncertainty, it may be prudent to assume a safe, conservative value. This was the case in the demonstration projects where capacity loss was taken as 20%.

As noted already, at the present time there is a lack of reliable information on transverse expansion caused by corrosion. In view of this, a simplified procedure[8,11] was developed. For design, the expansion strain was assumed as 0.1% (approximately 3 times the ultimate concrete tensile strain assumed to be 10% of the ultimate concrete failure strain).

A simplified two step design procedure was developed in which the requirements for strength and corrosion expansion were uncoupled. The strength requirements were first met by strain compatibility analysis in which the contribution of the FRP to the tensile (but not compressive) capacity were determined through interaction diagrams for the assumed shortfall in strength. The confining lateral strain provided by this strengthening was then checked to ensure that it satisfied the maximum assumed expansion of 0.1%.

Results obtained were found to be reasonable when compared to solutions from competing methods developed for seismic retrofit[8,11]. For bi-directional carbon fiber, only two FRP layers were required. Twice as many layers were required for fiberglass because of its lower strength and stiffness.

FRP Materials

Two types of materials were used. One was a pre-preg, the other a wet lay up. The pre-preg system was from Air Logistics in which all the FRP material was cut to size, resin-saturated in the factory and sent to the site in hermetically sealed pouches. The wet lay up system required on-site impregnation and was from Fyfe excepting for the Allen Creek Bridge where an alternative system was used. As this repair was carried out under dry conditions inside a coffer dam[8] it is not included here. Both carbon and fiberglass were used. Details of the properties of the fiber and the resin as provided by the suppliers are summarized in Tables 1-2[12-13].

Table 1. Properties of Pre-Preg System [12]

Fibers	Tensile Strength (MPa)	Tensile Modulus (GPa)	Load per Ply (kN/m)
Uni-directional (GFRP)	590	36	420
Bi-directional (GFRP)	320	21	210
Uni-directional (CFRP)	830	76	596
Bi-directional (CFRP)	590	22	420

Table 2. Properties of Wet Layup System [13]

Properties	Quantities
Tensile Strength	3.24 GPa
Tensile Modulus	72.4 GPa
Ultimate Elongation	4.5 %
Laminate Thickness	0.127 cm
Dry fiber weight per sq. yd.	915 g/m^2
Dry fiber thickness	0.038 cm

Field Demonstration

Four field demonstration studies were conducted at two contrasting sites. The first site, Allen Creek Bridge, Clearwater, FL is located in shallow, relatively calm waters; the subsequent repairs were carried out on piles supporting the Friendship Trails Bridge and the Gandy Bridge that span Tampa Bay, Florida's largest estuary that flows into the Gulf of Mexico. The procedure for wrapping under water was identical to that above water. Usually, wrapping was scheduled for low tide when the submerged region could be readily accessed. On rare occasions when the water level was higher, diving masks were used.

Allen Creek Bridge

Allen Creek Bridge is located on the busy US 19 highway connecting Clearwater and St. Petersburg, FL. The original bridge built in 1950 was supported on reinforced concrete piles driven into Allen Creek. In 1982, the bridge was widened and the widened section was supported on 35 cm square prestressed piles.

The waters from Allen Creek flow east into Old Tampa Bay that in turn joins the Gulf of Mexico to the south. The environment is very aggressive; all the reinforced concrete piles from the original construction had been rehabilitated several times. At low tide, the water level in the deepest portion of the creek is about 0.8m. At maximum high tide it is about 1.9m. This shallow depth meant that the underwater wrap did not require divers and could be carried out on a ladder.

Instrumentation was installed to allow linear polarization and corrosion potential measurements to be made[4] without junction boxes and wiring. Cores were taken from various locations to determine the chloride content at the level of the prestressing strands. Results indicated that the strands were corroding though there were no visible signs of corrosion.

Surface Preparation

Pile surfaces were covered with marine growth that had to be scraped off. Additionally, two of the four corners that were not rounded but chamfered had to be ground using an air-powered grinder. This was a difficult operation particularly for sections that were below the water line. Quick-setting hydraulic cement was used to fill any depression, discontinuities and to provide a smooth surface. The entire surface was pressure washed using freshwater to remove all dust, marine algae just prior to wrapping.

Figure 2. Patching damage (left) and grinding surface (right).

FRP Wrap

Two different schemes using two different materials were evaluated. In each scheme four piles were wrapped with two other instrumented piles serving as controls. In the first scheme, cofferdam construction was used and the piles wrapped using a bi-directional FRP in wet layup under dry conditions. Complete details may be found in Ref. 8. As this was wrapped under 'perfect' conditions, its performance provided a means for evaluating piles that were directly wrapped in water using the water activated urethane resin.

Post-wrap corrosion measurements showed that the corrosion rate was smaller for the wrapped piles compared to the unwrapped controls[4,8]. However, rates are still small.

Tampa Bay Bridges

The Friendship Trails Bridge and the Gandy Bridge are two of four bridges that cross Tampa Bay. When it was first opened to traffic in the 1920's it was the longest bridge in the United States reducing the distance between Tampa and St. Petersburg by 72 km. The environment is very corrosive and the majority of the piles supporting these bridges have sustained severe corrosion damage and needed to be repaired. Three demonstration studies were carried out on these piles over the period 2004-06.

All piles repaired were 50 cm x 50 cm either reinforced or prestressed concrete piles. They were all located in waters that are approximately 4.9 m deep. This meant that ladders could no longer be used in this situation. An innovative scaffolding system (Fig. 2) was designed and fabricated for the study. It was lightweight, modular yet sufficiently rigid when assembled to support 4-6 people. The scaffolding was suspended from the pile cap and extended 2.74 m below the pile cap. Its mesh flooring provided a secure platform around the pile that allowed the wrap to be carried out unimpeded in knee deep waters without the need for diving gear.

Two different FRP systems were used. One was the water-activated, pre-preg urethane resin system used in the Allen Creek Bridge. The other system used a special epoxy that could cure in water. The first system was used to wrap four piles – two using carbon and two using glass. In the second system, the FRP material had to be pre-impregnated on site prior to the wrap. Two piles were wrapped with fiberglass using this system. Of the two, one was an experimental FRP system that combined wrapping with a sacrificial cathodic protection system. Two other unwrapped piles in a similar initial state of disrepair were used as controls to evaluate the performance of the wrapped piles.

Figure 3.   Installation vertical (left) transverse layer (right) typically at low tide

As for the piles in the Allen Creek Bridge, several piles were instrumented but using a special rebar probe developed by the Florida Department of Transportation[7]. This allowed measurement of the corrosion current between the locations where they were located. Reductions in current compared to the controls were expected to provide a measure of the efficacy of the FRP wrap. Unlike, Allen Creek Bridge, junction boxes and wiring were required to allow readings to be taken. Preliminary corrosion measurements were however inconclusive because the system had not yet stabilized following the installation of the probes.

The FRP material for the first system was identical to that used in Allen Creek Bridge, i.e. unidirectional longitudinal layer followed by transverse bi-directional material. For the second system, all material was uni-directional. The scaffolding system enabled the wrapping to be carried out with relative ease.

On average, it took between 30 minutes and 45 minutes to wrap a pile depending on the number of FRP layers. For the wet-lay-up system, it took additional time to impregnate the FRP material and transport it to the pile site. The total time taken averaged about 90 minutes in this case.

Fig. 3 shows the installation of the longitudinal layer at the corners followed by the transverse layer that was applied spirally with no overlap. Fig. 4 shows a view of the wrapped piles and a close-up view more than two years later showing marine deposition on the wrap.

Figure 4.   View of FRP wrapped piles when new (left) and after 2 years (right)

New Developments

The results from the initial repairs carried out on the piles supporting Allen Creek Bridge and the Tampa Bay bridges showed that it was feasible to wrap corroding piles under typical field conditions. Corrosion rate measurements also indicated that the FRP wrapping led to a reduction in the corrosion rate. However, the FRP-concrete bond was found to be poorer when compared to the bond that was obtained under laboratory conditions[8,14].

Laboratory studies were undertaken over the past year to evaluate alternative strategies used by the aerospace industry to improve the underwater FRP-concrete bond. Two candidate techniques widely used in aerospace (1) vacuum bagging and (2) pressure bagging were evaluated. Prototype systems were developed and the bond evaluated both non-destructively and destructively[9]. Both techniques worked but in the subsequent field demonstration, pressure bagging was used. Qualitatively, the bond appeared to be much improved though no on-site destructive bonds have yet been conducted. Such tests are planned in the future.

The USF research team has just received funding for implementing a pilot sacrificial cathodic protection system. This is expected to be implemented over the next several months at the same Friendship Trails Bridge site.

Conclusions

This paper provides a brief overview of demonstration projects in which instrumented reinforced and prestressed piles were wrapped using different FRP systems. Initially, wrapping was carried out directly under water using ladders. Subsequently, a lightweight scaffolding system was developed that could be suspended from the pile cap. With all-around access, FRP wrapping was conducted with ease.

Based on our experience, it is clear that it is feasible to use FRP for underwater corrosion repair. The performance of the FRP in slowing down the corrosion rate appears to be comparable to that determined from laboratory testing. Visual inspection suggests that the repairs have held up well with the longest over four years old at this time. More detailed examination may be possible in the future when the Allen Bridge is replaced. However, on-site pull-out tests on several piles repaired in the initial phase have shown that the FRP-concrete bond was poorer than that obtained under dry laboratory conditions.

A pressure bagging technique was developed and successfully implemented last year to address this variability in bond. While no on-site pullout tests have yet been conducted, visual examination suggests that the bond was better. Efforts are currently underway to incorporate a sacrifical cathodic protection (CP) system within the wrap that can stop corrosion. With all the development work carried out, the prospects of FRP emerging as an alternative cost effective pile repair system appear bright.

Acknowledgements

The demonstration studies reported were funded by the Florida/US Department of Transportation and Hillsborough County. The support and guidance of Mr. Pepe Garcia and Mr. Steve Womble, both from Florida Department of Transportation are gratefully acknowledged. We thank Ms. Mara Nelson, and Mr. Nils Olsson, both from Hillsborough County, and SDR Engineering, Tallahassee for their contribution. We thank Mr. Franz Worth of Air Logistics and Mr. Ed Fyfe, President, Fyfe for their support. This work could not have been carried out without the tireless effort of current and former graduate students Dr. K S Suh, Dr M. Stokes, Mr. I Gualtera, Mr. D. Winters, Mr. Julio Aguilar and Mr. Andy Schrader.

References

1. Sen, R. "Advances in the Application of FRP for Repairing Corrosion Damage", Progress in Structural Engineering and Materials, 2003, Vol. 5, No 2, pp. 99-113.
2. Bazinet, S., Cercone, L. and Worth, F. "Composite FRP Moves into Underwater Repair Applications", SAMPE Journal, 2003, Vol. 39, No. 3, pp. 8-16.
3. Watson, R.J. "The Use of Composites in the Rehabilitation of Civil Engineering Structures", ACI SP 215 (Ed. S. Rizkalla and A. Nanni), ACI, Farmington Hills, MI, 2003, pp. 291-302.
4. Mullins, G., Sen, R., Suh, K and Winters, D. "Underwater FRP Repair of Prestressed Piles in the Allen Creek Bridge", ASCE, Journal of Composites for Construction, 2005, Vol. 9, No. 2, pp.136-146.
5. Mullins, G., Sen, R., Suh, K. and Winters, D. "A Demonstration of Underwater FRP Repair", Concrete International, 2006, Vol. 28, No. 1, pp. 1-4.
6. Sen, R., Mullins, G., Suh, K. S. and Winters, D. "FRP Application in Underwater Repair of Corroded Piles". ACI SP 230 (Eds. C. Shield, J. Busel, S. Walkup, D. Gremel), 2005, Vol. 2, pp 1139-1156.
7. Mullins, G., Sen, R., Suh, K and Winters, D. "Underwater Pile Wrap of the Friendship Trails Bridge", Final Report, Hillsborough County, June 2004, 32 pp.
8. Mullins, G., Sen, R., Suh, K.S. and Winters, D. "Use of FRP for Corrosion Mitigation Applications in a Marine Environment", Final Report submitted to Florida / US Department of Transportation, Tallahassee, FL October 2005, 406 pp.
9. Mullins, G., Sen, R., Winters, D. and Schrader, "Innovative Pile Repair", Final Report submitted to Hillsborough County, FL, January 2007, 40 pp.
10. Emmons, P. Means, P. *Concrete Repair and Maintenance Illustrated*, RS Means, Kingston, MA, 1993.
11. Sen, R., Mullins, G. and Shahawy, M. (2007). "FRP Repair and Strengthening of Structurally Deficient Piles". To appear in Journal of Transportation Research Board.
12. Air Logistics Corporation. Aquawrap Repair System, Pasadena, CA.
13. Fyfo Co. LLC, http://www.fyfeco.com/
14. Sen, R. and Mullins, G. "Application of FRP for Underwater Pile Repair", Composites Part B, 2007, Vol. 38, No. 5-6, 751-758.

Rehabilitation Design of a Bascule Bridge – Including Jacking Up an 800-Ton Bascule Leaf

Yuhe Yang, PE, SE, PB Americas, Inc., Seattle, Washington, USA
William B. Iyall, PE, City of Tacoma, Washington, USA
Eric R. Kelley, PE, PB Americas, Inc., Seattle, Washington, USA
Kongsak Pugasap, PB Americas, Inc., Seattle, Washington, USA

Abstract

The 904-foot Hylebos Waterway Bridge was constructed in 1939 and spans the Hylebos Waterway in Tacoma, Washington (USA). It consists of east and west approaches and a 290-foot double-leaf trunnion bascule span crossing the waterway.

This bridge was inoperable after a drive shaft on the east bascule leaf failed in early 2001. In 2004, a fire damaged the main trunnion girder and bearings inside the east bascule pier.

This project includes replacing both approach structures and restoring the bascule main span to operational condition. Project components include:
- Replacing the fire-damaged trunnion bearing
- Modifying the pinion support system to meet current design code
- Seismically retrofitting the bascule leaf structures
- Overhauling the bascule mechanical driving and electrical control systems; and
- Replacing both approaches.

Environmental regulations restrained construction activities, and because the bascule leaf weighed approximately 800 tons, replacing the bearings was a challenge. A tight in-water construction window and weak structure layer in the bascule pier's lower portion further complicated the repair effort. The design team developed unique solutions to these challenges and gained valuable experience in rehabilitating movable bridges.

Project Description

The Hylebos Waterway Bridge (see Figure 1) is located in the East Blair Peninsula in the City of Tacoma, State of Washington, United States. This bridge crosses the Hylebos Waterway and provides a second access to and from the peninsula. In early 2001, a drive shaft on the east bascule leaf failed, causing the bridge to become inoperable. A fire in late 2004 inside the east bascule pier caused additional damage to the bridge, including its main trunnion girder and bearings. This project's objective is to provide peninsula residents with an emergency evacuation route. A key component of this route is the Hylebos Bridge. The project will replace both aging approach structures and restore the bascule main span back to operational condition.

Bridge design, construction and maintenance 2007, Thomas Telford, London

Existing Bridge and its Deficiencies

The existing Hylebos Bridge was constructed in 1939. It consists of three principle elements: the east approach, west approach, and the bascule span that crosses the Hylebos Waterway. The entire bridge is approximately 904 feet long (see Figure 2).

The bascule span is a double-leaf trunnion bascule structure. This span has an overall length of 290 feet, including 216 feet 10 inches from the east trunnion's centerline to the west trunnion's centerline, and 36 feet 7 inches of fixed deck over each bascule pier. The bridge deck width is 37 feet 6 inches.

Figure 1: Hylebos Waterway Bridge
Tacoma, Washington, USA

Each bascule leaf consists of two main trusses and a solid concrete deck. The east approach is approximately 344 feet long and constructed of wood piling that supports solid sawn timber stringers supporting a concrete deck. The west approach was reconstructed in 1990, is approximately 344 feet, and constructed of wood piling that supports glue-laminated timber floor beams supporting a wood deck.

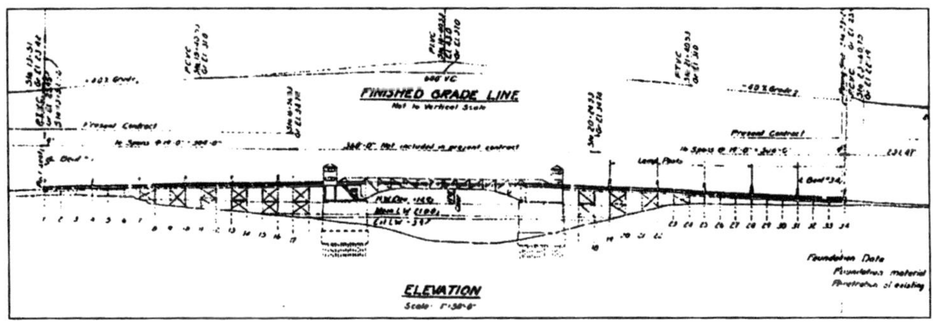

Figure 2: The Existing Hylebos Bridge

At the beginning of this project, a week-long on-site bridge inspection was conducted. Numerous structural deficiencies were either found or confirmed during this inspection. Some deficiencies had more severe impacts on the structure than others. The major deficiencies described in the inspection report follows.

East Approach Deficiencies

The east approach deck, timber stringers, sidewalk, and guardrails are all deteriorated. Pile caps from bents 23 to 34 are deteriorated. The approach is deficient for seismic forces.

West Approach Deficiencies

The west approach deck, sidewalk, guardrails, and pile caps need to be replaced.

Bascule Span Deficiencies

In addition to the broken drive shaft inside the east pier (see Figure 3), severe fire damage inside the east bascule pier was discovered and recorded during the inspection. The paint protecting the steel on the bascule span has numerous rust spots. The pier walls supporting the bascule have many spalls and cracks. The concrete deck has spalls and exposed rebar. The expansion joints on the deck are no longer functioning. The guard rails are deteriorated.

Figure 3: Broken Shaft of the East Bascule Leaf

Unique Challenges

The complexity of bridge rehabilitation projects varies. Rehabilitation of movable bridges is usually more complex than fixed-span bridges because of the more complex design load combinations, more strict requirements on the deflection control of certain structural elements, and the interaction between the moving structural members and the driving machinery.

This project presents a number of design challenges. Some of these challenges were not initially expected and discovered during the different design stages, which is not unusual for a complex bridge rehabilitation project.

Construction Restrictions

Environmental regulations put restraints on this project. Some main constraints include the following:

- Driving steel pipe piles is not allowed without a formal consultation process for permitting;
- The numbers of precast prestressed piles on each approach side are limited to approximately ten maximum;
- Open excavation in the water is not allowed without a formal consultation, and the possibility of approval is minimal; and
- Barge access during construction at the shallow water is limited, due to the environmental concern of the possibility of grounding.

Navigational requirements also limit the bascule leave-down time to nine days, including Labor Day weekend.

Seismic Retrofitting

Seismic retrofitting this bascule bridge poses the following unique challenges:

- The bascule counterweight adds significant weight to the superstructure;
- Connections at the bascule leaf supports are rigid without ductility, which does not provide the desired seismic force reduction through the structure member ductility behavior; and
- A hollow layout between the bascule pier and its footing provides very limited lateral resistance capacity.

Fire Damage

Severe fire damage inside the east bascule pier was discovered and recorded during the initial on-site inspection at the beginning of this project. The fire occurred in late 2004. This damage includes:

- Total loss of two electrical motors (see Figure 4);
- Bending or warping of steel members that support the machinery floor below (see Figure 5);
- Warping of the trunnion girder that supports the bascule leaf (see Figure 6);
- Misalignment of the trunnion bearings (see Figure 7); and
- Slight deformation of the bascule leaf truss members inside the bascule pier.

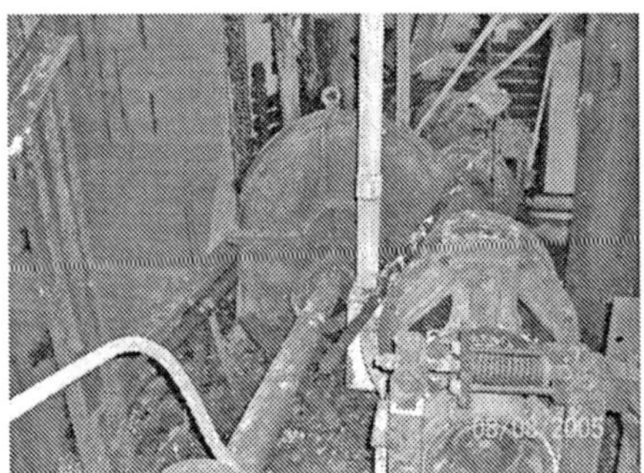

Figure 4: Damaged Motors

Figure 5: Fire Damage on Hangers

Figure 6: Warped Trunnion Girder

Figure 7: Bearing Fire Damage

Excessive Deflection

During the design process, a check by the engineers discovered that the machinery support system's existing design does not meet the current design requirement in the American Association of State Highway and Transportation Officials (AASHTO)'s Load and Resistance Factor Design (LRFD), "Movable Highway Bridge Design Specifications".

The existing bascule leaf driving machinery system is supported by a steel-framed floor slab inside the bascule pier. One side of the machinery floor slab is attached to the trunnion girder, and the other side is supported by steel hangers attached to the roadway deck above (see Figure 8).

Engineering calculations revealed that support by using hangers attached to the roadway deck stringers above is too flexible. Under the combined factored loading from operation of machinery and dynamic load allowance, the misalignment of the bearings and gearing that result from excessive deflection is beyond the tolerable limit. This is because it would compromise the machinery's operational life and result in premature failure (see Figure 9).

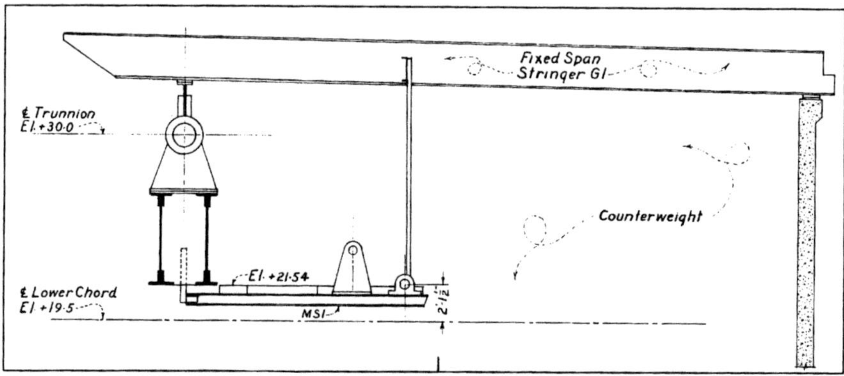

Figure 8: Existing Bascule Leaf Machinery Support

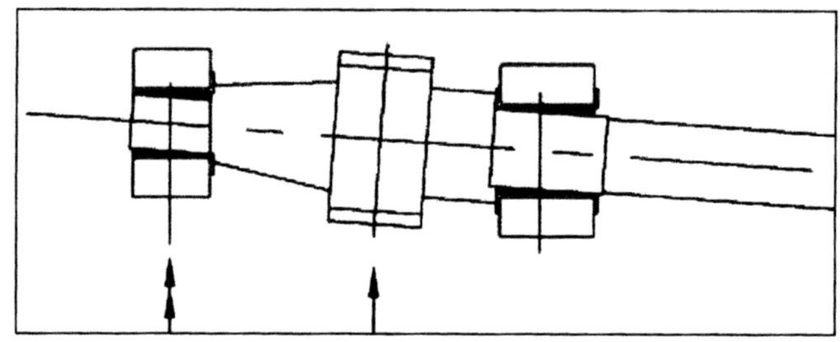

Figure 9: Excessive Deflection Causing Binding at the Bearings

Design Solutions

To rehabilitate a movable bridge that poses these unique challenges, unconventional design approaches and techniques were developed during the design process.

Communication

One of the major elements in the design process has been effective communication between the client, the project stakeholders, and the engineering design team. A good rehabilitation design should, at a minimum, consider the following:

- The client (user)'s functional needs;
- Applicable design code requirements;
- Available funding for the project;
- Constructability; and
- Construction schedule.

Throughout the design period, a series of design meetings were held between the clients and the design team to discuss needs vs. available funding, constructability and schedule, etc.

This communication helped identify the project constraints and established the client and the design team's common understanding of project needs and practical objectives based on the available funding. This set the project on the right course.

Using Existing Approaches as Construction Trestles

Early discussions with the environmental consultant and environmental permitting review agencies revealed that it would be difficult to obtain approval for pile driving in the water due to concerns about impacts on fisheries. The permitting review for pile driving would require a lengthy formal consultation process and would delay the project, consequently impacting other scheduled projects.

Avoiding pile driving meant that conventional construction trestles would not be a viable option for this project. Through a series constructability studies and discussions with construction specialists, a staged construction schedule was laid out. This schedule would allow the contractor to finish rehabilitation of the bascule span by using existing approaches as construction trestles for transporting construction material and personnel, and as part of temporary staging areas (before demolition).

Construction of the new approach foundations would also use the existing approach as construction platforms. Holes would be cut to open on the existing bridge deck, and the drilled shafts would be constructed through the open hole on the deck.

During demolition of the existing bridge deck, subsequent construction of the new bridge approaches, rehabilitation of the bascule piers, and construction to support areas adjacent to the bridge may be needed. To avoid pile driving that would have negative environmental impacts, the constructor could use floating construction platforms such as Flexifloat sectional barges.

Tailored Seismic Retrofit Approach

Because they are constrained by existing conditions, seismically retrofitting existing bridges is usually more challenging than designing new bridges. Given the unique difficulties described previously, seismically retrofitting movable bridges is even more of a challenge.

For bascule bridges, the bascule leaf trunnion's supports and connections are the critical components in the load path for seismic induced forces. From a seismic design point of view, the bascule leaf support structures are rigid and lack ductility. Therefore during a major earthquake event, the forces exerting on the trunnion bearing supports would be tremendously large which would cause difficulties in seismic retrofit design.

A hollow structural layer in the bascule pier, sandwiched between the counter-weight pit floor and the top of the footing, added another level of difficulty to the design. To seismically retrofit (strengthen) the bascule pier's substructure, extensive underwater soil excavation would be expected. This would certainly trigger a higher level of the environmental permitting review process, and consequently delay the start of the project. Retrofitting the bascule substructure's underwater portion would also cost more than the funding available to this project.

To solve these problems, the design team considered the possibility of retrofitting this bascule span in two phases. Current AASHTO design specifications require that new bridges be designed for an earthquake level with 475-year return period. After revaluating design options and considering this AASHTO requirement, the design team proposed the following approach:

- Replace the two bridge approach structures and design them for a seismic event of a 475-year return period, in accordance with AASHTO requirements.

- Partially seismically retrofit the bascule structure to a seismic design level of a 108-year return period, with the probability of exceeding 20% in 25 years. This 20% probability of exceedance is approximately equivalent to a seismic event of a 475-year return period in 100 years, which is a proper level for this 1930s-era bridge.

- Carry out the next phase of seismic retrofit on the bascule span when additional funding is available.

The project team proposed this design approach, which the client accepted. Figures 10 and 11 show the retrofit details that provide lateral support at the trunnion bearings.

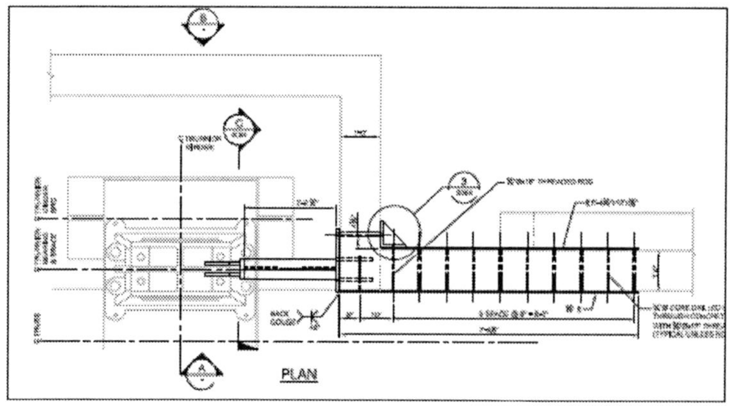

Figure 10: Plan View of the Seismic Restrainer Attached to Trunnion Bearing

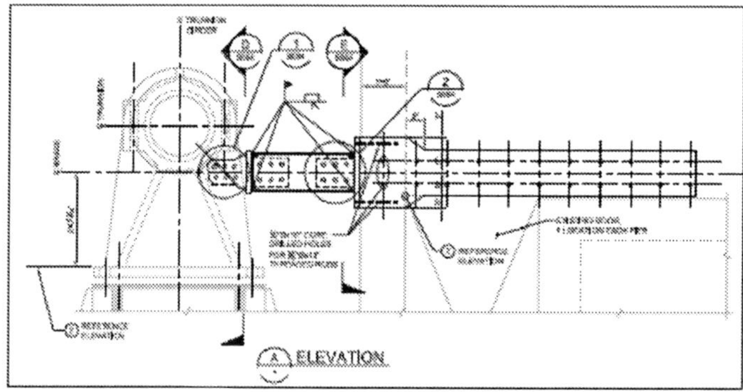

Figure 11: Elevation View of the Seismic Restrainer Attached to Trunnion Bearing

Replacing and Re-Aligning the Bascule Leaf Supporting Bearings

The fire mentioned previously caused severe damage inside the east bascule leaf. In addition to the motors and steel hangers needing to be replaced (see Figures 4 and 5), the fire caused a slight warping on the trunnion girder (see Figure 6). As a precautionary measure, small steel coupons were cut from the steel members once heated by the fire and tested in a lab. The results indicated that the fire had not changed the steel's metallurgical property.

Although the warping in the trunnion girder is small for such a large girder, it caused enough misalignment of the trunnion bearings (an approximately 0.1-degree horizontal offset between the two bearings) that the mechanical engineers on the team raised the "red flag". A field inspection of the bearings also indicated that the bearings' contacting surface may have been damaged by the fire and the water used to extinguish the fire. The conclusion was to replace and re-align the trunnion bearings.

In order to replace and re-align the bearings, three options were investigated:

- Heat straightening;
- Lifting the entire bascule leaf and replacing the bearings; and
- Jacking up the bascule leaf and replacing the bearings.

Heat-Straightening

After discussions with the potential heat-straightening contractor and experts and an on-site examination, the design team did not recommend the heat-straightening option. This was due to the concern that although this is generally a very cost effective method for repairing damaged steel structural members there are no previous examples that demonstrate how this process could be effectively applied to this project – a project that requires precision correction of the girder deformation in order to meet the mechanical engineer's requirements.

Lifting the Bascule Leaf

Lifting the entire bascule leaf off the bridge would have many benefits, including:

- Providing an un-obstructed working space for the replacing the bearings and for other seismic retrofit and rehabilitation activities inside the bascule piers; and

- Providing a factory facility for repairing and repainting the superstructure leaf in parallel with other construction activities on site.

Despite these benefits, further study indicated that this option is not feasible due to the shallow water around the bascule piers. Environmental concerns would also prohibit large vessels with heavy lifting equipment from entering the shallow water, for fear of the grounding.

Jacking the Bascule Leaf

More studies and the design were done in considering the option of jacking the bascule leaf. Many details had to be considered for the jacking design. To jack up the 800-ton bascule leaf and reposition the bearings while the jacks still support the leaf, the ideal jacking locations would be the spaces in the middle of the two bearings under each trunnion (see Figures 12 through 14). Jacking from this location would give the closest camber or deflection as the final permanent condition, which would facilitate the bearing re-alignment process.

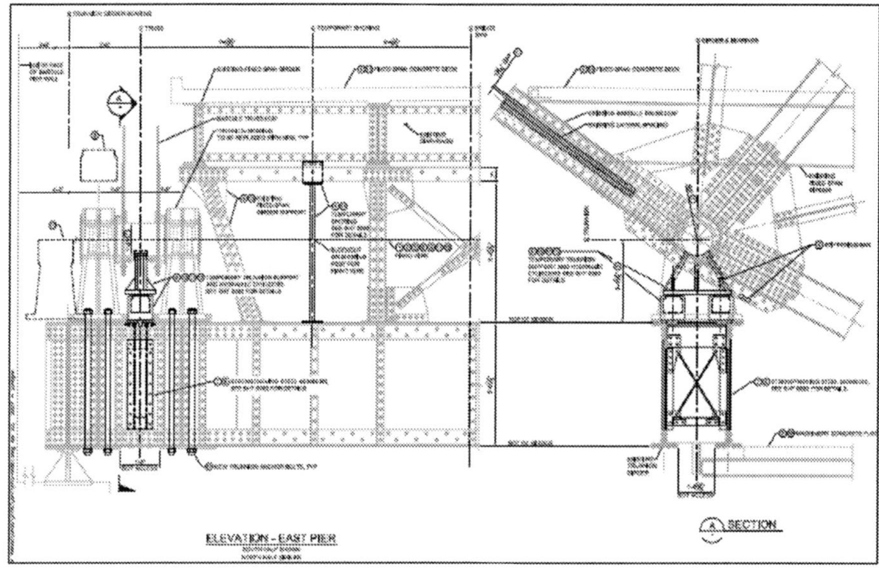

Figure 12: Jacking Design Details

Figure 13: Jacking Location
between the Bearings

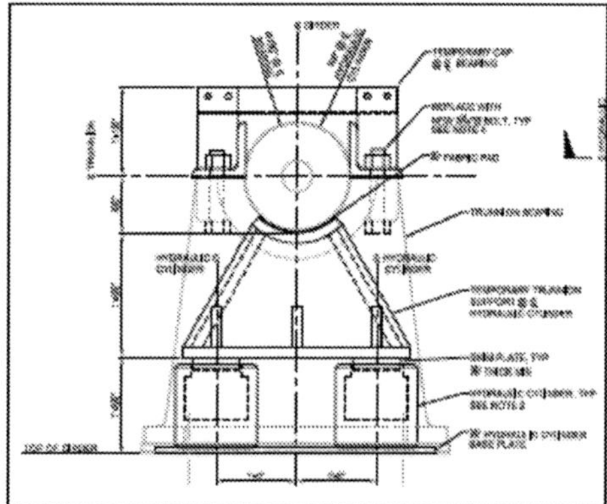

Figure 14: Jacking Detail and
Temporary Trunnion Support

The Enerpac CLS0-3002 hydraulic jack was chosen for the design. This type of jack is compact in size and can fit within the limited space between the bearings. Each jack is capable of lifting 300 tons. A total of four jacks (two under each trunnion) would provide sufficient jacking capacity to lift the 800-ton bascule leaf (see Figures 12 and 14). Two sections of the trunnion girder underneath these hydraulic jacks would also be strengthened, to provide support for the heavy lifting (see Figure 15).

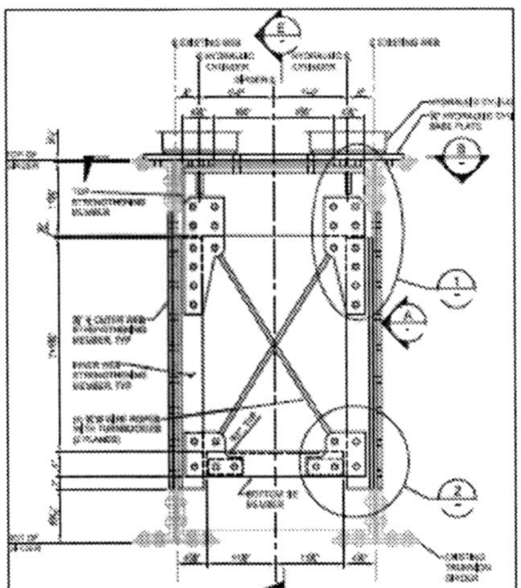

Figure 15: Strengthening Trunnion Girder under the Jacking Location

Modifying the Pinion Supports

As mentioned previously, the existing bascule drive pinions and their supports were discovered to have excessive deflections (see Figures 8 and 9), given the load combinations specified by current AASHTO LRFD "Movable Highway Bridge Design Specifications".

Investigations found that strengthening the existing support system configuration could not effectively reduce the amount of deflection. To control the deflection to within the mechanical designs' allowable limit, the entire bascule leaf pinion system would have to be re-designed and re-constructed.

The re-designed pinion is supported on the bascule pier wall. The original in-board supports of the pinions are removed from the re-designed support. Instead, the pinions are supported by steel frames that are directly anchored into the bascule pier walls, to provide firm and stable supports.

Future Consideration

For future consideration on similar projects, the engineering design team offers the following experience and "lessons-learned":

- "Expect the unexpected" for complex retrofit/rehabilitation projects;
- By communicating the project's objectives and design approaches in a timely manner, these objectives and approaches can be modified if necessary;
- It is important to evaluate existing structures thoroughly during a project's early stages;
- Seismic design criteria may vary for different structures;
- Projects will benefit from allow sufficient time for environmental permitting, if possible; and
- Raising an entire bascule leaf is rare, but the operation is possible.

An asymmetric Curve Cable Stayed Bridge in Parma (Italy)

P. Giorgio Malerba, Dept. Of Struct. Eng., Politecnico di Milano, Milan, Italy
Paolo Galli, Consultant engineer, Milan, Italy
Paolo Sorba, A.I.Erre Engineering S.r.l., Parma, Italy

Abstract

The De Gasperi Cable Stayed Bridge (Parma – Italy) was completed in 2005. It has an asymmetric geometry, with a single pylon inclined at an angle of 72° in respect to the horizontal plane, that stretches 75.00 m above deck level, for a total length of 79.00 m. The suspended curved deck is subdivided into a 40 m span (from the northern counterbalance to the pylon) and a 130 m span (from the pylon to the southern abutment). The deck, made up of two sections separated by a 5 m wide gap, has a composite steel - concrete cross section. The pylon is made up of two pre-stressed blades, transversally connected to one another by means of five couples of steel bracings and by reinforced concrete infill at the two ends. The foundations of the pylon and of the abutments are of the indirect type with piles.

The conception of the structure

Crossing over the river Parma, the cable stayed bridge named after Alcide De Gasperi connects the roundabout in Via Laghirano to the stretch of road which ends on Strada degli Argini, along the new Parma city beltway. When talking of bridges which are part of a urban viability network, we usually think of multi-span beams, resting on piles in the river bed and on concrete abutments at the ends. In this particular case, on considering the environmental impact of this simple and economical type of bridge, it was immediately clear that such a massive structure would not only block out the view, but even change the actual image of the river itself which, at this particular point, can be seen from different perspectives. In addition, the ever changing and often impetuous flow of this river would call for suitable foundations and piles into the river bed.

This last problem, together with the aesthetic issues, were solved by adopting a cable stayed system, in which the supporting action supplied by the piles is carried out from above by rows of cables gathered at the tip of a pylon which, in turn, directs the loads down to the foundations.

Due to the fact that the bridge is asymmetric, the single pylon has to be inclined in order to counterbalance the pull of the cables stretching from the suspended curved deck. This counterbalance action is provided by the weight of the inclined pylon itself and then by anchoring the tip of the pylon to a counterweight, which at the same time forms the abutment for the access road to Strada degli Argini.

General characteristics of the bridge

The bridge has asymmetric geometry, with one single pylon inclined at an angle of 72° in respect to the horizontal plane. The pylon has an inclined length of 79m and is 75m high (Figure 1). The suspended curved deck is subdivided into two spans. The first one stretches for 40m from the northern abutment to the pylon, while the second covers a 130m span from the pylon to the southern abutment.

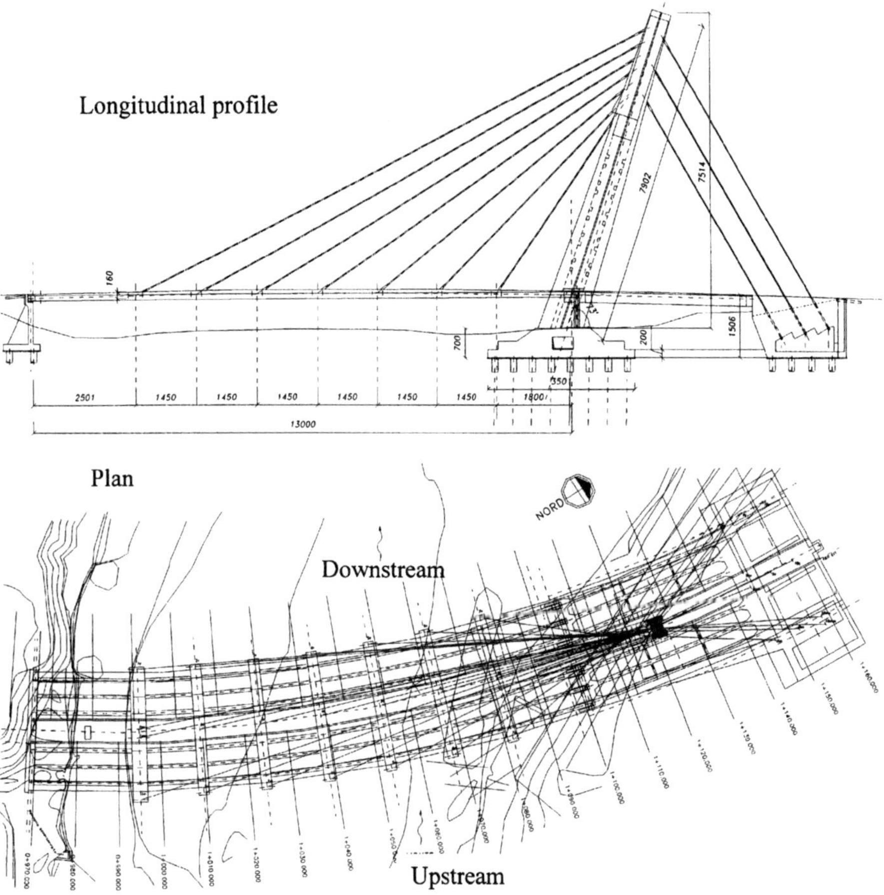

Figure 1. General characteristics of the bridge

Transversally, the deck is made up of two separate structures, each one 11.40m wide, having a composite steel – concrete section (Figure 2). The two parts are separated by a 5m gap, thus the total width is 27.8 m. In plan, the bridge axis is circular, having an average radius of 350.53 m. The vertical section shows a constant curvature with a radius of 3000m. The two sections of the deck are aligned transversally and slope at a gradient of 4.5%. Due to the overall geometry of the bridge, the cables are arranged in a half-fan pattern and lie in different planes. The 130m suspended span is supported by 7 sets of cables (1x4 + 6x3 = 22 cables). Another 3 sets of 11 cables connect the pylon to the counterweight.

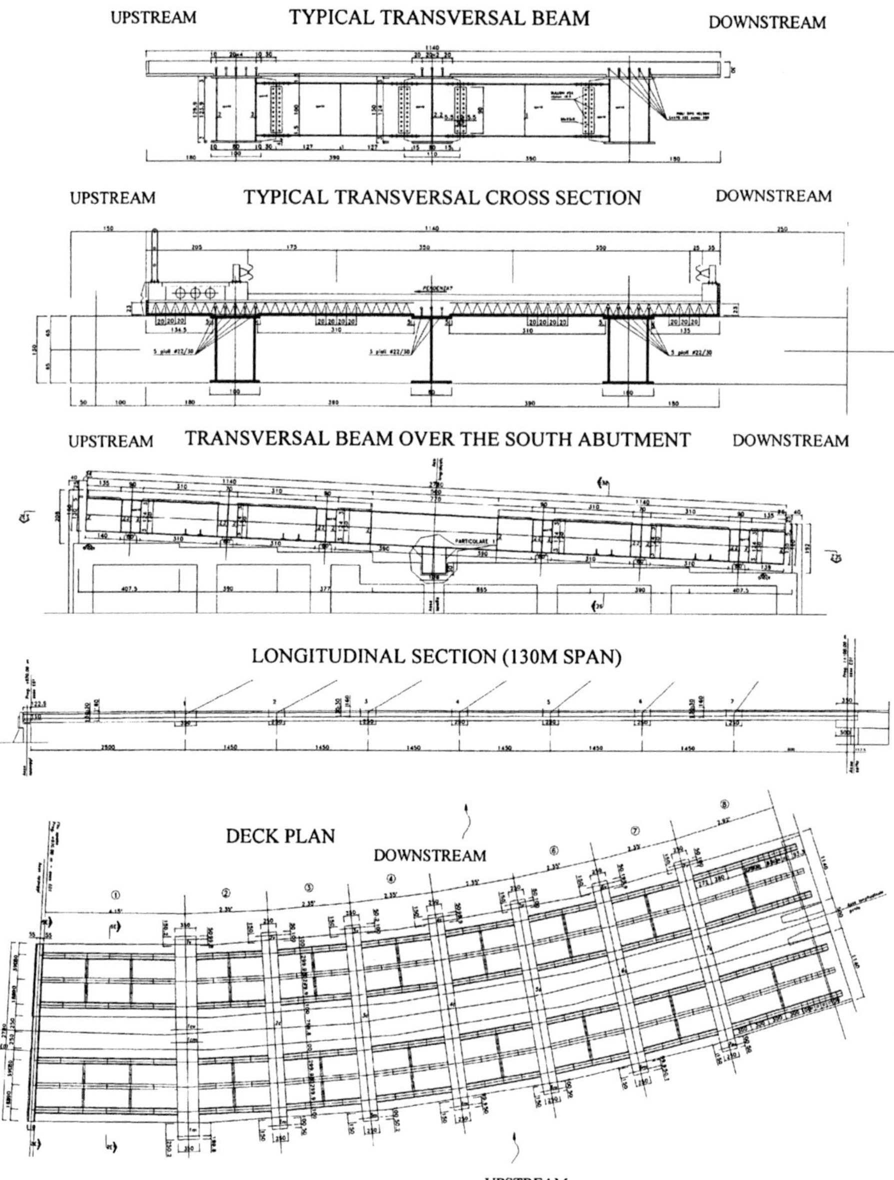

Figure 2. Plan and cross section of the suspended deck

The foundations

The southern abutment

The southern abutment sits on two rows of 7 foundation piles having a 150cm diameter and 37.40 m long. The front wall sits on a 1.50 m thick base and its height varies according to the transversal inclination of the structure. The wall also acts as a support for the embankment. In order to counterbalance the earth's thrust, the wall has been reinforced by means of 5 internal triangular buttresses.

The upper part of the wall is shaped so as to form large horizontal steps where 6 supports provide vertical constraint to the deck. At the top of the abutment, in a central position, a hollow hosts a vertical sliding bearing which constrains the radial movements of the deck.

The Foundations of the Pylon

The foundations of the pylon are formed by a grid made up of 6 x 8 = 48 foundation piles, of 150cm diameter and 37.40m length. They are arranged with a relatively regular spacing of 4.50m, so as to form a square mesh. The base of the foundations has a rectangular shape, 35.00 m long and 27.00m wide. The longitudinal section tapers out into large steps. The foundation block in the central area is made lighter by two prismatic voids, and then, at the riverside, by a duct which leads to the room where the cable heads which pre-stress the pylon are located.

The Structure of the counterweight

The counterweight is a vast structure for which great care was taken in calibrating the balance between the solid and hollow parts. The functions of the counterweight structure are: a) anchoring the mooring cables, which elastically constrain the vertex of the pylon; b) supporting the deck girders; c) acting as a connection to the northern road network.

The foundations are made up of 24 foundation piles of Ø 150cm and are 37.40m length. The foundation plinth has a trapezium layout and is 1.50m thick.

Vertically, the counterweight can be considered in two distinct parts having different widths. The lower one is the widest, due to the need to fan out the run of the cables towards the basis, so as to obtain an efficient contrasting action against lateral loads on the pylon and, and at the same time not cause interference with the roadway traffic. The upper part, which only involves the central group of 4 cables and the structure which connects it to the deck, is as wide as the deck itself.

Internally, the counterweight is subdivided into 5 hollow rooms by 4 walls. The ceiling of the empty rooms are perpendicular to the cables direction. The ceilings are reasonably thick (from 6,5 to 2,0m) so as to carry out a ballast function and, at the same time, to provide adequate lateral surfaces capable of absorbing the high shear actions conveyed by the cables.

The connecting structure

An asymmetric stayed system, having an inclined pylon which is sturdily constrained by mooring stays, conveys intense horizontal forces to the ground. In order to limit any movement between the counterweight and the pylon, two pre-stressed stringers connect to the base slabs of either foundation.

The deck

The structure has mixed composite steel – concrete sections and is made up of 3 longitudinal girders. The two edge ones are box girders while the central one has "I" section (Figures 2 and 3). The steel is of CORTEN type.

The 130m suspended span girders are the same depth (1.30m) for the length of 120m from the southern abutment, then their depth varies up to 1.75m, corresponding to the intermediate bearing that rests on the transversal walls at the sides of the pylon. In the suspended length, the longitudinal beams are connected to one another by 7 box girders. The transverse girders also house the anchorage devices of the stay cables.

In the short span, the girder height varies from 1.75m, at the axis of the pylon to 2.75m, at the northern abutment. In this section, the longitudinal girders are connected by 3 transversal trusses (Figure 3).

<table>
<tr><td>(a)</td><td>(b)</td></tr>
</table>

Figure 3. The deck during erection. a) the suspended span; b) the northern span behind the pylon.

THE Pylon

The Pylon structure

The pylon is divided vertically into two parts (Figure 4). The lower section is made up of 7 segments of reinforced concrete. In the upper section a steel box, that houses the stay anchorage devices, has been inserted in the gap between the two lateral pre-stressed concrete walls rising from the lower section, so that the overall image of the antenna is streamlined both from an upstream and a downstream perspective.

The two walls, here called the "blades", are braced by transverse elements. At the bottom, the bracing action is carried out by two reinforced concrete walls, 1m thick and 9.50m high, which run perpendicular to the blades. At the top, there is another reinforced concrete infill, on which the steel case has been fixed: it is as wide as the gap between the "blades" and 5.00m high. In the intermediate area, the transversal connection is by means of cross-shaped bracings, evenly spaced.

The anchoring system at the top of the pylon

The steel box placed at the top of the pylon houses the cables anchorage devices (Figure 5). The suspension system consists of cables of different diameters, having a minimum of 3 to a maximum of 91 strands. The forces acting in the stays reach a maximum of 12,000 kN, while the medium value is of 7,500 kN. Suitable distribution mechanisms have to be exploited in order to transfer the concentrated load to the concrete body of the pylon with acceptable stress rates.

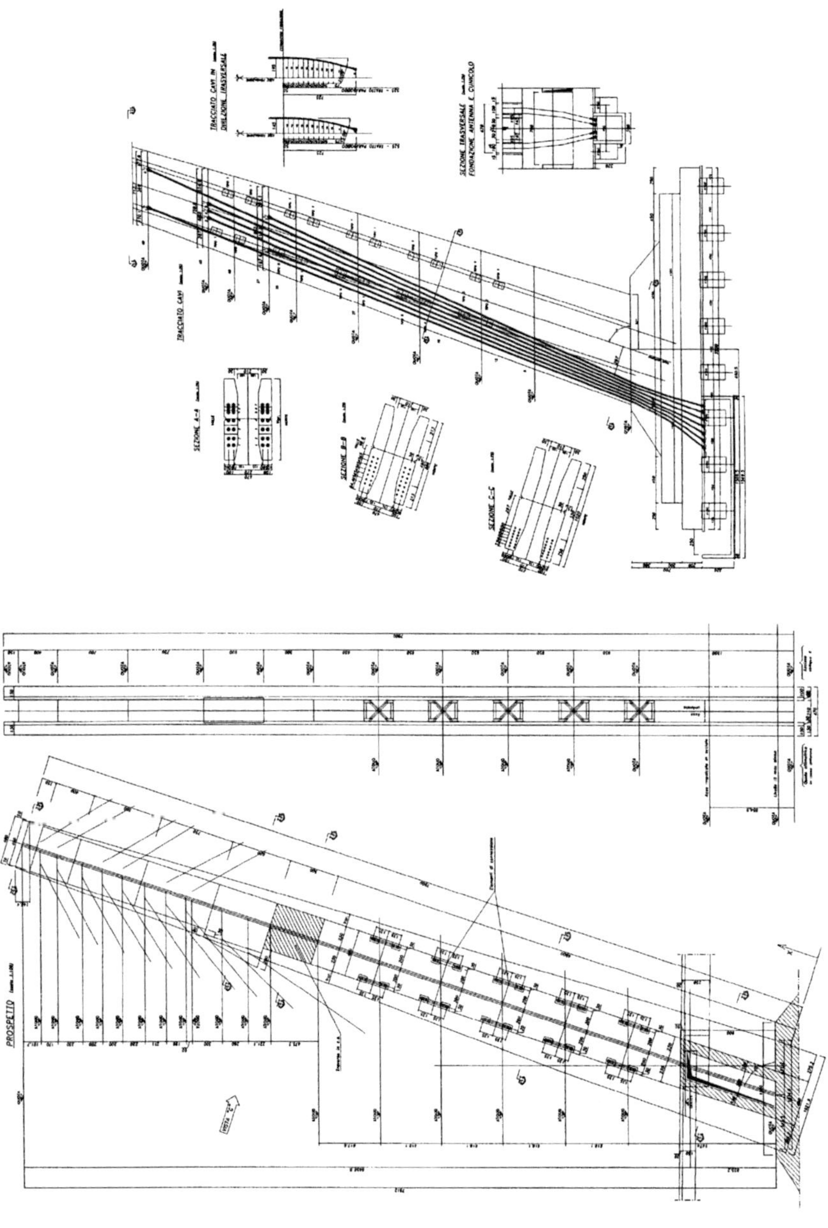

Figure 4. The antenna.

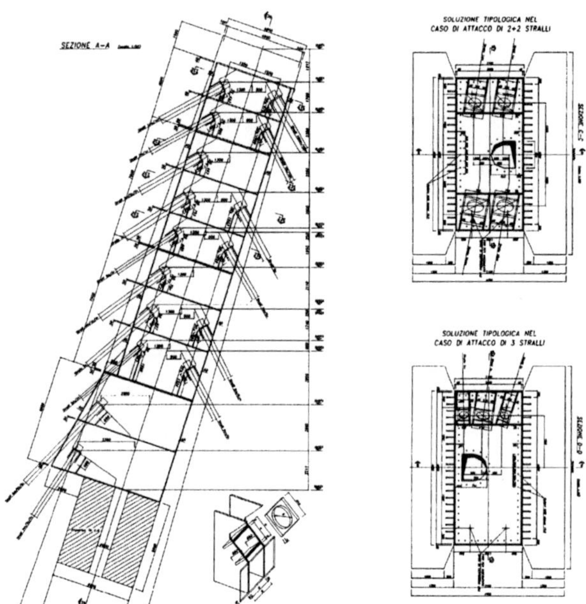

Figure 5. The steel anchoring system.

The first level of distribution is provided by the steel anchorage blocks, that support the cable pull forces on thick steel plates. At a second stage, the resultant force, each one having its own intensity and eccentricity, that act on these plates, have to be conveyed to the axis of the antenna, by means of stiff and robust systems. The transfer is carried out by the steel box, which can bear high local stresses with sizes and weights smaller than the ones that would have been required if the anchoring structure was made of concrete. Moreover, a steel structure can be manufactured in a workshop, thus allowing to build with great precision the complex arrangement of diaphragms and local stiffeners which provide to grade the critical distribution mechanism. In this case, the steel case, built by specialized manufacturers is made up of 4 segments which were assembled onto the concrete traverse at the top of the pylon.

It must be mentioned that, for a temporary stage of several months, the pylon had to stand alone, fixed only by means of its connection to the foundation. Because of the slope, the vertical forces had some eccentricity, this introduced a flexural component into the pylon. As a consequence, the section at the basis would have remained fully compressed only during the first erection phases. In the following stages, strong ordinary reinforcements would have been needed in the area facing the river, in order to adsorb the intense tensile stresses. These stresses would have become higher and higher as the erection of the pylon went on. Instead of introducing a great amount of ordinary reinforcement, the eccentricity due to the self weight has been counterbalanced by means of 28 pre-stressing cables, each one made up of 27 strands having a 0.6" diameter. These cables, starting with low eccentricity from the top, gradually converge towards the frontal sides of the two blades (Figure 4). Once the pylon was completed, the section at the base was fully compressed, with the southern edge (the one facing the river) experiencing stresses a little higher than the ones at the opposite edge. The subsequent action of the suspension stays enabled the force resultant to move towards the axis of the antenna.

The lateral walls.
At the side of the pylon, at a 15cm distance from the its lateral faces, there are two transversal walls which bear the vertical loads transmitted by the deck in the axis of the pylon, and the horizontal thrusts, due to the pull of the stays onto the curved deck of the bridge.

The stay system
The suspension system is made up of 33 stays. They all converge to the steel box and are slightly offset.

The 130m span is supported by means of seven sets of stays. The bottom anchoring devices are housed by the transverse beams which connect the two sections of the deck (Fig. 7). The first transverse beam, coming from the southern abutment, is supported by four stays, two at the ends and two at the midspan of the transverse beam. The other six transverse beams are supported by three stays each. Northward, the head of the pylon is directly connected to the counterbalance block by eleven anchoring stays.

The Bearing and Restraint System
The particular geometry of the bridge led to the following unusual restraint system for the deck:
-   in the northern section, behind the pylon, the deck is built into the counterbalance block;
-   the deck in axis with the transverse walls at the sides of the pylon lie on bidirectional bearing supports;
-   at the same section, the radial displacements are constrained by two shear keys, one for each half of the deck, sliding on vertical supports placed at the middle of the two walls;
-   the deck at the southern abutment lies on six bidirectional bearing supports; its radial displacements are constrained by a single shear key sliding over a vertical support and placed in the middle of the abutment upper surface.

Structural Analyses
The bridge had been studied by means of several numerical models. Displacements and stresses due to dead, permanent and live loads, combined according to several load combinations, have been systematically studied.
Specific analyses have been carried out to define the pre-tensioning of the stays and the tensioning sequence.
Other specific analyses have dealt with the pylon stability, the effects due to stays pre-tensioning during temporary erection stages, the details of the anchoring zones, the forces and the displacements regarding the bearing devices.

Construction Phases
In addition to usual engineering problems, bridge construction has also to deal with the solution of all issues regarding erection stages and temporary configuration, until the completion of the works.
The bridge erection has been developed through the following phases:
-   Foundations and abutment at south side.
-   Foundations of the pylon. Foundations and counterbalance at north side.
-   Placement of temporary piers in the riverbed.
-   Erection of the pylon and connection of the two blades through cross bracings.
-   Pre-stressing of the pylon.
-   Casting of the traverse at the top of the pylon.
-   Assembly of the blocks of the steel anchoring system at the top of the pylon and simultaneous advance of the two reinforced concrete "blades", until completion.

- Assembly of deck beams; welding of transverse beams to longitudinal beams.
- Casting of segments of reinforced concrete deck slabs. The strips adjacent to the transverses, where negative moments were expected, were cast after the first pre-tensioning and the application of all d ead loads. The gaps were then filled b y fibre reinforced concrete.
- Final tensioning of the stays, with complete recovery of the vertical displacements due to permanent loads (roadbed, sidewalks, guard rails)
- Finishing works (lighting, painting).

Monitoring and Field Measurements

During the erection stage, all the controls on specimens of materials and on the building procedures have been carried out. Welded parts have been checked through magnetoscopic and ultrasonic tests. A specimen of a cable made of 55/0,6" strands have been put under a pulse fatigue test for $2x10^6$ cycles, and then stretched to breaking load.

Systematic topographic surveys checked the pylon alignment and the correctness of the altimetric profile. The tensioning of the stays took about ten days for each of the two phases (first and second tensioning). The effects on the deck, due to pre-tensioning, have been indirectly controlled by means of topographical surveys. A direct control of the tension in stays was carried out during the second tensioning phase, by checking the tension reached in each strand at the end of the first one.

During its service life the structural behaviour of the bridge is monitored through a set of transducers which measure the main mechanical and environmental parameters. Data acquisition is carried out through a dedicated computer. Fig. 8 shows a view, taken from an upstream perspective, of the finished bridge.

Contractors

The bridge was built by a temporary association of contractors formed by: Nino Ferrari (main contractor), La Spezia – Italy; Cordioli & C. Costruzioni Metalliche, Valeggio sul Mincio (Vr) – Italy. Pre-tensioning and pre-stressing systems and bearing supports were supplied by ALGA S.p.A., Milan – Italy.

References

Malerba P.G. (Editor), (2006), Photos by Luca Piola. *Ponte De Gasperi, dai bozzetti all'opera finita (De Gasperi Bridge, from the sketches to the finished work)* – in Italian. Monte Università Parma Editore – info@mupeditore.it – www.mupeditore.it.

Acknowledgements

This paper is dedicated to memory of Francesco Martinez y Cabrera, Professor of "Bridge analysis and design" at the Politecnico di Milano, who conceived this bridge, and to the memory of Doct. Eng. Fabrizio Fabbri, who began this work with passion and competence. The photos of this article were taken by Mr. Luca Piola and Eng. Paolo Galli.

Figure 6. The bridge after completion, seen from an upstream perspective.

Bridge Widening – Stitching a steel girder reinforced concrete composite bridge to a post-tensioned prestressed voided deck structure

Peter Chong, B Sc (Hons), M Sc, C Eng, FIStructE, MICE, MIHT, P Eng, MIEM
Divisional Director, Peter Brett Associates, United Kingdom

Abstract

Converting a two lane bridge to a four lane structure by constructing a new structure and subsequently stitching it to the existing bridge is not a new concept. What made this structure unique is that the new structure is a steel-concrete composite and it was stitched to the existing post-tensioned voided slab bridge deck. Design challenges faced and overcame included the analysis, devising the connection details both for the deck and the abutments, the construction phasing and seeking to apply sustainable and health and safety principles wherever practicable and at every opportunity.

Introduction

Carriageway widening from 2 lanes to 4 lanes on the existing A228/M20 Western Motorway Overbridge necessitated a wider bridge structure. Bridge widening by constructing a new structure adjacent to and subsequently stitching it to the side of the existing bridge was chosen as the most appropriate in terms of minimizing disruption to the travelling public, its sustainability credentials and overall value for money.

The existing bridge comprises a single-span post-tensioned prestressed concrete voided deck slab flanked with edge cantilevers. This 37 year old structure is curved in plan and is simply supported on abutments which are inclined forward. It has an ultra slim constant depth deck with a depth to span ratio of approximately 1:30. Span from centreline bearings to centreline bearings is approximately 38 metres. The edge cantilever on the side to be widened is demolished to facilitate the stitching detail that was devised to connect the two parts of the widened bridge and transform it into a monolithic bridge deck.

Figure 1. Elevation of existing bridge

The widened portion comprises 3 pairs of braced steel girders made composite with a combination of in-situ and precast reinforced concrete deck slab and edge parapet elements. The steel girder nearest to the existing bridge is curved in sympathy with the demolished edge of the existing bridge whilst the rest of the girders are straight in plan. The design of the widening presented several interesting challenges. The new deck has to conform to the geometrical constraints and its flexural stiffness and deflection characteristics have been fine tuned to be comparable to that of the existing bridge. The use of both precast and insitu concrete have necessitated the consideration of additional construction sequence load cases. Vehicle collision load criteria (bridge strike) have influenced the choice of cross bracings for the steel concrete composite half of the widened structure. The construction of new abutments adjacent to the existing has to be carefully considered to ensure that the co-joined abutments behave monolithically and do not inadvertently overstress each other. The end result is a cost-effective and sustainable solution and the principles employed and lessons learnt could be beneficial for bridge designers considering bridge widening.

The M20 motorway is a key road corridor between the United Kingdom (UK) and the Continent of Europe. It has particular economic significance due to the large volume of Heavy Goods Vehicles using the route to access the Channel ports. Located within the English County of Kent, the M20 at Junction 4 links to the A228 – a key north-south strategic highway linking a number of major settlements from the Medway Towns, through the Medway Valley and on to Tunbridge Wells to the South.

The hybrid widened bridge, designed by Consultants Peter Brett Associates, is part of the £3.2m M20 Junction 4 Improvement scheme for client Kent County Council, under an Agreement with the Highways Agency (HA).

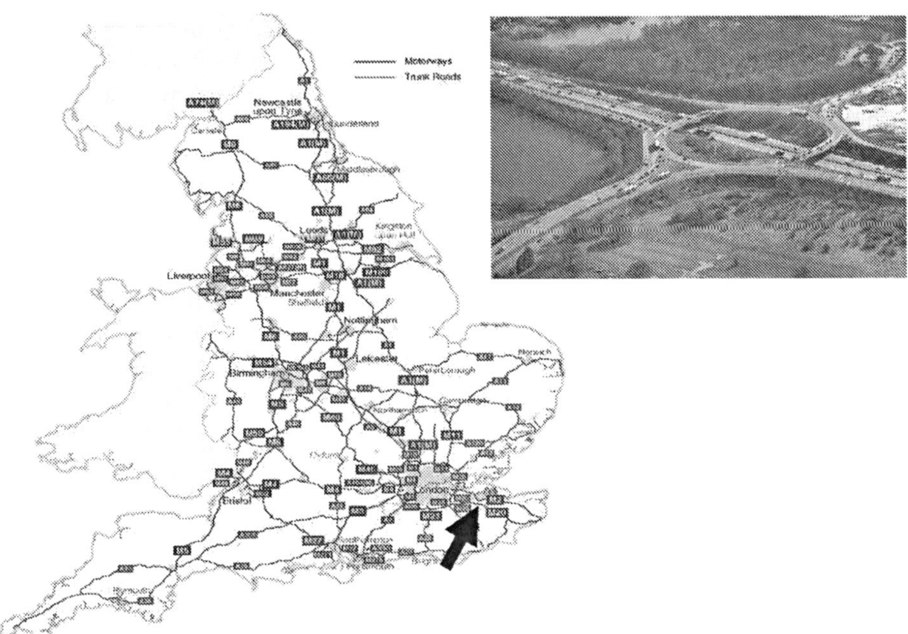

Figure 2. Network Map of England showing Motorways and Truck Roads for which the UK Highways Agency are responsible for managing, maintaining and improving (Arrow shows location of bridge along the M20 London to Dover Motorway)

Design
Structural aspects
As far as the author is concerned, he is not aware of any previous attempts to stitch a composite steel girder/reinforced concrete deck type of structure to an existing post-tensioned voided deck structure. Available Design Codes of Practice do not specifically cover unusual

situations like these either. Consequently the design philosophy devised is based on sound engineering first principles.

Central to the structural consideration is the question of how do you ensure that the two very different structural forms can be designed to behave as one both in the short term and throughout their design life?

Assessment of Existing Bridge

The existing bridge is assessed to be capable of sustaining 30 units of HB loading and the 40 tonne Assessment Live Load to current British Standards. The edge cantilever slabs (which carry the footways) however are substandard and bollards have been introduced to discourage vehicles from accessing onto them. The substandard cantilever on the side to be widened was therefore chosen to be demolished.

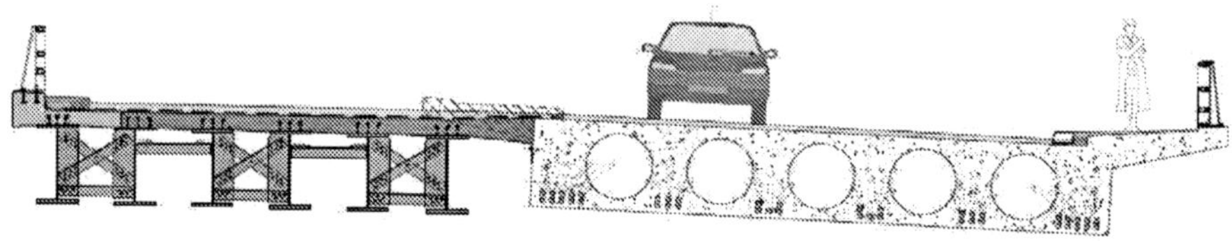

Figure 3. Cross section illustrating the steel concrete composite deck stitched to the voided pseudo-slab post-tensioned deck (with the substandard cantilever demolished)

Design Criteria
New portion of combined deck
The design load criterion for the new part of the widened deck is 37.5 units of HB (equivalent to a 150 tonne vehicle) to the UK Highways Agency Bridge Design document BD37/01. This has been interpreted to mean that if the two portions of the combined bridge deck were to be physically separated (in future) then the new portion should still be capable of sustaining the aforementioned design load.

Existing portion of combined deck
The existing deck has an assessed load capacity of 30 units of HB (equivalent to a 120 tonne vehicle). The design of the combined bridge should ensure that the older portion of the deck is not weakened and that its assessed load capacity is still valid on this portion of the deck.

Feasibility
Initial feasibility included complete demolition of the existing and replacing it with a brand new bridge. The governing criteria of minimising disruption to the travelling public, sustainability and overall value for money favour a lightweight prefabricated deck which is buildable and easily maintainable. Once the decision was taken to retain and incorporate the existing bridge into a wider structure the various structural forms considered included the use of super slim precast prestressed beams. These were subsequently rejected on the grounds of geometric constraints and costs.

Analysis
Feasibility of the choice of introducing a steel girders made composite with a reinforced concrete deck slab was confirmed based on the use of 2-Dimensional (2-D) grillage computer

analysis. The new and existing decks were modelled as discrete longitudinal and transverse grillage members. The 2-D model was subsequently used to undertake the detailed design for the out-of-plane effects.

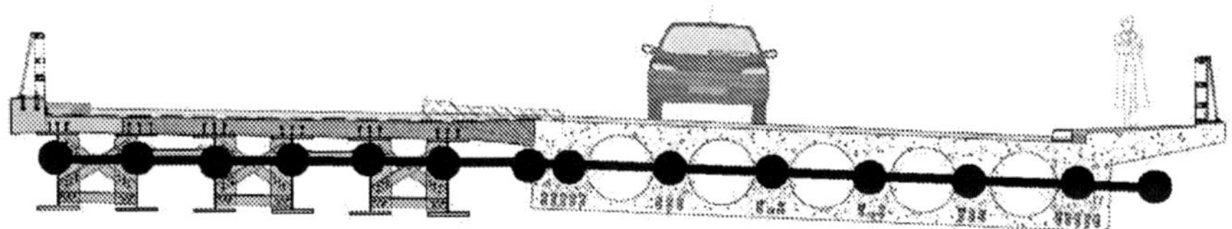

Figure 4. Cross section illustrating the 2-D Grillage (Longitudinal and Transverse) Members

The longitudinal members represent the discrete stiffness properties of the steel I-girders together with a section of the reinforced concrete deck for the widened part of the bridge and represent the discrete transformed pseudo-I and edge sections of the voided pseudo slab existing deck. Transverse grillage members represent the reinforced concrete deck slab (with associated X-bracings where it exists) on the widened part and represent the top and bottom concrete 'slabs' of the voided existing deck. At positions in the voided deck structure where solid transverse diaphragms exist, these were modelled as solid rectangular sections.

These grillage members are subjected to bending and shear under imposed load. Torsionless analysis has been adopted as permitted by the British Standard for the Design of Bridges BS 5400. The original intention was to closely match the flexural/deflection stiffness of the new deck to that of the existing deck so that the widened structure would behave monolithically as one as if they were constructed wide at the same time. It was soon apparent that this was not feasible because of the two different design load criteria that were dictated by the Highways Authority. The existing deck has an assessed load capacity of 30 units of HB. The design load criterion for the new part of the widened deck is 37.5 units of HB.

To accommodate the two different loading criteria, the final flexural stiffness for the new portion of the widened deck was fine tuned to 1.2 X the equivalent stiffness of the existing portion. This appeared to provide the 'best' compromise of relative stiffness which optimised material usage and load transfer across the interface zone which is the longitudinal strip of bridge deck between the new and the existing.

Special consideration was given to the shape of the steel girders in relation to the existing curved post-tensioned deck. Adopting straight girders would certainly simplify the analysis and construction details but causes challenges in terms of the transverse behaviour of the deck slab where it joins with the curved existing. The varying width of the new deck slab at the transition zone proved difficult to analyse, was considered to impose significant loads onto the existing deck and raised buildability issues.

The solution adopted was to design the steel girder nearest to the existing deck as a curved beam at a constant radial separation. The other girders are straight in plan and the varying width of the new deck slab was therefore moved away from the vulnerable transition/ interface zone between the new and existing bridge decks.

Load distribution (due to superimposed dead load and live loads) was analysed using the 2-D grillage model. Bending moments and shears from the global grillage analysis superimposed on the local analysis were utilised in the detailed design of the various structural components.

Load distribution (due self-weight of steelwork, permanent formwork and wet concrete) was analysed by hand and also with a 3-D space frame combined with Finite Element Method (FEM) model. A local FEM model (Figure 5) was also specifically created to model the effects of vehicular collision of the bottom flanges of the steel girders.

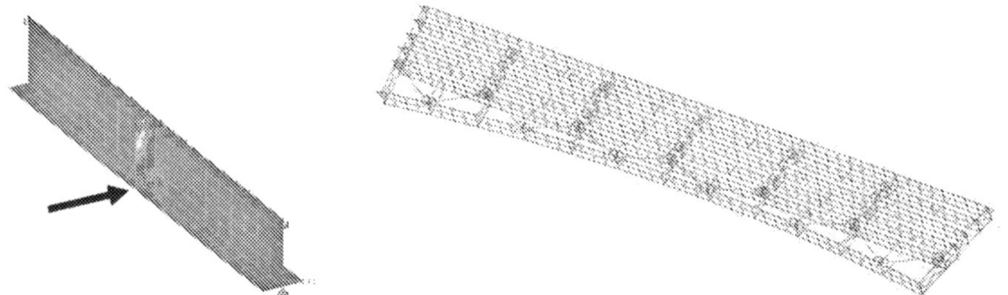

Figure 5. Local FEM model simulating vehicle collision and 3-D space frame combined with FEM with bracings

A global torsionless grillage system of analysis has been adopted. The torsional stiffness of the grillage members has not been taken into account in the analysis of the structure. Torsion certainly plays a minor role in the behaviour of the steel concrete composite part of the combined structure. The existing voided post-tensioned part of the combined structure is essentially a pseudo-slab with not insignificant torsional stiffness. As the analysis was intended to determine the general behaviour of the combined deck structure under load and to provide the basis of design for the steel concrete composite part only the author felt that the adoption of the torsionless grillage system is justified. The absence of torsional stiffness on the existing voided post-tensioned part of the combined structure would result in larger predicted bending moments and a safe lower bound assessed load carrying capacity.

Deck Stitch Detail

The interface zone between the new and existing bridge decks comprises two layers of 25mm diameter high yield (500N/mm^2) reinforcement bars dowelled into the side of the existing post-tensioned deck. Existing mild steel reinforcement (from the demolished cantilever footway) protruding from the side of the post-tensioned deck were carefully straightened by heat treatment and incorporated into the Stitch Detail. This Stitch Detail forms the connection of the two parts of the bridge deck making it a complete structure.

As the flexural longitudinal stiffnesses of the two parts have been optimised the longitudinal (in-plane) effects due to live loading were found not to be significant. The two layers of flexural steel provided catered for the worst possible transverse flexural effects due to eccentric live loads across the stitched interface zone. The dowelled steel (and the retained protruding steel from the existing deck) also resist the longitudinal shear caused by the differential shrinkage and residual creep effects at the interface zone between the new and existing structures. Differential shrinkage effects are dominant for the first year or so after the stitch detail was cast but with time the decreasing rate of shrinkage of the new slab coupled with the residual creep effects of the existing deck combine to reduce any further net increase in the interface shear loads.

Materials
Steel
Weathering steel grade S355J2W+N was specified to British Standard (BS) EN 10025-5:2004 (similar to Corten®) and the steel plate thicknesses include a 3mm sacrificial allowance per face for its 120 year design life. Weathering steel is not a rustproof material. If water is allowed to accumulate in pockets, those areas will experience higher corrosion rates and will rust as unprotected normal steel.

The use of weathering steel in construction therefore presented several challenges which were carefully addressed. Specific welding techniques ensure that the weld runs weather at the same rate as the parent material. The profile and details of connections of the girders were devised to minimize water ponding.

Weathering steel's normal surface weathering can lead to rust stains on nearby surfaces. Drip details were introduced to the bottom flanges of the steel girders near the concrete abutments to ensure discolouration staining does not occur.
To ensure a uniform appearance after fabrication, a blast clean to SA 2.5 quality was specified. It is envisaged that the uniform appearance will prevail as the weathering steel mellows. The bolted connections were secured with weather resistant grade High Strength Friction Grade bolts to BS 4395. Total tonnage of steel used is approximately 120 tonnes.

Concrete
The concrete specified to be used in the abutments and bridge deck is compressive strength class 40/50 to BS EN 206-1/ British Standard BS8500. In addition, silane treatment to HA BD43/03 to all exposed concrete surfaces has been specified to improve the durability of these surfaces that are vulnerable to salt spray (which is a consequence of the deposition of de-icing salts onto the road surfaces).

Precambering issues
Each steel girder is precambered to a parabolic profile to remove the effects of self weight of steel, the 'wet-concrete' construction stage (including the erection of the precast concrete parapet edge slabs) and superimposed dead load (which includes the road surfacing which is applied after the new deck has attained composite status). A notional precamber in respect of live load was also applied to ensure that the structure maintained a positive camber under normal loading conditions.

It was recognised that the curved girder and its adjacent straight girder would be subjected to lateral translations under load that cannot be predicted by the 2-D grillage. The 3-D space frame model was therefore specifically created to study the effects of lateral displacements under load. This gave confidence that the various construction stages would not produce excessive translational movements that could exceed dimensional tolerances. This 3-D frame model was also successfully used to study the effects of the loading of precast parapet edge members on the outermost pair of girders and subsequently provided the predictions on which remedial alignment to the edge members were successfully implemented.

Sustainability credentials
One key aspect of sustainability for bridges is about minimising the carbon footprint or the combination of embodied and operational energy over the full life cycle of the bridge. Although a quantitative embodied energy type of approach was not undertaken on the design of this bridge we can nonetheless qualitatively demonstrate many of its sustainability credentials.

Useful life of existing bridge

Demolishing the existing bridge to make way for a completely new replacement structure was an option. Demolish and replace with a new structure in simple terms could mean a doubling of embodied energy although the demolished concrete and reinforcement are recyclable to a certain extent. The existing bridge has a theoretical residual design life of 83 years and it makes economic sense to retain the structure if at all possible.

Use of weathering steel

Weathering steel (similar to Corten®) does not require painting and cuts down the need for regular maintenance which would involve traffic management and cause congestion.

Use of Glass Reinforced Plastic formwork

Use of GRP permanent formwork between the steel girders obviates the need for erection and the subsequent removal of temporary formwork – which requires traffic management and its attendant congestion related emissions.

The permanent formwork affords additional protection to the soffit of the reinforced concrete deck from salt spray from the traffic beneath – reducing maintenance (i.e. minimising operational energy) and the related traffic disruption effects.

Use of prefabricated components to reduce time for motorway closures / diversions

Prefabricated components offer a higher quality of manufacture and should enhance durability resulting in less maintenance and its attendant traffic disruption and emissions

Health and safety

United Kingdom (UK) legislation governing health and safety in relation to construction work (and it applied to the design of this structure) came under Statutory Instrument 1994 No. 3140 "The Construction (Design and Management) Regulations 1994". These Regulations, in particular Clause 13, Requirements on designer, required designers to design structures that are safe to build, safe in service and safe to decommission. The emphasis is to design out foreseeable risks to health and safety at source. To that end, risks associated with the construction, in-service and maintenance and ultimately decommissioning phases of the life of the structure are assessed.

There is a risk of working at height during the construction phase and especially over a busy motorway where the risk of components falling onto traffic below would have very serious implications. The edge parapet cantilever would have required falsework for insitu casting. The introduction of a precast element would mitigate the risk of working at height (at the edge). The precast elements were also designed to be stable when craned onto the steel girders and act as edge formwork for the subsequent insitu concrete deck pour.

The use of GRP permanent formwork between the steel girders creates a safe surface for the construction worker and protects the traffic below from falling debris.

Bracings

Bracings are required to facilitate the erection of the steel girders by craneage and prevent the lateral torsional buckling mode of failure.

They also continue to provide stability to the bare steel girders against wind and accidental loads and the subsequent construction sequence loads which include the placing of the precast edge parapet elements, permanent formwork and reinforcement steel fixing and the wet concrete stage of construction.

After the insitu deck has been constructed and the bridge deck is fully composite, the bracings then fulfil its load distribution function (in conjunction with the transverse property of the reinforced concrete slab) and additionally provides the necessary transfer load path for vehicular collision because the soffit of the widened portion of the bridge deck, although higher than the existing portion, is still less than the 5.7m headroom criterion, below which vehicle collision loads need be considered under UK bridge design rules.

The Finite Element Method form of analysis was utilised to study the effects of collision load at the bottom flanges of the steel girders (Figure 5).

Ground conditions
The new bridge abutments are founded in the geological stratum known as Folkestone Beds – predominately granular in nature (coarse sand) and thus would experience immediate settlement during the construction stage with little or no secondary long-term consolidation type of settlement.

Abutments
The existing abutments have been in-service for 37 years. It is anticipated that there is no further settlement. As a consequence, the spread footings of the widened abutments have been sized to keep the imposed bearing pressures to an acceptably low figure in order to minimise excessive settlements. A dowelled detail was adopted to connect the new and the existing abutments.

Precast parapet edge slabs

Figure 6. Precast parapet edge slabs installed on top of the edge pair of steel girders and details illustrating their interaction with the shear studs at the top of the girders

Partial Demolition
Demolition of cantilever to facilitate the stitch detail
Had the cantilever been structurally adequate it would have been incorporated into the widening scheme. As it was substandard, its removal was therefore necessary. The stitch detail comprised the retention of a significant amount of the existing cantilever (mild steel) reinforcing bars which is to be incorporated into the permanent works. Dowelled high yield reinforcing bars fixed into the side of the voided slab bridge deck via drilling and resin grouting completed the stitch detail. Various methods of removal were considered, hydro-

demolition was the obvious contender as the existing bridge is a highly stressed post-tensioned structure and hydro-demolition would not damage the existing reinforcing steel. However, the use of high pressure jetting would involve large quantities of water and would be a relatively slow process. The containment of the water and the debris above the busy motorway would be quite a challenge. The alternative of closing the motorway below to accommodate this relatively slow demolition technique was also considered to be unacceptable. It was therefore decided to consider and devise a mechanical form of demolition technique.

The edge cantilever of the 37 year old prestressed post-tensioned voided slab concrete bridge was demolished during a 13-hour weekend closure of the M20. This was the first activity of the construction sequence to construct the widened bridge comprising the existing post-tensioned bridge (with one cantilever taken out) and the new steel girder-reinforced concrete slab composite bridge stitched to its side.

Originally intended to perform saw cuts at right angles to the span, commencing at mid-span and perhaps several other saw-cuts at quarter and eighth spans. This procedure would gradually transfer the locked-in post-tensioned stress from the cantilever section back into the main body of the post-tensioned deck. Having removed the compressive stresses from the cantilever section completely, the cantilever would then behave principally as reinforced concrete sections which can be safely demolished by the hydraulic shears.

An alternative and quicker method of demolition would be to use two hydraulic shears to perform the initial and subsequent stress transfer starting at mid-span and working towards the supports simultaneously.

The main difference between the two techniques is the rate of stress transfer. The saw cutting is a more controlled method and fully achieves stress removal from the whole of the cantilever section prior to the subsequent demolition of what could be considered to be a transformed reinforced concrete section. It is recognised that the use of the tandem hydraulic shear only method would result in a relatively rapid release of locked-in prestress and create shock stress wave transfers into the main body of the deck. This was carefully checked out and calculations were performed which confirmed that the anticipated shock stress waves could be safely absorbed by the main deck.

Figure 7. Forty-five tonne excavators fitted with 360° rotating crunchers demolishing the cantilever section of the existing post-tensioned bridge

Construction
Substructure
Construction of the abutments was relatively straight forward and followed the traditional bottom up construction sequence of excavation, placing of blinding, fixing steel reinforcement bars, erection of falsework, concreting, striking of formwork and backfilling.

Abutment stitch detail
Dowel bars were drilled and grouted to the side of the existing abutment stem. The dowels will link the new abutment to the existing. Post drilled and grouted dowels are inherently inefficient to transfer shear forces, much less to resist pull-out. There are uncertainties associated with the existing strength and quality of the existing concrete. Pull out tests provided data on the behaviour of the grouted-in dowels and confirmed adequacy of the concrete substrate. Insitu shear testing is not easy to perform. Therefore it was decided to design out as much of the transfer forces as much as possible. To reduce the design loads on these dowels, a gap detail has been introduced between the stem of the new and existing abutment. This gap will only be grouted after all the dead loads have been transferred to supporting ground. In addition, the settlement and rotational behaviour of the new abutments were monitored throughout the construction process to ensure that as much of the anticipated movements have ceased prior to connecting the new abutment to the existing by grouting up the gap detail. Grouting up the gap detail effectively made the two adjacent abutments (new and existing) monolithic.

Superstructure
Erection of steelwork
The erection of steelwork is in strict accordance with the prescribed and designed Construction Sequence. The bridge girders were paired and lifted in by high capacity craneage. The first girder pair when secured became the 'anchor' pair of girders providing stability for the subsequent girders. Girders were design checked for lateral torsional buckling instability and were lifted in pairs. Pairs of girders are inherently stable under its own weight and under normal climatic conditions. Unless restrained to adjacent pairs, they could however become unstable in adverse conditions including accidental impact or exceptional windy conditions.

The first pair to be erected comprises the curved girder with its adjacent straight girder and the pair is connected with both vertical and plan bracings. This is devised as the 'anchor' pair which is able to provide the support necessary for the temporary stability of all the subsequently erected steelwork. Crane lifting straps are kept in place as a precaution until each pair is further secured to the anchor pair either directly or via a previously secured pair.

Figure 8. Paired lifting of steel girders using high capacity craneage

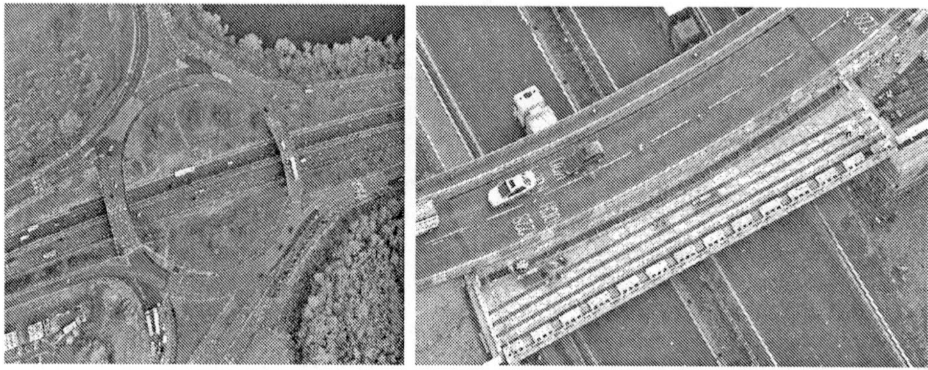

Figure 9. Aerial view of completed widened structure (left) and during construction

Figure 10. Elevation of completed structure

Design and Construction of the Pasteur-Cikapayang-Surapati Elevated Road and Bridge Project in Bandung, Indonesia

Andrew Yeoward, Halcrow Group Ltd, Shanghai, China
Innes Flett, Halcrow Group Ltd, Swindon, UK
Paul White, Halcrow Group Ltd, Swindon, UK

Abstract

Opened in June 2005, the Pasteur-Cikapayang-Surapati Elevated Road and Bridge Project in Bandung, Indonesia provides a vital transport corridor through the heart of Bandung, West Java.

The project comprises 2.1km of elevated viaduct and a 160m long cables stayed bridge. When designed it was one of the first examples in the country of the forms of construction employed, bringing innovative construction techniques to Indonesia's construction industry and developing its technical capabilities. The East and West elevated viaducts were constructed using precast concrete segmental balanced cantilever techniques, whilst the main cable stayed bridge was cast in-situ with parallel strand cables attached and stressed between the deck and the single central pylon. Bored cast in situ piles supported the cast insitu concrete piers. The entire project was given careful architectural treatment resulting in a high quality and elegant structure for the city of Bandung.

This paper gives a summary description of the design of the project and elaborates further on the precast segmental and cast in situ erection methods employed.

Background

In the early 1990's studies carried out by the Directorate General of Highways (Bina Marga) of the Government of Indonesia, as part of the Government's development plan, indicated that priority should be given to providing a new East-West elevated arterial road in the northern part of Bandung. At the time, traffic travelling through this part of the city needed to descend into and cross the valley of the Cikapundung river causing considerable delay in journey times when travelling east-west across the city. In 1996 Bina Marga appointed Halcrow Group Ltd (then Sir William Halcrow and Partners Ltd) in association with INCO of Kuwait, INDEC and Associates of Indonesia and LAPI-ITB to carry out the design development, detailed design and construction supervision of the project. Funding for the scheme was provided through a loan by the Government of Kuwait.

Design was completed at the end of 1997, and competitive tendering procedures were carried out in 1998 on FIDIC-based contract documents. A contract was signed with Indonesian-Kuwaiti

Joint Venture Wika-Waskita-CGC in 1999 for the construction of this challenging project over the congested streets of Bandung. During that period, the Asian financial crisis contributed to delays in starting construction on the project, and work finally commenced in earnest on 1 October 2001, with completion and opening of the finished structure in June 2005.

Various aspects of the design were described in a paper presented in Singapore in 1999[1], and the paper presented here seeks to provide an update on design modifications, and to describe the construction of the project.

Project Description

The plan of the project is shown on figure 1 below. The structure comprises a west viaduct 1284m in length, and east viaduct 840m in length and an asymmetric cable-stayed bridge 161m in length with a main span of 106m. The viaducts carried a dual-two lane highway and generally ran directly above the existing local road network, whilst the cable stayed bridge section was located in an area of domestic settlements and the valley of the Cikapundung river and carried a dual three lane highway. On-off ramp structures were located at each end of the cable stayed bridge, additionally providing the third traffic lane in each direction. The third lane over the valley was deemed necessary to carry the additional traffic that was expected to use the valley section of the highway only.

Selected design details are described in more detail elsewhere[1], but a summary of the main design features is given below

Viaducts

The viaducts comprised single precast concrete box girders supported by single piers at 44.5m centres located in the median of the local road network. Figure 1 shows a typical cross-section of the viaduct. Movement joints were provided at midspan by means of a 'mortice-and-tenon' joint at every 6 spans, giving an expansion length of 267m and allowing a consistent span length throughout the length of the viaducts

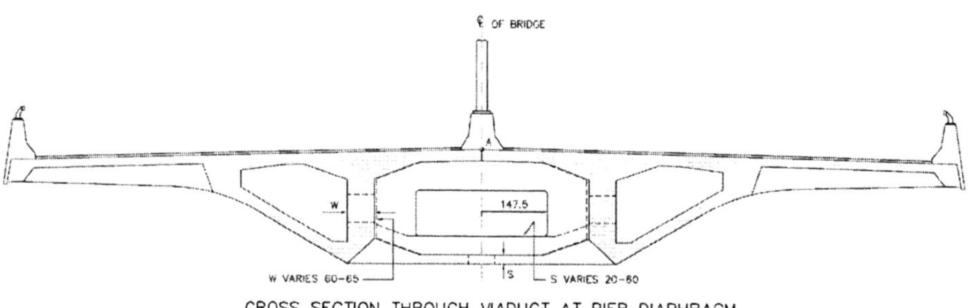

Figure 1 – Viaduct Superstructure

The Viaduct substructure and foundations comprise typically six bored cast in situ 1200m dia concrete piles supporting the single leaf pier. The piles were founded in the weathered breccia layer occurring typically at depths of 12-16 metres. The viaduct piers were designed to resist the high seismic loads by the provision of plastic hinges at their base. This resulted in a characteristic 'waisted' section below ground level and was design specifically to limit the amount of bending experienced by the foundations during a seismic event. This follows the Indonesian and ASSHTO recommendations and is designed to ensure that any repair works to the structure following such an event are not required to be carried out below ground level. Furthermore the seismic load was shared between the piers by means of lockup devices provided within the deck (to prevent vandalism) at each pier location.

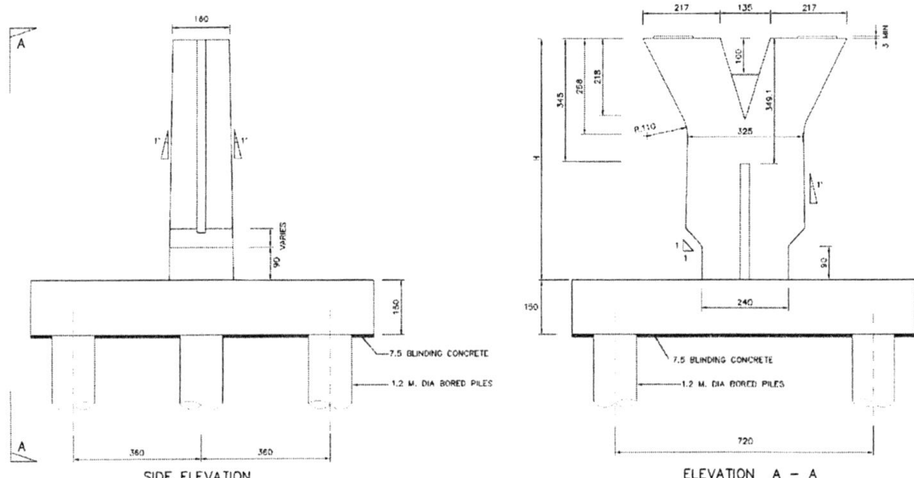

Figure 2 – Viaduct Substructure

Cable-stayed Bridge

Following careful considerable of topographic, aesthetic and other considerations, a main span length of 106m and sidespan length of 55m were selected. One of the key restrictions was the close proximity of the Bandung city airport, which limited the height of the tower. The considerable width of the deck at the tower location, 29m, also limited the choice of tower form to that of a single pylon leg, other alternatives such as A-frame or inverted Y-frame not giving the requisite high quality appearance.

The relatively stiff deck and the angle of the longer stay cables enabled the superstructure to be part stayed and part spanning longitudinally between the main pier and the first viaduct pier.

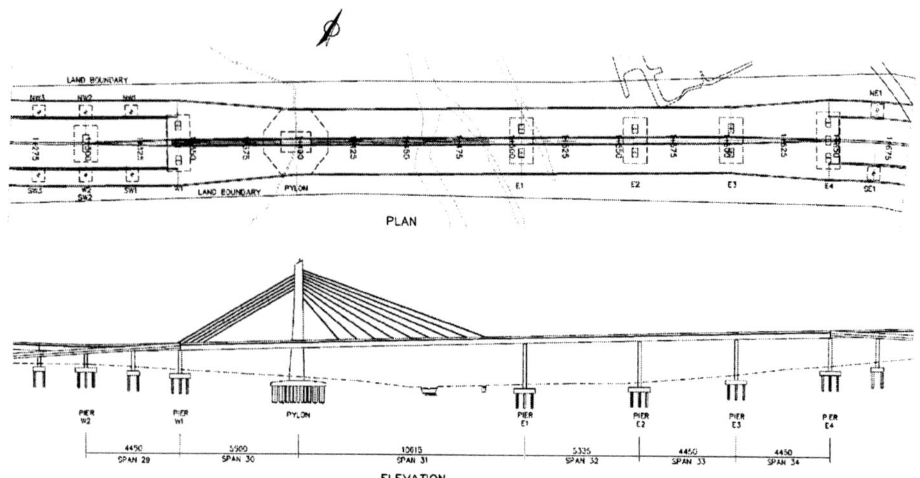

Figure 3 - Cable-stayed Bridge General Arrangement

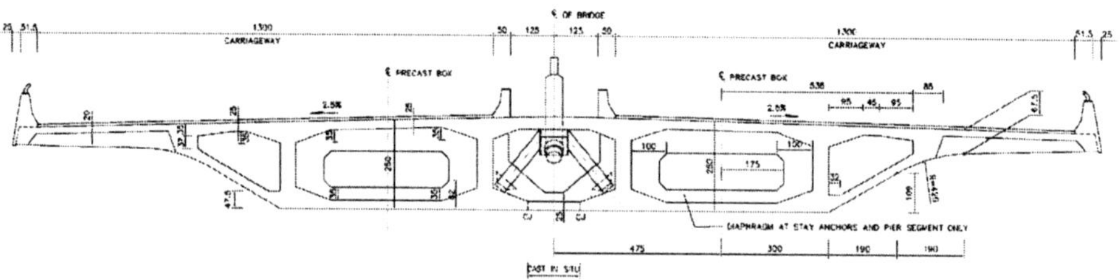

Figure 4 – Cable stayed Bridge Superstructure cross-sections

Ramps

Access ramps were provided at three locations:

- on/off-ramps at each side of the valley providing access to/from Jalan Cihampelas and Jalan Tamansari
- an on-ramp close to the western end of the project at Jalan Pasir Kaliki.

These comprised multispan cast-is-situ voided concrete decks supported from reinforced piers with bored cast in-situ piled foundations similar to the main viaduct. The span of the continuous ramp structure was 22.25m, following the 44.5m module of the viaducts.

Design Development

The design development of the project is described elsewhere[1], but a summary of the key design innovations and developments for the project and the region included

- design of piers to resist volcanic 'lahar' (mud) flow
- design of piers with plastic hinges to protect foundations in an extreme seismic event
- introduction of lock-up device connections between superstructure and substructure to transmit seismic forces
- introduction of precast segmental balanced cantilever construction to Bandung
- midspan movement joints
- development of cable-stayed bridge construction techniques

Construction

Construction commenced in earnest in October 2001 when the contractor Joint Venture Wika-Waskita-CGC mobilised on site. The following gives a summary of various aspects of the construction.

Foundations

Following successful completion of trial pile tests, construction of the bored cast-in-situ foundations commenced on the western viaduct. The founding material of weathered volcanic brecchia proved to be as predicted by the soils investigation and constructed pile lengths were generally as designed.

Piers

Figure 5 below show the reinforcement for the piers, including the 'waisted' area at ground level forming the plastic hinge. The density of the reinforcement was quite high, and following a review of the design it was found that some of the reinforcement could be curtailed, improving the constructability.

Figure 5 – Pier Reinforcement

Figure 6 – Finished Pier

Despite the density of reinforcement, concrete was placed in the steel forms without difficulty, and the finished product provided a high quality finish (figure 6).

Precasting Yard

The precasting yard was set up at Baros, a short distance away and adjacent to the tollroad which extended beyond the western end of the scheme. This would subsequently enable precast viaduct units to be brought along the toll road at night and taken around the toll booths to be delivered at the western end of the site.

Shortline casting methods were used and a total of two casting bays with four moulds per bay were eventually in use on the project. Figures 7 and 8 show general views of the casting yard.

Figure 7 General View of casting yard   Figure 8 Precast Units at the casting yard

A total of 658 units were produced in the yard. In span units 3.2m in length weighed approximately 85 tonnes, whilst the pier head segments (including diaphragms) weighed about 120 tonnes. These were delivered to site on multiwheeled low-loaders.

Viaduct Construction

The viaducts were constructed by the precast glued segmental balanced cantilever method. Segments arrived from the precasting yard and were delivered directly to the point of erection. A mobile crane was used to lift the pier units on to the piers and to lift the first pair of adjacent span units into position. Out-of-balance moments for these were taken by a steel supporting frame. A moveable erection gantry, and supported by the piers, was used to lift the subsequent precast units to their final position. Maximum out of balance was limited to one segment, and when a pair of balanced segments was in position they were stressed back to the cantilevered deck already constructed. Once a balanced cantilever pair was completed, the trailing cantilever was connected by an in-situ stitch to the adjacent previously constructed cantilever. Figures 9 and 10 show erection of the viaduct units.

During the construction of the first few of the 44.5m spans it was noticed that there was a tendency for the horizontal alignment to 'creep'. This was as a result of an initial lack of stiffness in the formwork frames. This was rectified, but only after a number of units had already been cast. The alignment error was rectified by introducing a small number of short cast in situ stitch elements

Figure 9 Erection Gantry

Figure 10 Erection of the viaduct units

Cable-stayed Bridge construction

Overall, the construction of the cable-stayed bridge was carried out in parallel with the construction of the viaducts. However, the largest foundation on the project was that of the main tower, which comprised 34 1200m dia piles connected by a 2760 m^3 reinforced concrete pilecap. This foundation was located in the Cikapundung valley, where ground conditions comprised alluvial material and conglomerates instead of the volcanic brecchia. Considerable care was therefore necessary to ensure that the assumed design parameters for the foundation piles was achieved. That, combined with the large pilecap and tower construction resulted in the cable-stayed bridge remaining on the construction programme critical path.

The cable-stayed bridge deck was originally designed to be constructed using precast elements which would take advantage of the precasting facilities already established for the viaducts. The same external forms would have been used to cast the profile of the superstructure which, for aesthetic reasons was maintained at a constant shape throughout the length of the project. The sidespan would have been constructed first on falsework, and the mainspan would then have been cantilevered out in the conventional manner for a cable-stayed bridge. This would have provided further innovation and technology development opportunities on the project. During the construction period however it was decided to construct the entire superstructure by cast-in-situ methods on falsework. This was possible due to the relatively short span of falsework required to bridge the Cikapundung river in the valley.

Thus the entire superstructure was cast-in-situ at a predefined level before the parallel strand cables supplied by Freysinnet were installed, lifting the entire mainspan to the correct final location (Figure 11). The counterbalancing force resisting the overturning moment of the mainspan is provided by the concrete side span deck, the on/off ramps and the approach viaduct.

Figure 11 - Construction progress on the cable stayed bridge

Conclusion

The objectives of the project were to relieve the traffic congestion in the city of Bandung, to provide a structure with a high aesthetic content and to provide technological innovation in the field of viaduct and cable-stayed bridge construction in Indonesia. All these objective were achieved, and the finished scheme has received a great deal of positive attention with visitors to the project in the first year after completion well into four figures. The figures below show various views of the completed bridge and viaduct.

Figure 12 Completed Viaduct

Figure 13 - Completed cable-stayed bridge

Acknowledgements
The authors wish to thank the Bina Marga, the Directorate General of Highways of the Government of Indonesia for the opportunity to present this paper.

References
(1) Flett ID and Yeoward AJ: Pasteur-Cikapayang-Surapati elevated Road and Bridge Project in Bandung, Indonesia. *Current and Future Trends in bridge design, construction and maintenance (Singapore) Thomas Telford, 1999*

Horizontally Curved Composite Plate Girders

M A Basher, Department of Civil and Structural Engineering, Universiti Kebangsaan Malaysia, 43600 UKM Bangi, Selangor Darul Ehsan, Malaysia
N E Shanmugam, Department of Civil and Structural Engineering, Universiti Kebangsaan Malaysia, 43600 UKM Bangi, Selangor Darul Ehsan, Malaysia
A K A Rashid, Department of Civil and Structural Engineering, Universiti Kebangsaan Malaysia, 43600 UKM Bangi, Selangor Darul Ehsan, Malaysia

Abstract

The paper is concerned with steel-concrete composite plate girders curved in plan. Finite element modeling of horizontally curved composite plate girders is presented in the paper. Details of the finite element modeling and the non-linear analysis are presented. Finite element results in respect of curved steel plate girders and straight composite plate girders tested by other researchers are presented first to assess the accuracy of the modeling. Preliminary studies on parameters such as curvature and steel flange width affecting the behaviour of composite girders are also reported herein.

Introduction

Plate girders curved in plan are frequently employed in the construction of modern highway bridges. Though the use of continuous curved girders has many advantages over conventional ways of using series of straight chords to form the curved alignment, engineers were reluctant to use them because of mathematical complexities associated with the analysis of such girder systems. The capacity of curved girders to resist external loads is notably decreased due to the existence of the initial curvature. At the design stage these girders are assumed to act independent of the deck slabs resting on them even though the deck slabs are connected to the girders by means of shear connectors. The advantage of composite action between the steel girders and concrete deck is not considered. The present study is concerned with such composite action in horizontally curved girders.

An experimental investigation on full-scale horizontally curved steel plate girders has been carried out to study their overall behaviour and to determine the shear strength by Zureick et al (2002). Shanmugam et al. (2003) have tested to failure medium size plate girders curved in plan and investigated the ultimate load behaviour and load carrying capacity. Web openings may need to be provided in these structural members. Therefore, Lian et al (2003) carried out experimental and finite element studies on horizontally curved plate girders containing centrally located circular web openings. Extensive studies considering a number of parameters have been carried out and simple design method proposed (Lian et al., 2004). Jung and White (2006) have reported recently results obtained from the finite element analyses of the curved girders tested by Zureick et al (2002). Both the elastic shear bucking and the full nonlinear maximum shear strength responses have been considered.

Bridge design, construction and maintenance 2007, Thomas Telford, London

The advantage of composite action between plate girders and concrete slab has been investigated recently in respect of straight girders. Allison et al (1982) carried out an experimental investigation on steel-concrete composite plate girders under combined shear and negative bending. Shanmugam and Baskar (2003a, 2003b, 2006) studied both experimentally and numerically the effects of combined shear and bending on composite plate girders. Finite element modeling was proposed, parametric studies carried out and design method to determine the shear capacity given. This paper is concerned with composite plate girders that are horizontally curved. Finite element method has been used to investigate the elastic and ultimate load behaviour and parametric studies carried out. Details of the finite element modeling and the results obtained from the analyses are presented herein.

Finite element analysis

Finite element software LUSAS version 13.7 has been used in the analysis presented in this paper. Three-dimensional models were developed by idealising the flange, web, and stiffener plates in plate girders using QSL8 semi-loof thin shell element in the LUSAS element library. Concrete slab was idealized by stress elements which are three dimensional hexahedral isoparametric solid continuum elements (HX20) with higher order models capable of modelling curved boundaries. This solid element has 20 nodes with three degrees of freedom at each node representing the three global directions in which it may move. Generally, the shell element can accommodate curved geometry with varying thickness and anisotropic and composite material properties. The element formulation takes account of both membrane and flexural deformations. Semi-loof shell element (QSL8) is a quadrilateral thin, doubly curved, isoparametric element formed by applying Kirchoff constraint at discrete points to a three-dimensional degenerated thick shell element. This degenerated element has three degrees of freedom (U, V, W) at corner nodes and five degrees of freedom (U, V, W, and rotations θ_1, θ_2 at loof points) at mid-side nodes. Complying with Kirchoff thin shell theory, transverse shearing deformations are excluded. The plate girders were modelled as simply supported girders by choosing Z-axes as the vertical axis. The pin-supported condition is simulated by fixing translation in X, Y and Z direction, whilst the rotation about X, Y and Z-axis is left free. Roller conditions were simulated by allowing translations in the X direction only and rotation about the axis being left free.

Material properties and Initial imperfections

In the modelling, steel plate girders are modelled using ungraded Mild Steel in LUSAS material library. This material has the properties of Young's Modulus 209 kN/mm^2 and Poisson's ratio of 0.3. The nonlinear material model used Von Mises yield criterion, an associated flow rule and isotropic hardening, giving three distinct regimes - elastic, perfectly plastic and multilinear strain hardening respectively.

A perfectly straight and undeformed model will provide different answers from a model with imperfect geometry. In LUSAS an imperfection can be built into the initial model by manually defining the appropriate geometry or it can be arrived at by loading a previous results file, selecting the load case of interest and choosing the LUSAS Data file command and selecting the deformed mesh factor option to create a model having deformed geometry appropriate to the eigenvalue or load case chosen. In this study, the initial imperfection of the modelled girders is obtained from the buckling analysis whereby the deformed mesh from the first eigenvalue is being used for the nonlinear analysis.

Finite element mesh

Regular mesh with element size of 80mm x 80mm defined in the Attributes options tabs are assigned to all surfaces of the modeled plate girders. The mesh suitable to the structural modeling shown in Figure 1 was chosen based on convergence studies carried out to determine the optimal mesh that gives a relatively accurate solution and one that has low computational time. It has been found that the mesh chosen is capable of producing results close to the actual behaviour of the girder.

Loading

Feature based loads are assigned to the model geometry and are effective over the whole of the feature to which they are assigned. In LUSAS, the feature based loads are (i) Structural: concentrated, body force, distributed, face, temperature, stress/strain, and beam loads and (ii) Prescribed: used to specify initial values for displacement, velocity or acceleration at a node. Prescribed loads with total prescribed displacement instead were being used in this analysis. A Prescribed Displacement defines a nodal movement by either a total or incremental prescribed displacement in global (or transformed) axis directions. Variables that are loaded with a nonzero prescribed variable will automatically be restrained in the required direction.

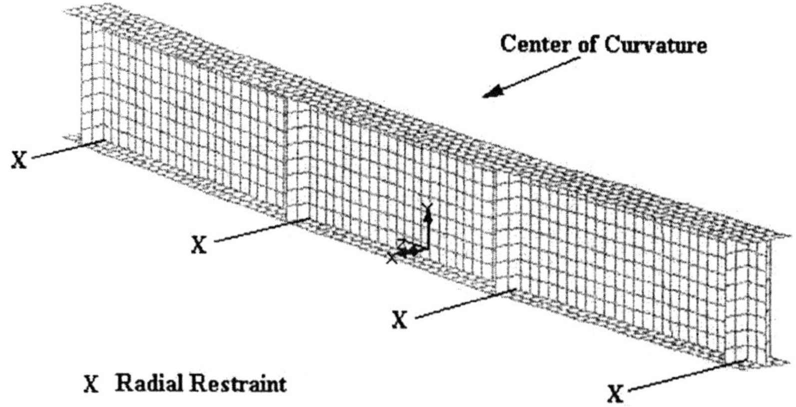

Figure 1 Typical finite element modelling used in the analyses

In *LUSAS*, incremental-iterative solution procedure is used. In this procedure, the total required load is applied in a number of increments. With each increment, a linear prediction of the nonlinear response is made, and subsequent iterative corrections are performed in order to restore equilibrium by the elimination of residual forces. The iterative corrections are referred to some form of convergence criteria that indicate to what extent an equilibrium state has been achieved. The nonlinear solution is based on the Newton-Raphson procedure. The details of the solution procedure are controlled using the Nonlinear Control properties assigned to load case. For the analysis of nonlinear problems, the solution procedure adopted may be of significance to the results obtained. In order to reduce this dependence, wherever possible, nonlinear control properties incorporate a series of generally applicable default settings, and automatically activated facilities. The incremental-iterative solution is based on Newton-Raphson iterations. In the Newton-Raphson procedure an initial prediction of the incremental solution is based on the tangent stiffness from which incremental displacements and their iterative corrections may be derived. For the Newton-Raphson solution procedures it is assumed that a displacement solution may be found for a given load increment and that, within each load increment, the load level remains constant. Such methods are therefore often referred to as constant load level incrementation procedures.

Accuracy of the finite element modelling

It is important to establish the accuracy of the finite element modeling before undertaking the analysis of the horizontally curved composite plate girders. Neither experimental nor analytical results could be found for comparison in the literature for such girders. Horizontally curved steel plate girder tested by Zureick et al. (2002) and straight composite plate girders tested by Shanmugam and Baskar (2003a, 2003b) were, therefore, considered for comparison and to assess the finite element modeling using LUSAS. Finite element modeling was made and analyses carried out to study the elastic and ultimate load behaviour of typical girders.

Zureick et al (2002) carried out full-scale tests on four steel plate girders identified in the text as S1, S2, S1-S and S2-S of 11.58 m chord length. Transverse stiffeners along the girder length were positioned such that the panel aspect ratio was 3 in the case of S1 and S2 and 1.5 in the case of S1-S and S2-S. The overall depth of the girders was around 1.22 m whilst the top and bottom flanges of around 22.9 mm thick varied in width from 546.6 mm as in the case of S1 and S1-S and, 556.3 mm for S2 and S2-S. Web slenderness (D/t_w) in all the four girders was kept 154, approximately equal to the minimum value of 150 permitted for transversely stiffened plate girders by AASHTO (2004). The flange slenderness ($b_f/2t_f$) varied between 11.92 and 12.19. The girders S1 and S1-S had a nominal radius of 63.63 m whilst the corresponding value for the girders S2 and S2-S was 36.58 m. The cross-sectional dimensions of the girders are shown in Figure 2 and the overall geometry of the test girders in Figure 3. Test set-up and applied loading in the tests are shown in Figure 4. Also shown in Figure 4 are shear force and bending moment diagrams corresponding to the applied loading. Typical girder, S1-S was analyzed using LUSAS and the finite element mesh shown in Figure 1 in order to establish the accuracy of the finite element modeling and, load-deflection plots in respect of the girder is shown in Figure 5. In the figure, experimental results along with the ABAQUS results given by Jung and White (2006) are also plotted for comparison. It can be seen from the figures that the load-deflection plots obtained from the analyses by LUSAS lie very close to the corresponding experimental and ABAQUS curves thus establishing the accuracy of the computer package LUSAS.

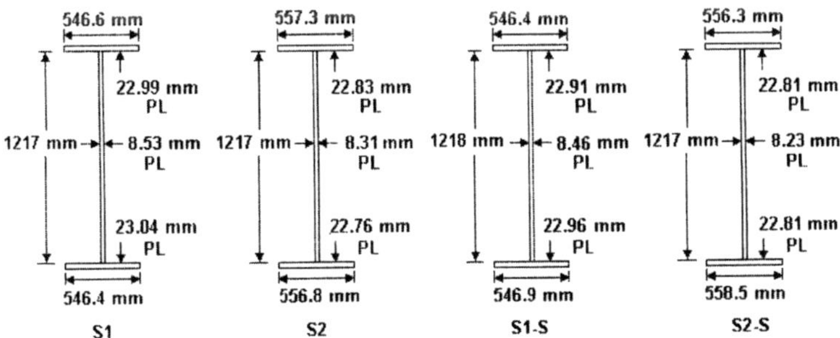

Figure 2 Cross-sectional dimensions of the test girders (Jung & White, 2006)

Studies similar to the above were carried out on straight composite plate girders reported by Shanmugam and Baskar (2003a, 2003b). Ten medium-scale composite girders were tested to failure and the results were presented along with those obtained from finite element analyses using ABAQUS. The tests were carried out in two phases; in the first phase, girders were subject to shear loading and in the second phase to the combined action of shear and bending. Two different web-depth to thickness (d/t) ratios viz. 250 and 150 and two different

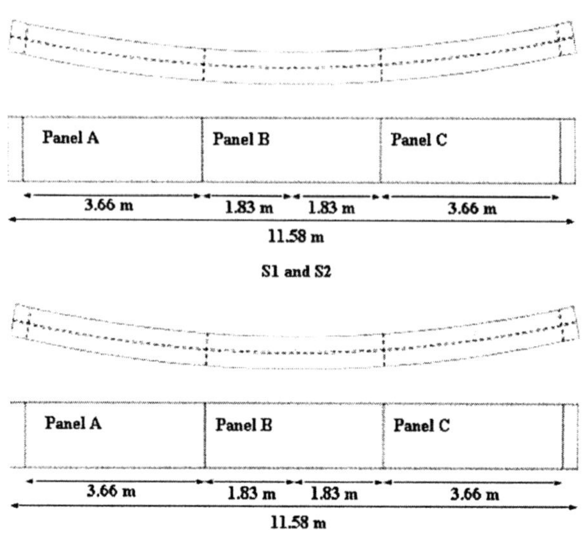

Figure 3 Overall Geometry of the test girders (Jung & White, 2006)

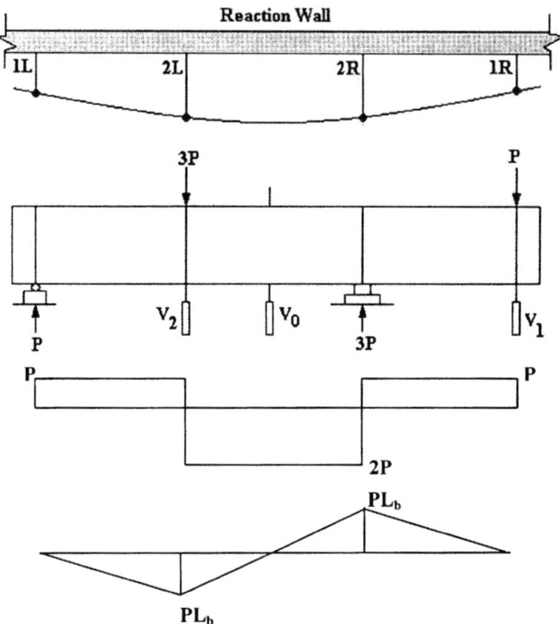

Figure 4 Test Set-up and applied loading (Jung & White, 2006)

moment/shear ratios were considered. 3mm thick web plate was chosen for girders with *d/t* ratio of 250 and, 5mm thick web plate for girders with *d/t* ratio 150. The panel aspect ratio of the web was restricted to 1.5 in all girders. In all composite girders the bond between steel girder and deck slab was achieved by means of shear studs, 19mm dia, 100mm long, welded in two rows. The width of deck slab was taken as 1000mm with an overall depth of 150mm. The slab width was 1200mm in some girders. The plate slenderness ratio of the flange outstands in steel girders varied from 5 to 6.67. Two of these girders, CPG9 and CPG10

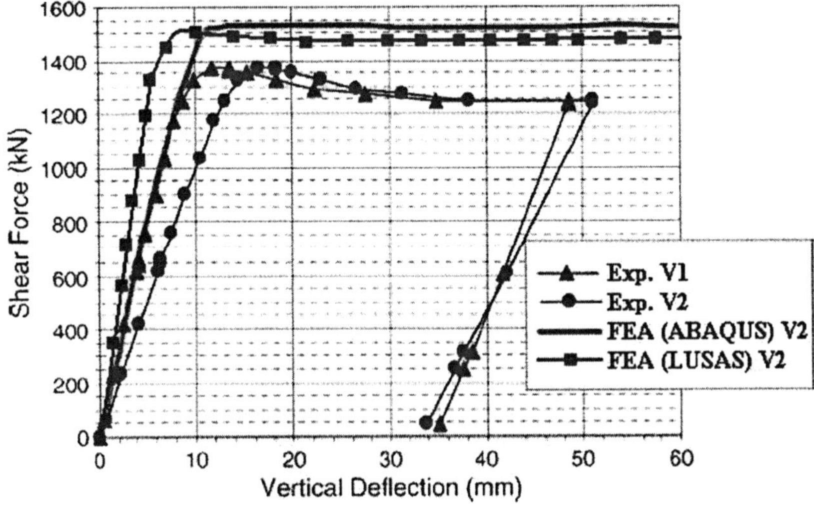

Figure 5 Load-deflection plots for steel girder S1-S (Jung & White, 2006)

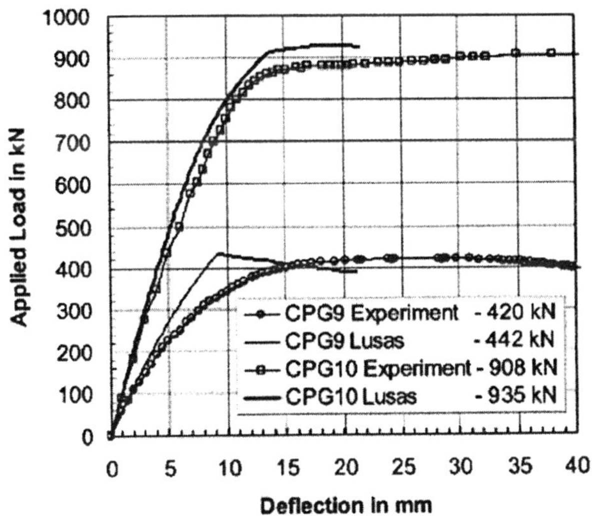

Figure 6 Load Deflection Plots for CPG9 and CPG10

were analyzed using LUSAS and the corresponding load-deflection plots are given along with the experimental results in Figure 6. It can be seen from the figures that LUSAS predictions lie very close to the experimental results thus proving the validity of the LUSAS analyses of straight composite plate girders accurately.

The results given above in respect of horizontally curved steel plate girder and straight composite girders establish the accuracy of the finite element modeling using LUSAS and the ability of the analyses to account for the horizontal curvature as well as the composite action between the concrete slab and steel plate girders. Having thus established the accuracy of the modeling using LUSAS it was decided to use the computer package for the analyses of horizontally curved composite plate girders.

Analysis of horizontally curved composite plate girders

The four steel plate girders, S1, S2, S1-S and S2-S tested by Zureick et al. (2002) are taken as the steel part of the composite girders analysed in this study. To each of these four steel girders, concrete slab of 200 mm thick and 2400mm wide is added at the top flange to act compositely with the steel part. The dimensions of the slab were chosen as per AASHTO recommendations for a composite girder. Full interaction in the composite action is assumed. The resulting four composite girders are identified herein in the text as C1, C2, C1-C and C2-Ccorresponding to S1, S2, S1-S and S2-S, respectively.

Description of the composite girders

Girder C1 corresponds to the steel girder S1 tested originally by Zureick et al. (2002) to examine the shear strength of a curved web panel of aspect ratio equal to 3. The ratio the web panel length to the radius of curvature was kept as 0.0575; subtended angle of 0.0575 between the cross frame locations, slightly greater than one-half of the maximum value of 0.1 permitted by AASHTO in the unified provisions for design of straight and curved I-girders was adopted (Jung and White, 2006). There were only four bearing stiffeners and no intermediate stiffeners were provided. Concrete slab of 200 mm thick and 2400 mm wide was added to the steel girder. Girder C1-C was identical to C1 but had an additional intermediate stiffener located at the center of each panel between bearing stiffeners. The resulting web panel ratio in this case was, therefore, equal to 1.5. Girder C2 was similar to C1 but differed only in the radius taken as 36.58 m thus giving a panel length to radius of curvature ratio of about 0.1. This value is close to the AASHTO requirement $L_b / R \leq 0.10$. Girder C2 – C was identical to C1 – C but had a radius of 36.58 m and panel aspect ratio of

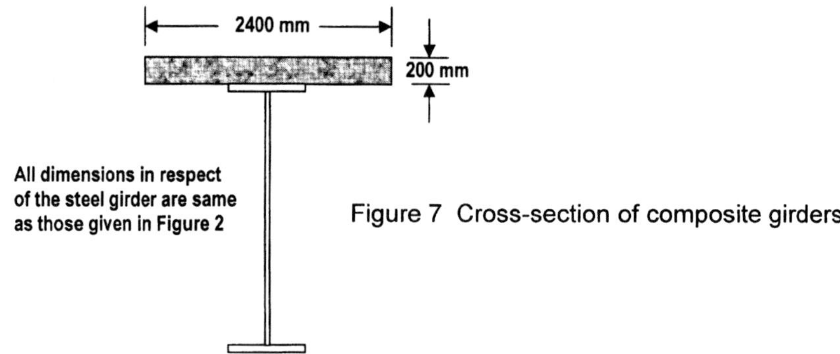

Figure 7 Cross-section of composite girders

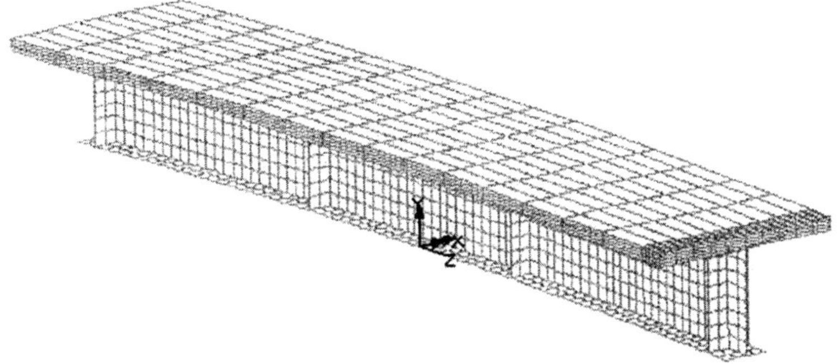

Figure 8 Typical finite element mesh for composite girders

1.5. Figure 7 shows the cross-section of a typical composite girder. A detailed finite element modeling was prepared for each of the girders; the finite element mesh of a typical girder is shown in Figure 8. The type of elements used for steel part of the girder and for concrete slab are same as those described earlier in this paper.

The support and loading conditions adopted for the analyses of composite girders are same as those used for steel girders by Zureick et al. (2002) and shown in Figure 4. The vertical support conditions provided for the physical model have been simulated in the finite element model by restraining the corresponding displacements at the nodes along a line across the width of the bottom flange. At the roller support, the girders are free to move along the tangential direction. Suitable restraints by means of truss elements are provided to the girder corresponding to the tube braces used to prevent the lateral torsional buckling of the girders as shown by 1L, 1R, 2L and 2R in Figure 4. Concentrated loads 3P and P as shown in Figure 4 are simulated in the analyses by beans of displacement control. Nominal residual stresses and imperfections for the steel part of the girder have been assumed in the analyses. The results obtained from the finite element analyses using LUSAS for all the four girders are presented in the following sections.

Results and discussion

The finite element analyses provided detailed output in terms of displacements, stresses, strains, moments and forces. However, for brevity only the most relevant results are presented herein for discussion. Figures 9-12 show the shear force-deflection plots for the four girders. Vertical deflection measured at the center of the supported span (V_2) is plotted against the corresponding shear force (P) in the span. In these figures, results corresponding to steel girders (Jung and White, 2006) are presented along with those for composite girders for comparison.

An elastic behaviour at the initial stages can be observed for all the girders and it becomes nonlinear soon after reaching the ultimate condition. Elastic buckling in the webs at the early stages of loading did not affect the overall behaviour of the girders. The behaviour of all the composite girders is same and it is similar to that of the steel girders and, enhancement in stiffness and ultimate load-carrying capacity compared to steel girders can be witnessed in all the composite girders. Composite girders exhibit significant gain in the ultimate load capacity, ranging from 25% to 35% over the corresponding values for steel girders. The gain should be attributed to the contribution due to the presence of concrete slab and the composite action. The increase in stiffness in respect of the girders C1, C1-C and C2 in particular is significant as shown in Figures 9-11. It is also clear from the results that the girders C1-C and C2-C having smaller web panel aspect ratios developed larger shear force compared to the corresponding girders C1 and C2, respectively with larger panel aspect ratios. Deformed shape after failure of a typical composite girder is shown in Figure 13. It can be seen that the webs developed tension field which was observed in the middle panel in which shear was maximum and the remaining panels also developed tension field as the load was increased. Similar behaviour has been reported for the corresponding steel girder also.

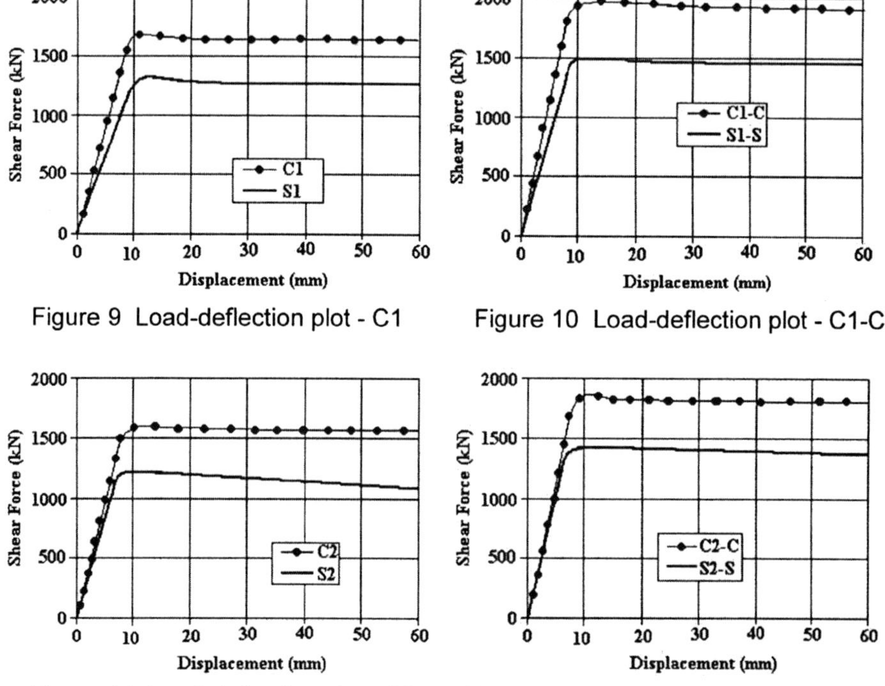

Figure 9 Load-deflection plot - C1   Figure 10 Load-deflection plot - C1-C

Figure 11 Load-deflection plot - C2   Figure 12 Load-deflection plot - C2 - C

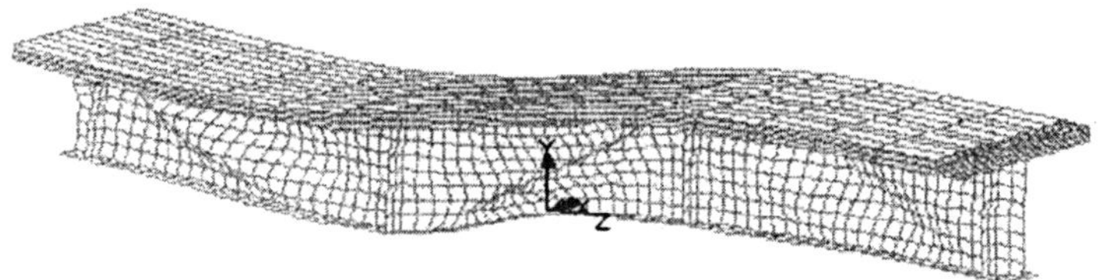

Figure 13 Deformed shape of the composite girder C1

Further analyses were carried out on composite girders to investigate the effect of radius of curvature and the flange width of the steel part of the girder. Three different radii of curvature viz. 50m, 75m and 100 m were considered in the analyses and the results are presented for the girder C1 in the form of shear force-deflection plots as shown in Figure 14. Corresponding results for straight girder are also presented for comparison. Similarly, analyses on composite girders were carried out considering four different flange widths viz. 203mm, 288mm, 460mm and 546mm for the plate girders keeping the thickness constant and giving plate slenderness for outstands ($b/2t_f$) of 4.6, 6.5, 10.5 and 12.4, respectively. Shear force-deflection plots from these analyses are given in Figure 15.

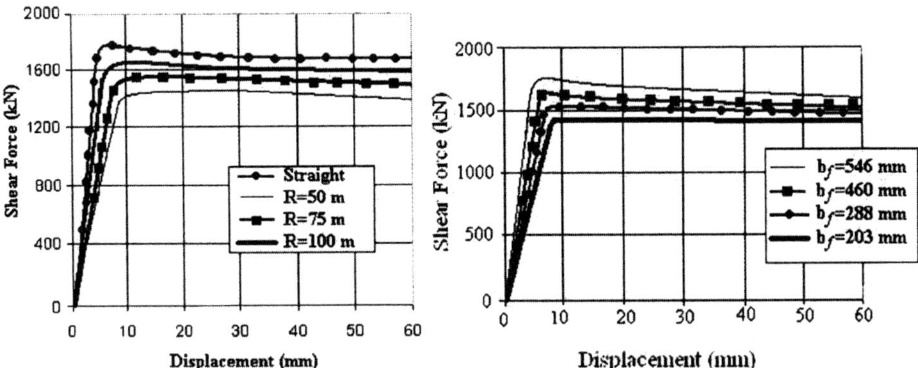

Figure 14 Effect of Radius of curvature   Figure 15 Effect of plate girder flange width

It is clear from Figure 14 that the straight girder resists the largest shear and its initial stiffness is greater compared to the curved girders. There is gradual drop in load resistance as the radius is decreased. Similar observations have been made in the results for other girders also. Larger flange width is found to result in larger load carrying capacity and stiffness though not significantly. It is necessary to study further the behaviour of composite girders and to investigate the influence of other parameters so that design methods could be formulated.

Conclusions

Finite element modeling using LUSAS computer package is proposed for the analysis of horizontally curved composite plate girders. The accuracy of the modeling is established by comparing the results thus obtained for test girders reported by other researchers. The behaviour of composite girders under the influence of parameters such as web panel aspect ratio, radius of curvature and flange width has been studied and the preliminary results are presented. It is necessary to carry out further analyses in order to obtain suitable design methods for horizontally curved composite plate girders.

References

AASHTO, LRFD bridge design specifications. 3rd ed. 2005 Interim provisions, Washington (DC): American Association of State and Highway Transportation Officials; 2004.

Allison, R.W., Johnson, R.P., and May, I.M., (1982). "Tension field action in composite plate girders." Proc. of the Instn. of Civil Engrs, London, V.73, part 2, June, 255- 276

Shanmugam, N.E., Mahendrakumar, M. and Thevendran, V., "Ultimate Load Behaviour of Horizontally Curved Plate Girders", Journal of Constructional Steel research, UK, Vol. 49, No. 4, 2003, 509-529.

Lian, V.T. and Shanmugam, N.E., "Openings in Horizontally Curved Plate Girders", Thin-Walled Structures, UK, Vol. 41, No. 2-3, 2003, 245-269.

Lian, V.T. and Shanmugam, N.E., "Design of Horizontally Curved Plate Girder Webs Containing Circular Openings", Thin Walled Structures, UK, Vol. 42, No. 5, 2004, 719-739.

Zureick, A. H., White, D. W., Phoawanich, N. and Park, J, "Shear strength of horizontally curved steel I girders – experimental tests.   Final report to Professional Services Industries, Inc. and Federal Highway Administration, March, 2002.

Jung, S.K. and White, D.W., "Shear strength of horizontally curved steel I girders – finite element analysis studies", Journal of Constructional Steel research, UK, Vol. 62, No. 4, 2006, 329-342.

Evans, H.R., Porter, D.M. and Rockey, K.C. (1978). "The collapse behavior of plate girders subjected to shear and bending." Int. Ass. Bridge Struct. Eng, periodical, proc. P-18/78

Baskar, K. and Shanmugam, N.E., (2003a). "Steel-concrete Composite Plate Girders Subject to Combined Shear and Bending." Journal of Constructional Steel Research, Vol. 59, n. 4., 531-557.

Shanmugam, N.E. and Baskar, K., (2003b). "Steel-concrete Composite Plate Girders Subject to Shear loading." Journal of Structural Engineering, ASCE, Sep., Vol. 129, n. 9, 1230-1242.

Shanmugam, N.E. and Baskar, K., "Design of Composite Plate Girders under Shear Loading", International Journal of Steel and Composite Structures, Vol. 6, No. 1, 2006, 1-14.

Assessment of stress conditions of a seismically damaged cable-stayed bridge after repair

C. C. Chen, Associate Professor, National Yunlin University of Science & Technology, Taiwan
K. C. Chang, Professor, National Taiwan University, Taiwan
Z. K. Lee, Ass. Research Fellow, National Center for Research on Earthquake Engineering, Taiwan.

Abstract

On 21 September, 1999, the Chi-Lu Bridge, just as its construction was near completion, was devastated by a strong earthquake. For safety concern, the bridge was repaired without making a detailed assessment of the structural conditions following the earthquake, so as to result in an unknown structural system. In order to verify the allowable live-load capacity, a two-stage load test was conducted before the official opening of the bridge. A modified structural system, in which the inherent statistical indeterminacy was purged by replacing all stay cables and the two side piers in the original structural system with a set of external forces, was then proposed to calculate the internal stress of the bridge subject under design load for the assessment of the structural safety of the bridge. The object of the paper is to present the load test and the modified analysis procedure in assessing the internal stress of the bridge.

Introduction

Chi-Lu Cable-stayed Bridge, crossing Juosheui River in the Central Taiwan, is a single-pylon symmetrical cable-stayed bridge with total length of 240m. The girder is a five-cell prestressed concrete structure, arranged in order of streamlining - 24m wide and 2.75m deep at the center. The hollow concrete pylon, 58m in height, is rigidly connected to the girder and the solid pylon pier of 15m in height, which is supported by bored piles. The cable system is designed in the semi-fan shape along the centerline of the girder, comprised of 17 pairs of stay cables in each side of the pylon. The movable bearings are set at both ends of the girder, with shear keys installed to provide for constraint in transverse direction. The general arrangement is shown in Fig. 1.

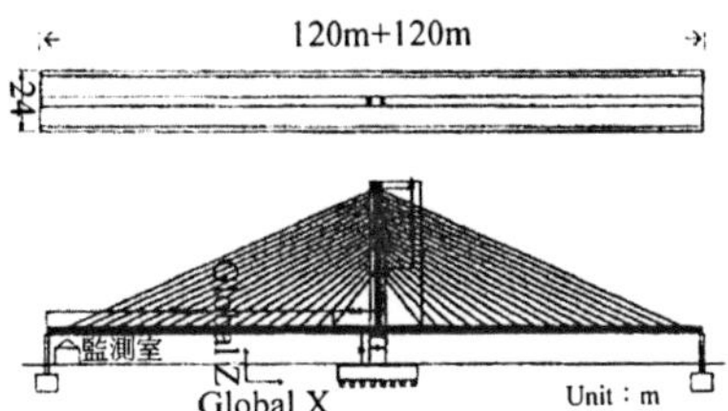

Fig. 1 General Arrangement of Chi-Lu Cable-stayed Bridge

Bridge design, construction and maintenance 2007, Thomas Telford, London

The bridge's construction commenced in 1996, and, just as nearly completed, was devastated by a strong earthquake of a magnitude of 7.3 on the Richter scale that occurred about three kilometers north of the bridge site in September 1999. Likely due to near-fault effect, most of the major structural components were seriously damaged. A large lateral shift of the bridge deck formed as a result of the strong transverse seismic force that also caused flexural-shear failure of prestressed concrete box girder at the juncture with the pylon. A very large torsional moment, induced by the transverse movement of the deck, caused excessive shear stress that led to a large vertical crack of 2 meters long stemming from the pylon bottom, and a large area of concrete pieces falling off. The pounding of the dislodged box girder gouged the parapet wall and caused the shear crack in the cross beams of both end piers. A thorough inspection of the whole cable system revealed that there was no serious damage to the major cable components, except that some stay cables had suffered serious deformation at the shim plates of the anchor heads, and that one stay cable during the stressing stage was extricated from its protection tube as the pulling rod was cut off by the shear force exerted by the heavy jack under the influence of the seismic motion. In accordance with the numerical analysis result of the bridge subjected to the recorded ground acceleration near bridge site, it was deduced that the girder lifted up in the initial stage of the earthquake and then moved just like a free cantilever beam during the seismic motion[1]

For safety concerns, without conducting a detailed evaluation of the structural system, the repairing work of the damaged structural components of the bridge, except its cable system, was underway right after the earthquake, and the entire prestressed concrete box girder was supported by several temporary towers located at the river bed. Two years later, after a complete retrofit plan was built up, the whole cable system was replaced and tensioned. In the meantime, in consideration of the public transportation needs of the nearby area, the bridge was expected to open temporarily for limited-size vehicles. Because the structural system of the bridge was changed to unknown status by the serious seismic damages, it has been a critical issue to confirm the correct stress condition and then assess the structural safety of the bridge, especially in service phase. Hence, a new method, which was based on a statically determinate structural system modified from the original one, was proposed to calculate the stress condition of the bridge subject under its own weight. Besides, load test was performed to check the allowable capacity of live load when the bridge was ready to open for traffic.

Load test

Besides an analytical study was carried out for the assessment of structural safety of the bridge, load tests were performed to check allowable capacity of live load when it was ready to open for traffic. Standard 30-ton trucks were adopted to be the source of test loading, which were deployed on the bridge deck according to the preset test loading patterns that simulated the design live loads prescribed in Taiwan's Bridge Code. It states that the bridge is designed for HS20-44 loading or the corresponding lane loading, and the sidewalk live load is 300 kg/m^2. The impact fraction was equal to 0.096, and the safety factor for overloading truck was revised to 30% in accordance with the guideline of the bridge. The bridge deck – 23.6 wide, consists of two traffic lanes and one sidewalk in each traffic direction. The reduction in intensity of load, 25% set for four traffic lanes according to the Code, was not considered in this design case. By taking all of the computational factors into account, the designed live load was approximately 6 tons/m, that was used for the reference value for the load test.

Fig. 2 shows the loading patterns implemented right before the official opening for traffic, including for static load test and dynamic load test. Uniform load test, unbalanced load test and torsional test were arranged in the static load test, and the maximum test load was set to be 5 tons/m, about 83% of the design lane loading. With inclusion of the impact load test and car-braking test, the dynamic load test was conducted for the purpose of identifying the natural frequencies of the bridge. The same trucks were employed to generate the dynamic responses for the tests. To obtain the structural information for the assessment of the structural safety of the bridge, several sensors were set up to collect the response signal induced by the specific loading patterns. The installed sensors, including inclinometer, displacement sensor and strain gauge, were organized to monitor pylon tilt and girder deflection. A number of velocimeters were applied to record the ambient vibration of the cable system, pylon and girder, for frequency identification.

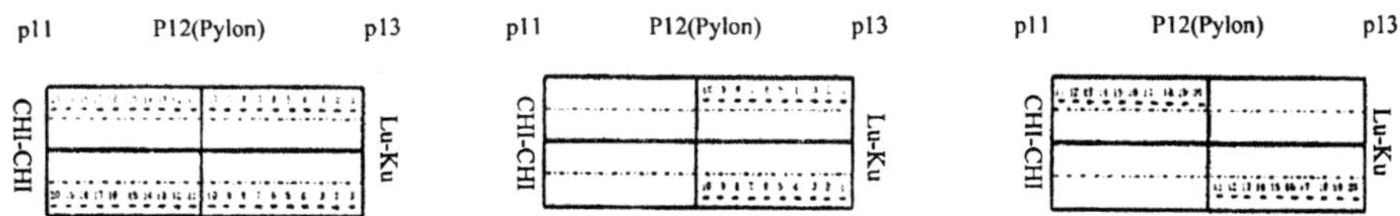

Fig. 2 Loading Patterns

When compared with the analyzed results, most of the measured deformations were smaller in the whole test loading patterns. Tab.1 lists the analyzed results and measured data of the girder deflection and pylon deformation in the uniform load test and unbalanced load test. It is clear that most of measured data corresponded to about 80% to 90% of the analyzed value, except the tilts at the pylon top. When the trucks were deployed on the deck one by one for the uniform load test, the increments of the concrete strain at the measured cross sections displayed an approximately linear relationship against truck number. In addition, while forty trucks were all positioned on the deck, the concrete surface inside box girder was inspected in detail, and no crack was found to have developed on the surface. The calculated variation of the cable forces, based on the change of natural frequencies of cables, closely matched the analyzed values from the bridge model subject under the assigned loading patterns. Following the static load test, the dynamic load test was undertaken that included the impact load test and car-braking test. According to the test plan, it was intended to identify the natural frequencies of the bridge from the measured data. In reality, the measured data was not clear for the frequencies identification due to certain error in test arrangement. By comparing the measured data with the analyzed result, it was concluded that the bridge was in good condition after repair and capable of providing enough capacity for the design live load.

Tab. 1 Measured Deformation in Each Load Test

Items		Analytical Results	Measured Data
Max. Deflection	Full Load on two Spans	4.6 cm	4.08 cm
	Full Load on One Span	8.8 cm	7.40 cm
Rotational Angle (Full Load on One Span)	Pylon Top	0.013°	0.0198°
	Pylon Post	0.041°	0.040°
	Pylon Bottom	0.032°	0.024°

Vibration test and model updating

For assessing the structural safety of the bridge after repair, a new numerical analysis procedure, which was based on a modified finite-element bridge model, was developed to calculate the internal stress of the repaired bridge subject under the design loads. The bridge model, which simulated the bridge structure in service phase, was updated in order to provide the more accurate output result for the safety assessment. The natural frequencies of the bridge were used as the reference value for the updating. In this study, after the load test and right before the official opening for traffic, a forced vibration test was performed to identify the natural frequency of the repaired structure. In the beginning of the test, the ambient vibration of the bridge was recorded.

Before the test, the original finite-element model, used in design phase, without updating, was used to make dynamical analysis for obtaining the mode shapes and corresponding natural frequencies. The direction of mode shapes is defined according to the calculated effective mass in three orthogonal directions. Figs. 3 to 5 show the first three mode shapes in the longitudinal direction, transverse direction and vertical direction, respectively. All of the longitudinal modes shown in Fig. 3 are in-plane and anti-symmetrical vibration of the bridge. Shown in Fig. 4, the first mode is mainly caused by the pylon vibration in the transverse direction and the remaining two modes are out-plane and symmetrical vibration of the bridge. All of three modes shown in Fig. 5 are in-plane and symmetrical vibration of the girder in combination with the still pylons. The corresponding natural frequencies, shown in Tab. 2, would be used to point out the possible range of the natural frequencies identified from the forced vibration test as well as ambient vibration test.

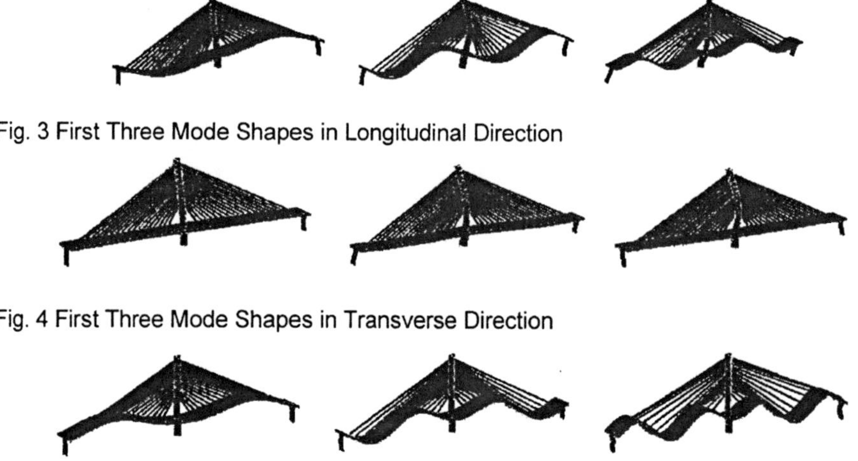

Fig. 3 First Three Mode Shapes in Longitudinal Direction

Fig. 4 First Three Mode Shapes in Transverse Direction

Fig. 5 First Three Mode Shapes in Vertical Direction

In the forced vibration test, a shaker was used to generate the exciting force. The excitation frequencies of the force were adjusted progressively from 0 Hz to 10 Hz in two perpendicular horizontal directions and from 1.5 Hz to 5Hz in vertical direction, at intervals of 0.05 Hz apart. Due to gravity effect, it was impossible to generate the exciting force with the excitation frequency below 1.5

Hz in vertical direction. The frequency-response curves were plotted for the identification of the natural frequencies of the bridge in three orthogonal directions. The X axis of the figure is the output exciting frequencies of the shaker and the Y axis is the amplitude of fast Fourier transform of the recorded vibration data at the point of the output excitation frequency divided by the exciting force of the shaker. The natural frequencies of the bridge should coincide with the output frequencies associated with the local maximum response in the figures.

For the description of the identified frequencies, the assigned vibration direction in following statements is defined by the measured direction. It is shown that the first two identified frequencies in longitudinal direction locate at 0.65 Hz and 1.65 Hz and no clear peak can be identified for higher modes. The first three identified frequencies in transverse direction are 0.55 Hz, 1.65 Hz and 2.65 Hz. The first three peaks in the figure locate at the output frequencies of 1.7Hz, 2.75Hz and 4.3Hz, which should be the natural frequencies associated with the second mode to fourth mode in vertical direction. As stated before, it was impossible to locate the first natural frequency because its value is less than the lowest excitation frequency generated by the shaker. According to the recorded ambient vibration data shown in frequency domain, one can say that the first natural frequency locates at 0.72 Hz in longitudinal direction and 0.51 Hz in transverse direction, and the second natural frequency is 1.94 Hz in vertical direction.

Tab. 2 lists the analytical natural frequencies and the identified frequencies from the forced vibration test as well as ambient vibration test. It is clear that there are some differences between the analytical values and the identified results. In addition, the identified frequencies are also different from the forced vibration test and ambient vibration test. As stated before, it was important to update the bridge model in order to provide the accurate stress condition for the assessment of the structural safety of the bridge after repair. A two-stage analysis process was adopted to update the bridge model in this study. At first, the sensitivity analysis was performed to select the decisive design parameters in changing the values of the natural frequencies shown in Tab.2. Then, a trial-and-error manner was applied to update the design parameters until the analytical frequencies close to the identified frequencies.

Tab.2 Analytical Frequencies and Identified Frequency

Mode	Longitudial Dir.			Transverse Dir.			Vertical Dir.		
	MIDAS	Forced Vib	Ambient Vib.	MIDAS	Forced Vib	Ambient Vib.	MIDAS	Forced Vib	Ambient Vib.
First Mode	0.56	0.72	0.65	0.47	0.51	0.55	1.02	-	-
Second Mode	1.59	-	1.65	1.81	-	1.65	1.90	1.94	1.70
Third Mode	2.17			2.63		2.65	3.30		2.75

The major design parameters considered in the sensitivity analysis were the flexural stiffness of the girder, pylon and end piers, the axial stiffness of the stay cables and the boundary condition at the both ends of the girder. The calculated results indicated that the decisive design parameters to the modes in transverse direction were the flexural stiffness of the pylon in weak axis, the flexural stiffness of the girder in strong axis and the flexural stiffness of the end piers in weak axis, but only the flexural stiffness of the pylon affects the first mode. The natural frequencies of the longitudinal modes and vertical modes were all affected by the axial stiffness of the stay cables, the flexural stiffness of the girder in weak direction and the boundary condition of the girder in longitudinal direction, but the last parameter was more decisive. The flexural stiffness of the pylon in strong axis only changed the natural frequencies of the vertical modes. The flexural stiffness of the end piers in weak direction was not important in all modes and would not be considered in the updating process. Besides the aforementioned design parameters, the mass of the girder was considered in the updating process.

In the beginning of the updating process, the Young's modulus of concrete structural components was adjusted from 2.8 E05 kg/cm^2 up to 3.1 E05 kg/cm^2 because the average concrete strength measured during the construction stage was closed to 420 kg/cm^2, larger than the design value of 350 kg/cm^2. During the trial-and-error updating process of the design parameters, it was found that the mass of the girder has to be increased, otherwise, it was not possible to update the other parameters. Therefore, an uniform load of 18.5 tons/m was added to the mass of the girder, which came from the pavement, parapet and so on, having been ignored in the original bridge model. The flexural stiffness of the pylon in weak axis was increased in order to enlarge the natural frequencies of the transverse modes. The flexural stiffness of the end piers in weak axis was decreased in order to reduce the natural frequency of the third transverse mode. The flexural stiffness of the girder in strong axis was not taken into account because the required adjustment is too large to be reasonable.

As the boundary condition of the girder was changed from roller to rigidity the natural frequencies of the longitudinal modes and vertical modes were increased to the intermediate values between that from the ambient vibration test and the forced vibration test. It was inferred that the boundary condition of the girder could be rigid during the ambient vibration test, because the external force was too small to move the girder from the bearing atop of the piers. The final calculated showed that there were still a big difference between the calculated value and identified value of the natural frequency of the second longitudinal mode. After updated, the bridge model was used to calculate the deflection for the unbalanced load case and uniform load case implemented in the load test right before the official opening again. When compared with the model before updating, the updated model gave the value of the deflection much close to the measured values in the load test, which also confirmed the updated model with higher accuracy.

Assessment of stress condition

If the internal stress of the bridge under its own weight was already determined, the updated bridge model could then be used to calculate the internal stress of the bridge subject under the design loads for the assessment of the structural safety of the bridge. As stated before, it was not possible to follow this procedure to assess the structural safety of Chi-Lu Bridge because the stress condition was dramatically changed by Chi-Chi earthquake. If the stress condition before the damage was known and the stress change induced by the damage and repair could be quantified, it was still possible to determine the stress condition after the damage and repair, by just releasing the quantified stress in the model before the damage. In reality, there was no any record about those items. Hence, another procedure was developed to calculate the internal stress of the bridge after repair, without the requirement of abovementioned information.

The reason why it was not possible to calculate the internal stress of the bridge, after repair, under its own weight directly from the loading analysis of the updated model was the bridge belonging to statically indeterminate system, of which the internal stress could not be calculated unless its construction, or damage, procedure was known. In contrast, without the information of the damage, the internal stress of the bridge after repair could be calculated if the structural system used for the calculation was a statically determinate system. Practically, the structural system of single-pylon cable-stay bridges could be modified to be statically determinate as long as the internal forces of the whole cable system and the reaction forces of the bearing atop both side piers could be obtained. Therefore, by replacing all stay cables and two end piers with a set of known external force, Chi-Lu Cable-stayed Bridge turned to be a statically determinate system, as shown in Fig.6, for the determination of the internal stress of the bridge, after repair, under its own weight[2].

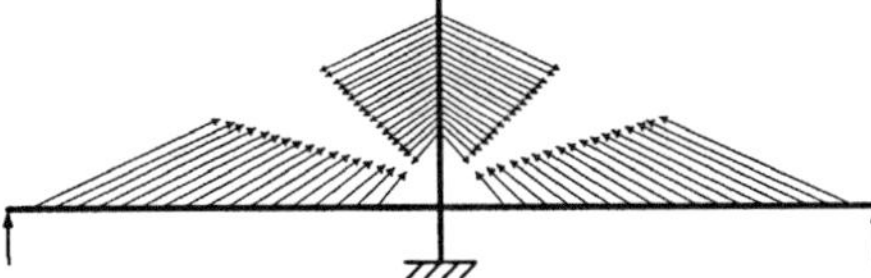

Fig. 6 Modified Structural Model

In order to obtain the external fore at the location of the piers, as shown in Fig. 6, a jack-up work was conducted after the repair was complete. The displacement between the pier top and the bottom of the girder was recorded by using dial gauge as the jack-up force was applied progressively to the end diaphragm of the girder. The measured displacement was plotted against the jack-up force in Fig.7. The displacement-force curve comprises of the first elastic range, transition zone and second elastic range. The measured displacement in the first elastic range should come from the released deformation of the compressed rubber inside the bearing pot, which implies the reaction force being transferred from the bearings to the hydraulic jacks. The measured displacement in the second elastic range should come from the elastic deflection of the girder, which means that the transfer of the reaction force was completed. Therefore, it could be realized that the reaction force of the pier would be 291 tons, locating at the starting point of the second elastic range.

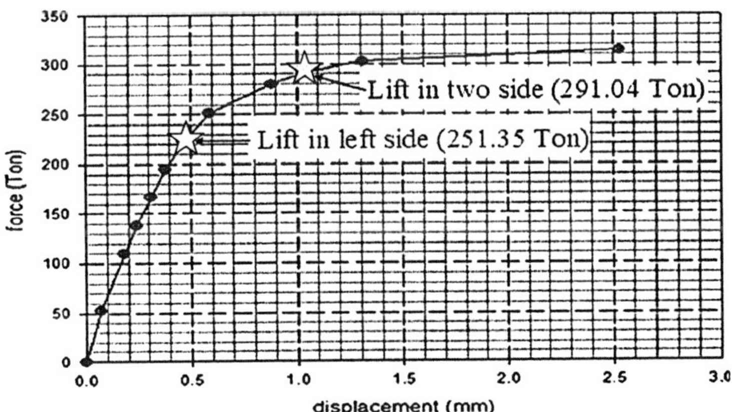

Fig.7 Force-Displacement Curve

The stay cables designed for Chi-Lu Cable-stayed Bridge are classified into three types, 33, 43 and 55 parallel strands, in combination with APS anchorage at both ends. The cables ranges from 28.3 m to 125.22 m long, with unit weight of about 50 kg/m. Based on a finite-element cable model, a search process, in which the analytical natural frequencies have to match the compared natural frequencies identified from the measured ambient vibration data, was developed to find out the optimal internal force, flexural rigidity and effective vibration length of the cables in a trial-and-error manner[3]. In this study, right before the official opening the ambient vibration data of all stay cables were recorded for the identification of their natural frequencies, which were used to be the compared frequencies in the search process of the cable forces. According to the analyzed results, it was not possible to determine the exact values of the effective vibration length of the cables, which might be equal to the free length of the cable plus 70% to 80% of the anchorage length. Hence, by taking two extreme values of 70% and 80%, two sets of cable forces were calculated to be the input external forces, as shown in Fig.6.

After the reaction forces at both end piers and the internal forces of the whole cable system were obtained for the bridge system after the repair and right before the official opening, the stress condition of the bridge was calculated by applying the obtained forces at the positions of both end piers and anchor blocks of the stay cables, as shown in Fig.6. Fig.8 shows the calculated top and bottom internal stresses of the girder. For the assessment of the structural safety of the bridge after repair, according to the design requirements prescribed in Taiwan's Bridge Code the complete bridge model was used to calculate the internal force of the bridge subject under each design load, such as live load, wind load, earthquake load and so on, and the combination values of the forces were calculated for the allowable stress check of the girder and the ultimate design check of the structural components. The result showed that the calculated stress of the girder was less than the allowable stress specified in the guideline of the bridge. Based on the result of the ultimate design check, the bridge after repair was proved to still possess enough capacity for the design loads.

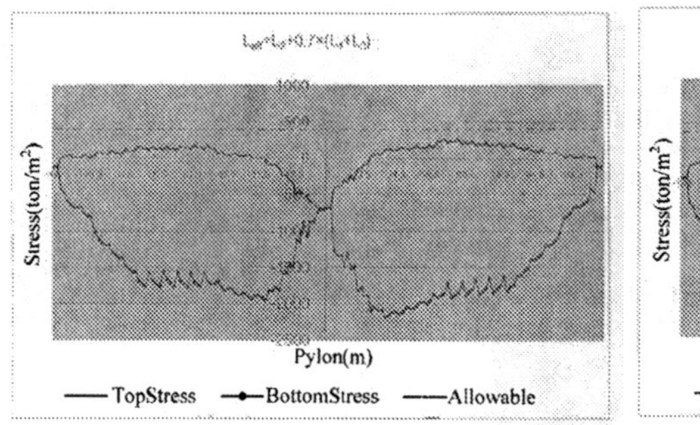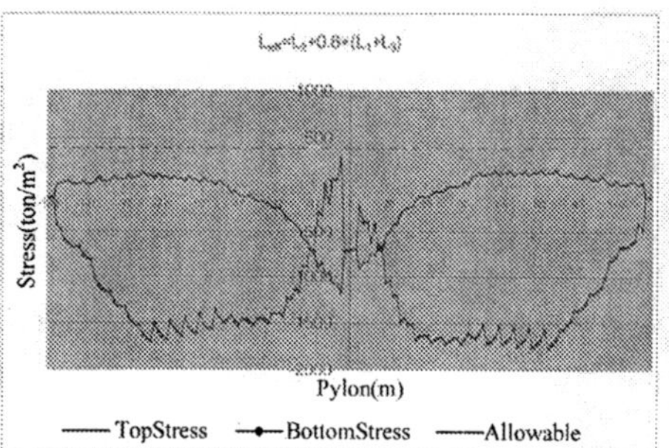

Fig. 8 Top and Bottom Stresses of the Girder.

Conclusion

Chi-Lu Cable-stayed Bridge was damaged in 1999 and officially opened for traffic in late 2004, after it was proved, experimentally and numerically, to be safe in service phase. So far, it has been monitored and remained in a good condition.

Reference

1.K. C. Chang, Y. L. Mo, C. C. Chen, L. C. Lai and C. C. Chou(2004), "Lessons Learned from the Damaged Chi-Lu Cable-Stayed Bridge", Journal of Bridge Engineering, ASCE, Vol. 9, No. 4, July/August.
2.Lee., Z. C. (2007). "Assessment of Stress Condition of the Seismically damaged Chi-Lu Cable-stayed Bridge after Repair." Thesis, NYUST, Taiwan.
3.Chen C. C., Lee, Z. K., Chou, C. C. and Chang, K. C. (2006). "Parametric Studies of Force Analysis of Stay Cable." Journal of Chinese Institute of Civil and Hydraulic Engineering, Vol.18, No.4, 555 - 563.

Taiwan High Speed Rail ：- The Structures and Quality Management System – A Case Study

Nigel Beaney Bsc CEng MICE, Taiwan High Speed Rail Corporation, Taiwan

Introduction
This paper has restricted its study to the types of structures and the methods of deck construction utilized on the THSR. It reviews some of the methods and the most common quality issues. The paper also provides a brief overview of the Quality Management System employed and provides some observations and suggestions on how the system could be improved to the benefit of the Employer and Contractors for projects of similar nature in the future.

The Project
Taiwan High Speed Rail Corporation (THSRC) was awarded a 35 year concession in 1998 to Build, and Operate the high speed railway between Taipei and Kaohsiung a total length of 345km. The HSR will be transferred to the government free at the end of the concession period. The railway is designed to operate at 300kph and incorporates the core systems from Japan using the Shinkansen 700T train sets. The alignment and the right of way were provided by the Government. There are 8 stations located along the HSR route with plans for a further 4.

The civil works were divided into 12 separate contracts and awarded to joint ventures of local and international companies on a design and construct basis to a basic design concept and specification provided by THSRC. The works were constructed between spring 2000 and autumn 2004.

The Contractors were responsible, amongst others, for the design, construction, safety, quality and environmental issues of the works.

Taiwan being a seismically active area and also prone to typhoons posed additional challenges.

THSRC carried out its own project management and also employed an independent Checking Engineer and Site Engineer team. An Independent Verification and Validation Engineer was also employed to certify that THSRC had complied with the terms of its contract with the Government and that the railway met the necessary safety and operational requirements.

The railway successfully entered revenue service at the beginning of 2007.

Key Statistics
In order to illustrate the scale of the construction works undertaken, figure 2 provides a

Bridge design, construction and maintenance 2007, Thomas Telford, London

summary of the project scope.

```
O/A length           345km
No. of Stations      8
Operating Speed      300kph
Length at grade        : -  31km
        Tunnels        : -  39km        N⁰ of Tunnels              : - 42
Cut & Cover Tunnels : -  24km
        Elevated       :-   251km       Max Continuous Length : - 157.316km
Major Quantities       : -
Structures
    Piles              : - ≅1400km      Total N⁰   30,880
    Spans              : - >8300
    Concrete           : - 8.5x10⁶m³     Rebar : - ≅ 2x10⁶te
    Prestress          : - ≅70,000te

Earthworks
    Excavation         : - 13x10⁶m³
Tunnels
    Excavation         : - ≅7x0 ⁶m³           Concrete : -≅2x0 ⁶m³
```

Figure 2 : -Summary of Project Scope

Structure types

In general the basic alignment proposed by the THSRC considered the use of a standard 30m span for the viaduct sections with longer spans for crossing freeways, railways & rivers. The Contractors were allowed to consider alternative span lengths and several adopted increased standard span lengths up to 40m for the viaduct sections.

Standard Structures

The standard structures were defined as the long continuous sections of viaduct consisting of a regular pattern of 30m or 35m spans with intermediate nonstandard bridges at infrequent intervals. It should be noted that the HSR route includes the longest section of continuous viaduct in the world at 157.316 km.

Fig. 3 : - view of viaduct section Paghuashan

The choice of construction method was determined according to the overall length of section, availability of resources, experience in precast works, local experience. In general where the JV included major overseas contractors the use of the FSPLM method was adopted. Where the dominant joint venture partner was Taiwanese they opted for the advanced shoring or multishoring systems. The FSPLM method was used on 5 contracts and advanced shoring was used on 6 contracts. Due to the high equipment and set up costs overall lengths in excess of approximately 10km are required for the FSPLM to be viable. For Advanced Shoring method minimum lengths of 3km are viable. The multishoring system being used for lengths less than 3km.

Full Span Precast Launch Method (FSPLM)

The FSPLM technique proved extremely reliable and efficient. Most contractors were able to maintain progress in excess of 1 deck unit placed per day per launcher with peak progress of approximately 40 units per month per launcher. The launchers proved versatile and were transported across earthwork sections to reach short lengths of viaduct, see Fig 4. They were turned through 180 degrees in place on the viaduct. They were dismantled, transported and re-erected Fig 5 and they were transversed sideways. With the exception of one contractor, the contractors used on line or parallel precast yards and similar launchers and carriers Figs 6, 7 & 8. However contract C215 was unable to obtain suitable area for their precast yard and adopted a yard off line. In this case they opted to use a combined launcher/carrier fig 9.

Fig.4 Traversing Earthworks

Fig.5 Transporting Launcher

Fig. 6 Precast Yard

Fig. 7 Carrier + Beam

Fig. 8 Launcher

Fig. 9 C215 Launcher

For precast yards the contractors adopted differing degrees of prefabrication and completion. C250 adopted on line factory methods and launched a complete deck unit complete with cable troughs, parapets and derailment walls. Other contractors opted to launch only the precast deck unit.

Within the precast yards the contractors adopted differing degrees of rebar prefabrication from total offline prefabrication including sub assemblies for diaphragms fig. 10 & 11 to in situ assembly within the precast forms. The quality achieved in all cases was to international standards with little or no remedial works necessary.

Fig. 10 Prefab deck rebar

Fig. 11 Rebar Prefab Diaphragm

The contractors not only demonstrated the versatility of the method they were also able to increase production through efficiencies in excess of the predicted 1 unit per day per launcher.

Advanced Shoring Method
Not all contractors adopted the FSPLM technique where the overall length was in excess of 10km. This was in part due to the local experience in the use of the Advanced Shoring method fig. 12.

Fig. 12 Advanced Shoring Method

Whereas the FSPLM technique exceeded the predicted progress rates the Advanced Shoring method fell short and most contractors were required to mobilize additional sets of equipment to meet their overall programme requirements.

The advanced shoring method requires the rebar to be fixed in place and quality issues did arise. The labour resources required were considerably greater than those needed for the FSPLM technique and this was a problem given that the guideway on all 12 contracts was being constructed simultaneously.

It is the authors opinion that if the length of viaduct justifies the use of the FSPLM technique, then the benefits outweigh the initial costs of setting up the precast facilities.

Non Standard Bridges

Non Standard Bridges are defined as short single or multispan bridges crossing highways, rivers, valleys or are located within longer sections of viaduct but of different construction.

In choosing the type of structure and construction method, the contractors took into consideration the span lengths, access, availability of equipment and resources, and local practice.

A brief description is provided of each method.

Multishoring or Birdcage Scaffolding Method

This method was used for individual bridges up to 5 spans crossing valleys, roads and where access was available, and is commonly used in Taiwan. Figure 13.

Fig. 13 Multishoring

Warren Truss

These bridges, whilst not the most aesthetically pleasing, have over time proved to be economical structures for railway bridges up to 150m span. These structures were successfully adopted on 5 contracts with the longest crossing highways, river and railway tracks on C250 with an overall length of 410m and individual maximum span of 140m. The choice between launching or in situ construction was made on the basis of access, physical restrictions and whether temporary supports were possible. Figure 14.

Fig. 14 Warren Truss

Balanced Cantilever Construction

This method is a common technique in Taiwan for highway bridge crossings of rivers, freeways, valleys etc. The common span combinations used are between 40/80/40, 20/40/20. The box depths were easily modified to suit the increased train loadings.

THSRC did however require the contractors to provide facilities for additional external prestress for future upgrading if it was deemed necessary. Figure 15.

Fig. 15 Balanced C'lever

Cast/Push Incremental

This techniques was used on only contract C260, despite the fact that it is relatively common form of construction in Taiwan. It was used in 4 locations where the access was difficult with a minimum number of spans of 6. A particular disadvantage of this system was the restriction to the max length of continuous deck not to exceed 105m, which was overcome by the use of temporary connections between each section.

Steel/Concrete Composite

This form of construction was only used in 2 locations, Contract C220 for a river crossing where steel plate girders were used with in situ concrete decking for spans up to 50m, and Contract C250 for a crossing of the freeway where the headroom was limited. In this case the contractor chose on open steel box with in situ concrete deck. The box being launched.

Quality Issues

The Contract required the Designer to take into consideration quality of construction and maintainability. Below are some examples of how the design impacted on the quality of the work and its long term maintenance. The two most common quality issues related to the concrete shear keys used for seismic stability, and the associated end diaphragm construction of the deck units. The typical arrangement most commonly used incorporated a deepened end diaphragm section to permit the seismic anchors to be fitted externally between the deck and pierhead.

This however creates several construction issues which if not properly addressed can lead to quality issues.

First, if the deck unit is constructed in situ, then the deepened diaphragm sections tend to attract all the debris and reinforcing tie wire. The area is heavily reinforced and also accommodates the bridge bearings and anchor bolts, anchorages for prestressing, and ducts for seismic anchor bars. It is very difficult to clean, place and compact the concrete, and as such this highly stressed area of the deck unit becomes prone to poor concrete quality with honey combing and voids a potential risk.

Secondly the purpose of the concrete shear key and its interaction with the bearings must be fully understood in determining the sequence and accuracy of construction of the concrete shear plinth on the pier heads. In some cases this was not fully understood and led to misaligned sliding surfaces and poorly fitting elastomeric pads. In certain areas considerable retrofitting of the shear keys was necessary.

An alternative adopted on some contracts was the introduction of mechanical shear keys rather than the cast in situ concrete type. These devices can be likened to an upside down pot bearing with vertical uplift clamps. These devices were aligned with the bearings before the deck was placed thus ensuring that the bearings and shear key acted as a composite unit. A further advantage was the lack of any tie bars vertical or horizontal, which reduced the congestion in the end diaphragms. Figures 16 and 17 illustrate the issues relating to shear keys and bearings.

Fig. 16 Concrete Shearkey

Fig. 17 Bearing + Mechanical Shearkey

Thirdly there is the difference between precast and in situ construction. Again it was evident that where precast construction was used notably in the FSPLM works or where the contractors utilized extensive prefabrication of rebar for the substructure works, Fig. 18 there were less quality issues and works requiring remediation. It is the authors opinion that some contractors and their designers did not recognize the value in precast or rebar prefabrication in respect of quality, progress and their impact on the overall costs of construction. For example had contractors prefabricated the diaphragm sections for in situ deck construction similar to the precast units Fig. 11, they would have been able to reduce the amount of accumulated debris, built in specific locations for concrete placement and vibration and thus reduce the risk of poor quality and reduce the overall period of construction for each deck unit. The choice as to whether to adopt prefabrication of the rebar or not was, in my opinion, determined by local practice and the benefits were not considered. Where defects were detected or suspected NDT techniques were used to determine the extent of the defect and to verify the remediation works. The types of techniques used included Visual, impulse test, Coring, Hammering, Penetration Radar, and Pull off tests (remediation works only).

The Remediation works required the removal of the poor concrete and recasting with approved mortars, or drilling and grouting using low viscous materials such as, Ama 600P, where voids were identified within the core of the concrete.

Fig. 18 Prefab Pierhead Rebar

Quality Management System (QMS)

The QMS required that the Contractors were fully responsible for their own quality systems of both design and construction by establishing a programme based upon the requirements of ISO9000.

This programme also required the Contractor to employ the services of an Independent Checking Engineer (CICE) for the design and geotechnical verification, however not for construction.

The Employer would monitor and audit the programmes by his own project management team, supplemented by an Independent Checking and Site Engineering team as illustrated in Fig. 19.

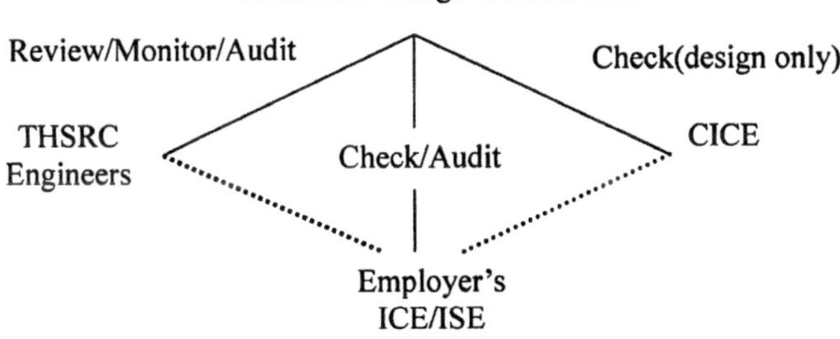

......denotes audits by ICE/ISE on THSRC.

Fig. 19 Checking/Monitoring Process

As noted in this paper, quality issues did arise and it is questionable whether on projects of this type a QMS controlled by the Contractor is effective. From a review of not only how the Contractor carried out his quality management but also how the Employer conducted his monitoring, improvements could be made which would have made the overall management of quality more effective.

The Employer's role was to review documents and monitor the Contractor's compliance with his quality procedures. There was considerable debate as to how involved the Employer's staff should become in the monitoring process. Should the monitoring be restricted strictly to a review of the Contractor records including Audits and Surveillances, "Reactive Monitoring", or should it also include a more detailed monitoring of the construction activities, "Proactive Monitoring?" Both types of monitoring by definition are correct. The challenge to those adopting the proactive approach was how to engage the Contractor in a quality partnership without encroaching on his overall obligations. This was achieved by having an open dialogue and demonstrating to him that the actions of the Employer provided benefits to himself as well as the Employer. A review of the quality issues by contract revealed that where the proactive approach was adopted, then the number of workmanship issues was greatly reduced.

For the most part the Contractor's QA team could not be considered independent of the site management teams and this resulted in undue pressures being exerted on the QA and QC inspectors to accept the status quo. The Contractors in general employed too few inspectors and with multiple work fronts operating in many areas on 24/7 basis their task was impossible. The Supervisors were competent and supported the inspectors. In terms of the overall project the types of remediation were acceptable and restricted in most parts to the types of defects noted above.

This all begs the question of who should be responsible for management of Quality. The Contractor or Employer? There is no simple answer to this question.

It is the authors opinion that whilst the QMS instigated by the Employer in general achieved the required results, Employers in the future should adopt a more proactive role in managing the quality of what is their infrastructure by perhaps : -

1) imposing hold points on the contractors inspection plans (the Employer having the option to exercise his right or not)
2) requiring the Contractor to demonstrate through full scale mock ups of critical areas that his methods will result in acceptable quality. (Fig. 20)
3) the role of the CICE should be extended to include random site inspection, since he is required to sign off on the as constructed works.
4) the Employer to hold regular meetings with the CICE and Contractor on matters of quality as well as design.
5) the Employer providing clear guidelines on the monitoring role.

Fig. 20 Grouting Bearing – notice the straps to remove air

Conclusions

This paper has by no means been able to review all topics relating to the structures constructed on the THSR project nor provide a detailed critique on all aspects affecting quality and progress.

However it is the author's hope that he has been able to highlight certain aspects that the methods adopted by the Contractors were successful in achieving the goals of quality, progress set by the Employer.

The study has highlighted that quality is not just about the Contractor instigating a Quality Management system, but more about a total commitment from the Employer, the Designer, the Contractor and independent checkers working in partnership to achieve the required quality which should be clearly set out in the Contract Specifications.

For an Employer to set out a schedule requiring 345km of guideway to be constructed simultaneously including some 249km of elevated structures within 4 years was ambitious. However through a partnership developed between the Employer and the Contractors it was achieved, to good standards of quality and safety and is a credit to all involved.

Acknowledgement

The author wishes to thank the colleagues of THSRC, who have assisted in providing information and in the preparation of this paper. He also is grateful to the management of THSRC for permitting this paper to published and presented at the conference.

Self-Consolidating Concrete for On-site Bridge Applications

Piotr Paczkowski, University of Nebraska, Lincoln, Nebraska, USA
Maria Kaszynska, Technical University of Szczecin, Szczecin, Poland
Andrzej S. Nowak, University of Nebraska, Lincoln, Nebraska, USA
Tomasz Lutomirski, University of Nebraska, Lincoln, Nebraska, USA

Abstract

The State of Nebraska is one of the national leaders in application of the self-consolidating concrete (SCC) for bridges. So far, it is mostly used in precasting plants where it is replacing ordinary concrete as SCC does not require vibration and the mixture flows filling also hard to reach corners within the formwork. Properly mixed SCC is not affected by segregation. It finds applications in precast concrete plants, due to excellent filling abilities, but also elimination of noise (no vibrators). However, SCC can also be used for cast-in-place applications (slabs, abutments, piers) and for repair of existing structures, in particular in hard to reach repair projects, with a restricted access for pumping concrete. In case of ordinary concrete, the major problem for on-site placement is assurance of good quality of material, proper degree of vibration, to avoid gaps and reduce shrinkage cracks. Self-consolidating concrete (SCC) offers an excellent solution. SCC will improve the quality of concrete structures by eliminating some of the potential for human error. It will replace manual compacting of fresh concrete with a modern semi-automatic placing technology. However, this type of concrete needs a more advanced mix design than ordinary concrete and a more careful quality assurance with more testing and checking at least during the startup period. Before any SCC is produced at the concrete plant or in a truck mixer, the mix has to be designed and tested to satisfy several key criteria. The cast-in-place applications of SCC require more attention to assure the flowing ability, passing ability, and resistance to segregation. The SCC is sensitive to water/cement ratio, placing temperature, delivery time, and admixture content. Major concerns are also related to the effectiveness of on-site quality control/quality assurance procedures, and high risk of form deformation and blowout due to excessive SCC pressure. The paper will present the requirements for cast-in-place application of the self-consolidating concrete, including procedures for testing fresh SCC on site to determine its key properties, such as filling ability, passing ability, and resistance to segregation, investigate the impact of the delivery time on the properties of fresh SCC, investigate the impact of the temperature on the properties of fresh SCC since SCC is more sensitive to temperature during hardening than conventional concrete.

Introduction

The current state-of-the-art in SCC was presented and discussed at the 1st International RILEM Symposium on SCC held in Stockholm (1999), then at the First North American Conference on the Design and Use of Self-Consolidating Concrete held at Northwestern

University (2002), RILEM Symposium in Iceland (2004), and recently at the SCC conference in Chicago (2005).

The cast-in-place applications of SCC require a special attention to achieve the primary SCC properties such as flowing ability, passing ability and resistance to segregation. These properties are very sensitive to the following parameters: the type of materials used, mix design (e.g. water/cement ratio and admixture content), placing temperature and delivery time. Therefore, prior to applying cast-in-place SCC, the following questions need to be answered:

- What are the proportions of the local materials that can be used for the development of SCC to be competitive to existing concrete mixes?
- What tests are needed to verify the quality of fresh SCC on site before casting? What are the acceptable ranges of test results?
- What is the acceptable range of delivery time for SCC? What actions are to be taken if this range is violated?
- What is the acceptable range of placing temperature for SCC? What actions are to be taken if this range is violated?

A European research project has been carried out to evaluate different SCC testing methods and classify SCC mixes based on test results (Schutter 2005). Japanese researchers have recommended a maximum of one hour delivery time and a temperature range from $5°$ C to $30°$ C for transporting and placing SCC on site (Ouchi 2003). Also, a group of researchers from Switzerland has studied experimentally the SCC pressure on formwork using strain gages for different mixes and casting methods (Leemann et al. 2006). High pressure values, even greater than the hydrostatic pressure, occurred when SCC pumped from the base, and high variation in pressure values occurred when SCC pumped from the top at different rates. These studies have concluded that more research is needed to gain a better understanding of SCC behavior for different site conditions.

Objectives

The main objective of this paper is to examine the influence of admixtures on rheological properties of SCC in terms of on-site bridge applications. In addition to that, the effectiveness of testing procedures for determining concrete key properties such as filling ability, passing ability, and resistance to segregation was checked. At last the impact of the delivery time on the properties of fresh SCC for the developed mix design was investigated.

Materials

The basic mix has been developed including specification of materials, mixing sequence and special requirements, and tolerances for proportioning of materials. The final product (mix design specifications) resulted in a concrete that satisfies the SCC requirements (flowability, passing ability, and resistance to segregation). These requirements were tested using the ASTM Slump Flow Test (C1611) and J-ring test (C1621). Base mix recipe (without admixtures) is shown in table 1.

Materials	Weight [kg/m^3]
1 PF CEMENT	480.3
Limestone 1/2"	416.3
Sand 47B	1238.2
Total Water	181.5
Total Weight	2316.3
w/c (cement)	0.50
w/cm (cement + fly ash)	0.38

Table 1. Base recipe.

	Weight [kg]	%
Cement	360.2	18
Class F fly ash	120.1	6
Aggregate 0-2.3mm	671.4	34
Aggregate 2.3 - 4.7mm	283.2	14
Aggregate > 4.7mm	547.2	28

Table 2. Percentage listing of dry mix.

Recipes were to use 1PF cement with pre-blended 25% of Class F fly ash and locally available aggregates. Gradation curves for the aggregates are presented in Fig.1.

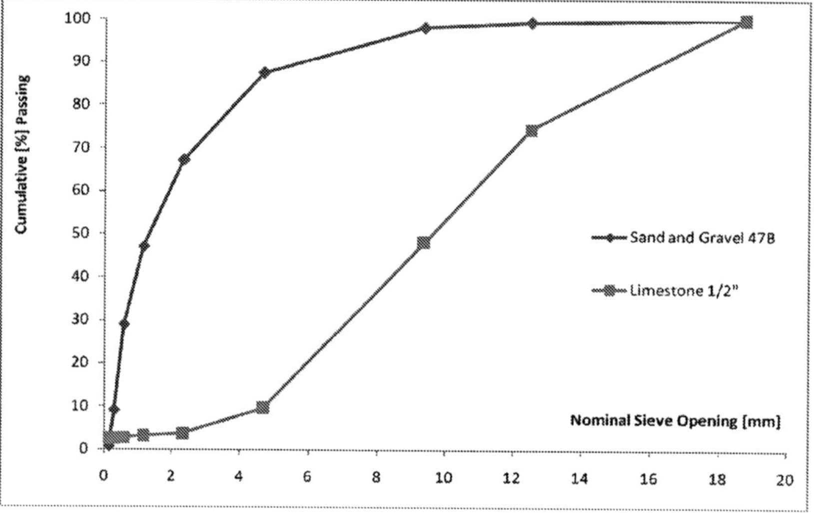

Fig.1. Aggregates gradation curves.

Table 2 shows the mix recipe with the percentage of dry constituents. To reduce water content, High-Range Water-Reducing Admixture Glenium 3400 NV was used. Viscosity-Modifying Admixture Rheomac VMA362 was included to increase the segregation resistance and to provide stability during transport and placement. To increase the workability times, Set Retarding Admixture Pozzolith 300R or Delvo Stabilizer was introduced into the mix. PAVE-AIR 90 Air-Entraining Admixture for Paving Concrete provided a control for air content in the mix.

Experimental Work

During the base mix development study it became clear that after performing Slump Flow (Fig. 2) and J-Ring (Fig. 3) tests with Visual Stability Index, it was not necessary to perform

V-funnel (Fig. 4) and L-box (Fig. 5) as they did not introduce any new information about properties of fresh mix. For all mixes, T_{50} was between 2 and 3 seconds just after the end of the mixing procedure, and 4 - 5 seconds when the spread circle decreased to 50 cm. Visual Stability Index was 0 to 1 for all considered mixes. The mixing procedure is shown in Figure 6.

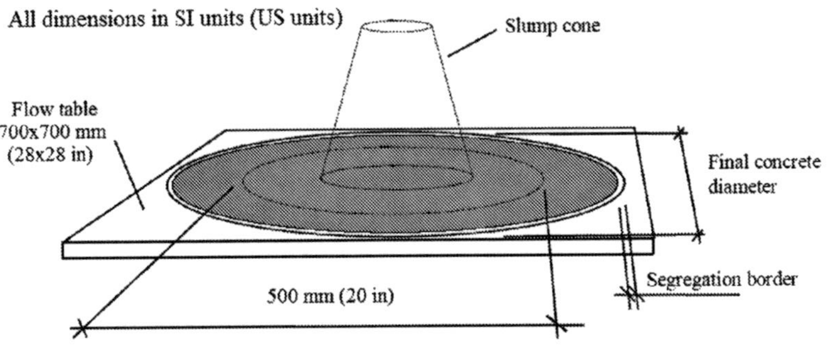

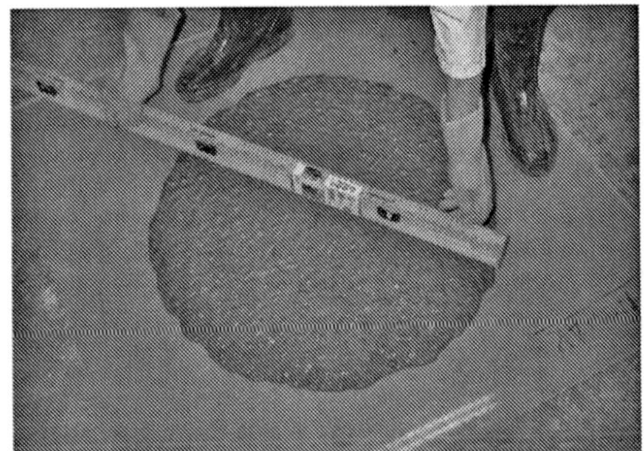

Fig.2. Slump Flow Test (Spread Circle).

Fig.3. J-Ring Test.

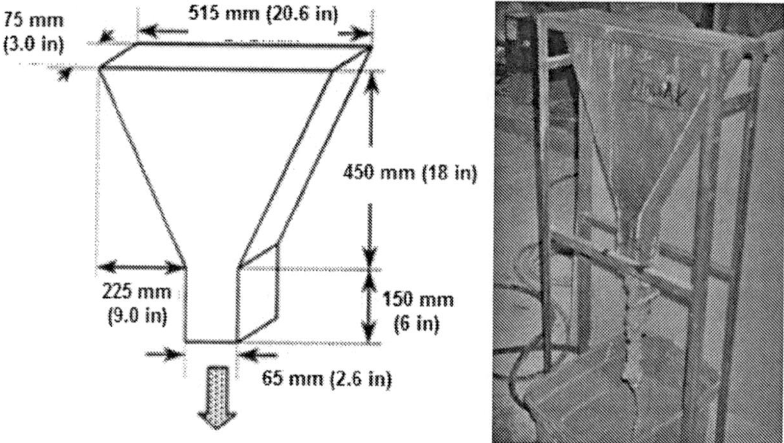

Fig.4. V-Funnel Test.

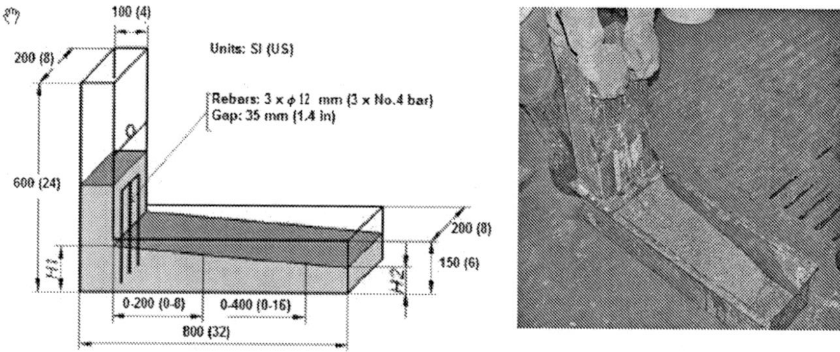

Fig.5. L-Box Test.

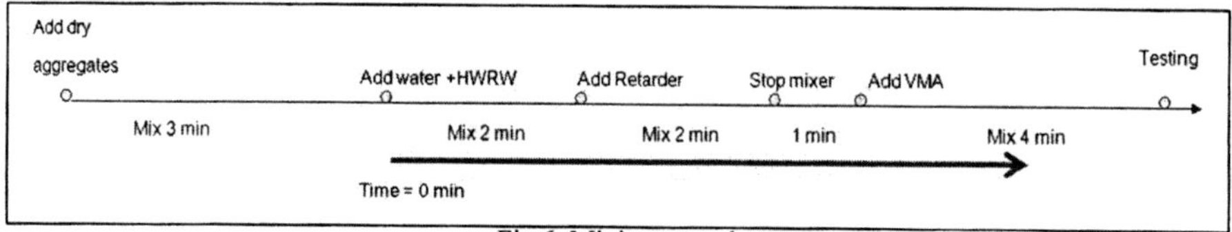

Fig.6. Mixing procedure.

Time Factor

To simulate delivery, agitation was performed in a mixer drum for up to 140 minutes. Every 10 to 15 minutes, Slump Flow and J-Ring tests were performed. Slump flow retention curves were obtained for different combinations of admixtures. It was found that VMA362 does not influence the workability time, so it is not considered as a variable (see Fig.7).

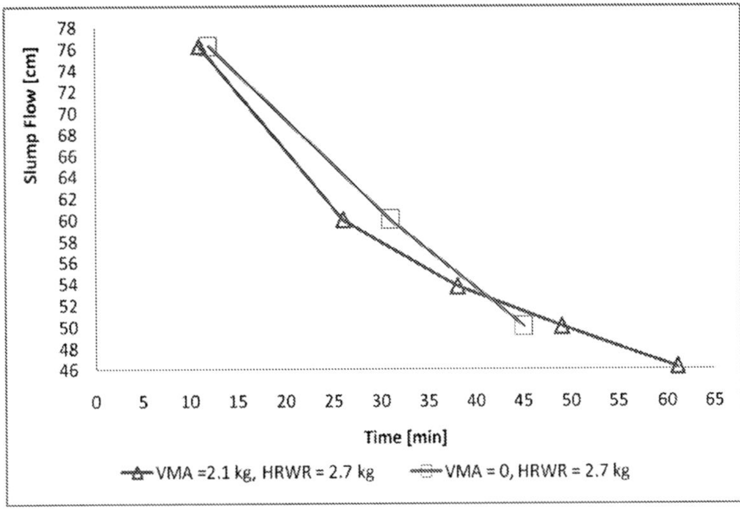

Fig.7. Slump Flow Retention Curve - Influence of VMA.

Several trials were performed to examine the possibility of extending the workability time with the usage of a retarder. At the end of every trial when the spread circle decreased below 50 cm, a secondary dosage of superplasticizer was applied. Figure 8 shows slump flow retention curves for different dosages of Pozzolith 300R. Numbers on the curves indicate the amount of superplasticizer that was added to restore the adequate SCC performance. There was a significant increase in workability time after the addition of retarding admixture. The amount of 2.7 kg per m^3 extended the workability time from 40 to 70 minutes, however, addition of more retarder did not help. Figure 9 presents the effect of Delvo Stabilizer on the workability range. This admixture works partially as a water reducer, and at the same time, it retards the setting time characteristics and acts as a stabilizer.

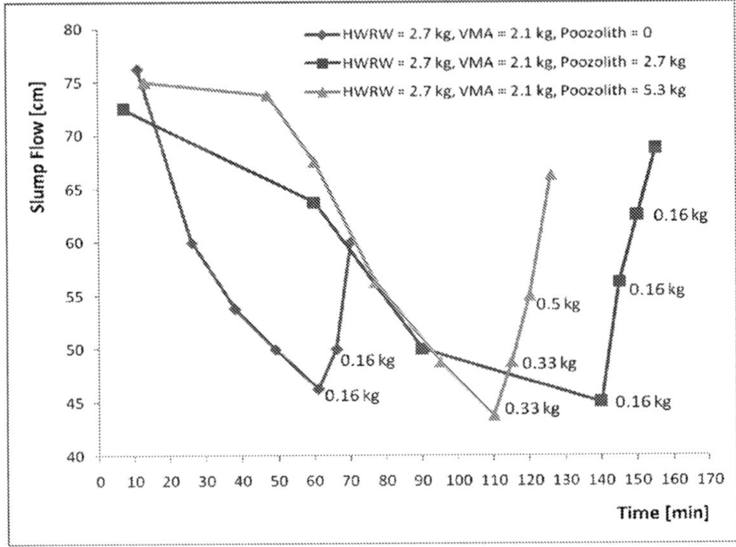

Fig.8. Slump Flow Retention Curve - Effect of retarder.

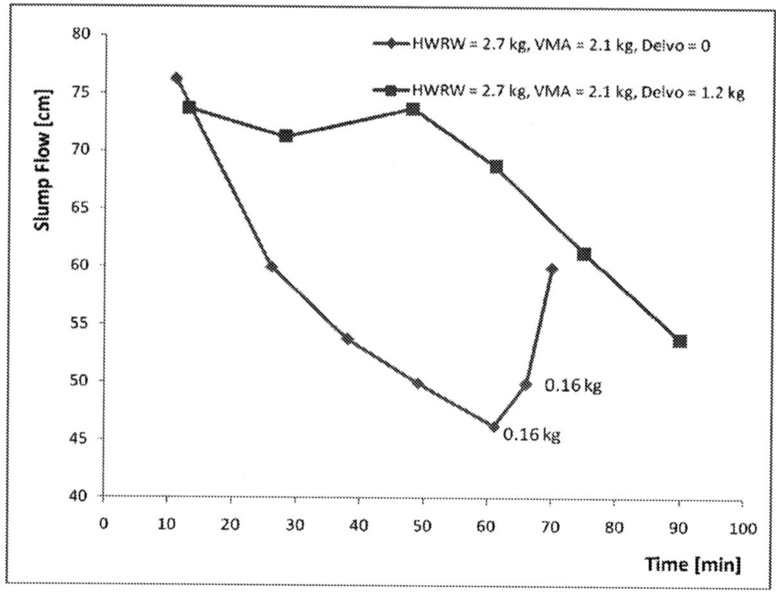

Fig.9. Slump Flow Retention Curve - Effect of Delvo Stabilizer.

Addition of a retarder can be necessary to extend workability time, but at the same time, it can effect the early compressive strength. Curves in Figure 10 present the SCC strength development for different amounts of the retarder. It can be clearly seen that Pozzolith 300R not only retards the setting, but it also increases the ultimate compressive strength.

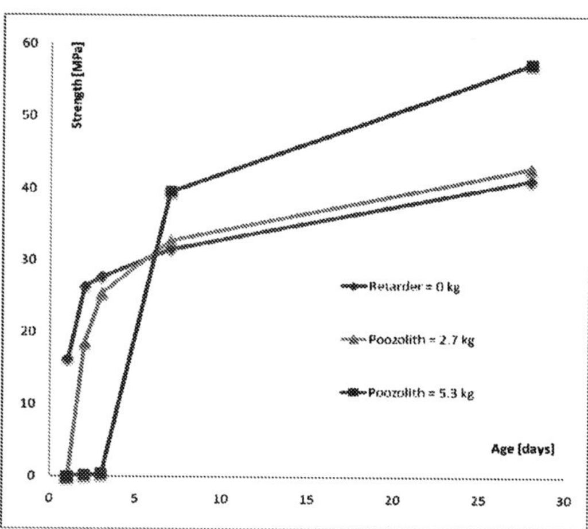

Fig.10. Strength Development Curves for different amounts of retarder.

Air Content

To ensure appropriate freeze and thaw resistance it is often recommended for concrete to have some plastic air. Depending on the environment and maximum aggregate size different levels of air entrainment are implied ("Design and Control of Concrete Mixtures" 14th Edition by PCA). For the considered recipe, the levels of air entrainment were examined in terms of the amount of Pave-Air 90 admixture. Modified Point-Count Method ASTM C457 was used in this test and then compared with the Pressure Method ASTM 231, which is more suitable for on-site applications.

Figure 11 (unpublished progress report by Morcous) presents the percentage amount of air measured for three different levels of air entraining admixture. Plastic air increases the resistance to freeze and thaw, but it has a significant impact on decreasing compressive strength. Figure 12 presents 7-day compressive strength for different levels of Air-Entraining Admixture. The amount of plastic air should be carefully selected depending on the required freeze and thaw resistance without decreasing compressive strength.

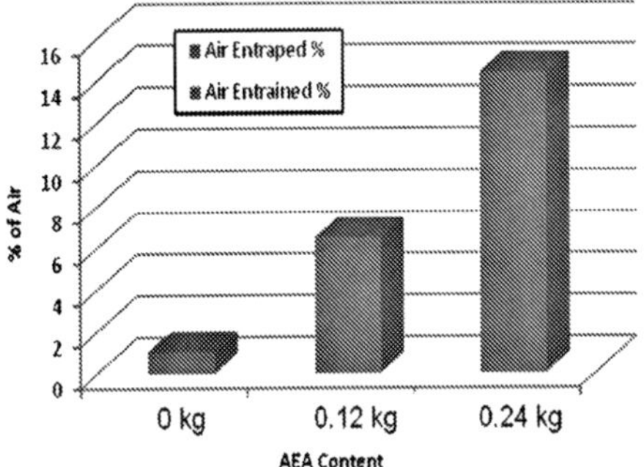

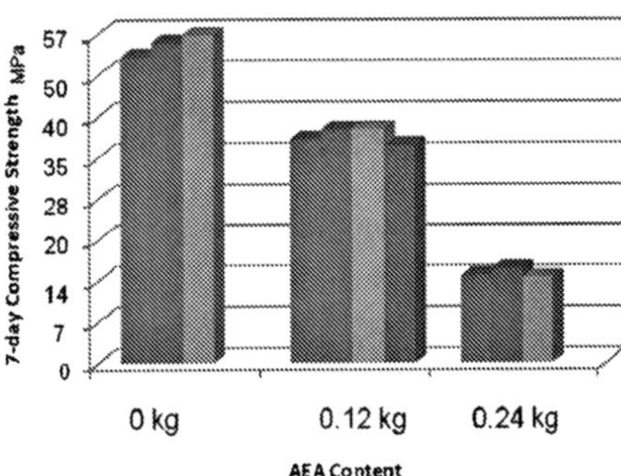

Figure 11. Air Entrainment as a function of AEA admixture.

Fig.12. 7-day compressive strengths as a function of Air Entraining Admixture

The air content was measured using the pressure method for three levels of air entrainment admixture, and the following results were obtained:

- for mix without air entrainment admixture – 4% air content
- for mix with 0.12 kg per m^3 of air entrainment admixture – 8% air content
- for mix with 0.24 kg per m^3 of air entrainment admixture – 16% air content

Conclusions

The basic mix recipe was developed for the maximum aggregate size of 0.5 in (12.5 mm). This helped to increase resistance to segregation and eliminated problems with blocking. It was found that for the on-site assessment of quality for the mix, it is sufficient to perform only J-ring test and Slump Flow test with Visual Stability Index. The analysis of delivery time effect on SCC properties showed that for on-site applications a retarder should be used and, if needed, an additional dosage of the superplasticizer. However, the amount of retarder should be carefully selected because an excessive amount does not increase workability time and it only affects the setting time.

Air content in concrete is currently the subject of intensive research. Comparison of the results from the pressure test and the Modified Point-Count Method ASTM C457 shows a good agreement. Therefore, the Pressure Method due to its simplicity can be successfully used on-site for a fast assessment of the air content. The larger plastic air content, the smaller compressive strength, therefore, practical quality control procedures are needed to ensure the adequate properties of concrete.

Acknowledgements

The presented research has been sponsored by the Nebraska Department of Roads. Thanks are due to the TAC members for their valuable comments, in particular Moe Jamshidi, Fouad Jaber and Lieska Halsey. We acknowledge with thanks the support provided by industrial sponsors: Mark Lafferty (Concrete Industries, Inc.), Terry Landon (Ready Mixed Concrete), and Norm Nelson (Lyman-Richey Corp.). Thanks are due to George Morcous (University of Nebraska) for his help in the testing program.

References

ASTM C136, 2004, "Standard Test Method for Sieve Analysis of Fine and Coarse Aggregates", ASTM Annual Book of Standards, Vol. 04-02.

ASTM C231, 2004, "Standard Test Method for Air Content of Freshly Mixed Concrete by the Pressure Method", ASTM Annual Book of Standards, Vol. 04-02.

ASTM C457, 2006, "Standard Test Method for Microscopical Determination of Parameters of the Air-Void System in Hardened Concrete", ASTM Annual Book of Standards, Vol. 04-02.

ASTM C1611, 2005, "Standard Test Method for Slump Flow of Self-Consolidating Concrete", ASTM Annual Book of Standards, Vol. 04-02.

ASTM C1621, 2006, "Standard Test Method for Passing Ability of Self-Consolidating Concrete by J-Ring", ASTM Annual Book of Standards, Vol. 04-02.

Corradi M., Khurama R., Magarotto R. "Self-Compacting Concrete: Some Significant Applications", Proceedings of the 4th International RILEM Symposium on Self-Consolidating Concrete, Northwestern University, Chicago, 2005, pp. 81-87

De Schutter, G. "Guidelines for Testing Fresh SCC" Report from European Research Project: Measurement of Properties of Fresh SCC. Sept. 2005.

European Federation of Concrete Admixtures (EPCA), May 2005, "The European Guidelines for Self-Compacting Concrete: Specifications, Production, and Use".

Kaszynska, M. and Nowak, A.S., "Effect of Mixing Tolerances on Performance of Self-Consolidating Concrete (SCC)", Proceedings of the 3rd International Symposium on High Performance Concrete, PCI, Orlando, FL, October 2003, CD ROM.

Laumet, P., Czarnecki, A.A. and Nowak, A. S., "Development of a US/Midwest Specific Self-Consolidating Concrete for Bridges," Proceedings of the Concrete Bridge Conference, Portland Cement Association, Reno, NV, May 2006.

Leeman, A., Hoffmann, C., and Winnefeld, F. "Pressure of Self-Consolidating Concrete on Formwork", ACI Concrete International, Feb. 2006.

Ouchi, M., Nakamura, S., Osterberg, T., Hallberg, S., and Lwin, M. "Applications of Self-Compacting Concrete in Japan, Europe, and the United States", ISHPC 2003.

Test Methods for Self-Consolidating Concrete, Technical Bulletin TB-1506, W.R. Grace & Co.-Conn., 2005

Walraven J. "Self-Compacting Concrete: Challenge for Designer and Researcher", Proceedings of the 4th International RILEM Symposium on Self-Consolidating Concrete, Northwestern University, Chicago, 2005, pp. 431-446

Design and Construction of Otanerua Eco-viaduct, Auckland, New Zealand

Moses Borland BEng. CEng. MICE, Northern Gateway Alliance, Auckland, New Zealand
Peter Lipscombe BE. PhD IntPE MIPENZ, Northern Gateway Alliance, Auckland, New Zealand
Rudolph Kotze BEng. BComm. CPEng. IntPE. MIPENZ,   Transit New Zealand

Abstract

Otanerua Eco-viaduct is the most challenging structure on the SH1: Northern Motorway extension project. The biodiversity of the area is of regional significance, and the challenge has been to construct a motorway in difficult terrain whilst minimising the impact on the environment.

The Otanerua Valley contains a diverse native forest that is home to the endangered fernbird and other native fauna. The 256m long eco-viaduct crosses over the Otanerua Valley, providing an ecological corridor between two breeding colonies of fernbird.

The design adopted an unconventional articulation as a result of the topography of the valley. In order to resist the seismic loading the deck is made integral at the abutments with an expansion joint at mid-length of the bridge.

The steep terrain and the requirement to maintain the native bush resulted in difficult access for conventional plant and prevented the deck elements being erected by crane.

The construction of this viaduct demonstrates the benefits of the alliance contract model in delivering economic structures in ecologically sensitive areas with challenging terrain.

Introduction

Otanerua Eco-viaduct is the most challenging structure on the Northern Motorway Extension (ALPURT B2) 30km north of Auckland, New Zealand (Fig.1). The project involves the construction of 7.5km of dual 2 lane motorway traversing a series of ridges and gullies, with a combination of viaducts, tunnels and earthworks.

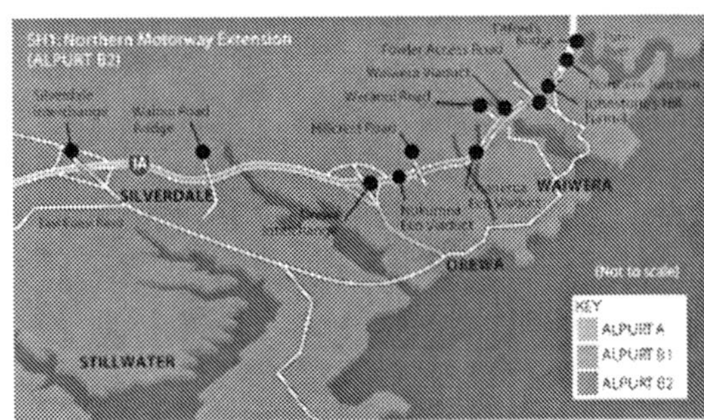

Fig. 1. Location Plan

Bridge design, construction and maintenance 2007, Thomas Telford, London

The project is being delivered by the Northern Gateway Alliance (NGA), which comprises Transit New Zealand, Fulton Hogan, Leighton Contractors, URS New Zealand, Tonkin & Taylor and Boffa Miskell. Dedicated staff and resources from each of the participant organisations have been brought together into one team, operating at a single office, for the duration of the project. All the participants in the Northern Gateway Alliance operate under a contractual framework where their commercial interests are aligned with actual project outcomes.

Under the alliance, all parties take collective responsibility for the successful completion of the project on time and within budget, and for achieving all the project's desired outcomes, including high construction standards, workplace safety, public communication and environmental improvements.

The project vision is to: "create a northern gateway that is a visual showcase of environmental and engineering excellence." Consistent with that vision the 256m long eco-viaduct has been provided specifically to create an ecological corridor for endangered native fauna in particular the semi flightless native New Zealand fernbird.

History and purpose of Otanerua Viaduct
When it was realised in 1996 that the route of the Northern Motorway extension (ALPURT B2) would cut through regionally significant bushland, the solution was to build a viaduct to traverse the area and create an ecological corridor beneath the viaduct that was suitable for the passage of aquatic, terrestrial and arboreal fauna.

In August 1996, a Heads of Agreement was signed between Transit New Zealand (Transit), the Department of Conservation and the Auckland Conservation Board for a viaduct to be constructed over the northern tributary of the Otanerua Stream.

The construction of the project was initiated by Transit in November 2003. Transit formed the Northern Gateway Alliance (NGA) in April 2004 and construction began in December the same year. Sustainability is at the heart of the project's design, and the Otanerua Eco-viaduct is the first major structure built in New Zealand solely to provide an ecological corridor.

Local Environment
Otanerua Valley is a heavily vegetated area of kauri and podocarp forest, interspersed with manuka gumland - a prime area of habitat for native and endangered flora and fauna. This area is classed as RAP-21 (Recommended Area for Protection 21), as listed in the Department of Conservation Management Strategy and Rodney District Council's District Plan. Resident species include the semi-flightless fernbird, north island robin, green and forest native geckos and at least seven species of native freshwater fish including koura (fresh water crayfish), inanga (whitebait), kokupu and banded kokupu.

Gumlands are areas that were once kauri forests and although long gone, the footprint of the forest remains. Kauri trees draw a great deal of nutrients from the soil and actually change its fertility to the extent that they can no longer survive there themselves. Kauri are gradually replaced with shrub species that can tolerate low fertility conditions such as manuka, kumarahou and waewaekaka. This is ideal habitat for the regionally endangered, semi flightless fern bird. Over many millennia fertility is returned to the soil and the kauri forest flourishes again. This cycle has been going on for many hundreds of thousands of years.

A key focus for the NGA was to preserve the ecological integrity of the area. This meant leaving the absolute minimum footprint on the ground so the bush could rehabilitate quickly. At the design stage this was achieved by creating a long bridge span and designing irrigation systems to maintain the valley's health and minimise the rain shadow effect under the bridge

Another sustainability driven decision was to use pulverised fly-ash (PFA) as a supplementary cementious material in all concrete used in the substructure. The concrete mixes typically used 30 to 40 percent cement replacement. The PFA used was a waste product from a local coal fired power station at Huntly, its use in concrete increases durability, workability and reduces cement demand.

The Otanerua Stream, with its pristine undeveloped catchment, is used by the Auckland Regional Council to benchmark the health and biodiversity of its waterways. As such, it was imperative to protect and monitor the health of the stream during construction. Baseline monitoring was undertaken before construction started and subsequently in February, May and September of each year. To date, with the bridge construction works complete, no adverse effects have been recorded.

Bridge Design

Concept Design

When the Northern Gateway Alliance team began working on the design in 2004, the design was essentially started from a concept design stage as there were few major constraints. The land had been acquired and the motorway corridor designation agreed with Rodney District Council and Auckland Regional Council. There were also some agreements with landowners and the Department of Conservation that the NGA had to comply with. The motorway designation and land take was for an 80 km/h design speed but the NGA, with integral support from Transit, targeted achieving a 100 km/h design speed. The challenge was to achieve an optimum solution that balanced engineering excellence, ecological protection and value for money.

The Specimen Design for the Otanerua Eco-viaduct showed a bridge 203 metres long by 23 metres wide on an eight percent gradient.

Given that the viaduct was to provide an ecological corridor, the NGA undertook a study of similar north-south orientated bridges to gauge the health of vegetation that could be expected under the proposed bridge. The study revealed no growth inboard of a line 43 degrees from the drip line of the bridge and only stunted growth inboard of a 27 degree line. This implied that over half the length of the Specimen Design bridge would have barren ground beneath it and unrestricted vegetation growth would be limited to a narrow 40 metre wide corridor.

The Specimen Design did not meet modern road geometric requirements. The new standards (the 'State Highway Geometric Design' manual for 100km/h design speed) stipulate flatter gradients, larger radii curves and shoulder widening to provide safe sight distances around curves in the road.

In designing the Otanerua Eco-viaduct for the new criteria, the NGA realigned the road, eased the gradient to six percent and both raised and lengthened the bridge. The combined effect of this was to increase the corridor of unrestricted vegetation growth to 75 metres wide and bring the bridge into line with the new standards.

A permanent irrigation system was established, utilising rain run-off from an uphill catchment, to increase the width of the effective ecological corridor to the full length of the bridge, some 256 metres. These provisions will maximise the potential of the ecological corridor the bridge is intended to provide.

The choice of bridge type, span arrangement, construction method and material was open to all options. A variety of bridge designs were considered at concept stage including long, medium and short span options. The longer span options had the advantage of minimising the disturbance to the valley but would have required specialist construction equipment and were more expensive than the shorter span options. Options considered at concept design stage were (Fig. 2):

- Single pylon cable stayed
- Balanced Cantilever Concrete Box Girder
- Incrementally Launched Concrete Box Girder
- Medium and short span Steel Composite and
- Short span Precast Concrete Beam

Fig. 2(a). Balanced Cantilever Option

- provides reduced ecological footprint
- can be constructed without temporary access across the stream

Fig. 2(b). Cable Stayed Bridge Option.

- smallest environmental footprint
- visually attractive bridge
- most expensive of the options considered

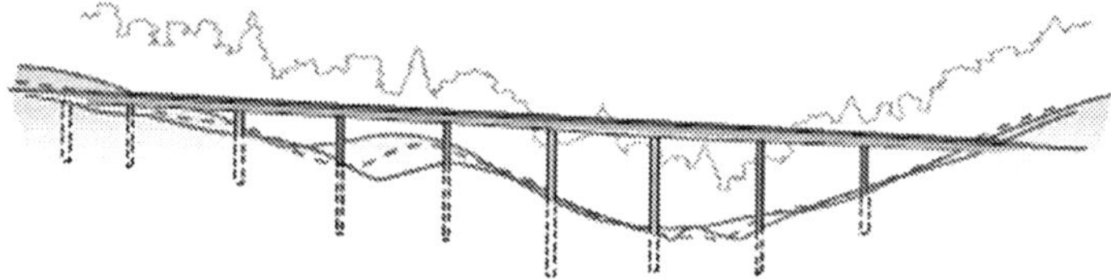

Fig. 2(c). Preferred option - Multi-span Viaduct using Precast Concrete Beams.

- Satisfies the functional requirements for the bridge
- Satisfies the environmental requirements for the ecological corridor.
- Most economic of the concept designs.

Preliminary Design

Preliminary designs for two bridges, one of precast concrete and one of steel composite construction, were developed to an advanced stage of detailed design to see if there was a clear difference in the estimated price. But with a cost difference of just two percent in favour of the concrete option, there was little differentiation between the two.

As the Land Transport Management Act (2003) requires a sustainable approach to the provision of transport infrastructure, sustainability principles played an important role in choosing between the two bridge designs.

The main factors taken into consideration when deciding on the final design of the viaduct were:

Aesthetics: Research concluded that the underside appearance of the steel design was considered inferior to the precast concrete option. However, since there would be little opportunity to view the eco-viaduct from below, it was considered a relatively minor issue.

Lifecycle: The ability to recycle steel economically favoured this option but although it is also possible to recycle concrete, current technology seldom makes this economically viable. This did not influence the decision as the cost of recycling 100 years from now, when discounted to present day values, would be very small.
The replacement of the steel corrosion protection system was analysed and expected to have a lifespan of 40 years. It would be necessary to replace the coating as microclimatic conditions caused by vegetation growth under the bridge would eventually deteriorate the coating. But when this cost was discounted to a net present value, the value was small and not a significant differentiator.

Supply Chain: Although there was deemed to be little distinction between the two structural systems in terms of their social, environmental and economic implications, it was the consideration of local supply versus international supply that influenced the decision.

Steel plate for the steel bridge would have had to be imported because the size of plate required is not readily manufactured in New Zealand. It was also likely that the beams would need to be fabricated overseas and imported as fully fabricated elements. This would require a longer lead time and there was concern this might cause potential programme delays.

In the concrete option, the components are made locally and from locally manufactured cement, aggregate and reinforcing steel. Therefore the concrete design had a sustainability and programme advantage.

Detailed Design

The bridge length was chosen to span the complete valley and avoid the need for approach embankments. The spans were chosen to suit Tee-roff prestressed concrete beams, provide equal spans and to ensure the pier positions avoided the Otanerua Stream which crosses the bridge site at an oblique angle between Piers F & G. (Fig. 3a)

The bridge foundations consist of 1800mm diameter bored concrete piles founded in the underlying rock.

The bridge columns are 1800mm diameter for the taller central piers and 1500mm diameter for the shorter piers. The columns are constructed directly on top of the piles. Reinforced concrete crosshead beams are constructed on top of the columns to support the deck.

The deck cross section consists of 10 Tee-roff prestressed concrete beams with an integral insitu reinforced concrete deck slab (Fig. 3b). The beams are simply supported on elastomeric bearings; this had advantages for construction over continuous construction as there was no need to support the beams temporarily until the ends were cast into the crosshead. A flexible linkage slab was cast over the pier to provide a continuous running surface. A grillage analysis of the deck was carried out to determine the forces in the beams and slab.

The bridge is located in a seismic zone and was designed to resist seismic loads. The geometry of the valley means that the central piers are tall and flexible while the piers closer to the abutments are relatively short and stiff. Conventional articulation with expansion joints at the ends of the deck would have resulted in high horizontal loads from seismic, shrinkage and thermal effects on the shorter piers requiring sleeving of the piles to provide the necessary articulation. To overcome this it was decided to provide an expansion joint at mid-length of the bridge (Pier E) and cast the deck integrally with the abutments. Resistance to longitudinal loads is provided by anchoring each half of the bridge back to a piled deadman beam behind each abutment.

The bridge is provided with reinforced concrete edge barriers topped with an oval steel rail which provide containment to US Federal Highway Authority Test Level 5.

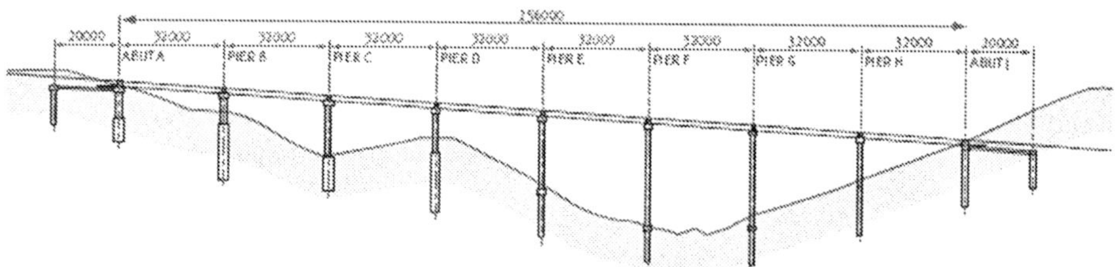

Fig. 3(a). Longitudinal Section

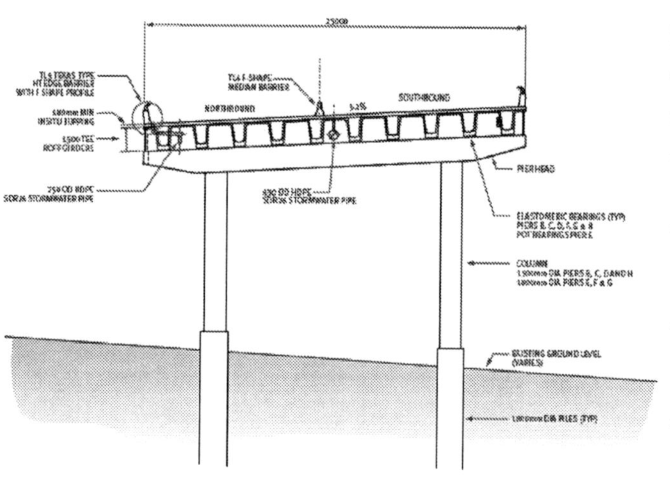

Fig. 3(b). Typical Cross Section

Key Details:

- 256 metres long
- 25 meters wide
- Eight spans of 32 metres
- Reinforced concrete piers and foundations
- Precast concrete Tee-roff (open top super-tee) beams
- Constant longitudinal gradient of 6% and 5 % crossfall

The Completed Design

The Otanerua Eco-viaduct is 256m long by 25m wide and towers 35m over the Otanerua Stream. It provides for two lanes of traffic in each direction with median separation. The height of the bridge is such that, over its middle section, it is just above the young forest canopy, while at the abutments it is at the forest floor. Detailing of the bridge barriers is intended to enhance this driving experience and allow appreciation of the Otanerua Valley from an elevated vantage point.

Construction Methodology

An advantage of an alliance is the early constructor involvement to ensure that the design is tailored to suit construction methods that best suit the site. The design and construction teams were based together in the same office and worked closely on the design. Formal design review meetings were supplemented with regular day to day discussions on various details. Construction methods used for the viaduct were largely dictated by the difficult terrain in the Otanerua Valley.

Access Tracks and Work Platforms

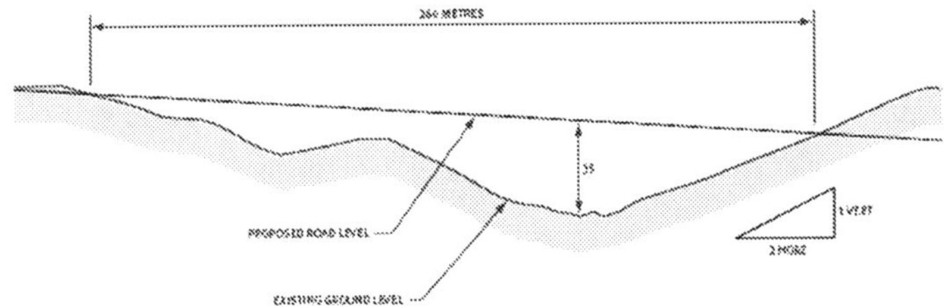

Fig. 4. Profile of Valley

The steep valley sides and the narrow motorway corridor, combined with the desire to protect the native vegetation, resulted in the use of very steep access tracks running straight up and down the valley slopes (Fig 4). The width of the tracks were kept to a minimum, this minimised the scar on the sensitive landscape and resulted in access being restricted to 4-wheel drive and tracked vehicles. A temporary bridge was constructed across the Otanerua Stream to avoid any disturbance to the stream bed (Fig. 5).

Fig. 5. Temporary Bridge over Stream

Individual work platforms were constructed at each pier position to provide a stable platform for piling and craneage and the size of these was also minimised. Bulldozers and winches were used to transport heavy plant up and down the steep slopes (Fig 6). Silt fences were erected around each work platform and on either side of the stream to control sediment runoff

Fig. 6. Moving Crane on Steep Slope

Piling

The geology of the site consisted of 3 – 10m of stiff residual clays overlying weak to moderately strong inter bedded Pakiri Formation siltstone and sandstone rock. The 1.8 metre diameter bored piles were constructed using a hydraulic rotary piling rig. Temporary casings were used to support the clay overburden and a six metre rock socket was bored using a rock auger. The rock sockets were then grooved to enhance side friction.

Drilling of the sockets in the soft rock conditions caused smearing of the socket sides, so the NGA developed a rotating jet wash attachment for the piling rig to clear the smearing from the pile sides. Concrete for piling was pumped for distances of up to 180 metres in areas that were inaccessible to concrete trucks.

Column Construction

Columns were constructed in four metre lifts using a steel jump form. When striking the formwork, the top section of the form was left in place to form the working platform for erecting the next lift. Permanent feature grooves were cast in the top lift of the column to receive the support clamps for the crosshead falsework.

Crosshead Construction

The crossheads were constructed insitu on a falsework system of beams spanning between the permanent columns. The system was developed as an economic solution to working with piers that are up to 31 metres high.

Clamps were stressed on to the top of the columns to provide support to the falsework beams. The beams were then supported on hydraulic jacks with lockable collars to allow easy adjustment of level and striking of the falsework.

The falsework system was assembled on the ground around the piers, complete with soffit formwork, work platforms and handrailing. This was done to minimise the amount of work carried out at height. The falsework was lifted into position with the support collars attached, using chain hoists mounted on top of the permanent columns (Fig. 7). The falsework was temporarily supported on dywidag ties until the support collars were stressed to the columns. The soffit formwork was then completed around the columns and the reinforcement put in place. Steel formwork was erected to the sides and the crosshead was concreted using a static concrete pump.

Fig. 7. Falsework Lifting

Support cones were cast into the crosshead to support access platforms for deck erection. These platforms were installed before the crosshead falsework was removed to minimise the amount of work carried out at height.

The falsework was then lowered to the ground using the chain hoists and split into three sections to allow it to be removed from around the piers (Fig. 8). The joints between sections were formed using prestressed bar to minimise the time taken to assemble and dismantle them.

Fig. 8. Falsework Lowering

Deck Construction

Each span of the bridge deck comprises ten 1.5 metre deep Tee-roff beams, covered by a continuous insitu concrete deck slab.

Erecting the beams conventionally with cranes from the valley floor was not possible due to the steep terrain and the requirement to minimise the clearance of vegetation. The chosen method of construction was to launch the beams from deck level using a self-launching overhead truss.

An existing launching truss was imported from Australia and modified to suit the varying skew angles of the bridge. The bridge was curved on a 900m radius and the skew at the supports varied from 14 to 30 degrees, the front and rear supports of the truss were fitted using sliding connections to allow the skew to be varied.

The truss was launched forward supported on 'Launching Bogies' running on rail track laid on the deck. Kentledge was fitted to the rear of the truss to rotate the truss prior to launching (Fig. 9). During beam erection the truss was supported on bogies running on transverse rails to allow the truss to move to each beam location..

Fig. 9. Truss Launching

The transverse rails were set level to minimise the forces required to move the truss. Two lifting trolleys equipped with lifting jacks were mounted on top of the truss to lift and move the Tee-roff beams.

The 31-metre long beams were delivered to site by a truck and rear powered jinker which provided the manoeuvrability required to access the work site. Each beam was reversed into the rear of the truss, which then picked up the front of the beam and lifted it clear of the jinker, which then drove out. The truck then reversed the rear of the beam under the rear trolley which lifted it clear of the truck. The beam was then pulled forward into the required position (Fig. 10). The truss was moved transversely to the required position, and lowered the beam onto the support bearings. Epoxy mortar was spread on top of the bearings before lowering the beam onto temporary supports which remained in place until the deck was cast. A temporary prop was installed under the down slope flange to stabilise the beam.

Once all the beams were installed the gaps between the beams were sealed and permanent formwork installed over the troughs in the beams. The deck reinforcement was then placed and the concrete deck pour completed.

Fig. 10. Beam Launching

Edge Barriers

The Test Level 5 edge barriers and median barrier were constructed insitu in steel formwork using a travelling gantry on the deck to move the formwork and provide access to the outer edge.

Environmental rehabilitation

Following completion of construction the work platforms and access platforms will be removed and the valley re-profiled to its original shape. The cleared areas will be revegetated to restore the habitat to its previous condition. In the Otanerua Valley eco-sourced seeds from within the gumlands are being used to grow plants for revegetation and mulch from removed vegetation will be used in the replanting process. This mulch contains dormant seeds from the original vegetation many of which will germinate and grow in conjunction with planted nursery seedlings.

Conclusion

The construction of Otanerua Viaduct required close co-operation between the environmental, design and construction teams from concept design to construction to achieve the desired outcomes. Construction of the viaduct demonstrates the benefits of the alliance contract model in fostering the integrated approach required to deliver economic structures in ecologically sensitive areas with challenging terrain (Fig. 11).

Fig. 11. Completed Structure

Achnowledgements

The authors wish to thank the Northern Gateway Alliance and Transit New Zealand for permission to publish this paper

Identification of an FRP Deck using Iterative Training of Neural Network

Doo Kie Kim, Professor, Department of Civil and Environmental Engineering, Kunsan National University, Kunsan, Jeonbuk, South Korea
Dong Hyawn Kim, Professor, Department of Ocean System Engineering, Kunsan National University, Kunsan, Jeonbuk, South Korea
Jintao Cui, Graduate, Department of Civil and Environmental Engineering, Kunsan National University, Kunsan, Jeonbuk, South Korea
Hyeong Yeol Seo, Graduate, Department of Civil and Environmental Engineering, Kunsan National University, Kunsan, Jeonbuk, South Korea

ABSTRACT

Fiber reinforced polymer (FRP) composite decks are new to bridge applications and hence not much literature exists on their structural and mechanical behavior. In studying this behavior, system identification (SI) – the process of determining a model of a dynamic system based on experimental data – is used. However, there are large discrepancies between the numerical displacement through static analysis of the primary model and the experimental displacement through a static load test. Therefore, this paper is presented to improve the efficiency of SI using neural networks (NN) integrated using an iteration method. It optimizes the FRP finite element (FE) model for SI. In the SI process, displacements are used as inputs while stiffness is the output. Two methods – the response surface method and iterative training of NN – are incorporated to optimize the estimated stiffness. Numerical displacements after identification are then compared with experimental displacements. Results are evaluated through the mean square value of the discrepancies between the numerical and the experimental displacements at 6 points. As iterations are performed, a slightly fluctuating yet decreasing trend in the discrepancies is observed. This shows that the process of optimization through NN is effective, and that the proposed method is efficient for system identification.

Bridge design, construction and maintenance 2007, Thomas Telford, London

Keywords: Fiber Reinforced Polymer (FRP); system identification; Neural Networks (NN); response surface method; iteration.

Introduction

The use of Fiber Reinforced Polymer (FRP) as a primary structural material is developing rapidly in the construction industry. FRP materials have considerable advantages in terms of weight, strength and corrosion resistance. They have been used for several decades in the aerospace, automobile and marine industries, where they have been very successful in very adverse environmental conditions. Although FRP composites are increasingly being considered for use in civil engineering, their widespread use is constrained due to current consideration of higher initial cost, lack of comprehensive design approaches and guidelines, and the predominant use of a one-to-one replacement methodology that often restricts the full utilization of the characteristics of the material. The development of such new FRP composite bridge systems raises concerns related to the dynamic response of the material to traffic loads and mass and stiffness characteristics, which are significantly different from those of conventional steel and structural concrete bridge structural components. Therefore, a technique for system identification is needed to update the finite element models of FRP decks.

Identification is the determination, based on input and output, of a system within a specified class of systems, to which the system under test is equivalent. System identification usually consists of two stages--model selection, and parameter estimation. There are many different ways of undertaking system identification, among which is the use of neural networks, which is the technique utilized in this paper. The identification of mathematical models of physical structures based on experimental measurements is a problem that has been receiving increasing attention recently. Numerous publications and methods are available on the subject of system identification of structures [1-3]. The approach that has been used most often is that of neural networks, which has been used in identification for health monitoring [4] and damage detection [5]. Feng et al [6] described a method of building baseline models for bridge performance monitoring using neural networks, and an improved approach for nonlinear system identification using neural networks was proposed by Gupta and N.K. Sinha [7].

In this paper, an attempt to build a baseline model for an FRP deck is described. To improve its accuracy, the NN methodology is integrated by iteration. A full size laboratory test was carried out to determine the static displacements of the structure. Then, a numerical model was constructed to investigate the effectiveness of the proposed method. Simulation results demonstrate that the proposed method is effective and is an efficient technique for system identification.

Theoretical background
Neural Networks

Rumelhart et al [8] reported the development of the back-propagation neural network (NN). NN is the most common of the self-learning models of artificial neural networks. A simple architecture of NN consists of an input layer, a hidden layer, an output layer, and connections between them (Fig. 1). Sigmoid functions are utilized as non-linear activation functions for all layers.

The corresponding architecture for back-propagation learning incorporates both the forward and the backward phases of the computations involved in the learning process. The learning mechanism of this back-propagation network is a generalized delta rule that performs a gradient descent on the error space to minimize the total error between the actual calculated values and the desired ones of an output layer during modification of the connection strengths. In other words, a least mean square procedure is carried out which finds the values of the connection strengths that minimize the error function by using a gradient descent method. The error function of the output layer is defined as:

$$E = \frac{1}{2}\sum_{k=1}(O_{pk} - T_k)^2 \tag{1}$$

where O_{pk} and T_k are the calculated values and desired values of the output layer, respectively.   Then:

$$O_{pk} = f_k(\sum_j W_{kj}O_{pj} + \theta_k) \tag{2}$$

where f_k, W_{kj}, O_{pi}, and θ_k are the activation function, connection strength, output values and biases of a hidden layer k respectively and:

$$O_{pj} = f_j(\sum_i W_{ji}O_{pi} + \theta_j) \tag{3}$$

where f_j, W_{ji}, O_{pi}, and θ_j are the activation function, connection strength, output values and biases of a hidden layer j.

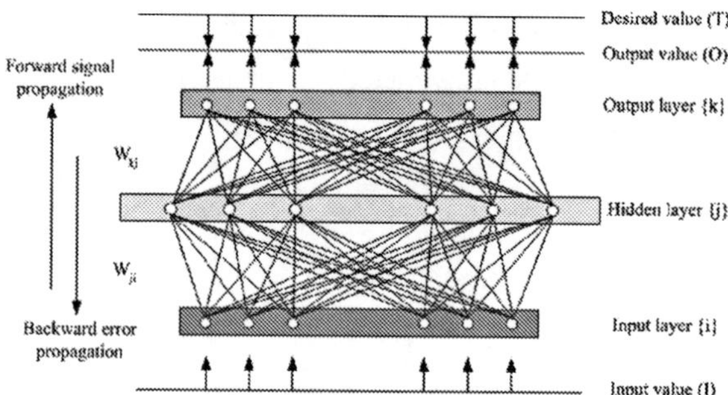

Fig.1. Structure of back-propagation neural network (Kim et al. [10])

In the back-propagation network, the error at the output neurons is propagated backward to hidden layer neurons, and then to input layer neurons modifying the connection weights and the biases between them by a generalized delta rule. The modification of the weights and the biases in such a generalized delta rule is by use of a gradient descent of the error.

Response surface method

Response surface methodology (RSM) is a set of statistical techniques designed to find the optimized value of the response or to examine the relationship between experimental responses and variations in the values of input variables. It is also used to optimize response quantities, which are influenced by several independent variables, as it provides simple models of complicated processes.

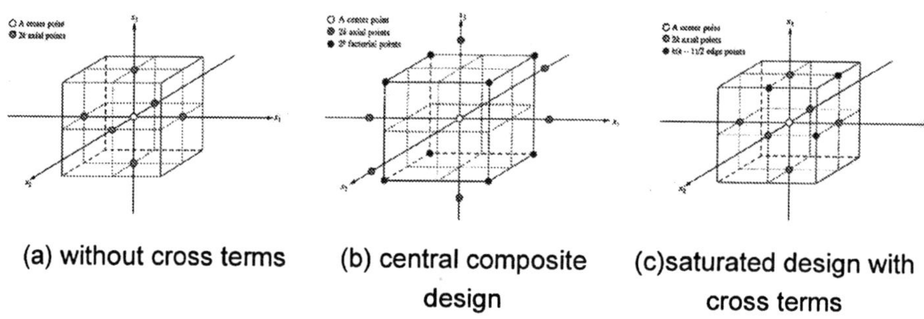

(a) without cross terms   (b) central composite design   (c)saturated design with cross terms

Fig.2. Design points for sampling [15]

Sampling point is one of the response surface methods used here to perfect the training database. First, in the design range of stiffness, only first-order approximations of g(X) (equation (1)) were used for sampling which resulted in 13 (6*2+1) evenly distributed design points. Then, six values of stiffness were changed simultaneously to form the upper and lower boundaries. Finally, two kinds of second-order polynomial (Fig. 2 (b), (c)) were added in succession to find the most effective and efficient mode. This can be described by equation (2).

$$g(X) = ax + b$$

(1)

$$g(X) = a + \sum_{i=1}^{n} b_i x_i + \sum_{i,j=1}^{n} c_i x_i x_j + \sum_{i,j,k=1}^{n} c_i x_i x_j x_k, (n \geq 1).$$

(2)

Application of the Proposed Method
Experimental tests

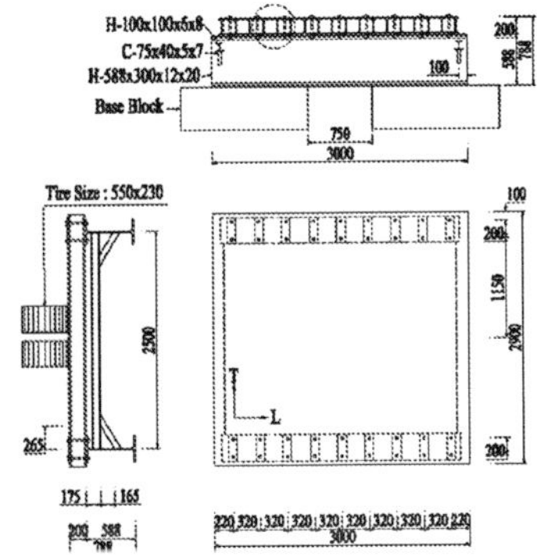

Fig.3. Schematic of FRP Test Mode   Fig.4. FRP Test Model

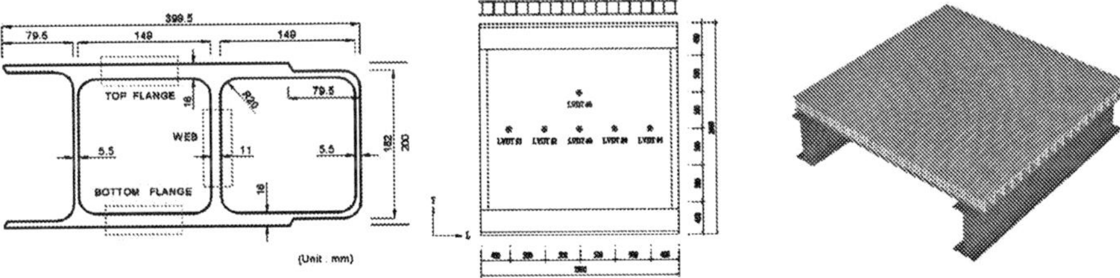

Fig.5. Unit module section   Fig.6. Schematic of   Fig.7. Finite element
                                 measured points   model

For the static displacement tests of the FRP composite deck, two steel girders were used along the longitudinal direction and two smaller steel girders along the transverse direction.   Four bracings were then placed between them so as to bind the test model as shown schematically in Fig. 3 and in the photograph in Fig. 4. Several FRP deck units shown in Fig. 5 comprise the FRP composite deck. The two longitudinal steel girders are supported by four base blocks at each end of the bottom flanges, in which the unsupported lengths are about a quarter of the steel girder lengths. For these two girders, H100*100*6*8 and C75*40*5*7 I-shape sections are used. The load of magnitude 50KN is exerted at the central area of the deck using two jacks. The static displacement experiment was performed three times, so that the displacements at 6 points were measured. The measured points and experimental results are shown in Fig. 6 and Table 1.

Table 1 Experimental displacements at 6 points (mm)

Items	P1	P2	P3	P4	P5	P6
Mean	0.1533	0.4900	2.2833	0.9033	0.2267	1.7367
Test 1	0.1600	0.5100	2.3300	0.9200	0.2200	1.7600
Test 2	0.1500	0.4800	2.2900	0.9100	0.2300	1.7500
Test 3	0.1500	0.4800	2.2300	0.8800	0.2300	1.7000

Primary finite element model

Strand7 [9] was selected as the finite element analysis tool. A total of 172 beam elements, 19,836 plate elements and 19,261 nodes were used to build the test model as shown in Fig. 7. Plate elements are extensively used since the deck is made up of FRP laminates. Table 2 shows the theoretical values of material properties and geometric parameters used to build the finite element model.

Table 2 Material properties and geometric parameters

Item	Thickness(mm)	$E_X (GPa)$	$E_Y (GPa)$	V_{XY}	$G_{XY} (GPa)$	$\rho (g/cm^3)$
Top Flange	18	15.83	14.86	0.253	4.457	1.9
Web	11	17.61	14.27	0.287	4.953	1.9
Bottom Flange	16	15.21	15.80	0.230	4.310	1.9

Process of system identification

During the process of system identification, displacements at 6 different points were selected as inputs, while the stiffnesses (E_X and E_Y) of the flanges and web were the outputs. This meant that there were a total of 6 input parameters and 6 output parameters. Here a NN that could be deployed in the MATLAB [13] was utilized with the collaboration of Strand7. The process of system identification can be summarized as the following:

1. Building of a primary finite element model
2. Establishment of training database

3. Training of Neural Network
4. Numerical verification
5. Estimation of stiffness using trained NN
6. Static analysis using estimated stiffness
7. Comparison of numerical displacements and experimental displacements
8. Method of iteration: repeat steps 2-7.

Numerical verification

Three cases of stiffness and displacements as shown in Table 3 were utilized as test patterns for the NN on which to perform the numerical verification. These were obtained by linear static analysis of the FE model in Strand7. Here case Ⅰ, Ⅱ and Ⅲ means that three cases of target stiffness were set through certain percentage changes of the theoretical stiffness.

Table 3 Test pattern to perform numerical verification

(a) Stiffness

Item	Top Flange		Web		Bottom Flange	
(unit:GPa)	E_X	E_Y	E_X	E_Y	E_X	E_Y
CASE Ⅰ	17.41	14.86	17.61	14.27	15.21	15.80
CASE Ⅱ	15.83	14.86	17.61	14.27	16.73	15.80
CASE Ⅲ	15.83	14.86	19.37	14.27	15.21	15.80

(b) Displacements

Items	P1	P2	P3	P4	P5	P6
CASE Ⅰ	0.2093	0.7237	2.7786	1.1864	0.3777	1.9700
CASE Ⅱ	0.2081	0.7229	2.7835	1.1853	0.3766	1.9735
CASE Ⅲ	0.2128	0.7238	2.7584	1.1808	0.3821	1.9537

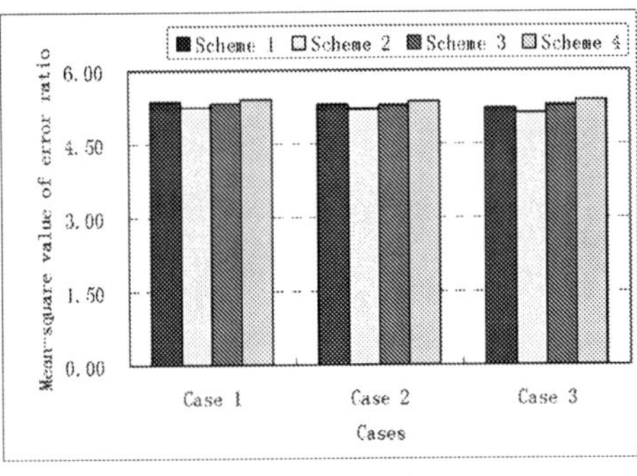

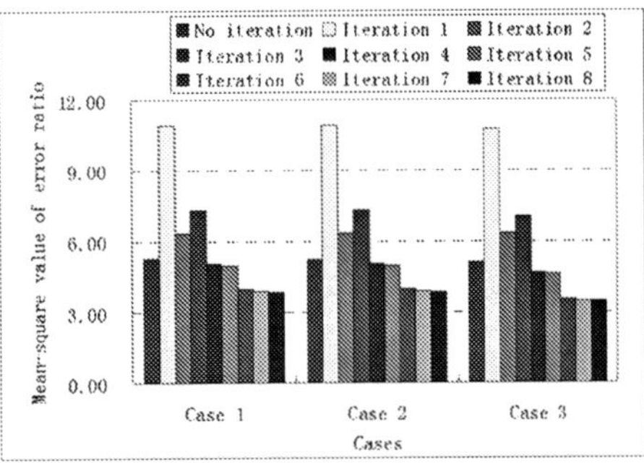

(a) Results of sampling point   (b) Results of iterations

Fig.8. Results of numerical verification

The results of the numerical verification were compared through the mean square value of errors between the estimated stiffness and the target stiffness as shown in Fig. 8. There are a total of 4 schemes of training database, which are coordinated with four types of sampling points as shown in Fig. 8 (a). Four experimental design models to select design points were considered in this section: Scheme 1 means saturated design using a second-order polymer without cross terms; scheme 2 means central composite design with a second-order polymer with cross terms; and schemes 3 and 4 mean saturated design using a second-order polymer with cross terms Ⅰ, Ⅱ respectively, as shown in Fig. 2. These four schemes use 13, 15, 27, and 43 cases of training database and Fig. 8 (b) shows the results when the iterations are performed.

Example studies

Here experimental displacements as shown in Table 1 were used as inputs while the stiffness is estimated by the NN. The model used here is built through simulating the tested FRP deck as described in section outlining the primary finite element model. Finally, the results are compared and discussed in the next section.

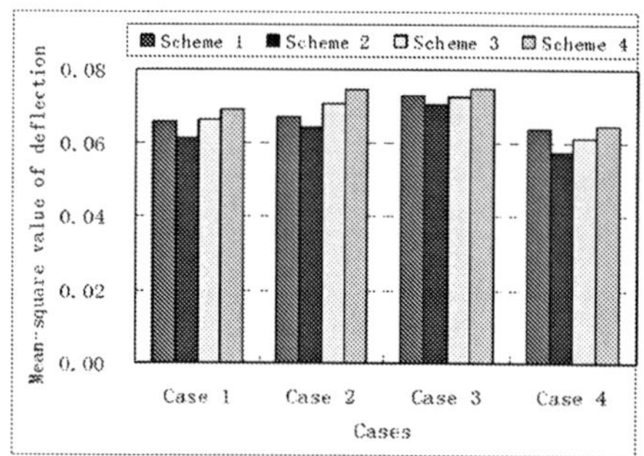

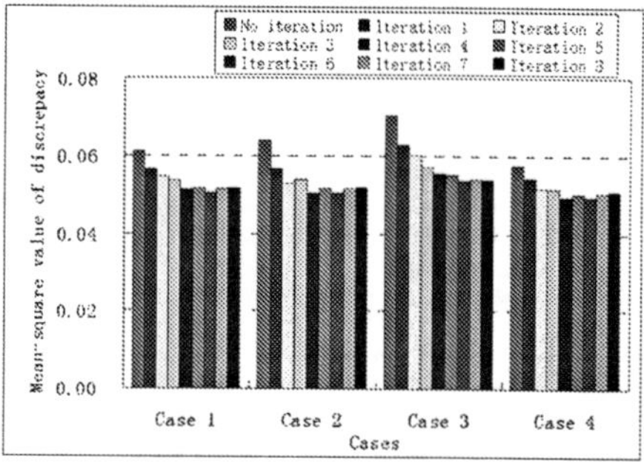

(a) Results of sampling point   (b) Results of iterations

Fig.9. Results of example studies

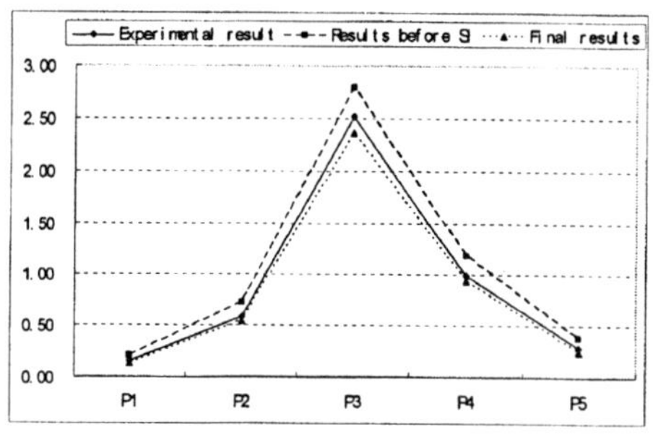

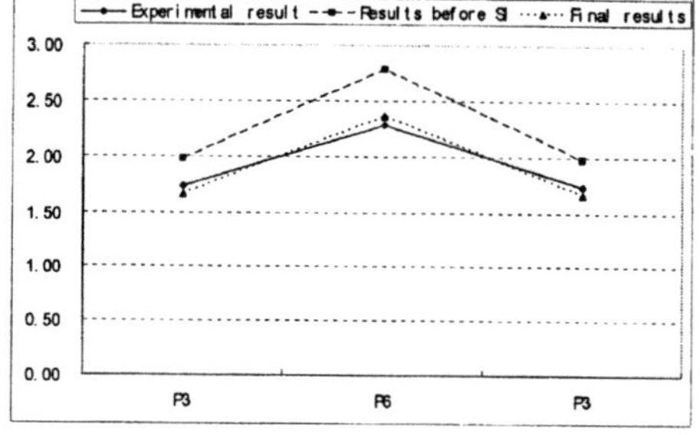

(a) Longitudinal direction   (b) Transverse direction

Fig.10. Comparison of numerical and experimental results

Results and discussion

As can be seen from the comparison in Fig. 8 (a), scheme 2 is the most effective in estimating stiffness, which means that 15 cases of training database were sufficient to obtain the optimum value. It can be observed that the accuracy may not be improved as the number of training data increases. To interpret this trend, a target line in the coordinate can be assumed, and the aim is to find the approximation line, which is closest to the target line. The approximation line is represented by linking the sampling points. In doing so, the accuracy is improved at the sampling points although in the median area (i.e. in between sampling points) this accuracy may be decreased.

To verify the accuracy of the proposed method, the mean square value of the discrepancies between numerical displacements and experimental displacements at 6 points were compared as shown in Fig. 9. The horizontal axes represent the cases of the numerical and experimental displacements respectively. From comparison of the iteration results, it can be noted that there is a decreasing trend in the discrepancy between these values as more iterations are performed, although this does fluctuate slightly. The results of the numerical displacements using the final estimated stiffness are compared with the experimental displacements as shown in Fig. 10.

Conclusions

This paper proposed the iterative training of Neural Networks together with a response surface method for the system identification of an FE modeled FRP deck. In this study, the following conclusions are drawn:

(1) An increased number of training data does not guarantee an improved accuracy in the method. However, an approximation very near the target accuracy can be sought by linking the sampling points. In this way, the precision is improved at the sampling points while that at the median area (i.e. in between sampling points) may be decreased.

(2) Although fluctuating, discrepancies in the numerical and experimental displacements show a decreasing trend. This means that in general, the proposed methodology is an effective procedure for system identification.

Acknowledgement

The research described in this paper was funded by the project for the development of long-life deck systems for bridges, a program of the Korea Institute of Construction Technology (KICT), Structure Research Department. The authors wish to express their gratitude for the financial support.

References

[1]   K.J. Astrom and P. Eykoff, System identification – a survey, Automatica 7(2) (1971) 123-162

[2]   L. Ljung, System Identification: Theory for the User, Prentice-Hall, Englewood Cliffs, NJ, 1987.

[3]   S.A. Billings, Identification of nonlinear systems--a survey, IEE Proc. 127 (1980) 272-285.

[4]   A.G. Chassiakos and S.F. Masri, Identification of structural systems by neural networks, Journal of Mathematics and Computers in Simulation 40 (1996) 637-656

[5]   J.J. Lee, J.W. Lee, J.H. Yi, C.B. Yun, H.Y. Jung, Neural networks-based damage detection for bridges considering errors in baseline finite element models, Journal of Sound and Vibration 280 (2005) 555-578

[6]   M.Q. Feng, D.K. Kim, J.H. Yi and Y. Chen, Baseline models for bridge performance monitoring. Journal of Engineering Mechanics, DOI: 10.1061/ (ASCE) 0733-9399(2004)130:5(562)

[7]   P. Gupta and N.K. Sinha, An improved approach for nonlinear system identification using neural networks, Journal of The Franklin Institute 336(1999) 721-734

[8]   Rumelhart, D.E., Hinton, G.E., Williams, R.J.: Learning representations by back-propagating errors. Vol.323. Nature (1986) 533-536

[9]   Strand7 Online Help, Release 2.4. Suite1, Level 5, 65 York Street, Sydney, 2000. Australia, 2005.

[10] Kim, D.H., and Park, W.S. (2005). "Neural network for design and reliability analysis of rubble mound breakwaters." *Ocean engineering*. 32(11/12), 1332-1349.

[11] Aizerman, M.A., Braverman, E.M., and Rozonoer, L.I. (1964), "Theoretical foundations of the potential function method in pattern recognition learning", Automation and Remote control, 25, 821-837.

[12] Yu, P.S., Chen, S.T., and Chang, I.F (2006), "Support vector regression for the real-time flood stage forecasting", Journal of Hydrology, 328(3-4), 704-716.

[13] Matlab user's guide, Release 7.1. The MathWorks, Inc. June, 2005.

[14] Hou, Sh. J, Li, Q, Long Sh. Y, Yang, X. J, Li, W. "Design optimization of regular hexagonal thin-walled columns with crashworthiness criteria", Finite elements in analysis and design, 43 (2007) 555 − 565.

[15] Haldar,A, Mahadevan,S. Reliability assessment using stochastic finite element analysis. USA, 2000.

STRUCTURAL OPTIMIZATION OF THE CABLE STAYED ARCH

P. G. Malerba, Dept. Of Struct. Eng., Politecnico di Milano, Milan, Italy
P. Galli, Consultant engineer, Milan, Italy
M. Di Domizio, Consultant engineer, Milan, Italy
G. Comaita, Consultant engineer, Milan, Italy

Abstract

A cable stayed arch tends to behave like a curved beam on an elastic foundation. With a proper arrangement, the stays act as internal constraints that may strongly improve the overall static behavior. This paper considers an arch constrained by stays radiating from a focus which lies on the arch's axis of symmetry and deals with the search for the optimal values of the vertical position of the focus, of the pre-tension of the internal stays and of the ratio between the section of the internal stays and that of the lower ones. A comparison with the corresponding behaviour of a simple tied arch and a recent application of the cable stayed arch in a wide industrial storage roof are shown.

Introduction

A correctly designed arch is a highly efficient structural typology, in which all sections tend to be fully compressed, the bearing structures' volumes tend to be minimized and the material's bearing capacity can be completely exploited.

In order to benefit from the above mentioned increment of efficiency, the ends of the arch have to be properly restrained by suitable vertical and horizontal reactions. In fact, the safety of these structure mainly depends on the reliability of the end supports, that is, on the stiffness and on the load bearing capacity of the soil-foundation system or of the substructures which support the arch. The previously recalled positive characteristics allowed ancients to build relatively large span structures by using only no tension materials. On the other hand, inefficient end supports has caused the loss of the majority of the structures of the past.

From the static point of view, an evolution of the arch concept arose from the possibility of avoiding the need from an external horizontal thrust, by introducing a tie along the chord. In bridges, the role of the tie is usually played directly by the structure of the deck. The tie counterbalances, in part or completely, the horizontal actions and, by means of imposed deformations (e.g. through a pre-tensioning action) it allows control of both the whole stress field along the length of the arch and the relative displacements between the two ends at the supports. The value of such displacements depends on the relative stiffnesses of the arch and of the tie. This improvement allows the removal of the limits due to the necessity for high horizontal thrust reactions and leads the whole system (arch plus tie) to behave like a simply supported beam.

The employment of tied arches in bridges (the so called "combined arch-beam systems", due to

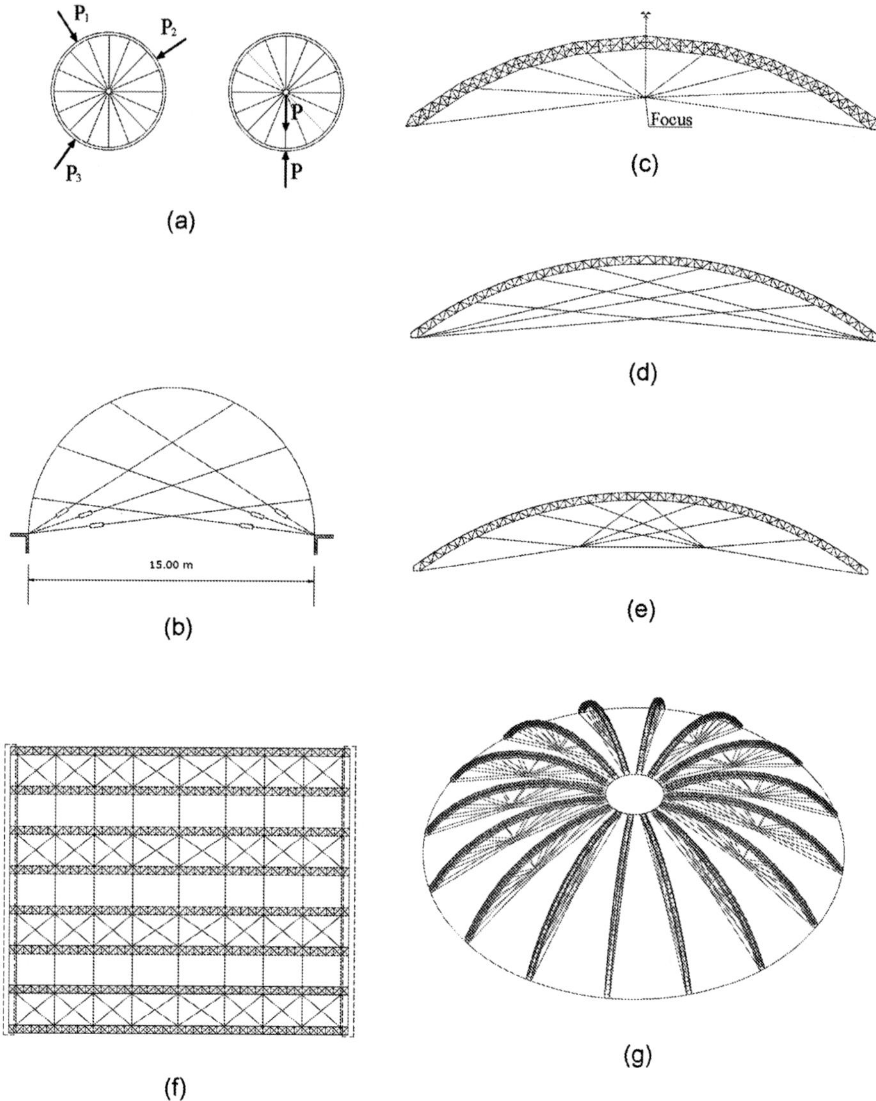

Figure 1. (a) Rings reinforced by closely spaced spokes (from Hetényi, Chap. IX, §48); (b) Cable stayed arches in GUM Department Store, Moscow (by V. G. Shukhov, from [1]). (c) Different assembly of stayed arches from [1]; (f) ÷ (g) Large roof plant and isometric view of a dome made of cable stayed arches [2].

Langer [3], Nielsen [6] and others) resulted in experimentation with many different arrangements so as to make the structure's overall behaviour more and more efficient. A wide range of literature deals with the problems connected with the definition of the arch shape, stiffness distribution, structural analysis and erection methods. At the current time, research tends to focus on further reductions in weight, while keeping the buckling behaviour under control.

Suggestions in this sense can be drawn from the examination of other structural systems. For instance, a current example of an efficient combined system of a curved beam and a collaborating stiffening addition is given by bicycle wheels (Fig. 1, a). A ring, which is reinforced by a large number of closely spaced spokes, behaves like a beam on elastic foundation. "If initial tension is put into the spokes they are able to resist compressive as well as tensile forces, and the force produced in the spokes by the external loading on the ring, will be proportional to the radial deflection of the ring at every point" (Hetényi, [5], Chap. IX, §48). The external loading is balanced by the resultant of the spoke forces, which is transmitted through the axle.

In an arch, the lack of symmetry means that it is not so intuitive to define the position of the point to which the resultant forces transmitted by the radiating members have to be conveyed. The present paper deals with this little used arch typology: the cable stayed arch (Fig. 1, b÷e).

The above mentioned tied arch, in which a single stay connects the two ends, can be considered the simplest form of cable stayed arch. This scheme can be enriched either by connecting the two ends to others suitable points along the arch through two fans of stays (Fig. 1, d), or by connecting the ends and others suitable points of the arch to a central pin ("the focus"), thus creating a radiating pattern of stays (Fig. 1,c). Other mixed stay configurations can be proposed (Fig. 1, e). In all cases, the contribution of the additional internal elastic restraints suggests new ways to condition the arch's structural behaviour and to increase its stiffness and the intensity of its critical loads. Assemblies of stayed arches can be adopted in large span roofs or in domes (Fig. 1, b, f, g).

The criteria that lead to the choice of the optimal shape for a cable stayed arch are neither simple nor immediate. For instance, the simple addition of a bundle of stays starting from the centre of the tie in a normal tied arch, gives little improvement in structural performance. In fact, the tie still attracts the highest force contribution in containing the arch thrust, while the others stays remain poorly engaged.

With reference to a symmetric radiating scheme, a significant evolution of the static behaviour and of the buckling performance under different load distributions, is determined by moving the pin position along the vertical segment that runs from the centre of the chord to the crown of the arch. Other potentially positive effects can be achieved by properly choosing the stiffness of the stays as a function of their position and by impressing a suitable pre-tension in the stays.

The best performances are investigated here by means of an optimization technique, focussed on finding under several load distribution; the optimal vertical position of the central focus, the optimal distribution of the stay stiffness and the optimal distribution of the pre-tension in the stays.

The results are measured in terms of the maximum band of eccentricity of the centre of thrust with respect to the arch axis, of the minimum relative displacement between the two ends of the arch, and of the value of the buckling multiplier associated with different load distributions. Comparisons with the corresponding behaviour of simple tied arches are shown.

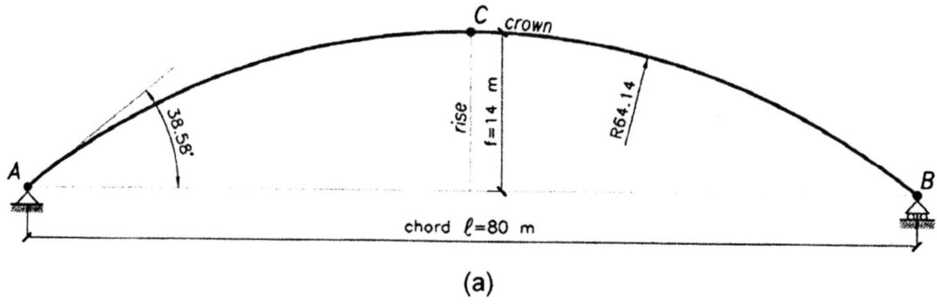

(a)

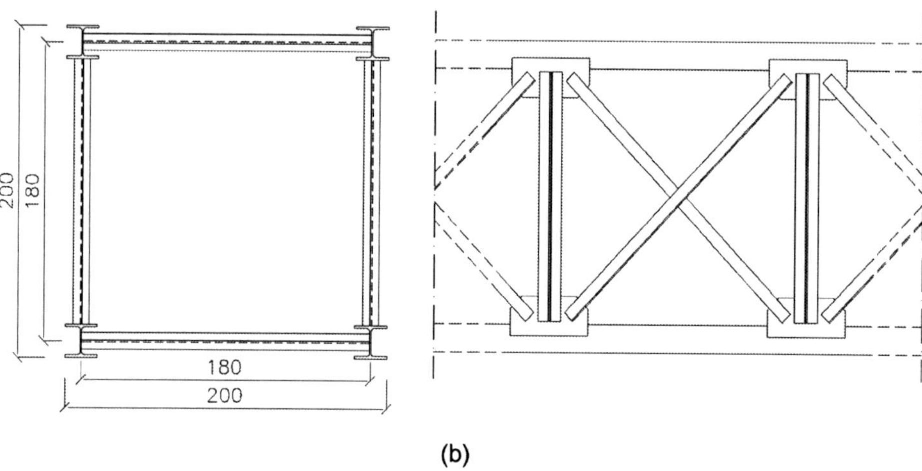

(b)

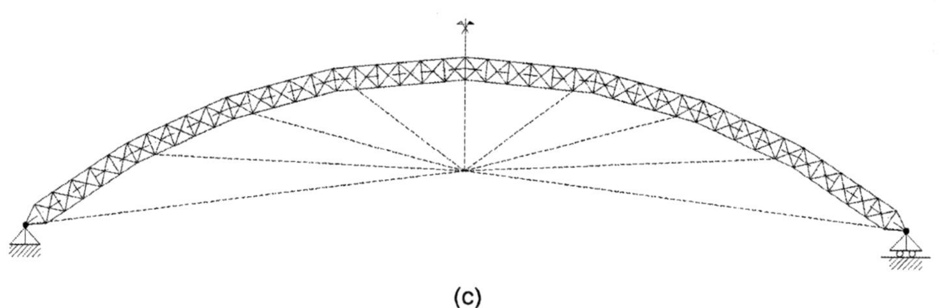

(c)

Figure 2. (a) Geometry of the arch. (b) Sectional and wall characteristics (c) View of the truss cable stayed arch.

The application of the cable stayed arch in a large span roof is then presented.

The Stayed Arch with Radiating Stays

We refer to the circular arch shown in Fig. 2. The arch has radius r = 63.14m, span l = 80m, overall length l_0 = 86.38m, rise f = 14m and end slope at the supports α_o = 38°,58 (Fig. 2 a).

The arch structure is made up of a truss composed of four HE200B stringers and L profile bracings. The section geometry is shown in Fig. 2. b.

The arch is restrained by a symmetrical set of nine stays, converging to a focus placed on the axis of symmetry. The lower stays are made of two full closed lock Ø80 ropes (A_{sl} = 3933mm^2). All interior stays have the same section, to be optimized as a suitable fraction $\rho_s = A_{s,i} / A_{s,l}$ of the lower ones. At the end of the optimizing process, the internal stays section will be determined as 2Ø44 ($2A_{si}$ = 2x1303mm2).

The pre-tension level of the lower stays is assumed to be the one that, under the combination of 100% of the dead load and 50% of the live load, reduces the relative displacement between the supports to zero. The pre-tension of the internal stays has to be optimized as a fraction of that of the lower stays.

The dead load is due to the self weight of the arch and to the loads on the roof. The intensity of the dead load is g_0 = 17.3 kN/m. For the sake of simplicity the dead load is assumed uniform over the span, instead of varying with the function $g = g_o /\cos \alpha$. The intensity of the live load is q = 32kN/m.

Four load combination were considered:
- Load Combination 1: Dead load, plus 50% of the live load.
- Load Combination 2: Dead load, plus 100% of the live load
- Load Combination 3: Dead load, without live load
- Load Combination 4: Dead load, plus 100% of the live load on only half of the span of the arch.

Optimization Problem

With reference to the structure shown in Fig. 2, we are searching for the focus position, the internal stays' area and their pre-tension. This should lead to a smaller band of internal eccentricity along the arch, or, in other words, it determines a thrust line as close as possible to the axis line of the arch. As collateral improvements, we also expect a structure that is stiffer than the simple tied arch and has higher critical loads.

<u>Objective Function</u>. The Objective Function to be minimized is the maximum internal eccentricity E_{max} along the arch, under the given load combination shown at top of Fig. 3. The Objective Function is expressed in terms of the following design variables [4], [7]:
- the vertical position y_D of the focus D, along the rise on the axis of symmetry. Preliminary analyses showed that such a measure has practical meaning in the interval $0 \le y_D \le 0,57f$;
- the pre-tensioning of the internal stays $\varepsilon_{s,i}$, which is assumed equal for all the stays and is measured as percentage of the pre-tensioning $\varepsilon_{s,l}$ of the two lower stays, which have the function of ties. The level of pre-tensioning may vary between 0 and 100% of the deformation imposed on to the lower stays;

- the ratio $\rho_s = A_{s,i} / A_{s,l}$ between the sectional area of the internal stays $A_{s,i}$ and that of the lower ones $A_{s,l}$. This ratio is assumed to be $0.1 \leq \rho_s \leq 0.9$.

<u>Displacement Constraints</u>. No displacement constraints are introduced. With reference to load condition 1 (dead loads, plus 50% of the live loads) a suitable value of the initial pre-tensioning $\varepsilon_{s,l}$ of the lower stays was chosen in order to limit the relative displacement between the two earth supports to nearly zero ($\delta_{A,B} \leq 0.1mm$). Relative displacements between the supports under the other load conditions are controlled during the solution process.

<u>Stress Constraints</u>. The following stress constraints were assumed:
- Stays: No compression. Maximum allowable stress $\sigma_s \leq 400N/mm^2$.
- Arch Members: Maximum allowable stress $|\sigma_a| \leq 160N/mm^2$.

<u>Synthesis</u>. In synthesis, the optimisation problem is stated in the form:
Objective Function:

$$E_{max} = E_{max}\left(\{y_D\}, \{\varepsilon_{s,i}\}, \{\rho_{s,i}\}\right) \rightarrow min \tag{1}$$

Design Variables:

$$0 \leq y_D \leq 0.57f \tag{2}$$
$$0 \leq \varepsilon_{s,i} \leq \varepsilon_{s,l} \tag{3}$$
$$0.1 \leq \rho_S \leq 0.9 \tag{4}$$

Stress Constraints:

$$0.0N/mm^2 \leq \sigma_s \leq 400N/mm^2 \tag{5}$$
$$-160N/mm^2 \leq \sigma_a \leq 160N/mm^2 \tag{6}$$

Search for the minimum

The search for the minimum of the Objective Function has been carried out with the Downhill Simplex Method [8], [2]. This method, originally proposed for the solution of unconstrained optimization problems, has been properly modified in order to handle the behavioural stress constraints.

As already said, Load Combination 1 had been used to define the initial pre-tension of the lower stays. Within the problem formulation of the previous paragraph, the last three load combinations were considered: (a) Load combination 2: Dead load, plus 100% of the live load. (b) Load combination 3: Dead load, without live load. (c) Load combination 4: Dead load, plus 100% of the live load on only half span of the arch.

During the analyses, in order to assess the sensitivity with respect to each design variable, partial results regarding:
- the focus height y_D
- the level of pre-tensioning of the internal stays $\varepsilon_{s,i}$
- the ratio between the sectional areas of the stays $\rho_{s,i}$
- the relative displacements between the supports $\delta_{A,B}$
- the actual tension in the stays/ties $\overline{\sigma}_{s,l}$
- the maximum eccentricity e_{max}
- the limit multiplier associated with the eulerian critical load

were recorded for each load combination.

Results

A synthesis of the results is given in Table 1. Fig. 3 shows the bending moments , the axial thrust and the eccentricity along the horizontal projection of the arch for each of the three load combinations.

Table 1. Main results regarding the behaviour of the cable stayed arch in the final configuration.

Span [m]	80	Relative Displac. $\delta_{A,B}$ Load Comb. 2: *[mm]*	63.5
Rise [m]	14	Relative Displac. $\delta_{A,B}$ Load Comb. 3: *[mm]*	-65.4
Height of the section [m]	2	Relative Displac. $\delta_{A,B}$ Load Comb. 4: *[mm]*	-1.4
Focus Coordinate y_D [m]	3.73	Thrust in the internal stays Load Comb. 2: *[kN]*	3139.5
Number of stays	9	Thrust in the lower stays Load Comb. 2: *[kN]*	1396.4
Lower stay section [mm^2]	80	Thrust in the lower stays Load Comb. 2: *[kN]*	1942.4
Internal stay section [mm^2]	44	Max eccentricity. Load Comb. 2: emax *[cm]*	28.55
Internal stays pre-tensioning [%]	44.65	Max eccentricity. Load Comb. 3: emax *[cm]*	28.70
Ratio between areas $\rho_s = A_{s,i} / A_{s,l}$	0.3	Max eccentricity. Load Comb. 4: emax *[cm]*	28.44

Table 2 shows a comparison between the values of axial thrust, bending moment, maximum eccentricity and minimum load multiplier obtained for Load Combination 4 (Dead load, plus 100% of the live load on half span of the arch) in a) a cable stayed arch, b) a tensioned tied arch, c) a non tensioned tied arch.

Table 2. Comparison between characteristics parameters of the cable stayed arch with no tensioned and tensioned tied arches (Load Combination 4).

	N_{max} *[kN]*	M_{max}*[kNm]*	e_{max} *[cm]*	k_{min}
a) Cable stayed arch	3311	854	28	37
b) Arch with horizontal tensioned tie	2117	3267	154	10,8
c) Arch with horizontal non tensioned tie	2039	3940	193	11,6

The main characteristics which emerge from the analyses can be summarized as follows:
-   The introduction and the pre-tension of an horizontal tie allows an arch to be conceived which does not need horizontal support reactions. This is a well known and experienced solution. A suitable pre-tension is the one which makes the relative displacement between the end supports equal to zero for a given load combination; for different load combinations, these displacements remain relatively small. From an external point of view, the tied arch works like a simply supported beam.
-   The addition of internal stays to the tied arch typology does not modify the overall behaviour if the stays start from the centre of the horizontal tie, and is practically useless.
-   Only by shifting the focus position upwards is it possible to activate a static distribution between the ties (which are no longer horizontal and will from now on be called lower stays), the internal stays and the arch segments. The focus position cannot be directly defined without a specific optimization process.
-   Although the introduction of the internal says brings with it a significant reduction in the internal eccentricity along the arch segments, a further reduction can be obtained by pre-tensioning these stays up to a certain fraction of the pre-tension imposed to the lower stays.

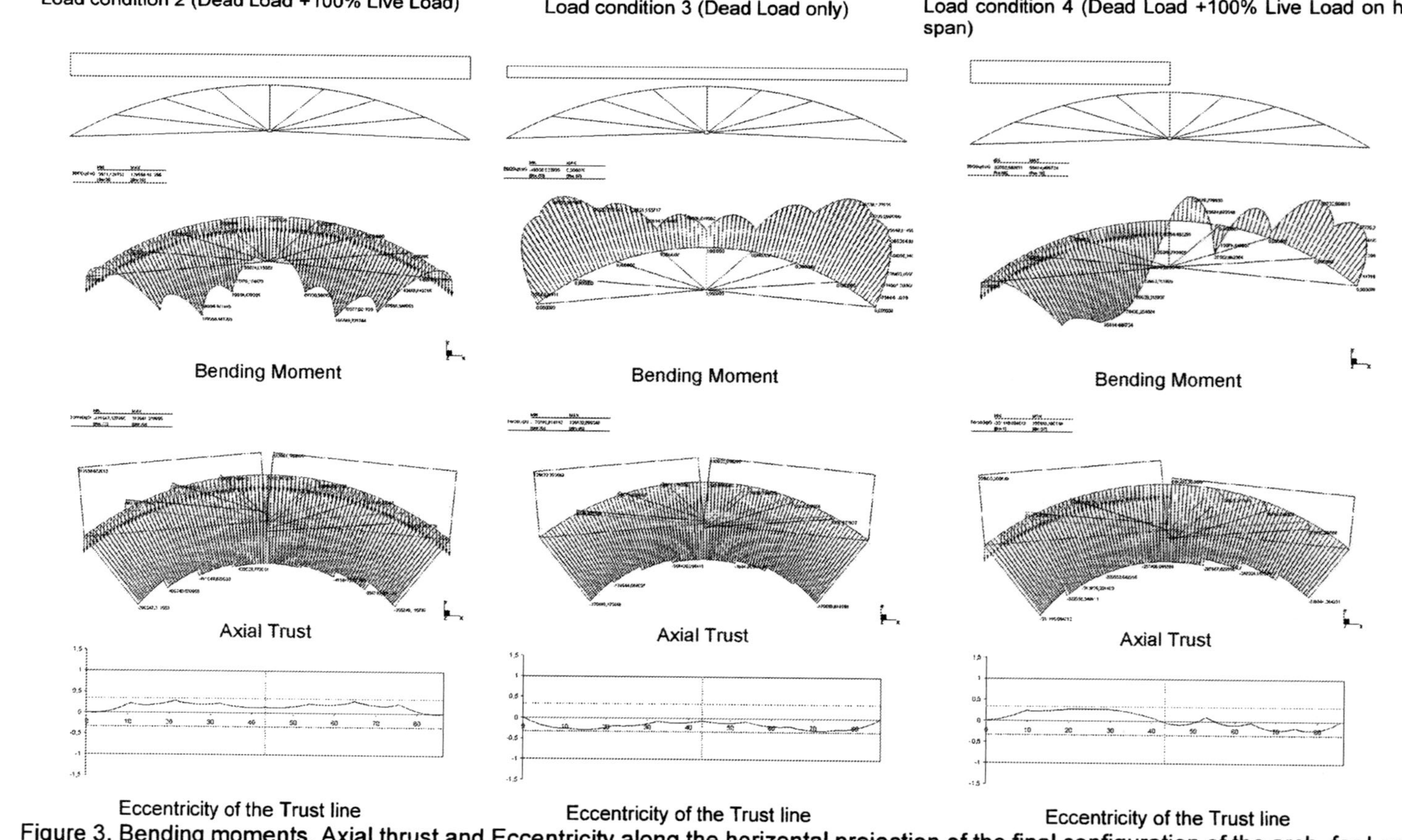

Figure 3. Bending moments, Axial thrust and Eccentricity along the horizontal projection of the final configuration of the arch, for Load Combinations 2, 3, 4.

- A further improvement can be achieved by adequately proportioning the internal stay section. The optimal pre-tension intensity and the optimal area of the internal stays can be defined by means of the same analysis which optimises the focus position.

An Engineering application

A cable stayed arch has been used in a large span roof, which covers an industrial storage of about 12,000 square metres.

The choice of the structure to be employed was submitted to an international competition. The evaluation criteria took into particular consideration the type of structure, the erection system, the level of interferences with the manufacturing activities and the costs. After some comparative reckonings, the unusual typology of the steel cable stayed arch was assumed as the main supporting system.

The storage plan has an irregular shape: it is made up of first a rectangular part, with a transverse span of 84 m and a length of 80 m, and continues with a curved sector, bent up to 90° with respect to the initial section and having different transverse spans (maximum 91 m, minimum 60m). Such a shape resulted from the operating characteristics of the handling system.

The roof is covered by means of 17 circular arches, pinned at the top of reinforced concrete columns. Each arch is restrained by nine stays, converging at a focus placed at the third point of the vertical segment which goes from the centre of the chord to the crown of the arches.

The arch section has a triangular shape and is 2.00m high and 2.00m wide. The upper stringers are made of HE200A profiles, the single lower stringer is made of HE300A profiles. The upper face and the lateral, inclined webs are braced by angle profiles: the upper face has L70x7 transverse elements and L80x8 cross bracing. The lateral faces have L80x8 transverse elements and L90x9 cross bracing.

The arches are connected to one another through longitudinal truss beams and are periodically stiffened by truss cross bracings. The general appearance is that of a large spatial truss roof. Fig. 4 shows a phase of the erection stage and an aerial view of the completed roof.

Conclusions

Arches are curved structures that carry loads primarily by developing axial compression. This typical behavior can be improved by activating a system of pre-tensioned stays converging to a focus in a suitable position.

This paper presented an introductory study on the structural performance of a cable stayed arch. From the structural point of view, the main differences with respect a traditional tied arch are:
- An increase in the thrust along the segments of the arch, mainly due to the pre-tensioning action introduced.
- A strong reduction in the bending moments and, as a consequence, of the internal eccentricity along the arch. The thrust line can be maintained very close to the axis of the arch.
- A general increase in the stiffness, with lower vertical displacements.
- A strong increase in the limit multiplier associated with the eulerian critical load.

The same structure can undergo further improvements by adequately shaping the axis of the arch and tuning the stiffness of the arch segments and of the stays.

New research is in progress in order to explore these aspects under a wide variety of loading conditions, like those due to the uplift and to lateral wind pressures. Another important issue regards the sequence of pre-tensioning in order to achieve the expected final configuration.

(a)

(b)

Figure 4. Roof supported by cable stayed arches and covering of a storage of 12000m². (a) View of the arches during an erection stage. (b) Aerial view of the completed roof.

References

1. Belenya, E., Prestressed Load-Bearing Metal Structures. Mir Publisher, Moscow, 1077.
2. Buraschi, L., Calabrò, P., Prestational Analysis of the Cable Stayed Arch. Dissertation, Politecnico di Milano, Milan, 2006 (In Italian).
3. De Miranda, F., Il Ponte Sistema Langer. Costruzioni Metalliche N. 1, Milano, 1955. (In Italian).
4. Kirsch, U., Optimum structural design. Concepts, methods and applications. McGraw-Hill, 1981.
5. M. Hetényi, Beams on Elastic Foundation. Ann Arbor: The University of Michigan Press, 1946, 7th Printing 1964.
6. Nielsen, O. F., Bogenträger mit schräg gestellten hängestangen. AIPC, Vol. IV, p. 429, 1936.
7. Rao, S. S., Engineering Optimisation, Theory and Practice. John Wiley and Sons, Inc., 1996.
8. Vetterling, Teukolsky, Flannery: Numerical recipes in Fortran 77. Cambridge University Press, 1996.

Under-slung Cable Structures-A Feasible Alternative?

Christos Christodoulou Cardiff School of Engineering, Cardiff University, U.K.
Dr. Robert J. Lark Cardiff School of Engineering, Cardiff University, U. K.

Abstract

This paper examines the behaviour and feasibility of under-slung cable structures. Consideration is given to their structural behaviour, their dynamics, their cost-effectiveness and the practicality of adopting such structures. The parameters that govern the behaviour and response of these structures are investigated and structural arrangements and details are proposed, which it is suggested would make the construction of such structures feasible. The aim of the paper is to generate discussion of this novel form of construction and to identify opportunities for the exploitation and development of such a concept.

This paper discusses how underslung structures, when combined with a trussed stiffening girder, can provide an innovative but more sympathetic structure, which is stiff, stable and economically viable for both medium and long-span structures. It is shown that when there is a need to minimise the number or limit the location of the foundations required, such structures provide a means of adopting relatively large spans, while restricting the height of the cable supports and hence minimising the impact of the structure.

The paper concludes by summarising the benefits that can be obtained from such structures both during construction and in use, and seeks to identify the opportunities and challenges that must still be met to facilitate the exploitation and development of such a concept.

Introduction

Suspension type bridges have been established from the early stages of primitive man, far before the first records of modern girder bridges ever existed. Bridges were generally built between trees, the cables consisted of hemp ropes, and typically the footway consisted of transverse timber boards and suspended by more or less vertical rope hangers. Such structures were common practice in early communities as it was a relatively straightforward form of construction without a need for any real understanding of the mode of behaviour. The so-called "Western Civilizations", only became interested in suspension bridges after the introduction of wrought iron, firstly in China in the form of chains. In the UK in particular, these were firstly forged on a large scale for use as anchor chains for ships, and bridges using such chains were often built near the early shipyards. However, most of these early types suffered oscillations in high winds and some collapsed as a result[1].

The economic utilization of materials for construction demands that, as far as it is possible, the predominating stresses in any structure should be those for which the material is best suited. Because of the great strength and reliability of materials like steel and high-tech composites in tension and the uncertainties and inefficiency involved in the design of massive compression members, for economic designs the form that should be adopted for long span bridges should generally be one that exploits these tensile properties. Indeed, this argument explains why cable structures dominate bridge design to such extent at all span levels - short to super span (>1200m) – and the extent of their application is such that nowadays, they can be used in virtually any situation. By nature, they are aesthetically stunning structures, renowned for their sensitivity to excitation and deformations, but extremely safe structures incorporating a high degree of redundancy. They are particularly flexible, cost efficient in terms of their weight/cost ratio and sustainable given their ease of maintenance, replacement of components and consequential longevity.

More recently, further structural developments have been achieved with regards to suspension and cable-stayed bridges with short to moderate spans.[2&3] These developments have incorporated the use of either concrete or steel as the main load bearing members, and the success of these projects has been assured by the use of simple but ingenious geometrical arrangements.

Concrete self-anchored suspension bridges have been introduced in China, with the construction of the Golden Bay Bridge, Lan Qi Bridge and Wan Xin Bridge. As described by Zhang et al[2] these projects, apart from being aesthetically pleasing, have resulted in significant savings in their construction costs. In particular, it is estimated that for the Lan Qi Bridge, the self-anchored suspension system resulted into a saving of nearly US\$ 4.17 million when compared with a gravity-anchored solution. These designs incorporate a closed loop cable system with local anchorages in the girders, thus making them fully pre-stressed structures for any loading condition.

Under-slung cable arrangements have not been widely used and examples are few and far between. Ruiz-Teran and Aparicio[3] report that only twenty such cable-stayed structures have been built in the past thirty years, although they do acknowledge that a number of this type of design have been proposed but not built. It is suggested that the reasons for this are that there is a lack of understanding of their structural behaviour, a scarcity of studies of these structural forms and a reluctance of public authorities to accept such highly unconventional structures.

The aim of this paper is to generate discussion of these novel forms of construction and to identify opportunities for the exploitation and development of such concepts. The paper discusses how underslung structures, when combined with a trussed stiffening girder, can provide an innovative but more sympathetic structure, which is stiff, stable and economically viable for both medium and long-span structures. It will be shown that when there is a need to minimise the number or limit the location of the foundations required, such structures provide a means of adopting relatively large spans, while restricting the height of the cable supports and hence minimising the impact of the structure.

The generic arrangement that is considered can be summarised as a deck, supported by vertical compression struts below the deck, which are themselves supported by the cables. Various layouts may be adopted according to the requirements of specific sites and Ruiz-Teran and Aparicio[3] describe a range of single and multi-span structures that are both externally and internally anchored.

Structural behaviour

For the purposes of this paper, a generic arrangement is discussed (Figure 1) that could be adapted to suit a range of situations by appropriate modification of the stiffening girder and the suspension system. The particular design that is presented was developed in response to an undergraduate design competition[4] for an 8m wide (an unusually narrow structure), 150m span structure that was required to carry an imposed loading of the order of 20 kN/m^2.

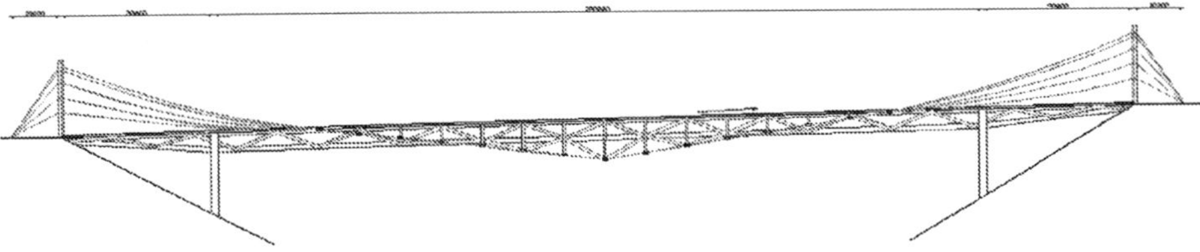

Figure 1: General Arrangement

Stiffening girder

Truss arrangements are excellent for dealing with high moments and shear forces and in addition they provide outstanding torsional stiffness (Figure 2). Because an under-slung arrangement is adopted, the top, compression chord will also carry and distribute local bending moments due to flexure between the nodes of the truss[1]. Treating the truss as one single beam, a simple static analysis can be performed. For the arrangement shown in Figure 1 in which the stiffening girder is supported at the piers this results in the maximum moments in the stiffening girder occurring over the piers and being of a hogging nature. As a result, the chords of the stiffening girder have to be designed for both tension and compression, a condition for which the type of truss proposed is again ideal because of its balanced configuration.

For the design proposed here it was suggested that rectangular hollow sections should be used for the top chords in order to increase the surface area through which the composite action with the deck can be achieved. Hollow sections are aesthetically pleasing but it is recognised that their use will increase the cost and complexity of the fabrication process. As such, it will probably be more economic to use open sections and, although this will require careful consideration of local buckling effects, the relatively short bay lengths that are possible with the proposed arrangement should relatively easily facilitate this change.

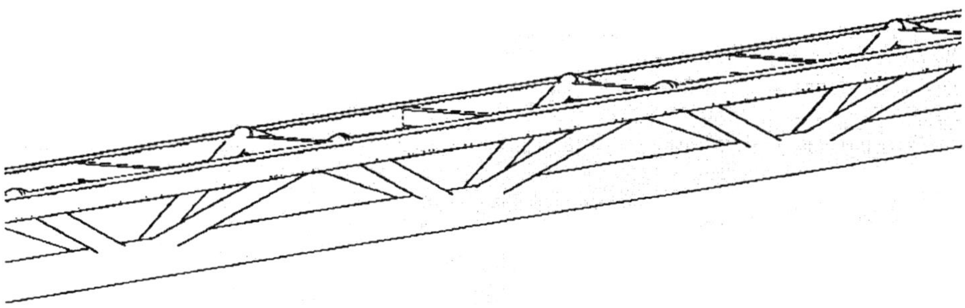

Figure 2: 3-D Model of the proposed truss system

For the bottom chord, a single member may be used. This layout will result in cost savings due the reduction in structural steelwork that is required. Additionally, this arrangement will improve both the aerodynamic behaviour of the structure and its aesthetics. Again, in the particular example presented here, circular hollow sections were proposed because of their small surface area, excellent torsional stiffness and high tensile and compressive capacity.

The induced axial forces in the chords are inversely dependent to the depth of the truss, although a balance needs to be struck between ensuring that it has sufficient stiffness to distribute the load as evenly as possible to the main suspension system that supports it, while not making it so stiff that it acts as a beam between the rigid supports at the fixed piers. To achieve this relatively shallow depths are preferable, as they will also reduce the impact of the structure on the surrounding environment, be less susceptible to adverse wind loading, and be easier to transport and assemble. For individual, large-scale projects the optimum depth will need to be determined. This will be a function of the axial forces, the amount of structural steelwork required to accommodate them and the resulting stiffness of the composite girder.

With regards to the elevational, inclined diagonals (Figure 1), it was proposed that they should also be circular hollow sections to reduce air drag and to enable them to withstand the large tensile and compressive axial forces that are induced in them, particularly near the fixed supports where the shear forces are large. To triangulate the structure in three dimensions these members were also inclined in cross-section (Figures 2 &3). This enhances the torsional stability of the structure and thus improves its sway resistance.

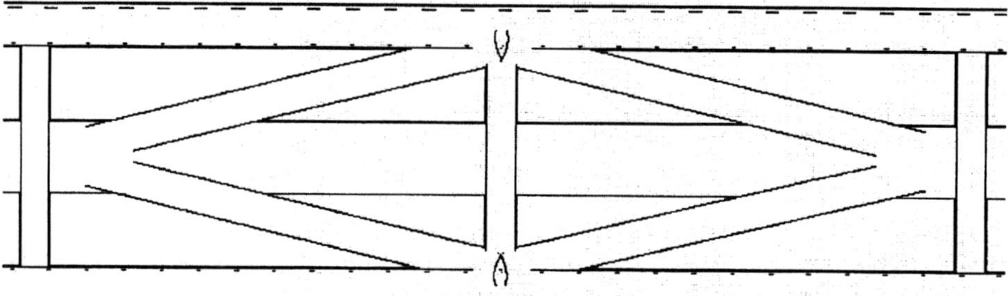

Figure 3: Top view of the truss arrangement, showing inclined diagonal members

Stability cross-bracing for the top chords can take one of two forms. They can be independent of the deck slab and used solely to provide restraint to the compression chords and therefore relatively slender, or they can also be part of the composite truss-deck system. If the latter approach is adopted the cross-members will also carry local moments. It is likely that such an approach is more appropriate when larger deck widths are required, therefore necessitating a thicker slab, although in this case fatigue of the cross-member / chord member connections due to torsional load effects may be a significant design criterion.

Cable system

The analysis of the proposed system is similar to that applicable to traditional suspension bridges; therefore the required sag of the cables determines the height of the pylons. For under-slung arrangements, the critical sag is the dip of the cable under the deck as it is over this length that the cables are loaded. Arguably, the optimum cable dip lies somewhere between 1.5 and 2.0 times the depth of the truss. However, what needs to be ensured is that the cables make best use of their excellent tensile resistance and hence cost effectiveness and that to do this they should be designed to carry the full imposed load and a significant proportion of the self-weight of the whole system.

Vertical struts are used to transfer the loads to the cables, and these struts are connected as shown in Figure 4. The main concern with this detail is the local bending that is likely to be induced in the struts because of the differential movement that will occur between their ends due to the difference in stiffness between the cables and the stiffening girder. This effect needs to be investigated in more detail.

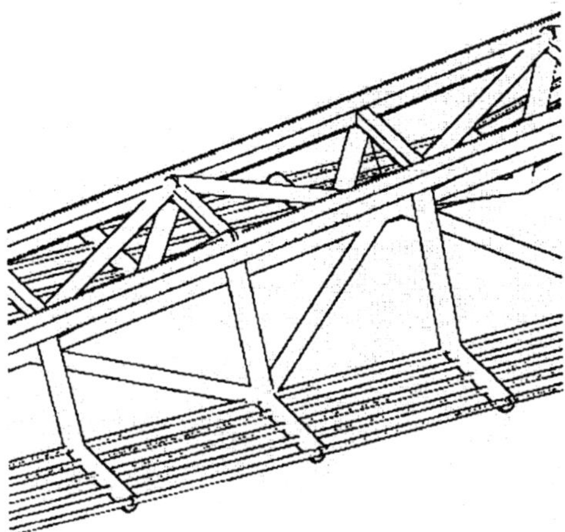

Figure 4: 3-D view of the whole structural system, emphasizing the cable connections

To provide transverse restraint to the cables additional diagonal members are added locally to enhance the transverse stiffness of the structure. These are again diagonal elements as shown in Figures 4 & 5. In order not to over-complicate the structure and to minimize the amount of additional steelwork required, it is proposed that these bracing members should be placed at every second position. The studies made to date suggest that such an arrangement is sufficient and indeed contributes to ensuring a uniform distribution of load between the stiffening girder and the cables.

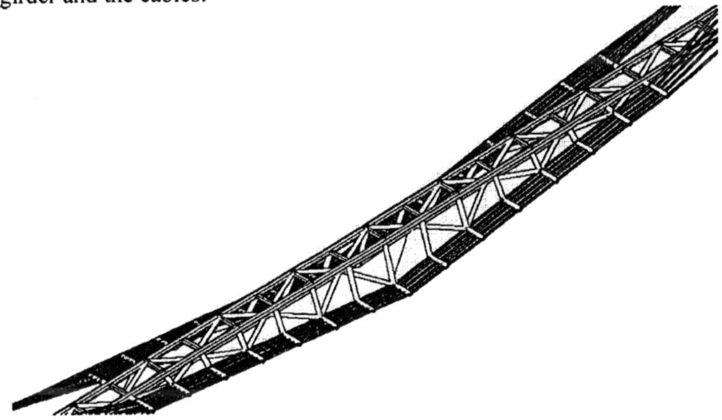

Figure 5: Isometric view of the structural arrangement

Pylons

The main purpose of the pylons is to accommodate the displacements and the vertical forces generated by the cables. Traditionally, the pylon's height has been dictated by the sag of the main cables. However, in this particular case, the critical sag is the one below the bridge deck and as a result the pylon's height is in theory of no great significance. Nevertheless, the lower this height the greater the resultant horizontal force in the cable and the less efficient the design. Also the greater these forces the larger the anchor block that is required, although if short pylons are required to minimise the visual impact then this design does facilitate this option.

Dynamic Response

The importance and significance of the dynamic response of suspended structures has been appreciated for many years. However, the detailed analysis of such effects is complex and very dependent on structure specific details. To date, time has not permitted such an analysis of the proposed design but the effects that need to be considered and the conceptual approach that has been adopted in order to limit the response of the structure to these effects are outlined here.

The flow of air over the slender deck of a any bridge structure (Figure 6) produces vertical bending and torsional oscillations of the deck. The response of the deck to these oscillations can take the following forms:-

Vortex shedding response

The periodic shedding of vortices from the upper and lower surfaces of the deck causes alternating aerodynamic forces to be applied to the structure. The response will decrease when structural damping is increased but will increase when the intensity of the turbulent flow is increased.[1]

These vortices create turbulences and consequently vibrations. The distance of the vortices from the structure is speed dependant thus the worst-case scenario is when the airflow is perpendicular to the cross-section. These vortices can also move both vertically and horizontally, thus exacerbating the vibrations, but it is believed that the cross section that has been adopted can be adjusted to limit these effects.

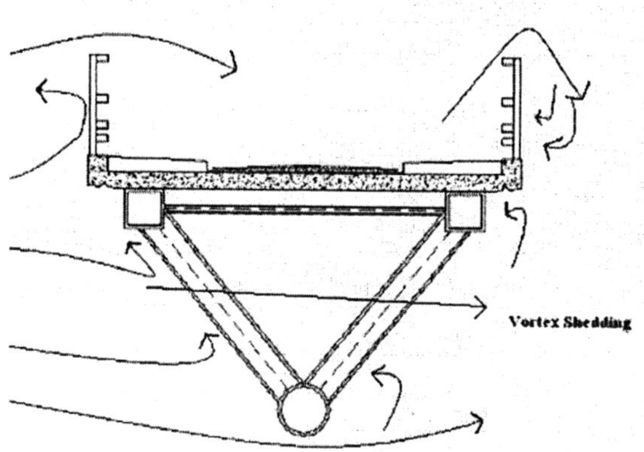

Figure 6: Airflows around the proposed structure

Turbulent or buffeting response

Wind loadings also varies due to the continuous fluctuation of the direction and velocity of the wind. If the energy of the turbulance generated by these fluctuations is sufficient then the resulting forced oscillations can in turn be sufficient to affect the serviceability of this type of structure. Such events are not uncommon and even the long-span Akashi Kaikyo Bridge has been closed six times because of such serviceability concerns. Further work will be required to address this issue for the proposed design.

Flutter response

Near what is defined as the critical wind speed for the cross-section, the vibrational response of the structure is amplified. This response can either be vertical or torsional, however for traditional deck arrangements, torsional instability is normally dominant.[4] Flutter occurs when the torsional and bending frequencies of the structure coincide. To prevent this structural damping of at least one of the frequencies may be required to ensure that this does not occur.

For the proposed design, the truss has excellent flexural and torsional stiffness and therefore should not be susceptible to this effect, provided that these stiffnesses are sufficiently different. Additionally, for moderate wind speeds the presence of such a complex configuration and the physical size of the elements will partially divert the flow of the air away from the deck resulting in reduced aerodynamic pressures on it. The scattered airflow will eventually produce turbulences and vortices but these will now favour stability, as they will reduce the fluttering effect.

Forms and Methods of Construction

In the form of cable structure proposed in this paper the aim has been to enhance the axial capacity while limiting the required flexural capacity, under both permanent and variable loading. The axial response is a function of the tension in the main cables and the compression capacity of the pylons, struts and deck. The flexural response is related to the flexural stiffness of the deck structure. In the proposed design, increasing the tensions in the cable can enhance the axial response and as a result the flexural response can be reduced, resulting in a more slender and hence less expensive structure. However, by adopting a truss for the main girder it is possible to balance these two responses and in the current design the truss has been deliberately sized to ensure that it can largely carry the permanent loads in flexure to aid both the proposed construction technique and the three span configuration that was required for this location.

In theory, using arrangements similar to the ones shown in figures 7 & 8 it is possible to compensate for 100% of the permanent loading, so that bending in the deck is limited to the local effects between the points of support provided by the cable system[3]. In this case the vertical struts will act as virtual piers and the effective clear span can be radically reduced, offering great savings both in terms of materials and costs.

For externally anchored designs such as that proposed in this paper, the disadvantage of this approach is that they require significant anchor blocks to accommodate the large tensile forces in the main cables. In the case of the competition for which this design was developed, this was not a problem as the ground conditions were such that the cables could be socketed directly into competent rock. However, where this is not the case self-anchored solutions[2] should provide an appropriate alternative and using one or other of these approaches, such designs are applicable in theory to both single span and multi-span bridges.

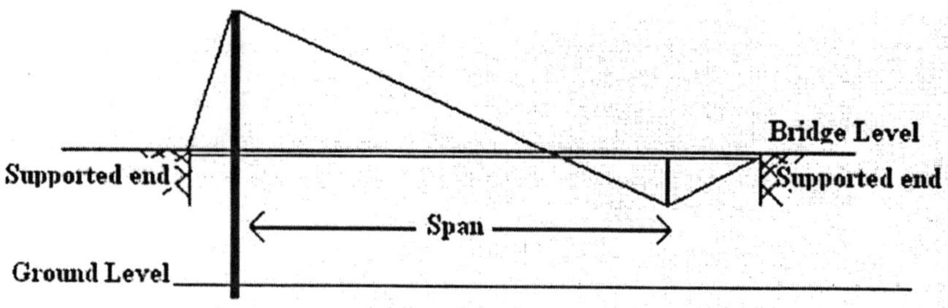

Figure 7: Idealised unsymmetrical mono-under-slung arrangement

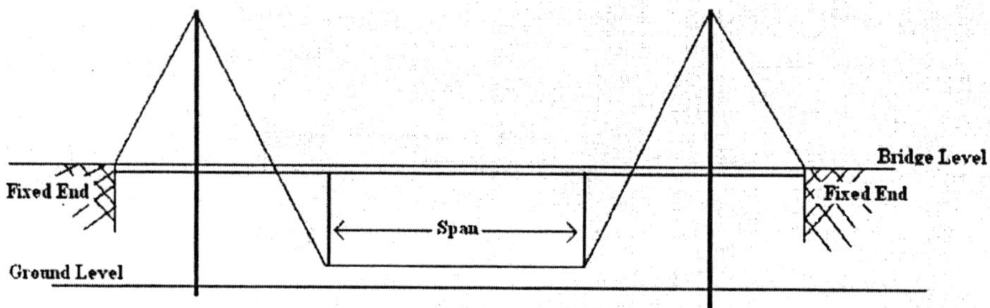

Figure 8: Idealised symmetrical under-slung arrangement

In the competition design the cable system that was proposed comprised ten preformed, spiral strands. Readily manufactured, they can be loosely attached to the deck either prior to erection if the latter is launched into position, or immediately after erection if it is lifted into position. Once in position, these cables can then be stressed in a predetermined sequence to ensure an even distribution of load both between each of the cables and between the deck and the cables. Preformed spiral strand is typically designed with a factor of safety in excess of 2.0 because of the high local fatigue effects generated within the strand as individual wires move and flex over one another due to the helical, spring like structure of the strand. Despite this, because of the high tensile strength of the drawn wire from which they are constructed, such strands are extremely efficient and are nowadays the favoured form of cable for both cable-stayed and suspension bridges.

Of course, in the form of structure proposed here, as in a trampoline, the greater the load in the cable the greater the stiffness of the structure. Indeed, such systems can be so efficient that, under live loading, fatigue of the connections between the cables and the deck can be a significant issue. Again further work is required to both quantify this effect in relation to the balance that is struck between the axial stiffness of the cables and the flexural stiffness of the deck and the detailed design of the cable clamps to accommodate this fatigue.

The construction of this type of design could be achieved in a number of ways. The main components of the truss are relatively lightweight and, as such, the elements could be assembled on site or could be prefabricated into larger parts and then transported to site for erection. This, in turn, could then be achieved either by cantilevering the sections from the supports using temporary stays or by launching the structure from one or both sides of the crossing, again either staying the cantilevers with temporary supports or relying on the inherent flexural stiffness of the deck if the cantilever is not too great. Launching sequences are ideal where site constraints are of crucial importance, such as bridges over deep gorges, where the availability of intermediate supports is limited or in cases where intermediate pylons cannot be used because of unstable ground conditions. In the competition design the truss was deliberately chosen and proportioned to enable it to carry its self-weight in flexure so that it could be launched without the need for any special temporary works other than a launch nose. This made the truss heavier than it needed to be for the final condition but did enable the specific site constraints to be satisfied and the final tension that was required in the cables to be reduced.

Conclusions

In response to an undergraduate design competition a novel, under-slung cable structure was proposed. This paper has considered the parameters that govern the behaviour and response of such a structure and structural arrangements and details have been proposed which it is suggested would make the construction of such a structure feasible.

The paper has discussed how underslung structures, when combined with a radical, trussed stiffening girder, can provide an efficient and innovative but more sympathetic structure, which is stiff, stable and economically viable for both medium and long-span structures. It has been shown that when there is a need to minimise the number or limit the location of the foundations required, such structures provide a means of adopting relatively large spans by using 'virtual' supports, while restricting the height of the cable pylons and hence minimising the impact of the structure.

The paper has addressed the benefits that can be obtained from such structures both during construction and in use, and has identified a number of opportunities and challenges that must still be met to facilitate the exploitation and development of such a concept. These include the need for a detailed consideration of the wind, vibrational and fatigue response of what will be an inherently flexible structure and a closer look at the design of details such as the cable clamps, saddles and connections, which will be subject to significant static and fatigue load effects.

References

1. Christos Christodoulou, *Suspension bridges and tying up of main cables underneath the bridge deck,* I.C.E. Wales Students and Graduates Papers Competition 2006, Cardiff 2006.

2. Z. Zhang, Q.-J. Teng and W.-L. Qiu, *Recent concrete, self-anchored suspension bridges in China,* Bridge Engineering Journal 159, Paper 14286, December 2006.

3. A.M. Ruiz-Teran and A.C. Aparicio, *General overview of cable-stayed bridges: Conventional cable-stayed bridges and new types,* Proceedings of the First International Conference on Advances in Bridge Engineering, Bridges – Past, Present and Future, Volume 2, Brunel University, West London, 26 – 28 June, 2006.

4. Christos Christodoulou, *2005/06 Undergraduate Prize Awards for Steel Bridge Design,* MEng Thesis, Cardiff University, Cardiff 2006.

5. M.J. Ryall et al. 2000, *Manual of bridge engineering,* Thomas Telford, London.

Six signature bridges in the People's Republic of China

Joost Meyboom, Dr.sc.tech., P.Eng., Delcan Infrastructure, Canada
Thijs Verburg, Architect, Verburg Hoogendijk Architects, Netherlands
Hugh Hawk, M.A.Sc., P.Eng., Delcan Infrastructure, Canada

Abstract

This paper describes the design of six signature bridges in the People's Republic of China. The bridge concepts were developed by DHV Engineering Consultancy (Shanghai) Co., Ltd., Verburg Hoogendijk Architects (VHA) and Delcan Infrastructure. Detailed design of two of the bridges was provided by the Tianjin Municipal Engineering & Design Institute.

The evolution of several designs and the corresponding interaction between architect and bridge engineer is described. This paper presents the design development and structural arrangement of these bridges and highlights some of the key technical considerations made during preliminary design. The bridges discussed are:

- The Tongnan Bridge (Tianjin) – a spine-like truss with a cantilevered orthotropic steel deck;
- The Liulin Bridge (Tianjin) – a two span cable supported bridge consisting of a deck suspended from three, half arches;
- The Phoenix Bridges (Guanzhou) – three unique arch structures with spans ranging from 100 m to 400 m and with helix-like and inclined arch arrangements ;
- Tuanbo Bridge (Tianjin) – A suspension bridge with a 500 m main span and inclined, centrally arranged towers.

The Design Process

Haven't we all at some time laid a beam across a ditch and tried to cross it without falling in? In this way we have all gained some experience in bridge engineering. Building bridges is in our genes – it is the instinct to get to the other side. A bridge, however, can be much more than a plank over water.

In many assignments, the architect is only required to provide some ornamentation in the name of aesthetics and allowed to participate with the preconceived idea that they only increase the cost of a project with designs that cannot be constructed. Of course there are architects who would rather ignore issues of constructability but then again this is not limited to architects. VHA's approach has been to develop designs in parallel with considerations of constructability and, conversely Delcan Infrastructure has always been an enthusiastic participant in considering a project's aesthetics.

Architects and Engineers

Typically the bridge industry includes bridge contractors, practicing engineers, academics and researchers, bridge owners and administrators, facility managers, design and assessing engineers, managers and operators, standards and guidance experts, legal professionals, safety regulators – but not architects.

Many people believe that an engineer is sufficient to achieve a good bridge design. We believe that this cannot be further from the truth. In fact where engineers and architects have cooperated as partners and equals, the concept, design, detailing and the eventual product has been extraordinary. Other fields of expertise could equally be added to the team with equally remarkable results. The artificial separation between the architect and engineer has been the subject of debate over the years and is a reflection of the industrial era's societal requirement for specialization. This specialization has brought society enormous benefits but it has also left its scars.

Five years ago during a conference on steel construction, VHA was asked the question why architects use so little steel in infrastructure projects. The answer was because of unfamiliarity with steel construction methods and a lack of understanding of the fundamentals of steel behaviour. At that time the educational streams for architects and engineers at major universities such as the Technical University of Delft were completely separate.

As a consequence of this discussion a so-called "Structural Steel Master Class" was set up in Rotterdam. This initiative aimed at providing young architects and engineers with guidance from recognized "Masters" in the fields of engineering and architecture and the opportunity to work together for a week to develop a sample project. The design results of these "Master Classes" were without fail, inspirational for both the young architects and engineers and the "Masters". It reinforced the notion that direct cooperation between architects and engineers leads to better, more appropriate and complete designs.

A fully integrated design approach between engineers and architects has been the most successful approach taken by VHA in achieving satisfactory and meaningful infrastructure projects. Delcan Infrastructure has developed a similar philosophy. To demonstrate the validity and value of this approach, six bridge designs that were developed by an engineer/architect team in the People's Republic of China are discussed. The bridges are:

- The Tongnan Bridge (Tianjin) – a spine-like truss with a cantilevered orthotropic steel deck;
- The Liulin Bridge (Tianjin) – a two span cable supported bridge consisting of a deck suspended from three, half arches;
- The Phoenix Bridges (Guanzhou) – three unique arch structures with spans ranging from 100 m to 400 m and with helix-like and inclined arch arrangements;
- Tuanbo Bridge (Tianjin) – A suspension bridge with a 500 m main span and inclined, centrally arranged towers.

Before providing a description of these bridge designs, the evolution of our design team should be mentioned. In 2004 DHV Shanghai contacted VHA in Amsterdam to explore a partnership aimed at winning infrastructure design competitions in China. Delcan Infrastructure in Canada, a member of the DHV group of companies with considerable bridge engineering expertise, was similarly approached and thus the bridge design team arose. This team has been together and cooperated successfully for several years in undertaking several design competitions in China.

Tongnan bridge

This bridge has a main span of 128 m and a total continuous length of 176 m. The bridge will carry four lanes of traffic and 2-4.5m wide sidewalks. This will give an out-to-out deck width 29.2 m. The bridge is shown in Fig. 1.

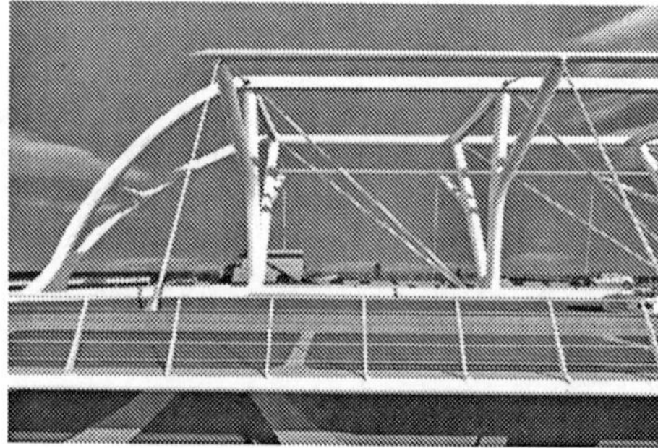

Figure 1. Tongnan bridge.

The bridge consists of a spine-like central truss that supports two parallel steel box girders separated by a 3 m wide opening along the length of the bridge. The boxes are 8.6 m wide with a maximum depth of 1.35 m. The boxes are provided with internal webs and the top plate will consist of an orthotropic plate.

Transverse ribs spaced at 4 m connect the two boxes and extend to the edge of the sidewalk. A central, 13 m deep truss will be engaged by the ribs to limit main span deflections. The truss consists of tree-shaped vertical posts. Hangers are provided to connect the edge of the deck with the top of the truss posts. Truss panels are 16 m long, see Fig. 2.

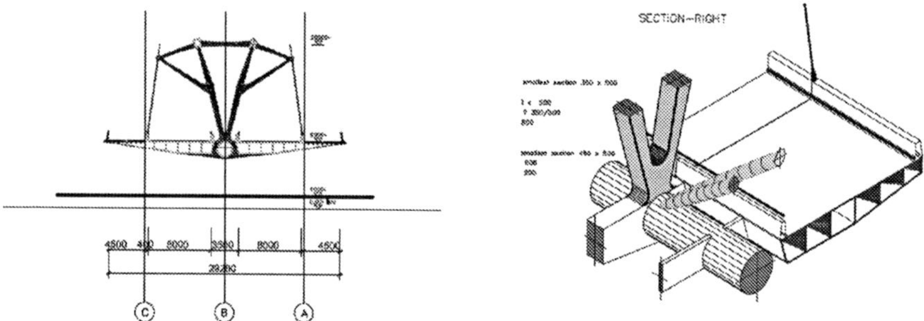

Figure 2. Cross section and details of Tongnan Bridge.

Continuity between the main and side spans is provided to control deflections and improve durability by eliminating joints. This will require hold-down bearings at the abutments where unfactored uplift forces in the order of 210 t and 95 t per bearing can be expected for dead load and live load, respectively. Options considered to provide uplift capacity included:
- Hold-down cables,
- Hold-down bearings,
- The introduction of additional dead load in the end spans. This could be achieved by the introduction of kentledge such as structural steel and/or concrete, for example, or (as is the case in moveable bridges) by the introduction of cast iron deadweight.

Luilin bridge

The Luilin Bridge has of two 87m spans. The form of the bridge was developed to mimic a dragonfly and consists of an arch which has been cut in half and separated. The ends of the arch are supported by props and the arch ribs are referred to as "flying girders", see Fig. 3 and 4.

Load on the deck is carried to transverse ribs that are suspended from cables that carry the load into the "flying girders". Cables are spaced at 5m and discontinued 20m from the centerline of the central pier. The flying girders are propped at one end and fixed at the central pier to give them a span of about 77m. The flying girders extend past the prop by about 10 m and cables are provided in the extension.

Figure 3. Luilin Bridge.

The deck is supported at the pier with corbels attached to the flying girders and lateral bearings are required between the deck and the props. The cables are stressed to carry the entire dead load of the deck. Stress is introduced in the cables by having them cut to predetermined lengths and hanging individual deck elements from the cables prior to connecting the deck elements into a continuous system.

In general the deck spans between cables. Because there are no cables within 20 m of either side of the central pier, the deck will be required to act as a 40 m long, two span continuous beam over the pier.

Figure 4. Luilin Bridge.

The architecture of the flying girders requires a constantly changing section. This results in an eccentricity of as much as 200mm between the cable and the center of gravity of the flying girder section. This eccentricity produces a twist of about 2mm in the cross section.

An important consideration in the design of the Liulin Bridge is unbalanced live load. By making every effort to minimize the moment at the central pier caused by live load located in only one span, a simple framing arrangement can be achieved over the pier resulting in a more constructible bridge and a reduction in steel. Based on preliminary calculations the unbalanced live load moment in the edge and central flying girders are about 18 MN-m and 52 MN-m. Positive moments in the span are about 11 MN-m. The maximum moment in the flying girder occurs at the pier and is 280 MN-m in the center flying girder.

The Phoenix Bridges

The Phoenix Bridges were developed as a series of three arch structures over the Jiaomen and Lower Hengli waterways in Guongzhou. In Chinese mythology the phoenix symbolizes the feminine and as such elegant, flowing arches were chosen for these bridges. The bridges are the Oscillation Bridge over the Jiaomen Waterway, The Kinetic Bridge over the Upper Hengli Waterway and the Inseparable Bridge over the Lower Hengli Waterway.

One goal of the design was to ensure that the separate functions and characters of the bridges would be clearly identifiable yet harmonious with one another. The main structural element of each bridge is an elegant and "light" arch. To make sure that this character is conveyed it was proposed to use steel coated with white paint. The deck is prefabricated concrete to enhance quality and smoothness and give the bridges a solid and sturdy appearance.

The bridges will carry a light rail system (LRT), highway traffic and bicycle/pedestrian users. A description of each of the three bridges is provided below.

The Oscillation Bridge

The bridge consists of a series of arches that are diagonally arranged over a 600 m length and form the main support structure of the bridge. This construction gives the Oscillation Bridge a unique and different appearance from every angle. In total there are eight arches – five small ones on each side of the two central larger arches that are arranged to emphasize the navigation channel. The smaller arches support the light rail and traffic lanes while the larger arches also support the footpaths. The required roadway width of 58 m allowed the deck to be divided into five segments – two pedestrian decks, two car decks and one LRT deck. This arrangement allows daylight to penetrate to the areas under the bridge and especially at both the river banks, see Figs. 5 and 6.

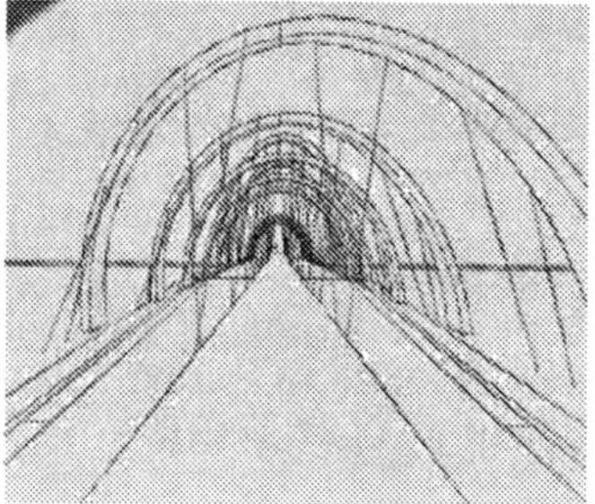

Figure 5. Sketches of the Oscillation Bridge.

The arches are skewed at 40 degrees to the bridge centre line. The two central larger arches span 107.62m and have a rise of 47 m. Due to some overlap of the skewed spans, the arch foundations of the two large arches are separated longitudinally by 70m. The arch rib consists of a steel box section with a variable section ranging from 4.5m x 2.0m at the top of the arch to 4.5 m x 4.0 m deep at the base of the arch. The smaller arches span 86.5m.

The floor system consists of 3-2m deep multi-cell continuous prestressed concrete box girders. Top flange, bottom flange and web thicknesses of 22cm, 18cm, and 18cm respectively were used. Transverse deck ribs are used to transfer deck loads to cables suspended from the arch ribs. The cable hangers are 90mm in diameter and are spaced at 6m. Cables consist of galvanized steel strand wrapped with polyethylene and protective steel pipe.

The flexural stiffness of the bridge is provided by the interaction between the deck system and the arches through the cable hangers. The cable hangers effectively reduce the bending moment in the deck system. The transverse deck ribs help to increase the transverse stiffness of the deck system.

Eccentric and unbalanced lateral loads on the arch will cause the arch to twist. This twist is resisted at the arch base by fixity to the foundations.

Figure 6. The Oscillation Bridge.

The Kinetic Bridge

Kinetic Bridge is the central phoenix bridge and, being a gateway between the Pearl Delta and Guangzhou, it symbolizes the entrance into the future and an exit from the past. It crosses the 225 m wide Upper Hengli Waterway.

The Kinetic Bridge consists of 4-200m long arches that form a structural sculpture held with tension. The arches lean on each other, they cross each other under an angle and are of different heights and lengths. This gives the structure a unified appearance that is different from every angle. The outer arches support the pedestrian and vehicle deck while the inner arches support the vehicle and LRT deck, see Fig. 7.

Figure 7. Arrangement and detail sketches of the Kinetic Bridge.

The shorter arches have a rise to span ratio of 1:3, and consist of a trianglular steel box sections with dimensions in the order of 2.1m x 2.7m high. The longer arches have a span to rise ratio of 1:4, and triangular steel box cross sections with dimensions in the order of 1.5m x 1.8m high. 3-2m deep prestressed concrete continuous box girders are used as floor beams. Transverse ribs spaced at 8m are used to transfer deck loads into cables suspended from the arch ribs. 90mm diameter cable hangers are used.

The Inseparable Bridge

When coming from the Guangzhou direction, this bridge will be the last bridge crossed and is intended to provide a "unifying" quality. The Inseparable Bridge crosses the busy Lower Hengli Waterway which is 420 m wide at the crossing. This bridge will be the highest of the three new bridges, making it a prominent viewpoint. From this perspective, it was decided that the shape of the bridge would be light and elegant, yet strong. The result is a combination of arches, reminiscent of a stone being skipped over the water or concentric rings expanding in the water. The arches are inseparable and are a symbol of unison, See Fig. 8.

The bridge is 540m long and consists of a series of concrete-filled steel pipe arches. The floor system consists of 3 continuous multi-cell prestressed concrete box girders. Lateral ribs are spaced at 10m and transfer load from the deck into cables suspended from the arch ribs. Lateral connecting elements are also arranged between each arch pair. The main bridge includes two different sets of arch spans - a central composition consisting of three continuous arch pairs spanning 140m, 260m and 140m and side compositions consisting of 4 continuous arches spanning 75m, 195m, 195m and 75m. The central and side arches thus support the three longitudinal box girders through transverse ribs that are attached by hangers to the arches. All of the arches are separated in cross section and elevation.

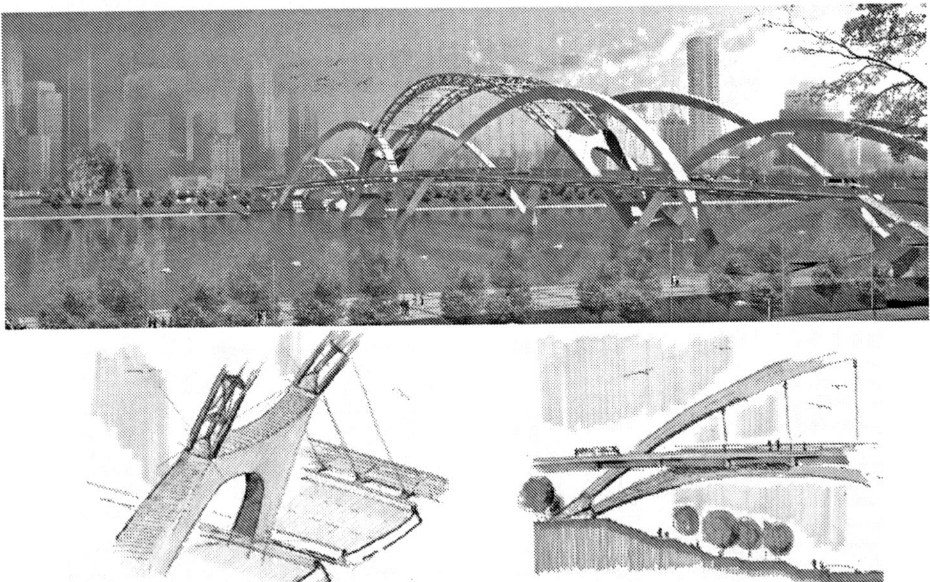

Figure 8. Arrangement and detail sketches for the Inseparable Bridge.

There are 8 lateral connecting elements for each pair of central arches. All of the connecting elements will consist of concrete filled steel pipe. The floor system consists of three 2m deep continuous multi-cell pre-stressed concrete box girders. Thicknesses for top flange, bottom flange and web are 22cm, 18cm, and 18cm respectively. Transverse ribs carry deck loads into cables that are suspended from the arch ribs. Transverse ribs are constructed as pre-stressed concrete box girders that are 1.6m deep and 1.6m wide.

Tuanbo Bridge

The Tuanbo Bridge was designed to carry vehicular traffic, bicycle/pedestrian traffic as well as LRT loads. The main bridge consists of 2 towers, a 3 span continuous precast box girder, two main cables, two side stabilizer cables, saddles, ground anchors and cable hangers, see Fig. 9.

The towers are built-up using arch structures and the cross section of each tower leg is a nominal 5.0m x 2.5m hollow rectangular section of reinforced concrete.

The two main bridge cables are 500 mm diameter each and are located at each side of the LRT line that is located along the centerline of the bridge. The cables are thus approximately 11m apart. The deck is suspended from the main cable with double cable hangers. These hangers are vertical and have a diameter of 80mm. The hangers are spaced at 25m in the longitudinal direction.

In order to enhance the torsional stiffness of the structure, stabilizing cables were added along the sides of the bridge. The internal force change in the cable hangers resulting from rotation of the superstructure provides some torsional stiffness because the force difference change can form a moment couple to resist longitudinal torsion. The central girder of the deck system, being a closed box system also provides a significant contribution to torsional stiffness. The stabilizing cables are 200mm in diameter. The hangers for the stabilizer cables are single cable hangers, each with a 60mm diameter and also spaced at 25m. The side stabilizer cables with hangers pull down on the deck and produce a stabilizing effect against rotations from eccentric loading.

The deck consists of a 3 span continuous prestressed concrete box beam. In cross section the deck is a 3 cell prestressed concrete box varying in depth from 7.5m at the center to 3.8m at the sides. The box supports transverse ribs that carry the deck cantilevers on the upper level and the bicycle and pedestrian lanes on a lower level. The light rail traffic and vehicle way are on the top level and the bicycle and pedestrian way are on the bottom level.

Figure 9. Tuanbo Bridge.

A gravity anchor block was adopted for the bridge and each side anchorage is estimated to require about 72,000 m^3 of mass concrete and fill.

The bridge requires of 8 main saddles and 4 cable distributing saddles, The 8 main saddles are symmetrically located at the 4 arch shaped main towers and the 4 cable distributing saddles are symmetrically located at the 2 ends of the bridge.

Conclusion

Tree-like structures, the dragonfly as metaphor, inclined and spiral arches and even the sun are all elements that were used to six signature bridges in the People's Republic of China. Historic and metal layers were found, analyzed and added to the site constraints to allow the architect/engineer bridge engineering team to develop designs that matched the project specific design criteria and also meet the client's needs.

Aerodynamic Study of Cable Oscillations and Mitigation Measures for Long-Span Bridges

Dr Ngai Yeung, Chief Technical Officer, Arup, Hong Kong SAR, China
Steve Kite, Associate, Arup, Hong Kong SAR, China
Dr Mingshui Li, Professor, Southwest Jiaotong University, China

Abstract

This paper describes the aerodynamic study of the cable oscillations induced by rain and wind and mitigation measures for long-span bridges. A series of wind tunnel testing have been carried out for the purpose of selecting appropriate cable types. The stay cable with different diameters and different surface configurations including smooth, with dimples and with helical fillets, were tested in the wind tunnel. Different azimuth angles of the cables, combined with different wind speed and rain intensities were also tested. The dampers were used to measure the mitigation effect of additional damping to cable oscillation. The results indicate that large oscillation of stay cable will occur at medium rain intensity and wind speed and the use of helical and dimpled surface can effectively reduce the oscillation although the drag loading is increased. Additional damping will further reduce the cable oscillation.

Introduction

Due to Typhoon climate in Hong Kong, the wind load is generally the control loading case for the design of cable-supported long-span bridges. The dynamic effects of wind load that needs to be catered for in the design includes buffeting due to the turbulent nature of wind, vortex shedding, galloping, flutter instabilities and interference effects. The oscillation of stay cables due to rain and wind, especially those long cables up to 500m for long-span cable-stayed bridges, is also a key dynamic wind effect to be considered and will dominant the choice of suitable cable types and surface texture.

Past experience shows that the rain and wind induced stay cable oscillations occur in a wind speed range of 10 - 20 m/s and at light to medium rain fall intensity. The magnitude of oscillation has been reported to be about 3 times the stay cable diameter. And the most severe oscillations are found to occur when the wind direction is at some specific azimuth angles, not perpendicular to the bridge span. Due to lack of proven analytical model for the rain-wind induced vibration of stay cable, a series of wind tunnel testing have been carried out to find the critical wind speed, rain fall intensity, and wind direction that induce the most severe oscillations in stay cables with selected diameters and surface textures. The mitigation measures have also been studied.

Full Scale Cable Models

The full scale cable models were used in the wind tunnel testing to ensure Reynolds Number similarity for the model and the prototype. Two cable diameters were selected for the study, i.e., 139mm and 169mm. Three types of surface configurations were studied. They were smooth surface, surface with dimple configurations, and surface with helical fillet (Figure 1).

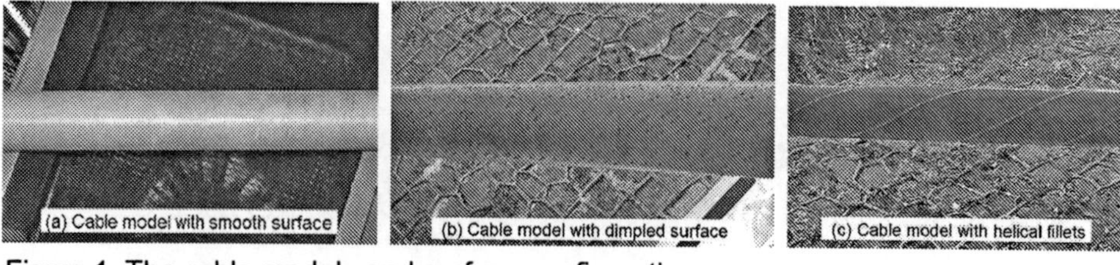

Figure 1. The cable models and surface configurations

The cable models were built from commercially available circular section HDPE cable ducts supported by stainless steel inner tubes. The dimpled surface was made by stamping the dimpled pattern on the surface of the smooth cable. The helical fillet was made by cutting a PE plate into 2mm wide fillets and attaching on to the smooth surface with 900mm step. All of the model surfaces were treated with dust and pollution to model the soiled surface condition of the stay cables in their service condition. The dimension of the dimple was 5.6mm x 4.0mm for the 139mm diameter cable and 6.7mm x 4.8mm for the 169mm diameter cable. The pattern of the helical fillet is illustrated in Figure 2.

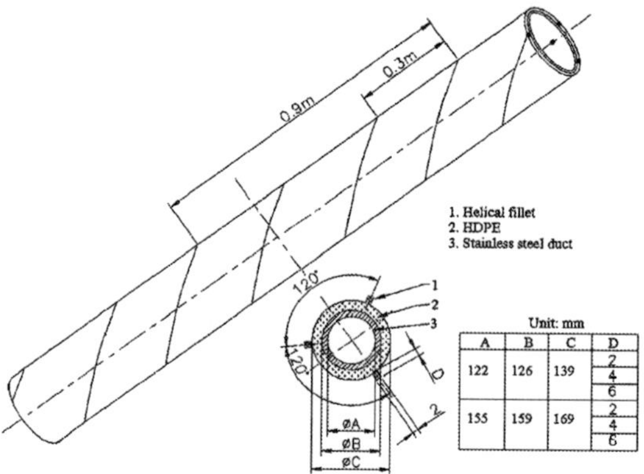

A	B	C	D
122	126	139	2 / 4 / 6
155	159	169	2 / 4 / 6

Figure 2. Pattern of helical fillet on cable surface

Similitude and Dynamic Parameters

The cable models were made to represent the exterior surface conditions of the prototype, no wires or strands were present inside the HDPE duct. Additional mass in the form of soft lead was installed inside the model to increase the total mass. With reference to PTI's recommendation [1], the cable models are not required to have identical mass / unit length as the prototype cables. It is sufficient for cable models to respect similitude for the Scruton Number (or mass-damping parameter).

The Scruton Number is defined as $Sc = \dfrac{m\zeta}{\rho D^2}$, in which m is mass / length, ζ is the damping ratio (relative to critical), ρ is the density of air, and D is the outer diameter of the cable tube. The Scruton Number represents the sensitivity of cables to rain and wind excitations and is smaller in model scale, thus the cable model can be more easily excited under rain and wind conditions than the prototype cable. The dynamic parameters of the cable model and the prototype cable are compared in Table 1. The inherent damping is 0.095% for cable diameter 139mm and 0.099% for cable diameter 169mm. The additional damping was supplied by installing an external fluid damper with adjustable damping ratios.

Table 1. Dynamic parameters of cables

Parameter	Outer diameter 139mm		Outer diameter 169mm	
	Prototype	Model	Prototype	Model
Frequency f (Hz)	0.47-1.41	0.91	0.22-0.66	0.63
Mass m (kg/m)	51.3	23.2	118.9	31.5
Damping ζ	0.1%	0.095%-1.01%	0.1%	0.099%-0.92%
Sc	2.1	0.93-9.9	3.4	0.90-8.3

Experimental Set-Up

The tests were carried out in an open circuit wind tunnel of down-blowing type with open test section located at Southwest Jiaotong University.   It can produce wind speeds ranging from 1 m/s to 20m/s.   The effective area of the test section is 1.3m wide and 1.5m high.   A specially designed test rig is situated at the test section to reproduce the required rain and wind conditions.   The cable model was supported by two pairs of springs at two ends and connected to the test rig (see Figure 3).

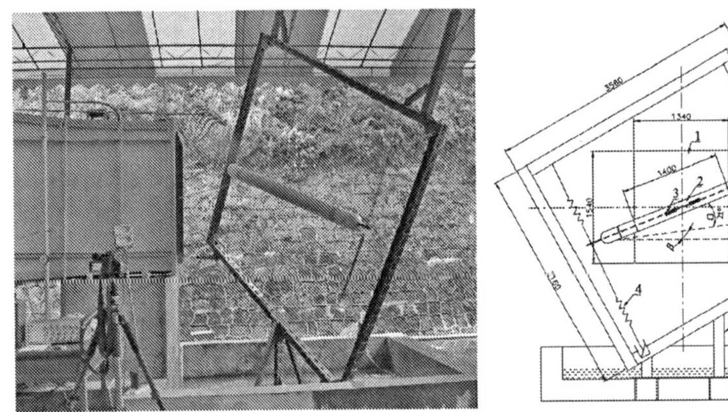

Figure 3. Experimental set-up

Every cable model was 2.72m long comprising an active model in the middle and two dummy models at the two ends to render end effects unimportant.   Two force transducers were installed at the top of the springs and calibrated to measure the vibrating displacement of the cable model induced by rain and wind.   A meteorological ombrometer was used to measure the intensity of the simulated rainfall.

Testing Procedure

The parameters that may affect the rain and wind induced cable oscillations include cable diameter, surface configuration and different patterns of the configuration, damping, wind speed and azimuth angle, rain intensity, etc.   As the exact physical processes involved in the rain and wind induced cable oscillations are difficult to define and quantify, a series of aerodynamic tests were programmed to investigate how those parameters affect the cable oscillation, the critical wind and rain condition to excite the maximum oscillation, and mitigation measures.

The cable model could be sensitive to buffeting or vortex-induced oscillations due to its small Scruton Numbers. To make sure that the effect of buffeting or vortex-induced oscillations were insignificant in the present study, the smooth cable models were tested on the condition without rain prior to the rain and wind oscillation tests. Then a series of tests on the smooth cable models were conducted in rain and wind to find out the critical azimuth angle for the two diameters. Each cable was tested at medium rain intensity of 80mm/hr with azimuth angle varying from 25 deg to 65 deg with increment of 5 deg. The wind speed varied from 5m/s to 10m/s with incremental speed of 1m/s and from 10 m/s to 15 m/s with incremental speed of 0.5 m/s when large vibration was observed.

Using the critical azimuth angle, wind speed and rain intensity identified above, the smooth cables with damping levels (relative to critical) 0.3%, 0.6% and 1.0% were tested to investigate the effect of additional damping provided by viscous dampers. These tests were repeated for the cable models with helical fillet and dimpled surfaces. Three depths of helical fillets and two depths of dimple patterns were studied. The azimuth angle, wind speed and rain intensity were varied to obtain the strongest excitations in the presence of helical fillets and dimples.

The duration of each test was 10 min. The first 8 min was for development of the rain and wind induced oscillations and the remaining 2 min was for data acquisition and record. The root-mean-square (rms) response and maximum amplitude of oscillation were derived from the time history data for further study.

Experimental Results

The effects of azimuth angles, wind speeds and rain intensities to the cable oscillations are investigated based on the wind tunnel testing results.

Testing of Cables with Smooth Surface Configuration

On the condition that the cables were not exposed to rain, the wind induced oscillations of two cable diameters 139mm and 169mm showed a maximum amplitude of less than 20mm, which indicated that the possibility of buffeting or vortex-induced oscillations are comparatively low.

Effects of Azimuth Angles and Wind Speeds for Smooth Cables

The rain and wind induced vibration tests of the smooth cables were conducted under medium rain intensities of 85mm/hr for the 139mm diameter cable and 80mm/hr for the 169mm diameter cable. For both cable diameters, it was observed that at a critical wind speed, the cables experienced a peak oscillation. It was also observed that water rivulet was formed on the top and bottom of the cable surface, which was considered to be contributable to the excitation of the cable oscillation (see Figure 4).

Figure 4. Rivulet formed along the cable during rain and wind induced oscillations

The rms response (Arms) and maximum amplitude (Amax) of rain and wind induced vibrations at different azimuth angles (β) are plotted against wind speed for cable diameter 139mm in Figure 5.   The measurements from the testing of cable diameter 169mm are plotted in Figure 6.

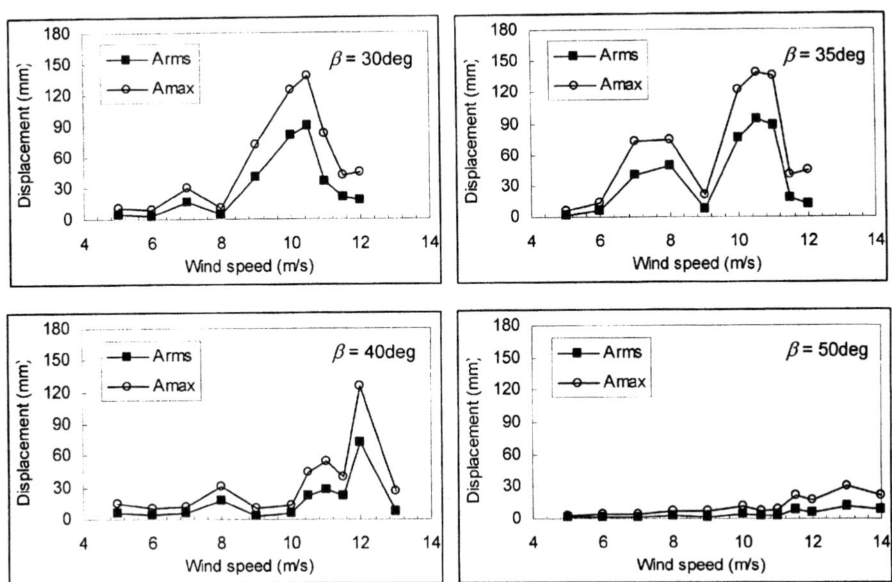

Figure 5. Amplitude of oscillation of cable dia. 139mm with rain intensity 85mm/hr

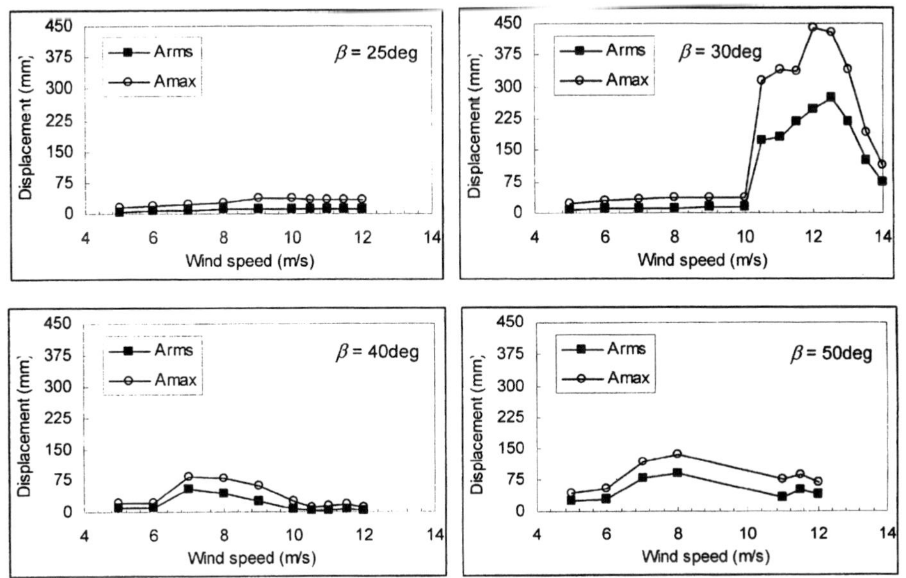

Figure 6. Amplitude of oscillation of cable dia. 169mm with rain intensity 80mm/hr

Figure 5 shows that the critical azimuth angle is β=35deg at wind speed V=10.5m/s for the 139mm diameter smooth cable. The maximum amplitude of oscillation is 140mm which equals the cable diameter. In the case of 169mm diameter cable, the critical azimuth angle is β=30deg at wind speed V=12m/s according to Figure 6. The cable experienced oscillation up to 440mm maximum which is about 2.5 times of its diameter.

Effects of Rain Intensity for Smooth Cables

With the critical azimuth angles identified above, the effects of rain intensity were tested and the results are plotted in Figure 7. This verified that rain and wind induced cable oscillations occur at medium rain intensity of around 80mm/hr. The use of viscous dampers to mitigate the excessive oscillation was then studied.

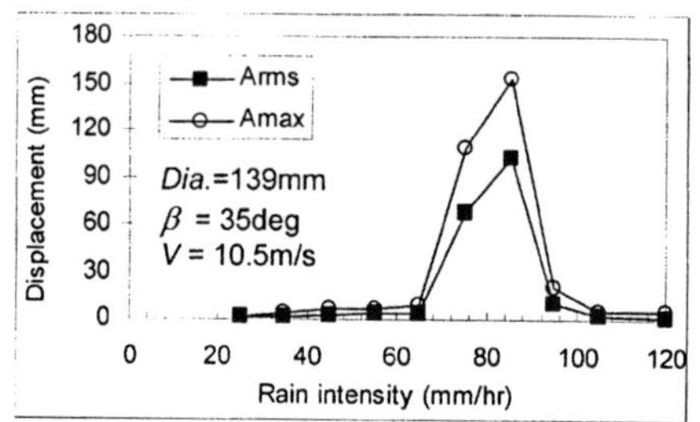

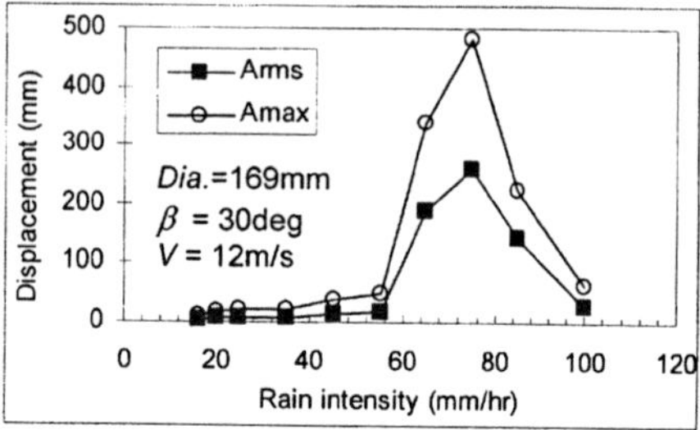

Figure 7. Rain and wind induced oscillation with varying rain intensities

Mitigation of Oscillation of Smooth Cables with Additional Damping

The effect of additional damping in suppressing the rain and wind induced cable oscillations is obvious as shown in Figure 8. The use of cable dampers, either internal dampers or external dampers, has also the benefit of mitigating the buffeting or vortex-induced vibrations.

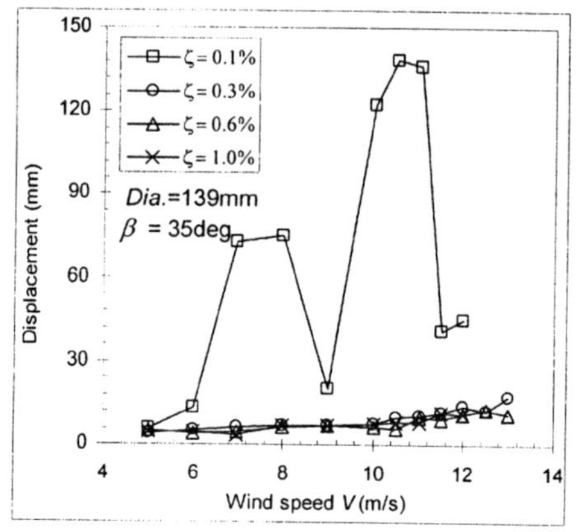

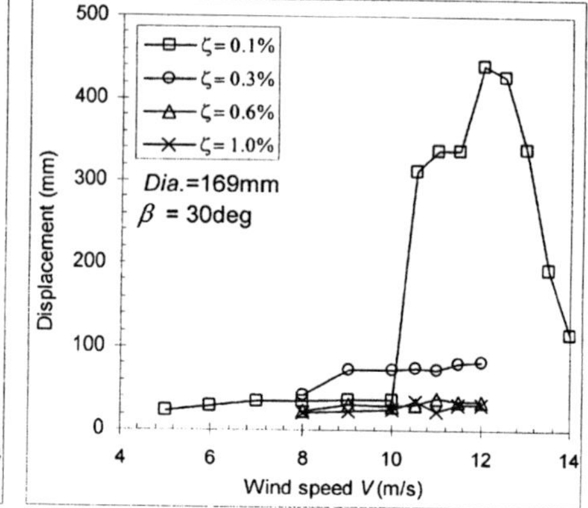

Figure 8. Rain and wind induced oscillation with varying damping ratios

Testing of Cables with Helical Fillet Surface Patterns

Three helical fillets of depth D = 2mm, 4mm and 6mm (see Figure 2) were tested. It was observed that no rivulet could be formed due to the presence of the helical fillets. The responses of the cable models under rain and wind excitation were comparatively small and plotted in Figure 9 for comparison. As the amplitude of vibration was very small in all situations under consideration, it was not necessary to add any more damping to the cable system with helical fillet surface patterns.

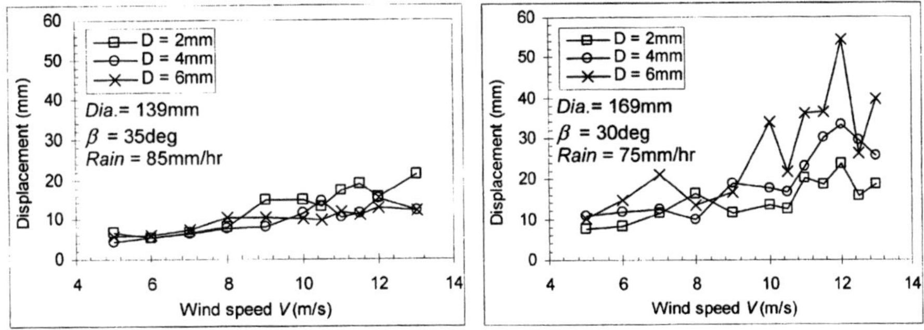

Figure 9. Rain and wind induced oscillation on cables with helical fillet

Testing of Cables with Dimpled Surface Pattern

The rms response (Arms) and maximum amplitude (Amax) of rain and wind induced vibrations of 139mm and 169mm diameter cables with dimpled surface pattern are plotted in Figure 10.   The 139mm diameter cable experience very small oscillations subject to rain and wind excitations.   The azimuth angle and rain intensity had no significant effect on the responses of the 139mm diameter cable.   However, strong vibration was observed for the 169mm diameter cable at an azimuth angle of 40deg and with rain fall intensity of 80mm/hr. The amplitude of the vibration was equal to the cable diameter.   Comparing the oscillations observed for the smooth cables, the mitigation effect of dimpled surface pattern is still obvious.

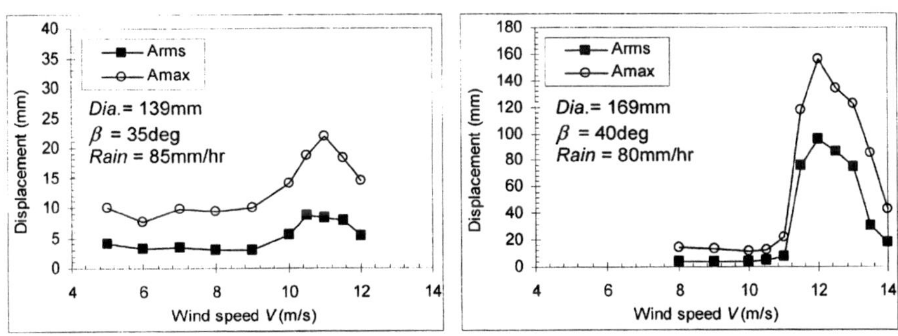

Figure 10. RMS and Max. vibration of cables with dimpled surface

Mitigation of Vibration of Dimpled Cables with Additional Damping

To further reduce the oscillation of 169mm diameter cable induced by rain and wind at the critical azimuth angle and rain intensity, additional damping was added to the dimpled cable model with viscous dampers.   The inherent damping of the 169mm diameter cable was measure to be about 0.1% (relative to critical).   Three higher damping levels, i.e., 0.3%, 0.6% and 1.0%, were achieved with viscous dampers.   The results from these tests with higher damping levels are compared in Figure 11, which indicate that additional damping is very effective in suppressing the rain and wind induced oscillations in stay cables.

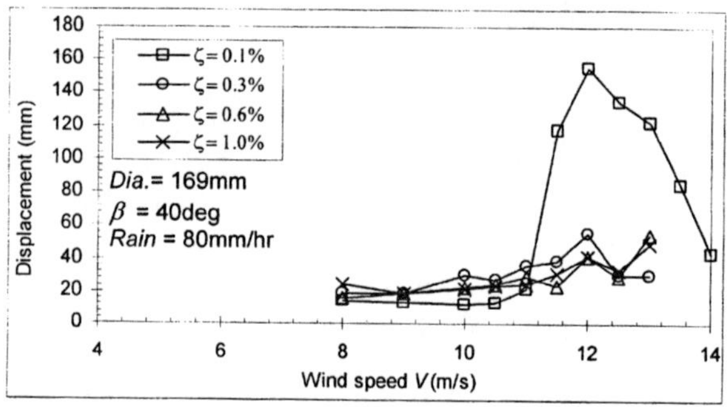

Figure 11. Rain and wind induced oscillation with varying damping ratios

Comparison of Three Cable Surface Configurations

The maximum rain and wind induced vibrations of the cables models with three types of surface configurations are compared in Figure 12, which shows that either the helical fillet configuration or the dimpled configuration is effective in suppressing the rain and wind induced cable oscillations.

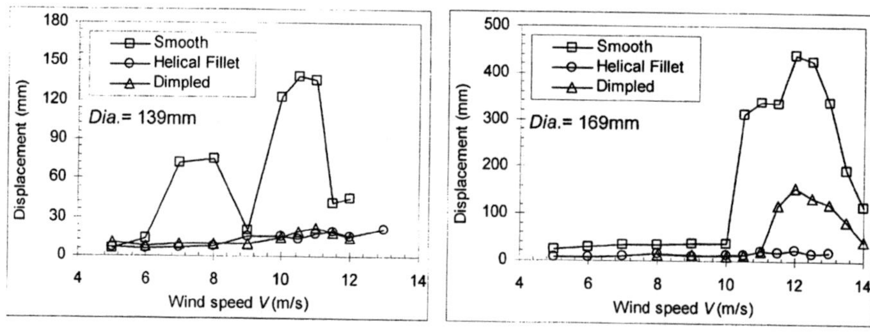

Figure 12. Comparison of cable surface configurations

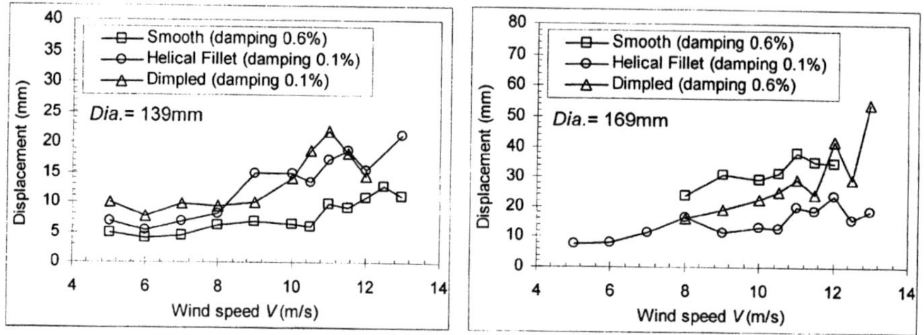

Figure 13. Comparison of cable surface configurations with added damping

When additional damping was applied to the cable system to suppress the excessive oscillations identified in Figure 12, the maximum amplitude of oscillation under all conditions could be reduced to an acceptable level as plotted in Figure 13. The helical fillet pattern and the dimpled pattern are effective in mitigation the rain and wind excited oscillations, but on the other hand, they cause an increase of drag loading in the stay cables. Higher drag coefficients are measured for greater depths of helical fillet and for deeper dimple indentations. Additional damping is extremely effective in suppressing the oscillations. Hence the most effective mitigation measure for long-span bridges is considered to a combination of cable surface texture and damper system, and the effectiveness of such measure should be verified by wind tunnel testing.

Conclusions

The design of stay cables plays an important role in the design of long-span bridges. The rain and wind induced oscillation of stay cables are known to the engineers but there have been no proven analytical models for tackling such issues. Hence experimental study is required. This leads to the aerodynamic study of the cable oscillations and mitigation measures presented in this paper.

A series of wind tunnel tests were conducted to investigate the cable oscillations induced by rain and wind, and the effects of helical fillets, dimpled surface configuration and additional damping on the oscillations. The testing results verify that the rain and wind induced oscillations occur in a wind speed range of 10 - 20 m/s and at light to medium rain fall intensity. Cables with smooth surface configuration are generally sensitive to the rain and wind excitation at their inherent damping level of about 0.1% (relative to critical). The additional damping is effective in mitigating the oscillations.

The surface configuration with helical fillet was found to be effective in suppressing the rain and wind induced oscillations. The dimpled configuration is also effective but the effectiveness may be dependent on the design of the dimple patterns. Larger oscillations were observed for the dimpled cable with diameter 169mm than that for the dimpled cable with a smaller diameter of 139mm. For a more efficient vibration control system, other mitigation measures including internal dampers, external dampers and cross ties for stay cables could be adopted.

Acknowledgement

This paper is submitted with the permission of the Highways Department, the Government of the Hong Kong Special Administrative Region.

References

[1] PTI Publication, "Recommendations for Stay Cable Design, Testing and Installation," Post-Tensioning Institute, Fourth Edition, March 2001.

The lightweight aluminium panel for bridge redecking

Dr T.W. Siwowski, Rzeszow University of Technology, Poland

Abstract

Replacing the deteriorated bridge decks to today's design standards and durability requirements are more economical alternatives to replacing the bridges themselves. Reasons such as this provide significant impetus for the development of bridge decks made of new advanced materials, that are durable, light and easy to install. Recently particularly effective is the use of FRP composites and aluminium alloys for lighweight bridge decks used for redecking of existing bridges. The research program has been undertaken to develop and implement an innovative aluminium bridge deck system, which would be applicable and realizable in domestic conditions. The several service load, ultimate load and dynamic tests have been carried out on the prefabricated 2.10 x 3.20 m deck panels, in order to examine and evaluate the panel behaviour under standard truck load and when loaded to failure. The main results of the panel tests are presented in the paper.

Introduction

The most vulnerable element of a bridge is its deck. Bridge deck deterioration of older bridges is a significant problem in aging of the highway system. The common problem of aged RC bridge decks is the spalling caused mainly by deicing salt. Therefore every ten to fifteen years RC bridge decks have to be replaced. Use of an advanced material bridge deck systems is viewed as a potential long-term solution for the concrete deck deterioration problem. Modern bridge redecking systems must meet the criteria of ease of construction, lightweight, economiaclly acceptable schemes, relatively short construction time and long service life expectancy. Most systems are prefabricated elements (panels) developed to achive economy through the repeated use of fabrication facility and to reduce on-site construction time and labor. Lightweight deck replacement is also a recognized strengthening technique for bridges that have structural limitations on the load carrying capacity, but have nonetheless sound steel stringers, cross-beams and main girders. Composite action is possible with most modern lighweight deck systems and can improve further the live-load capacity.

The recently developed redecking systems can be grouped according to material used. The groups are: (1) conventional materials as concrete, steel and timber and (2) modern advanced materials as: engineered cement composite, glulam timber, aluminium alloys and FRP composites. The contemporary progress of metal engineering, which led to the development of new generation aluminium alloys with excellent strength and durability, had let to wider utilisation of this material in civil and transportation engineering (Mazzolani 2004). Particularly effective is the use of aluminium alloys in bridge redecking (Höglund 1994,

Matteo 1997, Soetens et al. 2001, Siwowski & Żółtowski 2003, Okura et al. 2003). The removal of deteriorated heavy RC deck and the replacement with lighter one, engineered with aluminium, make possible to avoid the strengthening of the super- and substructure and thus cut the total cost of modernization. Furthermore the excellent corrosion resistance of aluminium alloys brings the saving of cost, spent for maintenance during service life of a bridge, eliminating also during that time a lot of environmental issues due to painting for corrosion protection. Additionally the application of aluminium deck shortens the closing time of the bridge, needed for carrying out the rehabilitation works. It reduces the social costs induced by traffic congestions (Herzog 1992).

Recognizing the potential benefits that aluminium could offer the transportation industry, the Department of Bridges at Rzeszów University of Technology has undertaken the research program to develop and implement an aluminium bridge deck system, which would be realizable and applicable in domestic conditions. The first phase of that study was to design aluminium extrusions, suitable to bridge decks. The second phase of the study, which is partially reported here, has comprised the experimental evaluation of the deck panel. The several service load, ultimate load and dynamic tests have been carried out on the prefabricated 2.10 x 3.20 m deck panels, in order to examine and evaluate the panel behaviour under standard truck load and when loaded to failure.

However, before the panel may be recommended for the use on deteriorated bridges, which need deck replacement, the next two phases of the research study must be completed. Phase three of the study will focus on the structural and environmental durability of a deck panel on the basis of fatigue testing in the laboratory and corrosion testing in the bridge environment. At the same time, the durability of the wearing surface will be assessed. The last, fourth phase of the study will involve a field evaluation of the deck system, which will replace deteriorated RC deck. The proposals of possible deck panel applications in replacement cases have been presented to road administration. After being approved, the preliminary replacement designs for two structurally deficient and functionally obsolete bridges have been also prepared. Long-term performance tests will be conducted after the bridges are open to the public in order to estimate service life of the aluminium bridge deck system. The final results of these tests are expected to be published soon.

Bridge deck panel description

The aluminium deck panel consists of the hollow extrusions with the cross-section shown in Figure 1. On the basis of analysis of the similar deck extrusions (Matteo 1997), a triangular one-voided section of profiles with the height of 0.16 m and the width of 0.12 m was accepted. These dimensions were limited by the recent capability of Polish aluminium extruder, which could fabricate extrusions with the section inscribed in a circle with the 0.2 m maximum diameter of piston. After the comprehensive material studies, the 6005A-T6 aluminium alloy was chosen to fabricate the extrusions, because of its optimal both the mechanical and anticorrosive properties (Siwowski 2002).

The tests were carried out on the deck panel 2.10 m wide and 3.2 m long, which comprised 16 extrusions welded together with the MIG butt welds. The dimensions of the individual panel were accepted with the assumption of its use in the redecked bridge structure. The linear support of the panel on steel beams was arranged in the experiment. The spacing of supports was equal to about 2.0 m, what suits the most frequently applied spacing of the main girders (or stringers) in the existing plate-beam bridges.

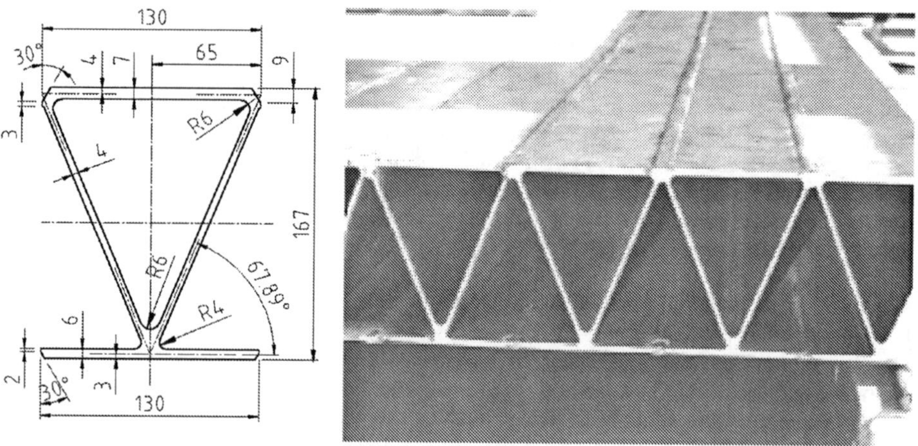

Figure 1. Cross-section of the aluminium deck extrusion

Although the resulting deck is geometrically orthotropic, the panel is typically oriented with extrusions parallel to the supporting girders and the direction of traffic. The deck constructed in this way cooperates very well with the girder (or stringer), creating a composite system (Matteo 1997, Dobmeier et al. 2001). When installed in this manner, stresses developed under loading can be generated by three different mechanisms, namely (a) longitudinal bending of composite girder - system I stresses; (b) transverse panel bending - system II stresses; (c) transverse bending of the panel top plate - system III stresses. The connections can be made for example by means of a galvanized bolts, carrying shear forces. The panel is suitable for the application on both an aluminium and steel girders. In the latter case the cover elements are applied on contact area of both metals in order to avoid the galvanic corrosion.

Service load evaluation

The main goal of static load tests was to obtain two basic sets of the physical parameters: i.e. strains and displacements, to be generated under the service load conditions and when loaded to failure. Instrumentation consists of 20 strain rosette gauges installed in strategic locations on the bottom surface and 25 rosette gauges installed on the top surface of the panel. More gauges were used on the top surface due to the presence of the load patch, which was expected to introduce the localized stresses. Seven deflection gauges were connected to the bottom surface to record displacement data. All together 52 discrete channels recorded the data.

Five service load tests used the same wheel load magnitude. However, they used different boundary conditions and load patch placement (Figure 2). These different boundary conditions and loading configurations were intended to replicate the scenarios typically encountered during actual service conditions. The magnitude and patch size of the service load were based upon the Polish bridge code. According to it, the dual S-truck wheel of the class B generates 150 kN load force. The tires distribute this force over the 0.20 x 0.60 m load patch. The laboratory tests used a pack of steel plates and hard rubber patch to simulate the contact area (Figure 3). In the service load tests loading was stopped once the magnitude reached the 150 kN mark, and the data were recorded. The panel was then unloaded at the same rate, and the data were recorded with the same frequency. Once unloaded, the gauges were scanned to see if any residual strains were present.

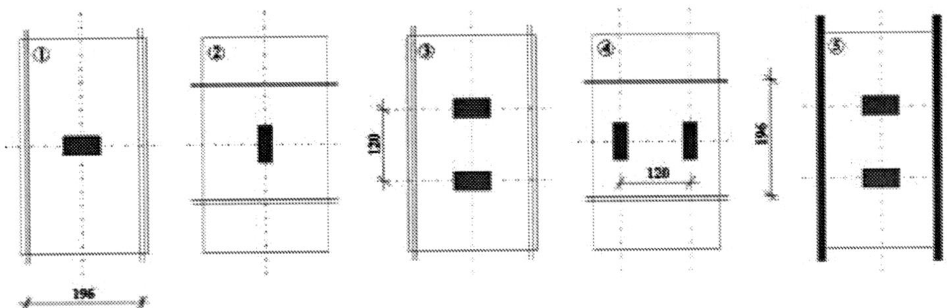

Figure 2. Service load configurations – static load cases

Figure 3. Experimental setup of the panel and load patch simulation

The results of the laboratory study indicated adequate strength and stiffness of the deck panel. The measured deflections in the middle of the bottom surface at service load P=150 kN and span/deflection ratios are presented in Table 1. There were nominally to be three repetitions of each test performed, which would then be averaged. The maximum deflection observed while testing occured under the load case III. The deflection at the middle point was 3.16 mm under a total load 2P=300 kN, which corresponds to a span to deflection ratio of 620. The span to deflection ratio value can be compared with the recommended value of 500 as given f.e. for FRP composite decks. This span to deflection ratio criterion is often used as a benchmark comparison for lightweight deck systems (Zureick et al. 2003).

Table 1. Deflections in the middle of the panel bottom plate

Load case	Applied service load (kN)	Single span deflection (mm)	Span/deflection ratio	Coefficient of determination R^2
I	150	2.15	912	0.999
II	150	2.07	947	0.9926
III	2x150	3.16	620	0.9955
IV	2x150	3.11	630	0.9968
V	2x150	2.79	702	0.9997

The load vs deflection curves for service loading for every load case are presented in Figure 4. The deflections from all tests exhibited a linear behaviour as seen here. The least coefficient of determination for a linear regression is 0.9926, which is relatively high and assures that the relationship between deflection and load is linear for the service load, although the material themselves is non-linear in nature.

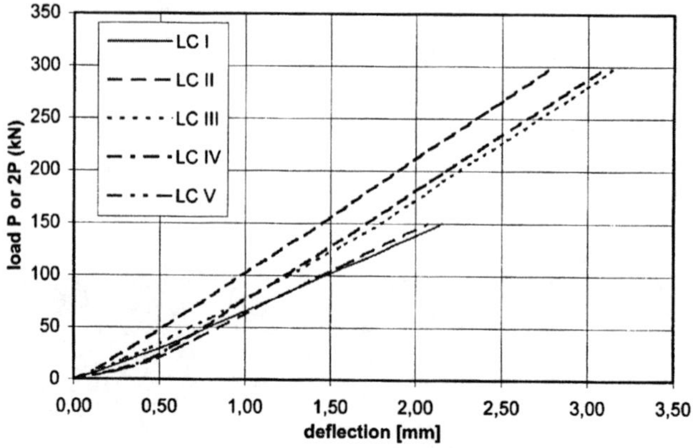

Figure 4. Load vs deflection curves for service load tests

The largest stresses obtained in the tests are summarized in Table 2. The maximum compressive stress recorded during testing was 171.6 MPa and occurred on the top surface under the load patch for load case I. These stresses resulted from localized bending of the top deck flange (system III stresses). The largest tensile stresses were also obtained on the top deck surface under the load patch. Again, localized bending of the top deck chord generated these large values – the maximum tensile stress was of 105.9 MPa for load case II. To understand the significance of these local stresses, one must first consider the global bending stresses developed on the bottom deck surface for load case III. The maximum tensile stress on the bottom deck surface was 51.8 MPa. This meant that local bending stresses were aproximatelly 2 – 3 times larger then the global bending stresses. Matteo et al. (1997) derived a ratio of 1.6 from experimental strain values. The importance of localized effects is showed in Figure 5, where the strains under the load patch along the main axis were recorded. Despite these high localized stresses, the magnitudes were still well within the allowable stresses for the design. All maximum stresses considered for strength analysis are lower then yield limit $f_{0.2}$ for aluminium alloy. There were 68%, 42% and 21% of $f_{0.2}$, respectivly. However, the compressive stresses in the HAZ were much closer to the allowable values for the design.

Table 2. Maximum stresses in the panel for the service load (P=150 kN)

Load case	Stresses (MPa)		bottom deck surface tensile	Tensile stress ratio
	top deck surface			
	compressive	tensile		
I	-171.57	66.08	31.78	2.08
II	-99.89	105.91	35.63	2.97
III	-109.13	45.22	51.8	0.87
IV	-46.55	13.37	49.98	0.27
V	-53.48	11.13	47.39	0.23

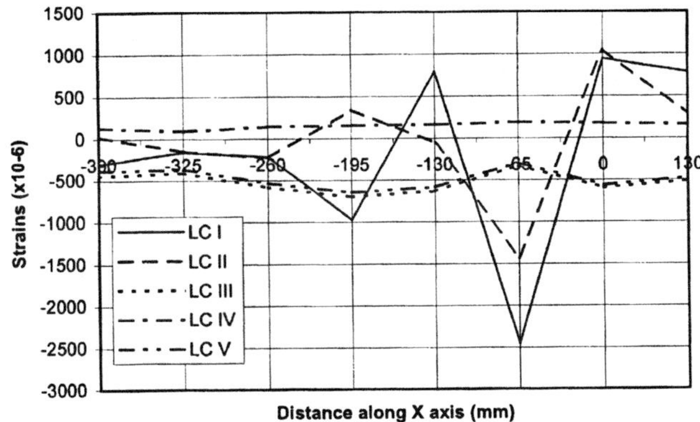

Figure 5. Strains under the load patch along the main axis

Failure testing

The ultimate load tests were conducted on the same deck panel as for service load testing, with instrumentation described above. The first failure test of the panel used the support and loading condition as for load case I. At a load of approximately 560.2 kN deck compliance changed and the sustainable load dropped to 538 kN. Testing was temporarily discontinued and closer inspection revealed that the stiffners near the load patch had buckled (Figure 6). Local yielding occured and the top deck surface under the load patch began to distort. To check if the deck possessed aditional capacity, loading was continued and the load was increased to 585 kN, when a pop was heard as a weld on the top surface began to fracture. The total deck load was imedietelly dropped to 572 kN. Although the deck was damaged, it still possessed substantial capacity. The deck finally achived a maximum load of 600 kN at which point the test was stopped (Figure 8).

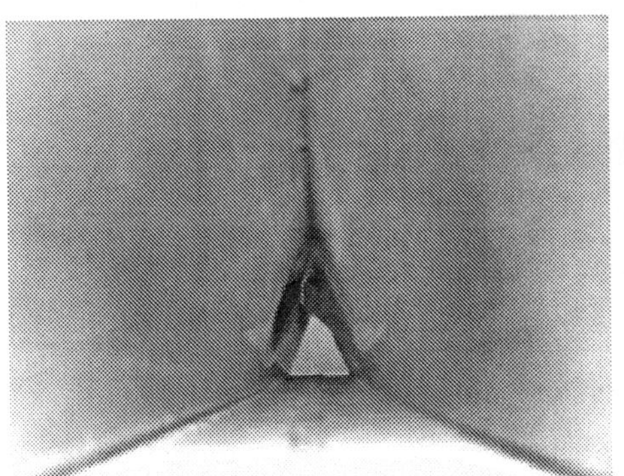

Figure 6. Failure mode for the first ultimate load test (load case I)

To further investigate the failure mechanism, a second ultimate-load test was conducted. This second test was similar to the first but used two load patches to apply the load (as for load case III). This allowed the total panel load to increase while simultaneously decreasing the localized force in each load patch. The dual load patches were positioned along the longitudinal centerline a distance of 1.20 m apart. Test protocol was similar to that of the first ultimate load test. At 2P=780 kN of total deck load a first pop was heard as a weld on the bottom surface began to fracture (Figure 7). The total deck load immediately dropped to approximately 750 kN. Although the deck was damaged, it still possessed capacity and loading of the structure continued from 750 kN up to maximum total load of 926 kN. Several fracturing bottom welds were heard during loading and the stiffeners began to distort. At the total load of 926 kN a violent pop was heard as a bottom weld fractured completely, resulting in total deck failure (Figure 8).

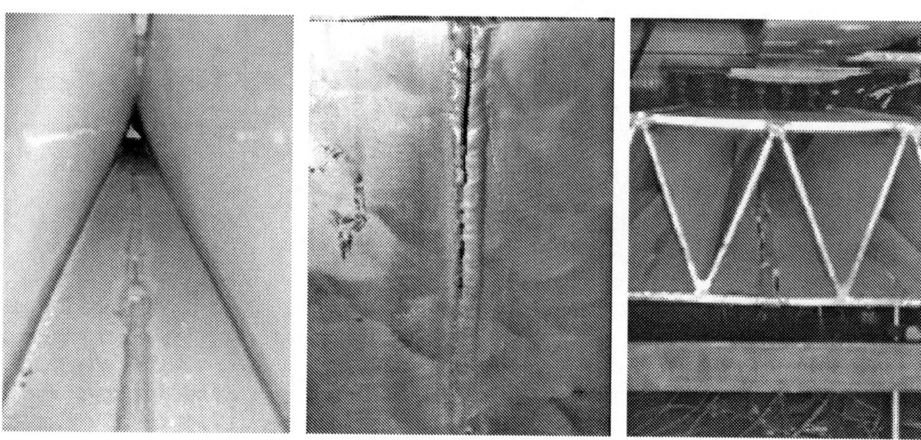

Figure 7. Failure mode the second ultimate load test (load case III)

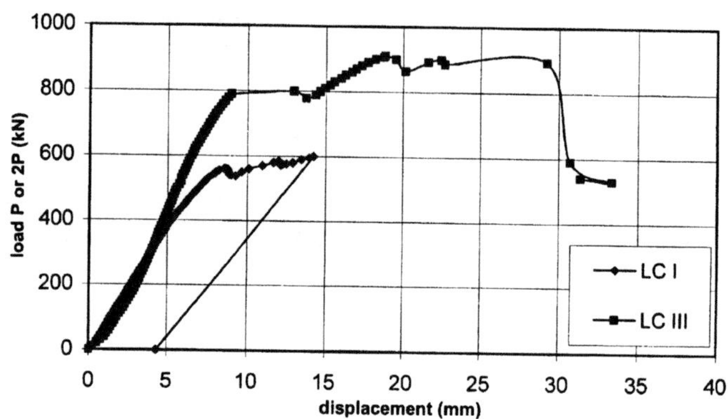

Figure 8. Load vs deflection curves for failure load tests

Dynamic response analysis

To determine the dynamic response of the deck panel the classical modal analysis was applied. This technique utilizes physical experiment to determine the response of the panel on given excitation force and thus predicting the dynamic behaviour of a deck. As a result the mathematical description, i.e. modal dynamic model of structure, is created. On the basis of the excitation force measurement and corresponding response determination, for example accelerations in particular points of structure, the frequency response function can be established. This function is a convenient description form of the dynamic characteristic of structure, because it includes the information about the relationship between input and output signals. Thanks to special curve fitting algorithm the modal parameters, describing respective mode shapes of vibration, can be determined from frequency response function. Modal parameters are identified for any individual neutral frequency or for frequency band with more then one frequency.

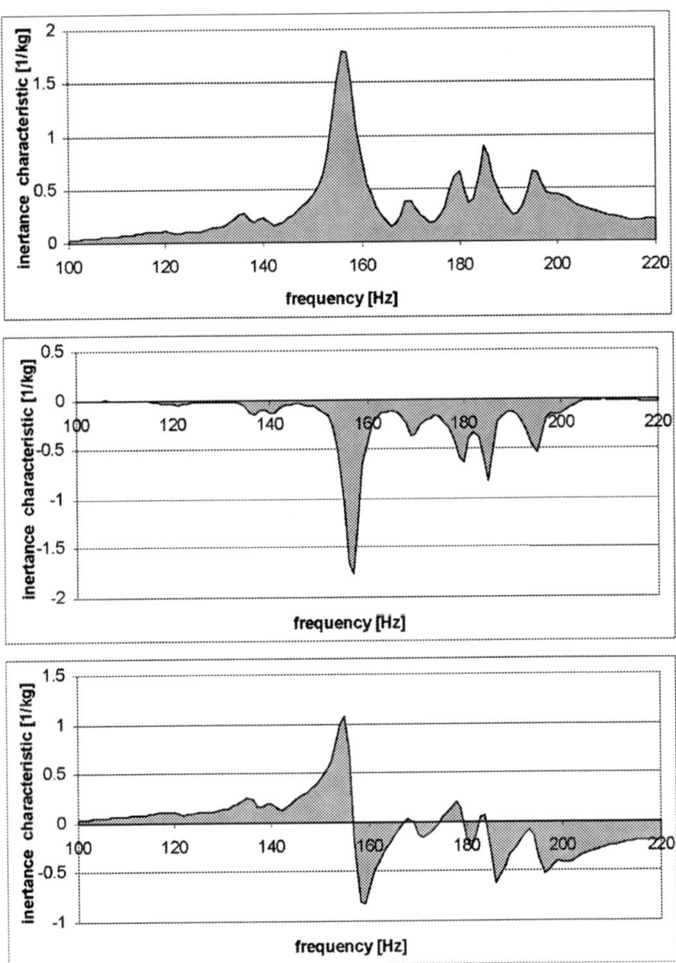

Figure 9. The inertance frequency characteristic enlarged in the band 100 - 220 Hz; From the top: B – absolute value, I – imaginary part, R – real part;

The modal parameters of the panel were determined on the basis of frequency characteristics. The identifying experiment comprised the random, sinusoidal and impulse excitation of panel vibration and the measurement of the corresponding response in the shape of accelerations of particular points in the function of frequency. In the test of panel the inertance frequency characteristic was used. The modal test with one reference signal was carried out on the panel. The structure was excited in one point by the inductor and the responses were recorded in over a dozen points by accelerometers. This is so called SIMO test, with the single input and multiple output. Instrumentation consists of the inductor TIRA W-I, power amplifier TIRA W-II and generator TG-100, which created the excitation system, force transducer 8201, charge amplifier 2651 and analyzer 2034, which created the force measurement system, and accelerometer PCB, charge amplifier GA-463, analyzer 2034 and analyzer LMS SCADAS III, which created system for acceleration recording and analysis (Łakota & Siwowski 2005).

The results of test analysis were obtained in the shape of frequency characteristic functions H(ω), from which the resonance frequencies and mode shapes were determined. Figure 9 shows the exemplary frequency characteristic function, given in the band of 100-220 Hz. On the basis of these figures the particular resonance frequencies were determined for values, in which imaginary part of inertance achieves maximum and, simultaneously, real part of inertance equals zero. Four first measured resonance frequencies of the panel are: 158, 174, 189, 206 Hz. The experimentally determined first mode shape is illustrated in Figure 10. This mode shape appears to correspond to the first longitudinal flexural mode and is basically compatible to the mode shape obtained numerically.

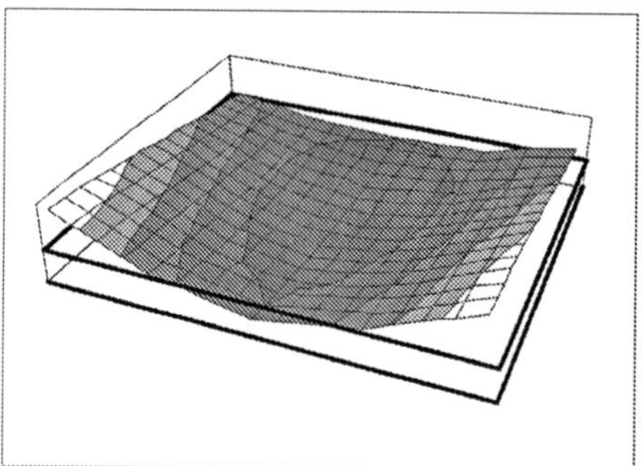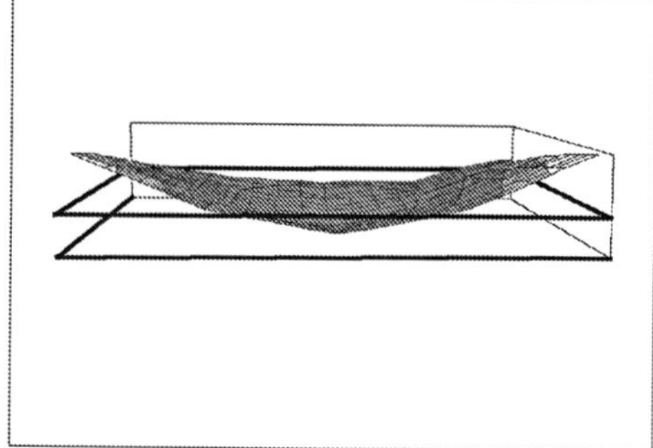

Figure 10. The first mode shape of panel for the frequency of 158 Hz

Conclusions

The results of the laboratory tests confirmed the adequate stiffness and strength of the aluminium bridge deck panel under service load. The stress state under the load patch is significantly influenced by localized bending. This influence is so strong that evolution of the failure regions is ultimately determined by these local effects. The welds are potential weak points. Strength in the HAZ is significantly reduced and fracture initiated at these locations. The failure mechanism and the failure load were identyfied during the tests, showing the adequate safety margins of the panel. In the first ultimate load test failure occurred at a load of about 560 kN by gross yielding and fracture underneath the load patch. In the second ultimate load test failure load was about 920 kN by fracture of welds on the bottom deck surface. The results of this research study were in good agreement with the model study carried out elsewhere. Although proprietary constraints prevented a specific comparison, the stress and displacement distribution and magnitudes were very similar.

The results of dynamic laboratory tests confirmed, that applied procedure – classical modal analysis - let to determine, with required accuracy, the inertancy frequency characteristics and thus the resonance frequencies and mode shapes of the panel. The adopted modal analysis gave rewarding results both in test data acquisition and the evaluation and interpretation of results. However, the findings prove the complexity of phenomena, happening with random excitation. From the practical point of view the very important finding of the dynamic test is the demonstration, that during the random excitation the resonance vibration of deck panel might be generated in certain bands of frequency. These bands were clearly determined in the test.

Results from the study clearly demonstrate that aluminium bridge deck panels are feasible alternative to RC decks from the standpoint of strength, servicebility and dynamic response. The panel may be recommended for the use on deteriorated bridges, which need deck replacement. An important question remains regarding the structural performance of the aluminium deck system. The long term behaviour under repeated loads and the fatigue resistance of longitudinal welds should be verified both in the laboratory and on-site. The above mentioned issues are the scope of the third phase of the study on aluminium bridge decks, which is still going on at the RUT Department of Bridges. The results of the third phase will be published soon elsewhere.

References

Dobmeier J.M., Barton F.W., Gomez J.P., Massarelli P.J., McKeel Jr. W.T., 2001. Failure study of an aluminum bridge deck panel. *Journal of Performance of Constructed Facilities* **15** (2): 68-75.

Herzog M.A.M., 1992. Cost effective aluminium decks for long span suspension bridges. In: *The proceedings of the 5th INALCO'92 International Conference on Aluminium Weldments, Munich, April, 1992.*

Höglund T., 1994. Bridges and bridge decks in aluminium. In: *Aluminum's Potential:Bridge construction; The proceedings of the bridge session of the 1994 ALUMITECH Conference. Atlanta, USA, October 1994.* Washington: The Aluminum Association.

Łakota W., Siwowski T., 2005. Numerical and experimental dynamic analysis of aluminium bridge deck panel. *Archives of Civil Engineering* **LI**(4): 557-607.

Matteo A.D., 1997. An Aluminum Bridge Deck Design for Highway Bridges. In: L.Kempner, C.B.Brown (ed). *Building to Last; The proceedings of Structures Congress XV, Portland, Oregon, USA, April, 1997.* New York: ASCE.

Mazzolani, F.M. 2004. Competing issues for aluminium alloys in structural engineering. *Progress in Structural Engineering and Materials* 6(4): 185-196.

Okura I., Hagisawa N., Naruo M., Toda H., 2003. Fatigue Behavior of Aluminum Deck Fabricated by Friction Stir Welding. *Structural Engineering/Earthquake Engineering, JSCE* **20** (1): 55-67.

Siwowski T., 2002. Static and stress analysis of aluminium bridge deck. In: *Current Issues of Civil and Environmental Engineering; The proceedings of the VII International Scientific Conference, Kosice, May 2002.* Kosice: Kosice Technical University.

Siwowski T., Żółtowski P., 2003. Analytical and experimental evaluation of aluminium bridge deck. In: Brisk Events (ed), *Lightweight Bridge Decks. The proceedings of European Bridge Engineering Conference, Rotterdam, March 2003.* Leusden: Brisk Events.

Soetens F., Van Hove B.W.E.M., Mennik J., 2001. Aluminium bridges in the Netherlands. In: Dimitris Kosteas, Menno Meyer – Sternberg (ed). *The proceedings of the 8th International Conference on Joints in Aluminium INALCO 2001, Munich, Germany, March 2001.*

Zureick A., Engindeniz M., Arnette J., Schneider C., 2003. Acceptance test specifications and guidelines for FRP bridg decks. *Structural Engineering, Mechanics and Materials Research Report No. 03-5.* Georgia Institute of Technology, School of Civil and Environmental Engineering, USA, January, 2003.

Influence of erection method on consumption of prestressing steel in superstructure of prestressed concrete bridges

A. Berger, MSc (Eng.), Freyssinet Polska Ltd., Milanówek, Poland
A. Ołdziejewska, MSc (Eng.), Freyssinet Polska Ltd., Milanówek, Poland
Prof. W. Radomski, Warsaw University of Technology, Warsaw, Poland

Abstract

Consumption of prestressing steel depends on a number of factors including the method used for the construction of the superstructures of prestressed concrete bridges. However, normally comparisons concerning this problem relate to bridge structures with various geometrical and material characteristics. The case when the bridge superstructure with the same characteristics is erected using two different construction methods is unusual. Such a case occurred in 2002, during construction of one of the bridge structures located at 'Czerniakowska Interchange' in Warsaw. The bridge structure with a total length of 800m was executed using both the incremental launching method and the cast-in-place method by means of stationary full-height scaffolding. The use of these two methods required the application of different types of prestressing, affecting the consumption of prestressing steel. Analysis of this problem is presented and discussed in the paper, taking into account both technological and economical aspects. Some conclusions of a more general nature are also presented.

Introduction

It is known that the consumption of prestressing steel depends on a number of factors including the method used for construction of the superstructures of prestressed concrete bridges. Therefore, the construction method should be analysed since it affects the total construction cost of a bridge.

Consumption of prestressing steel is normally defined by an index expressing the amount of steel per square meter of the bridge deck, [kg/m^2]. This index is usually used for economic comparison of various structural solutions of bridge structures constructed by means of different erection methods. In other words, the index is applicable to structures with different dimensions and shapes. The case when identical structures are erected with the use of two different erection methods is not often encountered in engineering practice. However, such a case occurred in 2002 during the construction of the 'Siekierkowska Route' in Warsaw, where one of its bridge structures is located at 'Czerniakowska Interchange' (Figure 1). This bridge, with a total length of about 800m, has been partly erected using the incremental launching method (for a length of about 590m) and partly using cast-in-place method with stationary full-height scaffolding (for the remaining length of about 210m). The use of these two erection methods has required the application of different types of prestressing, affecting the consumption of prestressing steel. In the case of the former method, centric tendons

Bridge design, construction and maintenance 2007, Thomas Telford, London

(centroidal prestressing with straight tendons), installed before launching, and final tendons have been applied, while in the case of the latter method only the final curved tendons have been used. The above difference in the tendon arrangements is characteristic of the two erection methods of bridge superstructures.

Apart from the analysis concerning the bridge structure at 'Czerniakowska Interchange', some more general remarks concerning the economic aspects of the use of the incremental launching method and cast-in-place method are reported. This analysis is based on the Polish economic condition. However, all the costs are expressed in euros (€) and, therefore, they can be used for comparison with costs in many other countries.

Description of the bridge structure

As reported earlier, the bridge structure presented in this paper is located at 'Czerniakowska Interchange', shown in Figure 1, where the relevant parts of the structure constructed by means of incremental launching method and cast-in-place method are denoted by CE-1A and CE-1B, respectively.

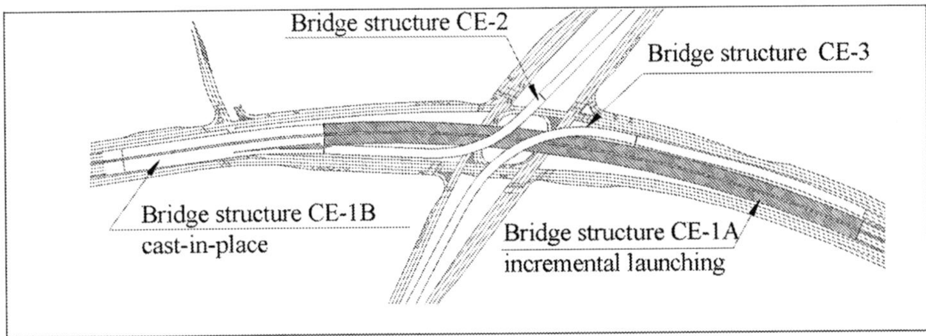

Figure 1. 'Czerniakowska Interchange' - plan

The bridge structure CE-1 consists of two independent, parallel carriageways (south carriageway and north carriageway). The bridge elevation is shown in Figure 2, while typical cross sections of the carriageways are presented in Figure 3. In Figure 3 the arrangements of the prestressing tendons are also shown, separately for the sections constructed with the use of the two erection methods.

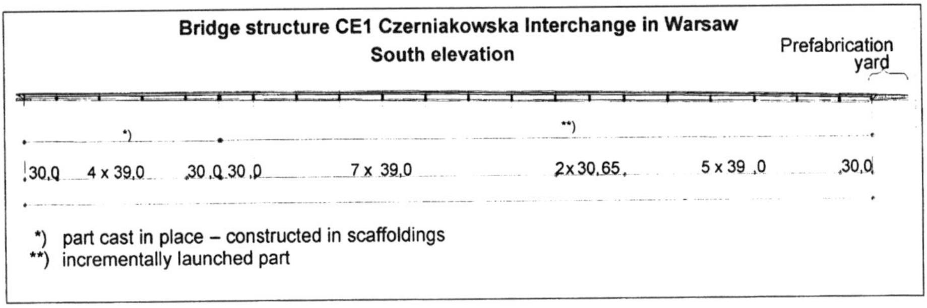

Figure 2. Elevation of south part of the bridge structure CE-1.

Each parallel carriageway of the structure consists of 16 spans with the lengths shown in Figure 2. The launching operations have been performed using 31 segments for each part of the structure. The total weight of each segment is 7.500 tons.

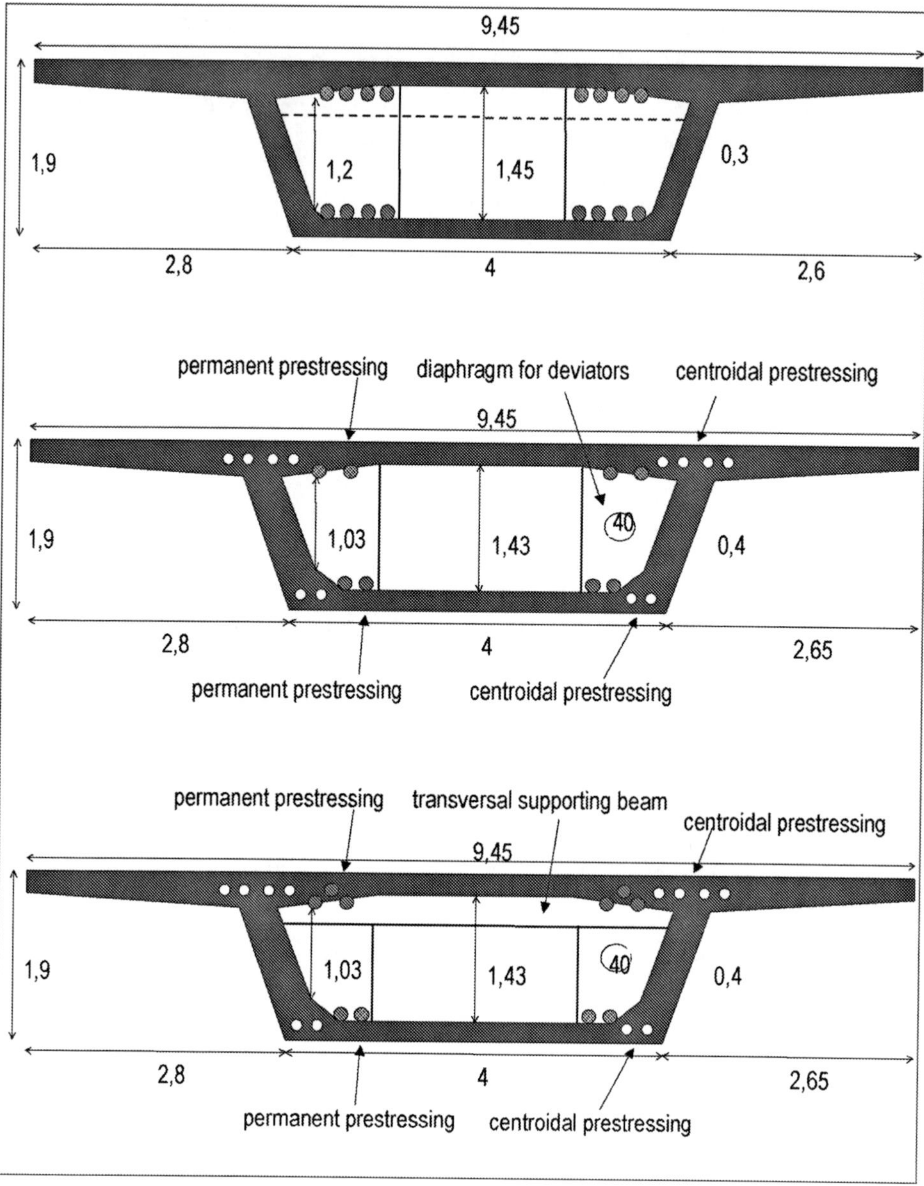

Figure 3. Cross sections of the bridge superstructure CE-1:
a) part of the bridge erected using cast-in-place method,
b) and c) part of the bridge erected using incremental launching method.

A view of the bridge structure during construction is shown in Figure 4, illustrating the junction zone for the two lengths constructed using the two different erection methods.

Figure 4. Bridge structure CE-1 during construction – junction zone of its two parts.

The use of two different erection methods resulted mainly from a change of geometrical parameters of the road interchange, one part was curved in plan and the other part was straight and had a connection with another bridge. After the cost analysis it was decided to use the launching method for the curved part and scaffolding method for the straight part.

The launching operations during bridge construction were difficult. This was partly due to the geometry of the structure, which is curved in plan with a radius of 1300m and curved in profile with a radius of 4500m, and partly due to the local conditions – the launching was carried out over an existing road with intensive traffic and also under existing bridge structures, constructed previously as part of the road interchange (see Figure 1). The last condition demanded a very high precision during the launching operation. For instance, the distance between some of the piers of the existing structures and the launched structure was only 8cm. However, although the launching process itself is interesting because of its non-routine application, it is beyond the scope of this paper and, therefore, it will not be described in detail herein.

Material consumption and costs

The required amounts of prestressing steel, anchorages and other equipment have been determined based on the relevant design calculations and technical characteristics of the two erection methods used. For each of them, it was possible to select various types of prestressing, e.g., bonded or unbonded prestressing, internal or external tendons etc. The results of a comparative analysis for the use of incremental launching are presented in Tables 1 and 2.

Table 1. Incremental launching of CE-1A bridge structure - Consumption of the prestressing tendons and accessories depending on the type of prestressing

Prestressing tendons and accessories (all of them - Freyssinet system)	Launch (centroidal) prestressing – internal, bonded, with couplers; Permanent prestressing – internal, bonded.	Launch (centroidal) prestressing – internal without couplers (with blisters); Permanent prestressing – external
Mass of prestressing steel – cables L15,7; 1860 MPa, 150 mm^2	406 t	426 t
Number of anchorages:		
– type 13C15	48	816
– type 19C15	276	276
Number of couplers:		
– type 13CC15	384	-
Length of non-galvanized steel sheets:		
– d = 80 mm	14 112 m	15 264 m
– d = 95 mm	9 396 m	-
Length of HPDE sheets:		
– d=95 mm	-	9 346 m

Table 2. Incremental launching of CE-1A bridge structure – cost of prestressing depending of its type

Permanent prestressing	Launch (centroidal) Prestressing	Relative cost $[€/kg]$[1]	Total cost $[€]$[2]
Internal prestressing	With couplers	2.11	859 400
	Without couplers	1.95	832 300
External prestressing	With couplers	2.18	894 500
	Without couplers	2.02	863 800

[1] Cost of 1 kg of prestressing steel applied for the structure; [2] Cost of prestressing steel including anchorages and couplers.

Following the technical and economical analysis, the General Contractor decided to use internal launch (centroidal) prestressing and external permanent prestressing for the CE-1A structure, and external permanent prestressing for the CE-1B structure. The continuity of all prestressing tendons (cables) was provided by jacking zones and not by couplers.

It should be emphasized that the selection of the type of prestressing has been made taking into account the total construction cost, not simply the prestressing cost itself. In this particular case, the prestressing cost is of the order of 3% - 4% of the total construction cost of the structure CE-1.

Finally, in the bridge structure CE-1A the following prestressing has been applied:
- launch (centroidal) prestressing: 12 cables 13L15,7 with jacking zones;
  cable length – two segments (2 · 19.5 m);
  at the end of each segment a half of the total number of cables are anchored;
- permanent prestressing: 4 cables 19L15.7 with jacking zones;
  cable length – two neighbouring bridge spans;
  anchorages – over the piers.

An example of the tendon arrangement in the structure CE-1A is shown in Figure 5, where some additional information is also provided.

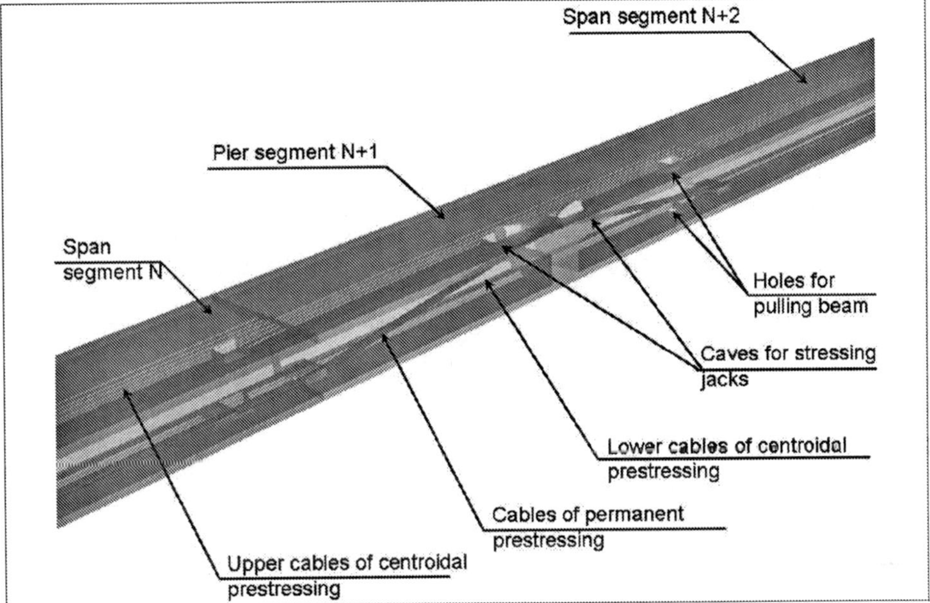

Figure 5. Prestressing in the bridge structure CE-1A (constructed with the use of incremental launching)

The average consumption of the prestressing materials in the structure CE-1A is 29.1 kg/m^2, where m^2 refers to the unit area of the bridge deck.

Finally, in bridge structure CE-1B (i.e., constructed with the use of cast-in-place method) the following prestressing has been applied:
- permanent prestressing (only): 8 cables 19L15.7 with jacking zones;
  cable length – two neighbuoring bridge spans;
  anchorages – over the piers.

The average consumption of the prestressing materials in the structure CE-1B is reduced to 19.0 kg/m^2. Therefore, it can be concluded that in the case of the incremental launching method, consumption of prestressing materials is about 35% higher than that in the case of the cast-in-place method. It should be noted that the bridge spans of the structure CE-1A and CE-1B are the same (see Figure 2).

The technical and economic analyses have shown that the unit costs of prestressing, using the two construction methods are as follows:

- incremental launching – 57.90 €/m^2;
- cast-in-place – 38.40 €/m^2.

Therefore, it can be concluded that taking into account the prestressing costs themselves, for this particular structure the cast-in-place method was evidently less expensive. This conclusion is in accordance with that expected resulting from the features of the method itself, requiring the use of launch tendons. However, it should be noted that the prestressing costs are not in general the decisive factor. Many other factors, such as scaffolding costs, construction time, possibility of usage of the terrain under the bridge superstructure, other social costs, etc, normally have a much greater influence on the economy of the erection method applied.

Additional analysis

Evidently the higher cost of prestressing in the case of the use of the incremental launching method justified a more general analysis showing the scope of the economic profits of the method. Such an analysis has been performed in the case of the bridge structure CE-1. The following four variants have been considered:

- variant A - incremental launching performed separately and consecutively (one part after another) for each carriageway (south and north) of the bridge structure CE-1;
- variant B - incremental launching performed at the same time for the two carriageways of the structure;
- variant C – stationary scaffolding for the two carriageways of the structure when the height of the bridge superstructure over the terrain is 5m;
- variant D – stationary scaffolding for two carriageways of the structure when the height of the bridge superstructure over the terrain is 11m.

The results of the analysis are presented in Figure 6, where the cost differences between the variants A, B, C, D are plotted for a range of bridge deck areas.

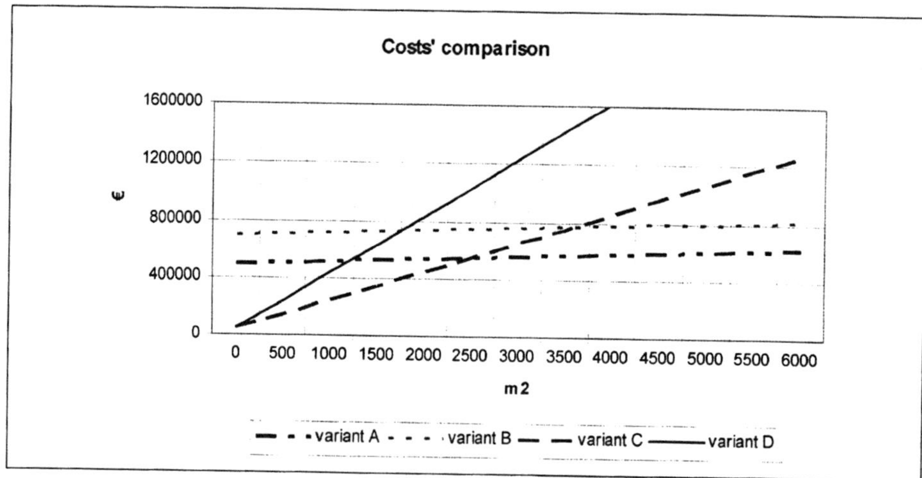

Figure 6. Comparison of the costs concerning the variants A, B, C, and D depending on the area of bridge deck

It can be seen that when the bridge deck area does not exceed about 1200 m^2, the use of stationary scaffolding (i.e , the use of cast-in-place method) is more economical than the use of the incremental launching method. The last method is evidently more economical than the former one when the area of bridge deck is greater than $3600m^2$. In the range of bridge deck area between $1200m^2$ and $3600m^2$, some other factors should be taken into account to select one of the construction options as more economically profitable, e.g. the time of construction.

Furthermore, it should be noticed that the cost in the case of the use of incremental launching method depends much less on the bridge deck area than the cost of cast-in-place method. Although it is known, on the other hand, that this problem is not commonly analysed to date in many situations.

Clearly, the diagram shown in Figure 6 can be considered only as the first approximation of the costs. The selection of the proper erection method should always be individually performed, based on the local technical and economic conditions

Final remarks

1. The prestressing cost itself is not, in general, a decisive factor influencing selection of the erection method for construction of bridge superstructures. In the particular case presented in this paper, the above cost is about 3% to 4% of the total construction cost. However, especially in the case of major bridges, prestressing cost in not negligible and, therefore, it should be analysed in particular.

2. In contemporary bridge engineering, the time of construction is often a decisive factor influencing the selection of the erection methods. The higher cost of the prestressing in the case of incremental launching method does not decrease its other economical values. However, prestressing cost analysis can be useful in many situations to determine more correctly all the criteria for selection of the proper erection method in the given technical and economic conditions.

3. The analysis presented above concerns only one case and is related to the Polish economic conditions. However, the information and conclusions seem to be of a more general nature and, therefore, can be helpful for similar considerations in many other countries.

Strengthening of Arched Masonry Bridges with Retrofitting Reinforcement - Research and Practice

Prof. Petr Stepanek, Faculty of Civil Engineering, Brno University of Technology, Brno, Czech Republic

Abstract

An experimental validation of additional strengthening consisting of
- pull-out (anchorage) tests of the strengthening system in masonry made from Czech burnt bricks,
- load tests of beams made from bricks "London Brick" and Czech bricks,
- load tests of vaults made from Czech bricks,

were carried out. The strengthening was realised by common steel reinforcement rebars, special glass fibber rebars and by special helix shaped rebars. Static and dynamic tests were performed. Aim of the work was
- evaluation, testing a new method for the masonry strengthening by retrofit reinforcement using non-prestressed steel and FRP rebars to resist the bending moment or eccentric axial force, respectively,
- development of a design methodology for strengthening – set up of mathematical model, parametric study, establish practice guide, development of new software or new modulus into existing program (for example Ring).

General reference to some existing practical realizations will be included into oral presentation.

Introduction

Faults and failures in these structures originate due to poor or insufficient maintenance, or inappropriate usage. The structures deteriorate due to age, climatic conditions, traffic load enlargement, dynamic load action and insufficient or neglected maintenance. Besides these reasons, some failures can occur as a consequence of imperfect structural design or faulty realization.

The load-bearing capacity of an arch can be augmented by its strengthening. A variety of strengthening methods are available. These vary in effectiveness. Generally this is done by forming a new arch over the existing one, bridging the arch with a framed structure without its loading at the existing vertical alignment of a transferring rod, or by inserting additional reinforcement. We can distinguish following strengthening methods:
- Saddling is one of the simplest and most popular of the traditional methods. This involves removal of the fill to expose the extrados of the vault. A reinforced or mass concrete flat or curved slab is subsequently cast in place over the original barrel. Original vault could be interlocked with new load bearing structure. Saddling increases (with minimal change to the external appearance) the load bearing capacity of the bridge. But it is expensive and will cause considerable disruption to traffic. The bridge could be also in a temporarily

vulnerable state once the fill has been removed, unless the barrel is supported from underneath which can be a costly process.

- Sprayed concrete applied to the intrados of the arch is another traditional method. This may be combined with a reinforcing mesh. Whilst this negates the need for drilling, the intrados is the part of the barrel exposed to weathering, resulting in friability. Applying sprayed concrete can cause moisture to be locked into the barrel. Other problems include poor composite behaviour and incompatible materials.
- Other strengthening systems combine the application of additional reinforcement and do not require the removal of the fill to expose the extrados of the barrel. Those systems (e.g. Archtec system, [1]) involve drilling through the road surface to install internal reinforcement into the arch, without causing any change to the existing appearance of the bridge and with minimal environmental disruption.
- Surface Reinforcement. There are proprietary systems available using a network of steel bars located in slots cut into the intrados or extrados and bonded using special adhesives. Such systems have been shown to increase the strength of the bridge. Access to the arch intrados is not always easy or possible.
- It is possible to have combined strengthening systems too. More importantly, the application of sprayed concrete and/or externally bonded or slotted reinforcement will have a detrimental effect on the appearance of the intrados. This is un-desirable in many situations and unacceptable for many structures of historic importance.

The ideal strengthening system would meet the following requirements: minimal change to the appearance of the bridge, minimal interruption to traffic and other road users, minimal impact on existing services in the bridge, provide an adequate increase in load carrying capacity, exhibit a ductile mechanism of failure, exhibit long term durability, be cost effective.

A method of additionally inserted non-prestressed reinforcement installed form the face side enables the additional strengthening of a masonry structure without serious intervention in the arch itself. The system is capable of distributing in a structure any newly originating stresses from the loading that start to act on a strengthened structure. The aim of reinforcement is to
- restrict the development of existing cracks and optionally to limit the onset of new ones,
- increase the load-bearing capacity of a bridge structure.

This system is applied from the arch face side into pre-prepared chases. For reinforcement it is possible to use conventional reinforcing steel bars, steel bars with anticorrosive treatment, stainless steel bars, high-strength reinforcement of different special shapes with anticorrosive treatment, FRP reinforcement (glass, carbon and aramid). For the best cohesion with the original materials, special cements and glues are used for the fixation of reinforcement and to ensure its coherence.

Performed tests of retrofitting reinforcement included into slots

Pull-out tests

The anchorage tests of different strengthening systems created by special reinforcement included into slots have been carried out on trial brick samples of dimensions 300/300/300, 375 and 450mm (width/height/length) brick up from solid burnt bricks of the strength class CPP P15 (MPa) and low quality cement-lime mortar. The embedment lengths of the Helifix ([2], rebar of diameter 6mm), normal reinforcing steel class 10505(R) rebars of diameter 6 mm and "home developed" GFRP reinforcement of diameter 6 mm have been tested. This GFRP reinforcement was developed by solution of research task 1HPK2/57 – [3].

In every anchorage sample either HeliBar of diameter 6mm or rebar 10505(R) or GFRP rebar of diameter 6mm was placed alternatively in an open slot of 10mm height and 30mm depth. The bars were fixed in place with HeliBond mortar, which filled the slot from the face of a block with the loaded end of a rebar to the distance equal to the certain embedment length (0). The specimens were installed in special steel testing frame prepared for the pull-out experiments and were fastened between rigid steel columns support (0). For the pull-out tests of specimens with HeliBar was used a mechanic jack and for the pull-out tests with reinforcement 10505(R) and GFRP the mechanic jack was replaced by a hydraulic jack.

A summary of performed pull-out tests of reinforcement anchored in masonry blocks, material of specimens, type of used reinforcement, number of tested embedment lengths and labeling of specimens are assembled in Table 1.

Table 1: List of Pull-out Tests Results – reinforcement HeliBar, GFRP and normal steel rebar 10505(R)

Specimen	Reinforce-ment	Embedment length [mm]	Nr. of specimens	Tensile force N_u [kN]	Stress in rebar σ_u [MPa]
H–150/i	Helibar	150	3	3,66	513
H–225/i	Helibar	225	3	5,73	803
H–300/i	Helibar	300	3	6,68	936
H–375/i	Helibar	375	3	6,60	924
H–450/i	Helibar	450	3	7,09	993
G-225/i	GFRP	225	3	8,05	285
G-300/i	GFRP	300	3	9,44	334
G-375/i	GFRP	375	3	18,60	658
G-450/i	GFRP	450	3	19,61	694
B–225/i	10505 (R)	225	3	10,43	369
B–300/i	10505 (R)	300	3	11,28	399
B–375/i	10505 (R)	375	3	16,22	574
B–450/i	10505 (R)	450	3	17,28	611

Fig. 1: Example of block of Series H and setup of pull-out tests

The results of carried-out pull/out tests measurements are assembled in Table 1 and are graphically presented in Fig. 2.

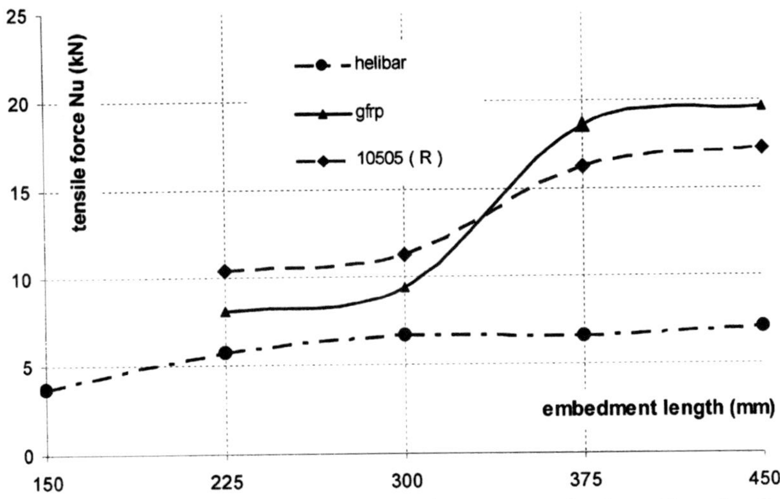

Fig. 2: Pull out test: ultimate axial force dependence on embedment length of rebar

Strengthened vaults – static tests

Within experimental parts of testing of strengthening three sets of masonry vaults for various loading types were manufactured. For the distinction of individual vaults is used notation jKi, where „j" corresponds to series number (1-3) and „i" to the strengthening method (1-3). The vaults were symmetrically loaded in ½ of the span - 1.series (j=1), asymmetrically in ¼ of the span - 2.series and symmetrically in both quarters of the span - 3.series (j=3) – Fig. 3.. Each series consists of three vaults: non-strengthened one – comparative (i=1), a vault reinforced in two chases (i=2) and a vault reinforced in three chases (i=3). The vaults were bricked up from full burnt bricks on lime-cement mortar of the width 890 mm, span 2600 mm, deflection 750 mm and radius 1500 mm. Into every reinforcing chases were embedded 2 bars HeliBar of special helical shape of diameter 8 mm or 1 bar of GFRP reinforcement with diameter 6mm. For provision of static boundary conditions the vaults were bricked up into the steel frames, in order to prevent their horizontal and vertical displacement in the bedding.

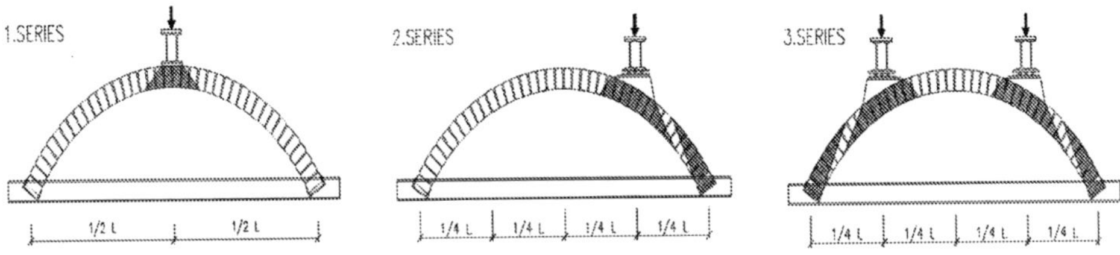

Fig. 3: Loading schemes of tested masonry vaults

The result values of the loading and corresponding deformations for all series for stainless steel reinforcement are presented in Tab. 2.

From the comparison of the load-bearing capacity of the individual vaults in the series it results that essential growth of the load-bearing capacity was achieved especially in the case of 1st series and 2nd series of the vaults, namely more than eight multiple growth. It was related to the vaults stressed by either concentrated or one-sided load, at which the vaults were loaded by the interaction of normal forces and bending moments. In the case of 3rd series the experiments did not prove the effects of strengthening by additionally inserted reinforcement on the vaults load-bearing capacity; no effects of reinforcement demonstrated themselves because the vaults were mainly compressed.

In the case of non-strengthened vaults of 1st and 2nd series the failure was acute, main crack was opened and the vault ruptured. In the case of the strengthened vaults of 1st and 2nd series came to the gradual opening of separate cracks until the failure, which was accompanied by the rapture of reinforcement from the chases. In the case of 3rd series of the vaults the failure of strengthened vaults was analogous to the failure of non-strengthened ones. The vaults in these series were mainly compressed, failure wasn't accompanied by the crack origin and the loss of stability was caused by very sudden crush of mortar in joints.

Table 2: Comparison of load-caring capacities and deformations of vaults (Helifix reinforcement)

j	i	Vault nr.	Maximal force $F_{max, jKi}$ [kN]	Maximal deformation w_{jKi} at $F_{max, jKi}$ [mm]	Comparison of deformations w_{jKi} at $F_{max, jK1}$ [mm]	F_{jKi}/F_{jK1}
11		1K1	4,737	3,233	3,233	1
	2	1K2	30,508	19,578	0,463	6,44
	3	1K3	39,98	18,614	0,309	8,44
21		2K1	4,933	1,678	1,687	1
	2	2K2	30,201	17,726	0,28	6,12
	3	2K3	43,756	16,884	0,331	8,87
31		3K1	368,584	8,975	8,975	1
	2	3K2	370,239	10,369	10,117	1
	3	3K3	439,772	8,9665	6,801	1,19

The results of the tests for the second series with GFRP reinforcement are presented in Tab. 3.

Table 3: Comparison of load-caring capacities and deformations of vaults (GFRP reinforcement)

j	i	Vault nr.	Maximal force $F_{max, jKi}$ [kN]	Maximal deformation w_{jKi} při $F_{max, jKi}$ [mm]	Comparison of deformations w_{jKi} při $F_{max, jK1}$ [mm]	F_{2Ki}/F_{2K1}
2	1	2K1	4,772	2,47	2,47	1
	2	2K2	30,61	17,512	0,3	6,41
	3	2K3	40,111	14,9	0,087	8,41

Comparison of behaviour of non-strengthened vaults (reference samples) is drawn on the Fig. 4. Influence of strengthening with stainless steel and GFRP is shown on the Fig. 5 and 6. The deformation under acting load (denoted S ¼ under) and on the other side unloaded half of vault (denoted S ¼ unloaded) are drawn.

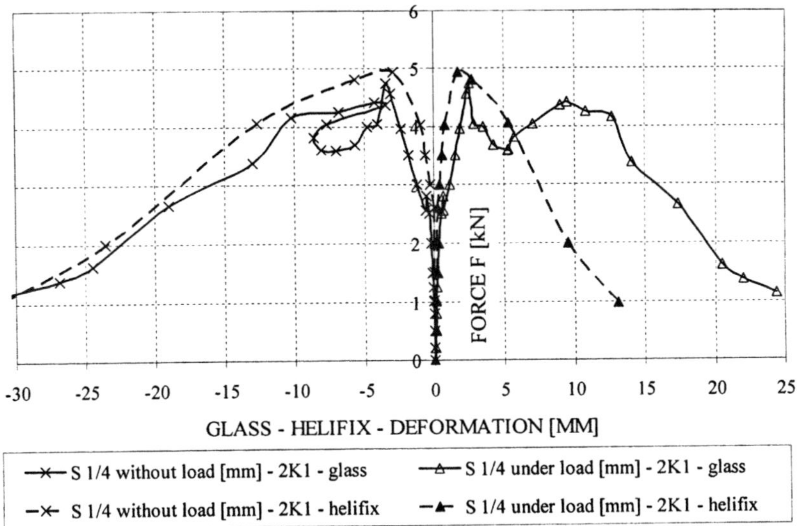

Fig. 4: Working diagrams of non-strengthened vaults (reference vault for Helifix and GFRP strengthening), loaded at ¼ of span

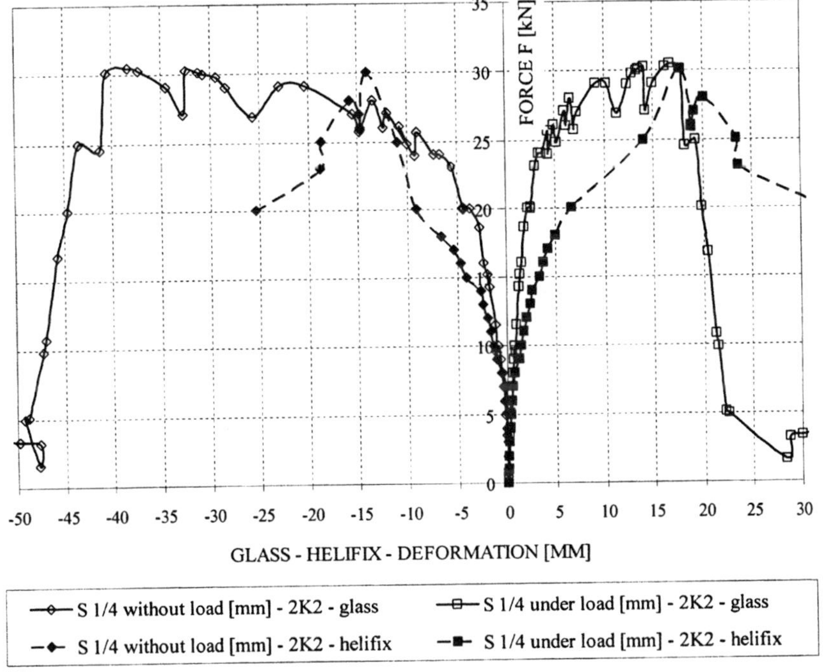

Fig. 5: Working diagrams of vaults strengthened with Helifix stainless steel and GFRP reinforcement in 2 chases (vaults are loaded at ¼ of span)

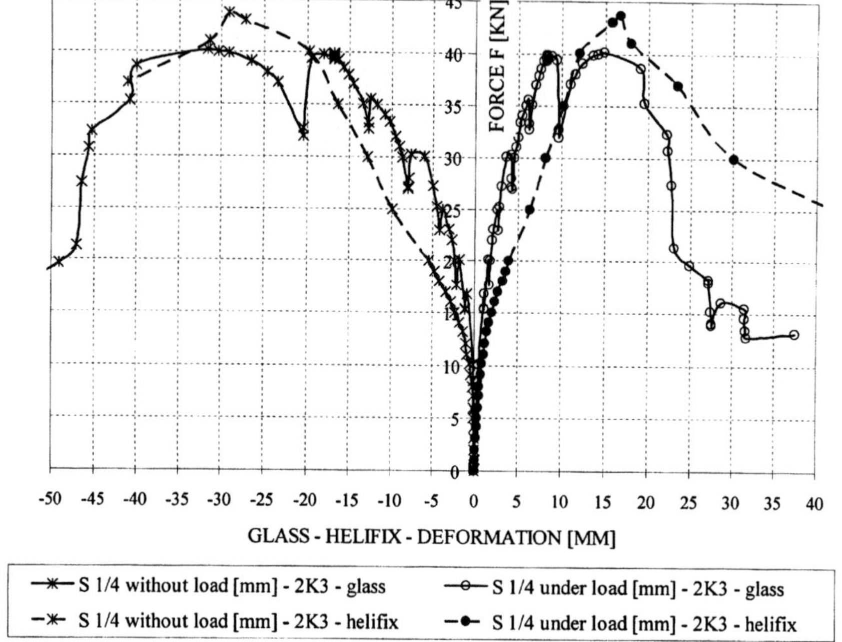

Fig. 6: Working diagrams of vaults strengthened with Helifix stainless steel and GFRP reinforcement in 3 chases (vaults are loaded at ¼ of span)

Strengthened vaults – dynamic tests

The reference (un-strengthened) vault collapse was reached with ultimate limit force 7 kN after 180000 cycles with frequency 4 Hz; loading force in ¼ of span fluctuated from 0,5 kN to 7,0 kN.

The strengthened vaults 2K2 and 2K3 were tested on the following way: static loading was increased up to cracking level of vault. After reaching of cracking force F_{cr} the dynamic loading was performed: frequency 4 Hz, loading force in ¼ of span changed from 2,5 kN to highest value F. This force was increased from F_{cr} with 1,0 kN increment after stabilization of deformation during last 5000 cycles. For the 2K2 vault was F_{cr} = 10 kN, for 2K3 vault was F_{cr} = 12 kN. Scheme of behavior by the test is shown on the Fig. 7. The deformation under acting load (denoted S ¼ under) and on the other side unloaded half of vault (denoted S ¼ unloaded) and at half span (denoted S ½) are drawn.

Mathematical modeling

The mathematical model of vault was made in ATENA software - [4]. It enables simulate the behaviour both of un-reinforced and reinforced structures with different types of reinforcing. The most convenient model for describing orthotropic non-contiguous character of masonry is a micro-model. This model can describe not only the materials characteristic of individual materials (bricks, mortar), but also their co-acting that is in the mathematical model of masonry. For first computation the 2D model was used.

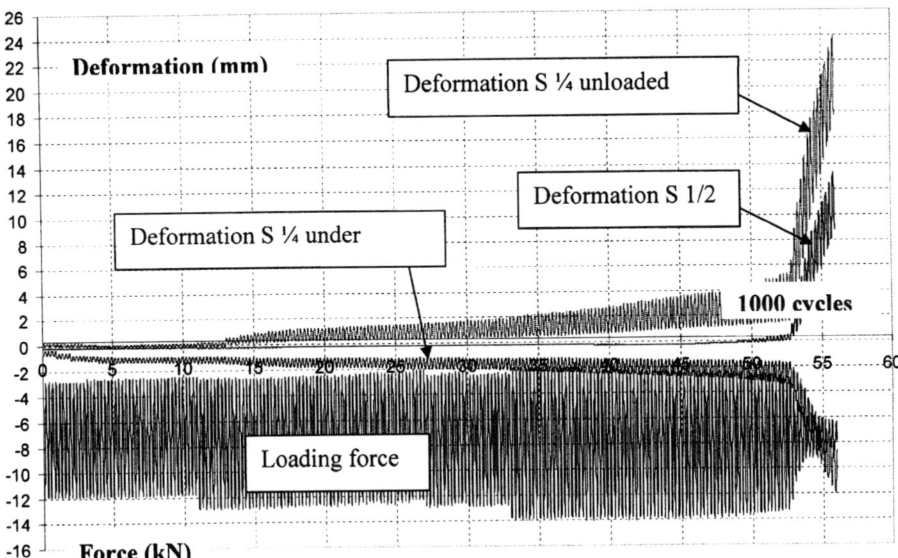

Fig. 7: Dynamic test of GFRP strengthened vault (2K2, reinforcement in 2 chases)

The contact model was based on a model of the dry friction (Mohr-Coulomb) defined by the shear cohesion c and by the friction factor φ (1). Maximum shearing stress is restricted by a linear relation

$$\tau = c + \sigma \cdot tg\varphi, \tag{1}$$

where σ is a magnitude of the contacting pressure stress (positive value). The ATENA contact model has two types of parameters. First set of parameters is describing the real physical properties of interface: tensile strength f_t, shear cohesion c and friction coefficient φ. They must correspond to real material properties. The second set of parameters are stiffness coefficients, which serve purely for numerical purpose. There are two stiffness coefficients, K_{nn} (normal), K_{tt} (tangential) and each has two values: basic and minimal. The basic stiffness represents the stiffness of the interface model in close state. Instead of rigid connection, the interface in closed state undergoes displacements according to the stiffness. The minimal stiffness serves only for numerical purposes as "predictor" within the nonlinear interactive solution method.

Models of un-reinforced vaults were used especially to checkout parameters of the contact. In the case of reinforced vaults is in the structural model in ATENA program also included reinforcement inserted into the chases. 2D model is unable to model precisely the reinforcement behaviour in the chases, unfortunately. For reinforcement is only implemented a presumption about its behaviour, namely by the bi-linear or tri-linear working diagram of steel. Into the calculation is also possible to implement a presumption about the reinforcement coherence with ambient material (bond-slip relation). The presumption about the reinforcement coherence with ambient environmental is possible to express on the bases of performed pull out tests.

In case of un-reinforced vaults the computational mathematical models approximate to the behaviour of real vaults. From comparison of tests and mathematical modelling is possible to obtain/identify the physically mechanically characteristics of masonry (mortar, bricks, contacts characteristics).

In the mathematical model of additionally reinforced vault is very important part of input data of the mathematical model the reinforcement mechanical bond with surrounding environment (mortar).

If in mathematical model is full mechanical bond of reinforcement to mortar and mortar to bricks under consideration, thus there is not redistribution of tension in rebar. Reinforcement is under stress only in regions where will get to open of contacts. In other section of reinforcement is tension subsequently approximately zero - Fig. 8a. With this it is also connected practically the same stiffness of a construction during loading.

From experiments is however evident a fall of stiffness at the load rising. As to the mathematical model corresponds to the performed experiments this bond strength have to be included into calculation. We used the bond slip model with partial bond strength according CEB-FIB Model Code 1990. Then the tension is redistributed in reinforcement how is displayed in next diagram – Fig. 8b.

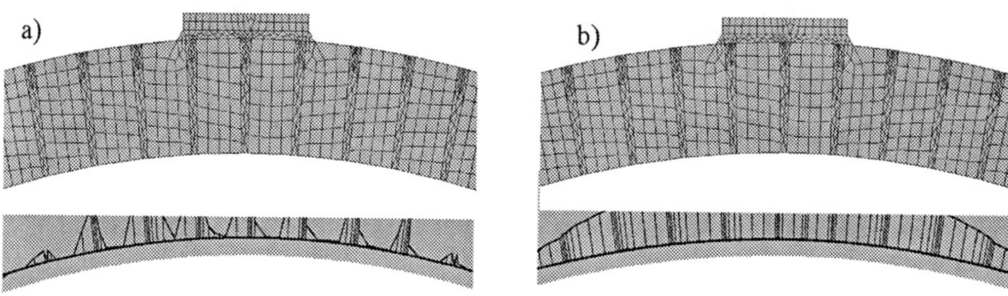

Fig. 8: Tension stress in rebar (MPa, vault 1K2, Helifix reinforcement): a) full bond strength, b) bond-slip model contact

Conclusion

The masonry strengthening was realised by common steel reinforcement rebars, special glass fibber rebars and by special helix shaped rebars included into masonry slots. In spite of performed tests it is possible to claim following conclusions:
- The anchorage length must be considered as such embedment length of reinforcement, when the ultimate force carried-out by a rebar does not further increase with rising embedment length. The anchorage length can be then determined according to mortar a masonry quality in the range between 300 and 550mm.
- The significant difference between the behaviour of special and concrete rebars has appeared when the peak tensile force was achieved. The special helix reinforcement is "screwed out" from the mortar but the rebar still carry a tensile force. In the case of concrete reinforcement and GFRP reinforcement with the groined surface is a failure of the anchorage zone brittle.
- Until cracks in the masonry develop, all materials behave elastically according to Hook's law. As cracks develop in the masonry, the bending stiffness of the masonry element decreases. After the cracking has developed in the grout, the stiffness of the beam in bending abruptly decreases rapidly.
- It is obvious that there is significant difference between the experimentally obtained ultimate load-bearing capacity and the load-bearing capacity calculated in accordance with design standards.

- Masonry with the retrofit reinforcement satisfies the load bearing function even after cracking has occurred. The ultimate load can be achieved by a number of causes, either individually or in combination:
  o failure of the masonry in its compressed area through crushing,
  o rupture of the reinforcement in tension (this can appear only after the masonry has failed through cracking in its tensioned area),
  o failure of the anchorage of the tensile reinforcement (collapse of the anchorage area),
  o failure of the masonry element from the development of shear cracks near the obvious supporting joints.
- The method has been approved by Heritage Authorities (not only) in Czech Republic.
- The described method of arch strengthening is an efficient and practical method that has been demonstrated in a full scale laboratory model.
- The firs series of dynamic tests showed possibility of stiffening of masonry vaults; the test loading was very high and umber of cycles very low. By normal structures the varying of loading between zero and maximum force does not occur at the way as was simulated in test.
- To this date the described system has been used to strengthen some masonry arch bridges constructed in different materials and geometries and for buildings masonry structures too. The repair operation is carried out without causing significant interruption to traffic. The strengthened bridges can normally be kept open to traffic throughout.

Examples of realization

Realization examples will be included into oral presentation.

Acknowledgement

This contribution has been prepared with the financial support of Ministry of Education, Youth and Sports, project No. MSM0021630519 "Progressive reliable and durable load bearing structures" and of a project of the Ministry of industry and trade No. 1HPK2/57 – „New generation of durable structures with increased resistance against aggressive actions". Part of experiments was performed under the sponsor support of firm Helifix UK and CZ.

References

[1] http://www.tfhrc.gov/pubrds/05mar/07.htm
[2] http://www.helifix.co.uk
[3] New generation of durable structures with increased resistance against aggressive actions. Annual report, BUT, FCE, 01/2007
[4] ČERVENKA, V., JENDELE, L.: ATENA Program Documentation, Theory and User's manual for ATENA 2D, Prague, Feb. 5, 2004

Construction of the Precast Segmental Approach Structures for Sutong Bridge

Xian Peng LIU, CHEC Wuhan Port Construction Corp, China
Zong Ping QIN, CHEC Wuhan Port Construction Corp, China
Y.W. LEUNG, YWL Engineering Pte Ltd, Singapore
Dr. Lingling YUE, YWL Engineering Pte Ltd, Singapore

Abstract

The precast segmental approach structure of the Sutong Bridge with 75m span length was constructed using balanced cantilever method. This precast segmental viaduct was characterized by deep foundation in the riverbed, 60m high columns connecting the main bridge and relatively long end spans in the balanced cantilever deck.

This paper covers the construction methods and logistics adopted to suit the particular site conditions. It discusses the casting yard selection and set up in the vicinity of the river bank, the design of the special equipment such as the short-line match casting moulds and two 160m long launching girders which were tailor-made for this project. The value engineering exercise in the post Contract award stage for facilitating construction is detailed. The geometry control method used in both the casting yard and erection front, the temporary works and stabilizing system are delineated. This paper also highlights the construction difficulties encountered and solutions to the problems.

Introduction

The Sutong Bridge is part of a traffic truck connecting cities Nantong and Suzhou (Changshu) across the Yangtze River in the south of Jiangsu Province (Figure 1). The total length of the highway is 32.4km. It consists of three parts: north bank link, central crossing and south bank link. The central crossing consists of a total length of 8.2km bridge structures with 6-lane dual carriageway. The main navigation channel is a cable-stayed bridge that has a central span of 1088m. Both the north and south approach structures consist of 30m, 50m or 75m long span prestressed concrete continuous single-cell box girder structure to be constructed by either cast in-situ or precast segmental method. The construction of the precast segmental approach structures commenced in April 2004 and completed in early 2007.

Figure 1: Location Plan

The Sutong Bridge project is located at the downstream of the Yangtze River close to the estuary with strong tidal effects at about 110 km from the river mouth. The site conditions were characterized by snow, storm, typhoon, strong tide, variable riverbed and rapid flow rate. The observed 30-year returned wind speed was 35.5m/s. The maximum difference between high tide and low tide observed was more than 5m. The maximum water depth was about 12m and the section flow rate was over 4m/s. These environmental factors posed constraints and challenges to many construction activities in this project.

Bridge Characteristics

Being the first precast segmental bridge built using short-line match-casting method in China, the approach structure consisted of 52 spans of single cell box girders. The typical bridge units were formed by 5-span or 10-span continuous deck supported on the pot bearings (Figure 2). The total length of the box girder measured along its centerline was approximate 3.7km. The deck width of a single carriageway was 17.5m (Figure 3). Other than one end span that was adjacent to the in-situ box girder at Natong side was a 50m unit, all other spans had an equal span length of 75m. The reinforced concrete columns with 4.5m x 6.5m rectangular hollow section were adopted for the maximum height of 60m and supported on pile caps with bored pile foundation.

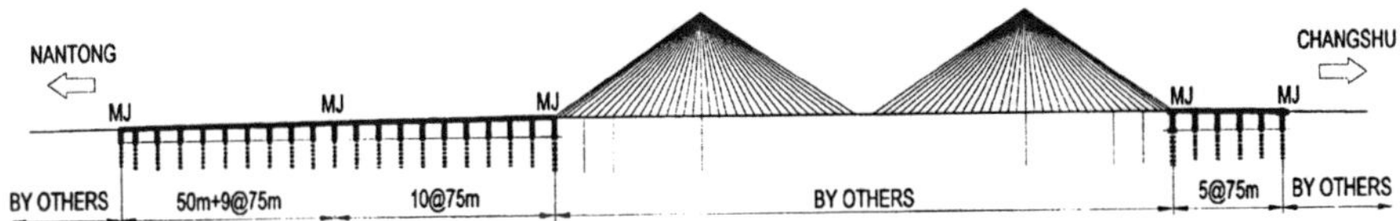

Figure 2: Bridge Elevation

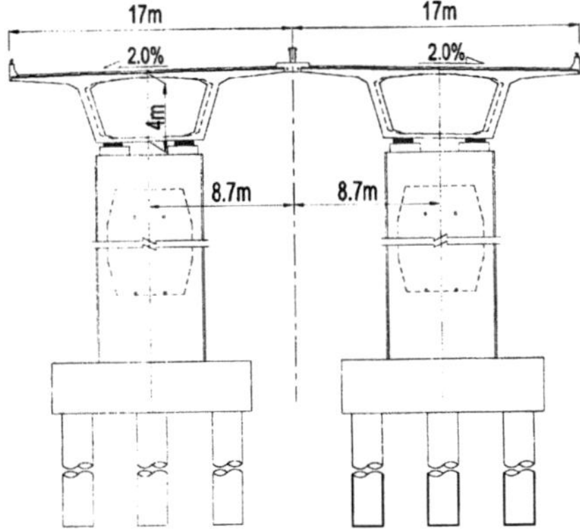

Figure 3: Bridge Sectional View

Value Engineering

The Engineer's original design of the precast segmental bridge deck was based on the balanced cantilever construction technique using long-line casting method with epoxy joints. A mixed prestressing system was employed. The cantilever and span tendons were internally prestressed while the continuity tendons were external.

During the Tender stage, YWL Engineering was engaged by the Contractor (CHEC) to provide an alternative concept design for segment production using short-line match-casting method together with the associated value engineering exercise focusing on improving constructability. The detailed design was later executed by the original designer after award of the Contract. Some key aspects of the value engineering included the following:

i)   Setting out concept of the box girder – the box girder soffit was always horizontal in the Engineer's design; thus the cross section of the box was not symmetric about its centerline. The depth of two webs varied with respect to changes in super-elevation. The alternative setting out concept employed a constant depth section by allowing rotation in the box axis (Refer to Figure 4). This arrangement would simplify the mould design and casting operation in the production.

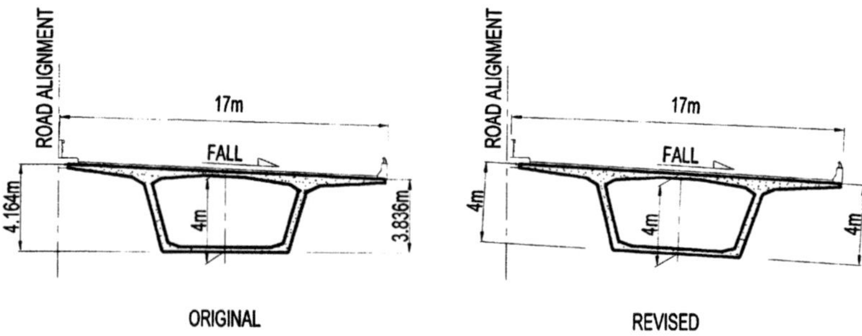

Figure 4: Box Girder Setting-Out Concept

ii)   Web transition details – near the support, the web thickness was increased for shear resistance. The original design adopted a detail with continuous transition. The proposed alternative made use of a "step" discontinuity at segment joints so that the inner core form for the segment mould could be standardized and thus the casting operation would be simplified. See Figure 5.

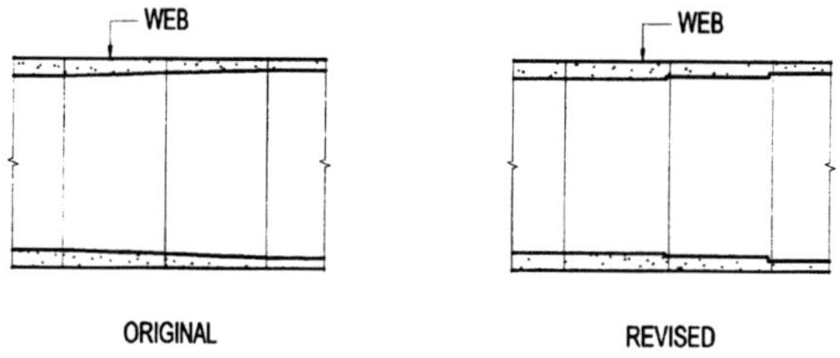

Figure 5: Box Girder Web Variation

iii)   Second stage casting of segment – there was a significant difference in self-weight between the typical and diaphragm segments (285 – 150 = 136 ton). The heavier diaphragm would demand extra capacity for all construction plants and equipment e.g. launching girder. The alternative construction concept involved re-design of the pier segment so that only a light weight shell segment was formed in the yard and the solid core was cast by insitu mean at a later stage after it was erected.

iv)   Standardization of structural elements – in the alternative proposal, many typical structural elements, such as, shear keys, blisters for cantilever tendons and diaphragm segments at end spans, were re-detailed and standardized in order to facilitate constructability.

v)   Closure segment at mid span – the Engineer's design at the mid span location consisted of a 3m long insitu cast closure segment. In the value engineering exercise, a precast option was proposed and adopted. The closure segment was 2.7m long with two 150mm insitu stitches.

Foundation

The project site is located in the alluvial plain of the Yangtze Delta, which is characterized by a thick layer of quaternary deposits. The bedrock is very deep ranging between 270m to 280m. In general, the upper layer of soil (-55m to -65m) consists of a loose to medium dense silty sand underlain by a layer of soft and compressible silty clay. Below the soft layer is the dense sandy deposits, which constitutes mainly the load bearing layers for the pile foundation.

The foundation of the precast segmental approach bridge was composed of 330 bored piles of diameter 1.8m. Piers were supported by either 8 or 9 numbers piles with 11m x 12m x 3m or 12m x 12m x 3.2m pilecaps respectively. All the piles were designed as friction piles with the founding levels ranging from -87m to -95m.

The temporary steel casings used for bored pile construction were formed by 12mm thick tubular section, and provided at the top portion from cap to -18m below seabed. The length varied from 21m to 33m depending on the geological condition. The casings were driven by a vibration hammer of 78 ton.

The drilling of the bored pile was carried out using reverse-circulation drilling rig mounted on the temporary staging. Global position system (GPS) was employed in the setting out of the piles. The foundation of the pile staging consisted of 12 numbers of 0.8m x 6mm or 1.0m x 10mm tubular piles of about 30m long that were tied together on top in 2 layers with I sections and bailey trusses to form the staging platform of 30m x 15m for one pile group construction. Bentonite slurry was employed to stabilize the bored holes. The characteristics for Bentonite suspensions are given in Table 1.

Table 1. Characteristics of the Proposed Bentonite Suspensions

Property	Fresh	Ready for re-use	Before concreting
PH	10~12	8~10	7~9
Density (G/ml)	<1.04	<1.08	1.06~1.10
Marsh viscosity (Sec)	26 ~ 35	25~26	20~24
Fluid loss (ml/30min)	<10	<15	<10
Sand content (%)	<0.3%	0.5~1.0%	<0.5%

The bored hole was filled with slurry before drilling of the portion below the casing commenced. The head of the drilling fluid was kept constant at 1.5m above the water level so as to ensure sufficient stabilizing pressure in the uncased shaft area.

After completion of drilling and initial base cleaning, the prefabricated reinforcement cages (8 numbers of 12m long per unit, 27 ton each) were installed by a 200 ton crane. Placement of concrete after final cleaning of the pile base was carried out using a 273mm diameter tremie pipe.   For 95m long piles, it required a 3-7 day construction cycle from commencement of drilling to completion of concreting works. See Figure 6 for the temporary platform and equipment for the piling works.

Figure 6: Staging Platform for Piling

Figure 7: Pilecap Construction

Pilecap construction was executed using heavy duty cofferdams which consisted of 4 side forms and a soffit with a total weight of 75 ton (See Figure 7). This falsework system was designed for constructability. The maximum weight of each module was limited to 15 ton to facilitate easy handling.   Initially, the cofferdams were overhang with temporary steelworks supported on the steel casing of the piles. After fine adjustment of cofferdam using hydraulic jacks, one meter thick layer of non-structural tremie concrete was poured in order to seal the base and stabilize the system prior to dewatering. The rest of the operations were carried out in dry. The typical cycle time for constructing a pile cap was 30-40 days.

Column

The reinforced concrete hollow columns were constructed using a self-climbing form with an integration of a working platform (See Figure 8). Hydraulic rams were employed for the climbing mechanism. A 70m high tower crane with a capacity of 120 ton-m was fixed on the pilecap for material handling. Concrete were supplied by a mixer barge and pumped via a pipe mounted on a scaffold system tied to the column. The slump of concrete was 150mm and each typical pour was 4m. The climbing form would be self-launched when the concrete reached a strength of 10 MPa. In general, a 4-day cycle time for a 4m pour height could be achieved.

Figure 8: Column Construction

Segment Casting

The segments were fabricated in a casting yard located at the north bank of the project site (Natong side). The yard was chosen at a favorable location to facilitate easy transportation of the heavy segments by barge. The loading point was connected by a 240m long temporary bridge and a jetty. See Figure 9 for the layout of the yard.

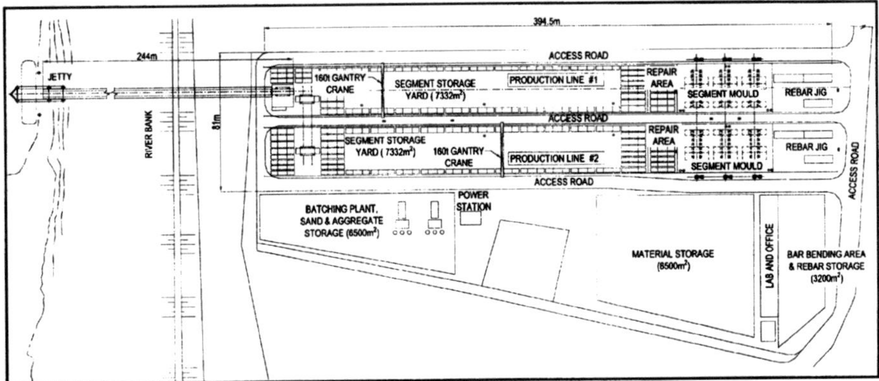

Figure 9: Casting Yard Layout Plan

In this project, a total of 1086 segments were fabricated in a period of 17 months. Due to the importance of the project, 6 sets of short-line casting cells were invested under a conservative assumption of 2.5-day casting cycle. The actual production characteristics, however, reached a typical cycle of 2 days for the standard segments after a brief initial learning period of 2 months. In the summer time, a 1-day cycle was achieved. The casting yard was about 40,000m² and equipped with 2 numbers of 160 ton gantry cranes for segment handling and 2 numbers of 16 ton gantry cranes for light duty tasks, such as, manipulating the reinforcement cage. The casting yard had a storage capacity of 354 segments based on two-layer stacking (Figure 10). Concrete was mixed in the casting yard by using 2 numbers of batching plants of 50m³ /h capacity each.

Figure 10: Segment Storage

Figure 11: Segment Curing Chamber

The bridge segments were produced by short-line match-casting method. It was a new technology used in China as earlier concrete segmental bridges had always been built with multi-insitu joints in order to deal with the geometry variations from the bridge alignment. In this project, the classical precast segmental method was employed in which the overall geometry of a bridge unit was captured in a casting cell in stages based on a given segmentation scheme. During each casting operation, the spatial relationship of a pair of conjugate segments in global coordinates was transformed to the local reference frame of the casting cell. Geometric errors that occurred during a casting operation were controlled and adjusted in the subsequent casting operations. Control points were fixed to the wet-cast segment before hardening of concrete for geometry controlling purposes. Survey was carried before and after casting a pair of conjugate segments. A proprietary computer software, GeomPro, developed by YWL, was used to control the geometry in the segment casting works. This software has a user-friendly interface equipped with a powerful databases and error detection checking system in order to minimize human errors during operation of the software.

The casting cell was approximately 90 ton each. In the design, the stiffness and flexibility of the mould were properly balanced with due consideration of the geometry of the viaduct. In winter seasons, the ambient temperature in the Yangtze River area would drop below -10°C. In order to achieve the target production rate, the air-conditioned brick houses were constructed for all casting cells to accelerate the gain of concrete strength of segments. In side these curing chambers, the temperature was maintained at above 10°C during winter. See Figure 11.

Segment Erection

The total erection period was about 15 months. Two overhead launching girders (Figure 12) were employed. Since the Sutong Bridge was a signature structure over the Yangtze River, the aesthetics of both the main bridge and approach structures was an important aspect of the design. The pier spacing and the proportion of deck and column were carefully considered and approved by a national bridge committee. With the even pier spacing, the 75m long balanced cantilever articulation led to an end span exceeding 30m (Refer to Figure 13). This imposed a taxing condition to the design of the launching girder. The 160m long overhead girders weighted about 1000 ton each and functioned both somewhat as a cantilever and span-by-span (end span) girder at the same time. Each girder had two winches with lifting capacities of 180 ton and 150 ton respectively. The lifting height of the winches was allowed for 70m so that segments could be picked up at sea level. The launching girder had been designed to resist a maximum wind speed of 58 m/s. However, the maximum wind speed was limited to16m/s during self-launching, and 22 m/s during segment erection. At out of service condition, no special tie down system was needed when the wind speed was below 30 m/s, otherwise the typhoon tie-down device was to be engaged.

Figure 12: LG Assembling

Figure 13: End Span Erection

The diaphragm segments of the initial two spans were erected using a barge crane of 800 ton in advance in order to facilitate the launching girder assembling and testing (Figure 12). The remaining diaphragms were erected either by the barge crane or launching girder. All other segments in the balanced cantilevers and end spans were constructed using the launching girders. During erection of the end span, a careful investigation of the deck-girder interaction was conducted in order to avoid overstressing of the hangers in the load transfer process.

The stabilization of the partially completed balanced cantilever was achieved by the use of vertical nailing consisted of 6 numbers of U shaped tendons (12@15.2mm) embedded in the column and preloaded to 50 ton each. The diaphragm segment was supported on 4 numbers of 100 ton temporary jacks (Figure 14). After geometry adjustment, the pier segment was fixed in position by concrete packers sandwiched with a layer (20mm) of sulphurous mortar that was later molten by the embedded electric arc for removal of the packers.

Figure 14: Erected Diaphragm Segment

After completion of the cantilever arms, the closure segment was lifted by the girder winch to the final position and supported by a pair of clamping beams mounted onto the erected cantilever tips. The launching girder could then be launched to the next span for continuing the erection works without completion of the stitches as all supports of the girder were rested on the pier segments. The insitu stitches were cast at the lowest temperature of the day. See The cantilever and span tendons were internally prestressed system consisted of 12, 15, 17 or 19 strands (15.2mm). The continuity tendons were externally prestressed system consisted of 25 strands and anchored at the diaphragm of every two spans.   Figure 15a and 15b for the launching girder in action.

The geometry control during erection was monitored at various stages. The theoretical profile of the deck, taking into consideration of the camber and stage effects, was compared with the observed results. Initially, some discrepancies were observed. After a thorough investigation, it was discovered that the problem was due to improper application of the temporary stressing system. Great improvement on the geometry accuracy was achieved for the remaining works.

Figure 15a, 15b: Partially Erected Bridge Deck

Conclusions

The precast segmental approach structure of the Sutong Bridge was a pilot project of this type of balanced cantilever construction in China. It was the result of the leadership and entrepreneur spirit of Jiangsu Provincial Sutong Bridge Commanding Department (STB). With close cooperation with many parties and experts, the project was successfully completed in a timely manner. The end product is a high quality signature structure over the Yangtze River and shall serve the public for many coming years.

Acknowledgment

For a project of this scale and complexity, there are always many individuals who have contributed to the works; their names, however, would be too long to be all listed. Special thanks are to those few who have immediate and significant help and influence on the authors' works, namely, Mr. Qingzhong YOU (STB), Mr. Yadong LIU (CHEC) and Ms Tujing (YWL). Their efforts made this project happen.

New Technologies Used in the Construction of a Cable-Stayed Bridge: Binzhou Yellow River Bridge in the Shandong Province In China

Zeying YANG, Civil and hydraulic engineering of Shandong university, Jinan, Shandong, China

Shenyuan LIU, Shandong Road and Bridge (group) CO., LTD. Jinan 250014, China

Jianbo QU, Highway Bureau of Shandong Province. Jinan 250014, China

Abstract

The Binzhou cable-stayed bridge, with three pylons over, the Yellow River in Shandong province in China was constructed between the years 2001 and 2004. The total length of this bridge is 1698 m. The width of a standard section of superstructure is 32.8m, which is the widest PC box girder bridge in China. Several new technologies were used during construction. This paper introduces these technologies and experiences to present the state of the art of bridge construction in China and give reference to the bridge engineers. The deepest drilled pile reached 120 m, which is the deepest drilled pile in China. A drilled hole can be washed in 15 minutes using a new technology. The depth of precipitation is lower than 15 centimeter in a cleaned hole. The safety and condition of this bridge remains good in the two years after its construction.

Introduction

Constructed between 2001 and 2004, the Binzhou highway bridge over the Yellow River in Shandong province in China is a 1698 m long cable-stayed bridge, with a main span of 768 m ($2 \times 42m + 2 \times 300m + 2 \times 42m$) (Fig. 1). A brief description of the cable-stayed bridge follows, but details are provided by reference [1]. The bridge comprises three cable-stayed cantilever sections, and has a 32.8m wide deck (which is the widest PC box girder bridge in China.) made up from a PC box girder.

In the back spans, two concrete piers provide additional restraints to the superstructure. Restraints to the deck are symmetrical about its longitudinal axis, and vertical and lateral movements are restrained at the piers and the pylons.

Bridge design, construction and maintenance 2007, Thomas Telford, London

Each pylon comprises two towers, and is supported with reinforced concrete piles. The piles are 1.5 m in diameter. 98 piles are 115 m deep and 24 piles are 80 m deep.

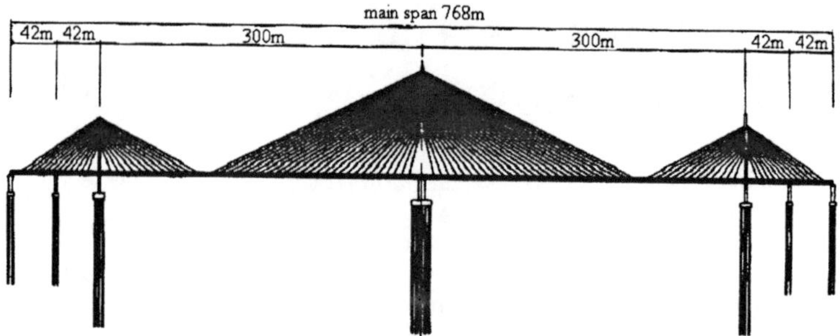

Fig. 1 Elevation of Binzhou highway bridge over yellow river

Erection of construction platform (new technology)

The main pylon (middle pylon in Fig. 1) is 49.5 m long and 18.5 m wide. Its foundation comprises 50 cast-in-place piles, which are 1.5 m in diameter and 115 m deep. Long-span pole type construction platform, which is shown in Fig.2, is adopted.

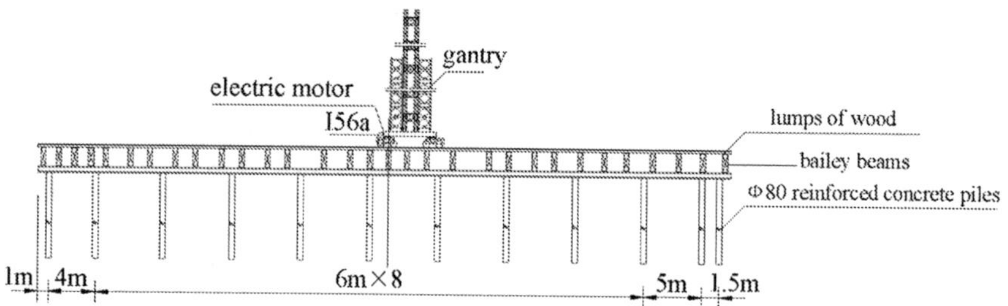

Fig. 2 Elevation of construction platform

The procedure for the erection of the platform was as follows. First, the cast-in-place piles, which are 0.8 m in diameter, were built at the longitudinal extremes of the pile cap of the main pylon. The span between the piles is 21 m. Bailey beams are put on the piles as longitudinal beams. Lastly, timber baulks ($0.2m \times 0.2m$) are put on the longitudinal beams. In order to make the lifting equipment easy to use and efficient a self-propelled gantry crane was installed on the platform. The gantry was 21 m high and had a 21 m span. The design lifting capacity of the gantry crane was 20 ton.

During construction, dislocation of the auger, lowering of the circular steel cages (which were 17 ton in weight), and erection of the auxiliary equipment were completed using the gantry crane. As a result, machine-time and the corresponding costs were saved. This new technology and the new type of construction platform, overcame the difficulties which a conventional construction platform could not solve.

Drilling

Auger type

At the location of the foundations the subsoil consisted of sandy clays, clayey sands (partial clay mingled with oyster shell), and clay. Due to the soil properties, the length of the piles was larger and their diameter smaller than normal and they had to be augered over their full length. Drilling in this type of soil demands that a carefully controlled, low torque technique is adopted. To achieve this, a normal-circulation rotary drilling method and reverse circulation washing of the borehole was adopted and using a KP-2000 type auger, the 122m deep boreholes could typically be finished in 5 days.

Verticality control

The soil conditions of the Yellow River are multi-layered and the soil properties of different layers vary greatly. The orientation of the drill bit is apt to deviate from the original direction when drilling through soil layers of different strengths and resistance and this can result in boreholes that are not vertical. To overcome this and to avoid the problems associated with piles that are not vertical, as well as installing a guide collar on the top of the auger and adding a counterweight to the drill bit, drilling speeds were increased to 20 revolutions per minute.

Reverse circulation washing (new technology)

From an analysis of the results of surveys of cast-in-place piles installed on previous projects throughout China, it was concluded that the majority of defective piles were caused by inclusions in the concrete and disruption of the interface between the concrete and the sides of the pile. These are typically the result of two effects: sediment in the mud fluid that is not properly extracted from the borehole, and clay blocks scraped from the sidewalls when lowering in the steel reinforcement cage. Because of the difficulty associated with thoroughly cleaning the boreholes using normal circulation washing, a reverse circulation washing technique was used in this case. An illustration of the reverse circulation washing technique is given in Figure 3.

The procedure for the construction of the piles was therefore as follows. Firstly, the boreholes were washed for 2 to 3 hours using a normal circulation washing technique. The steel cages and the conductor pipe for the concrete were then lowered into the borehole and the equipment for the reverse circulation process was installed. Thirdly, the borehole was washed for about a further 30 minutes using normal circulation washing to initially dilute the mud fluid and then finally the borehole was pulsed with compressed air at a pressure of between 0.6 MPa and 0.8 MPa at a rate of between 8 m^3/hr and 9 m^3/hr. This reverse circulation reduced the density of the mud fluid in the conductor pipe to about 0.6 t/m^3, which is less than that outside the conductor pipe. As a result, suction was generated in the pipe that forced the mud fluid and sediment at the bottom of borehole to rise up through the pipe and out of the borehole. Typical, 122 m deep boreholes could be fully washed in 2 hours using this reverse circulation technique as opposed to the 10 to 12 hours that would be required using the normal circulation method. Furthermore, the quality of the reverse circulation washing technique is significantly better than that of the normal method. All the pile boreholes on this bridge were washed using the reverse circulation method and all were classified as first quality after inspection.

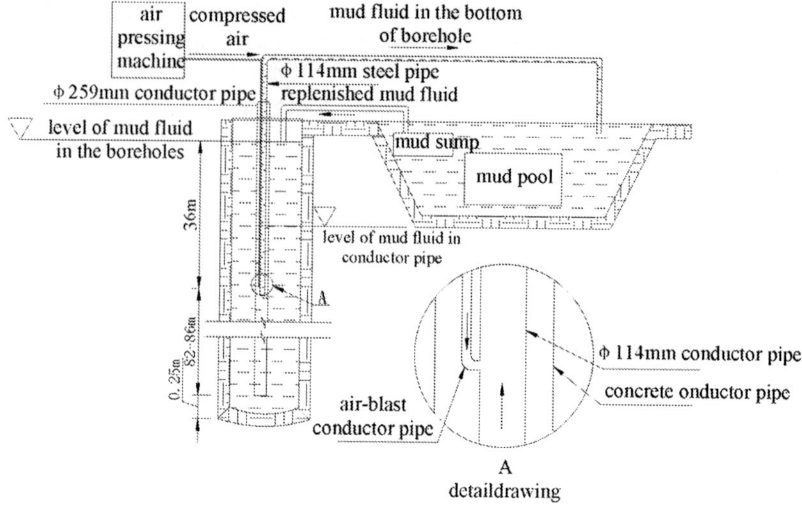

Fig.3 Schematic of reverse circulation washing

Placing of the liquid concrete

Conductor pipe

259 mm diameter steel pipes connected with spiral connecting sleeves were used as the conductor pipes to place the liquid concrete into the boreholes. Because of the volume of concrete that had to be placed in the first batch, if the distance between the lower end of the conductor pipe and the bottom of borehole was too small, difficulties were experienced with the conductor pipe choking with the liquid concrete. To avoid this, the distance between lower end of the conductor pipe and the bottom of the borehole was maintained between 40 cm and 60 cm, which was slightly greater than that required by the specification.

Concrete mix design

In order to increase the plasticity retention percentage (time to initial set) and workability of the concrete, a retarder and fly ash were added to the concrete mix. This both reduced the cost of the concrete and increased its workability.

Placing the liquid concrete

In order to increase the compaction of the concrete, as it was placed the conductor pipe was repeatedly lifted and lowered vertically by between 1 and 2 m by the crane. This had the effect of acting like a large poker vibrator ensuring that the concrete was properly distributed throughout the pile cross-section and that the presence of voids was minimised. Also to ensure that the top 50 cm of the pile was not disrupted by this procedure, redundant cement paste was discharged from the conductor pipe by physically pumping out the nozzle in this zone.

Conclusions

The 98 very long, slender piles required for the Binzhou cable-stayed bridge over Yellow River in China, were constructed to a first category quality in just 98 days by using novel techniques. This paper has described and reviewed these techniques. The bridge has now been open to traffic since 2004 and is still in good condition after three years of service.

References

[1] SUN Xianguo, ZHONG Yuan, AN Changjun. Design and construction of Binzhou Huanghe River Highway Bridge. Bridge building [J].2003.6, 41-43.

Analysis of the Influence of the Main Cable's Manufacturing Error on the Mechanical Properties of a Spatial Self-anchored Cable-net Suspension Bridge

Xia-Guo Ping, PhD student of Civil Engineering, Bridge Science Research Institute, Dalian University of Technology, Dalian, China.
Zhang-Zhe, Professor of Civil Engineering, Bridge Science Research Institute, Dalian University of Technology, Dalian, China.
Mei-Xiu Dao, Assistant engineer, Bridge Science Research Institute, China Zhongtie Major Bridge Engineering Group, Wuhan, China
Ye-Yi, PhD student of Civil Engineering, Bridge Science Research Institute, Dalian University of Technology, Dalian, China.

Abstract: The No.1 Xinghai Bay Bridge in Dalian is a self-anchored cable-net suspension bridge with a spatial net and a multi-transferred cable system, which is the most notable feature that differentiates it from normal self-anchored suspension bridges. The question that is addressed is, as the main component of the bridge, what kind of influence of will the main cable's manufacturing error have on the mechanical properties of the system? To answer this question, and with reference to the results of a specific experimental model, this paper describes the detailed analysis of the cable-net system using a non-linear FE model. The conclusion that is drawn is that the longer the length of cable 1, the larger the stiffness of the structure; while the shorter the length of cable 2, the smaller the stiffness of the structure.

Introduction

The No.1 Xinghai Bay Bridge in Dalian City is a pedestrian bridge located on the east side of the Xinghai Square which crosses the Malan River. It is a self-anchored cable-net suspension bridge with a main span of 134m and a width of 9m. The stiffening girder has a depth of 1.3m. The steel tower consists of two pipe section trusses, each of which has an inclination of 9.08°. In elevation the angle between the horizontal and the tower is 70°. The cable system is a spatial net cable with multi-transferred cables and hangers, as shown in Figure 1.

Bridge design, construction and maintenance 2007, Thomas Telford, London

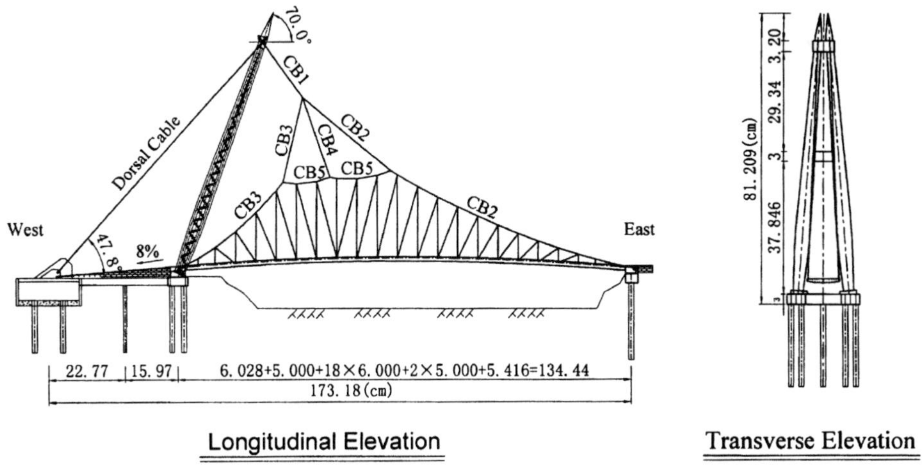

Figure 1   Transverse and longitudinal elevation of the bridge

As a self-anchored suspension bridge, its structural system can be divided into four main components as follows: the stiffening girder, the cable system supporting the stiffening girder, the towers supporting the cable system, and the anchor block supporting the cable system. Similar to other self-anchored suspension bridges, it is expected to exhibit a highly flexible and non-linear behaviour. One of the major concerns is its geometrical non-linear behaviour, as evidenced in the beam-column affect, the effect of large displacements, the cable sag effect and the influence of the initial stress in the system. The difference between this structure and the usual self-anchored suspension bridge is that its cable system is a spatial net cable. The hangers are connected to the main cable (cable 1) indirectly through cables 2 to 5. This type of multi-transferred cable system is the most notable characteristic of the bridge, but it also brings with it many difficulties associated with its design and construction. It is extremely hard to know exactly what its non-linear behaviour will be due to its unique spatial cable-net system. For such a non-linear spatial cable-net system, not only is there a need to conduct a detailed analysis using spatial finite element methods (FEM), but also a profound understanding of the structural behaviour is required, which is obtained by studying the conclusions of an experimental model. The results from these experimental studies are presented in this paper and are used to illustrate the comparative variations of the stiffness of stiffening girder because of variations in the length of the cables. A numerical analysis using the finite element program ANSYS has also been performed to verify the experimental results.

Experimental investigation

The necessity for experimental modelling

Experiments using models have been carried out for many years in a quest to discover and understand the behaviour and properties of bridges. The behaviour and response of large and complex engineering systems can often be predicted within predetermined accuracies from studies on small-scale replicas, the models, which can be produced economically and quickly. The problem can be more conveniently and economically studied with the help of scale models, provided that certain precautions are observed in the design and construction of such models.

There is no need for such models to be exact replicas on a smaller scale of the intended, prototype system. Indeed, it is one of the outstanding advantages of modelling that only those features of the prototype system need be reproduced that are essential in the study of certain selected aspects, which are the objectives of the investigation. It is necessary, however, that these prototype features are reproduced in the model in such a manner that a unique relationship between the prototype and model is provided in every aspect. This is achieved by establishing between the prototype and the model a similarity relationship. However, models may not always behave as might be expected from a theoretical analysis of the system. Even models that are constructed similarly in every aspect except for size, may display small—and occasionally not so small—differences in behaviour from model to model. These differences, referred to as scale effects, may well affect the test results and their interpretation and so, although not discussed further here (see Ref. (3) for this discussion), it is particularly important to be aware of these differences and the principles of model design if these interpretations are to be reliable.

Model design

According to the principle of similarity, a scaling factor of 1:25 was selected. The model consisted of the stiffening girder, the towers, the cable system (cables, hangers, dorsal cables, and rope grips), foundations, bearings and anchor blocks. For a more detailed description of the model the reader is referred to Ref. (1). The main targets of the experiment were to identify the coordinates of each key-point and to determine the forces in the hangers. The instrumentation used included a Total Station and measuring tapes for determining the geometry of the cable, a dial gauge for measuring the deflection of the stiffening girder, and transducers for measuring the forces in the cables. A number of points distributed along the length of the girder and the intersection points of the cables were selected as the measuring points.

Test set-up

As the most important load-carrying members of this bridge, it must be emphasized that the cable loads and lengths are fundamental in determining the ultimate geometry of the structure. However, it is particularly difficulty to choose the geometry of the cable in design and then to manufacture the cable system correctly during construction. Any error in the manufacture of the main cables will have an important influence on the construction of this bridge. The geometry and the force distribution between the suspension cables will change drastically due to any error in the length of the cables. At the same time, the deflection of the total system is to a large extent controlled by the length of the cables. Correspondingly, the forces in the cables and the hangers must be adjusted to ensure that the girder achieves its designed position. Only then can experimental tests be performed to obtain the vertical deflection of the stiffening girder under different load cases. It is these results that will be used to investigate the stiffness of the overall bridge.

Three tests on the model were designed to illustrate the structural behaviour of the bridge when subject to this type of error:
- Test 1, when the length of cable 1 is longer than its designed length;
- Test 2, when the lengths of all cables were equal to their designed length;
- Test 3, when the length of cable 2 is shorter than its designed length.

Observations and results

To study the influence of the manufacturing error on the whole stiffness, an equal uniform load was applied along the entire length of the girder for each of the different test cases. This was achieved in the experiments using steel plates stacked on the deck surface. The variation in stiffness was then derived by measuring the vertical deflections of the stiffening girder having first readjusted the forces in the cables and the hangers to achieve the designed bridge profile. The results of the experiments are displayed in Figures 2 and 3, and the readjusted geometry of the cables is shown in Figure 4.

In Figures 2 and 3 the node number-deflection curves of the stiffening girder are shown. The experimental curves for the correct length cables approximately match the curves obtained for the cable with an error in its length, however, the results for cable 1 are a little smaller than its results without a manufacturing error, and the results for cable 2 are a little larger than its results without a manufacturing error. The readjusted geometry of the cables differs from the design geometry as shown in Figure 4. These results indicate that the cable system is stiffer when the length of cable 1 is longer than designed, and, contrastingly, the cable system is more flexible when the length of cable 2 is shorter than designed.

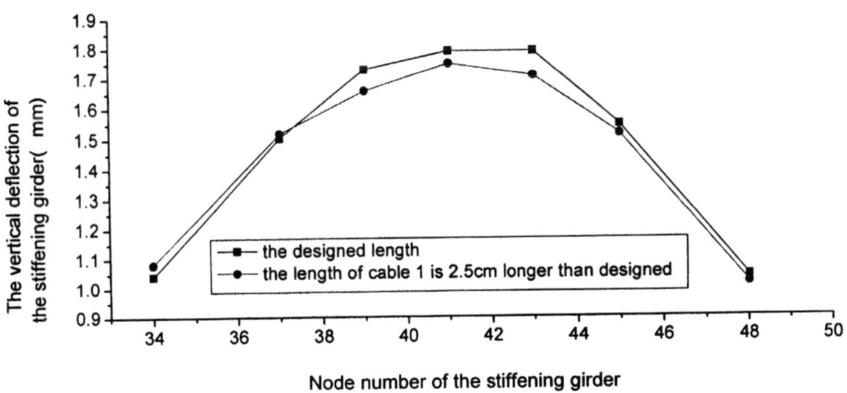

Figure 2   The node number-vertical deflection experimental curves for the stiffening girder when the length of cable 1 is 2.5cm longer than the design length

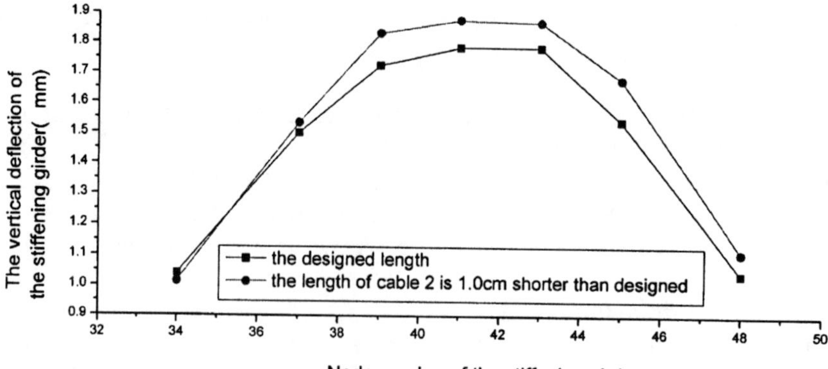

Figure 3   The node number-vertical deflection experimental curves for the stiffening girder when the length of cable 2 is 1.0cm shorter than the design length

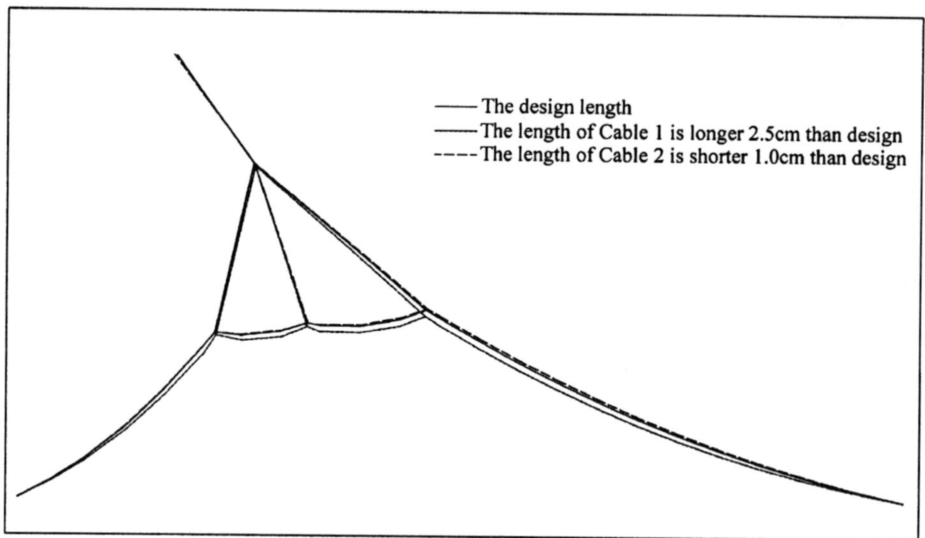

Figure 4   The geometry of the cable for the different test cases

Numerical analysis

For the numerical calculations, a finite element model of the bridge was developed using the commercial finite element program ANSYS, which has been used for the analysis of many structures in recent years. A three-dimensional FEM model was adopted for simulating the behaviour of the bridge. The results for the vertical deflection of the stiffening girder were obtained for the same load cases as used in the experiments, as shown in Figures 5 and 6.

These were:

    (1) Under the action of a uniform load when the length of cable 1 is 2.5cm longer than the designed length, and

    (2) Under the action of a uniform load when the length of cable 2 is 1.0cm shorter than the designed length.

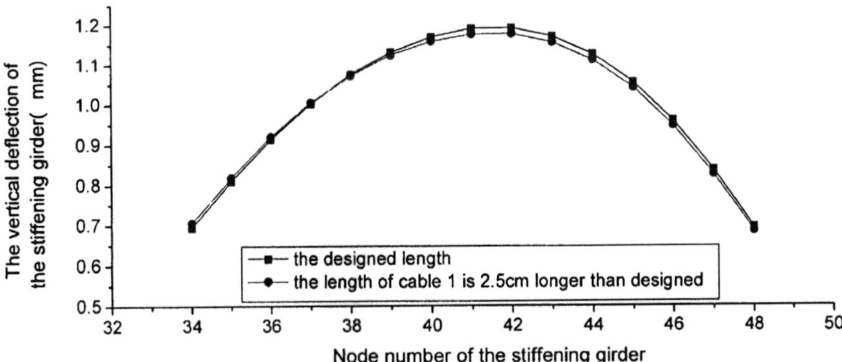

Figure 5   The node number-vertical deflection numerical curves for the stiffening girder when the length of cable 1 is 2.5cm longer than the designed length

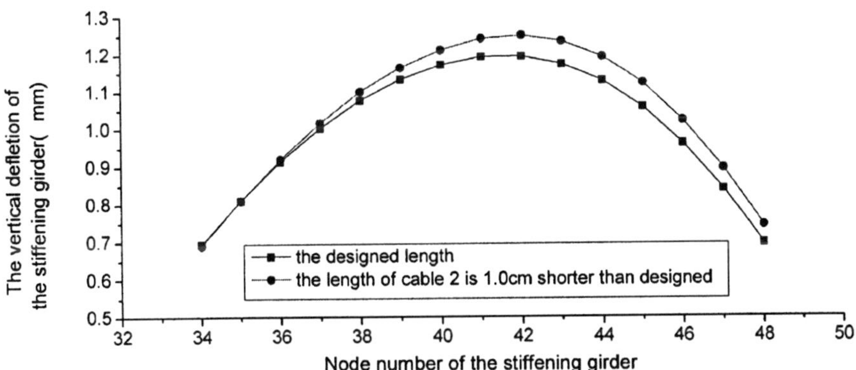

Figure 6   The node number-vertical deflection numerical curves for the stiffening girder when the length of cable 2 is 1.0cm shorter than the designed length

From Figures 5 and 6, the same conclusion as was obtained from the experimental results can be drawn; i.e. the longer the length of cable 1, the greater the stiffness of the structure, while the shorter the length of cable 2, the smaller the stiffness of the structure. However, it can also be seen that there is a considerable difference between the numerical results and the experimental results. This may be caused by the many uncertainties and imperfections in the experimental modelling, such as the material properties, geometry, assembly, and so on. However, it is also very difficult to obtain numerical models that give realistic results, in particular when strong non-linear effects are present, because of the many uncertainties involved when numerically modelling complex structures. Therefore, the conclusion is that these two methods actually give results that are very close and the numerical model can be used to carry out further analyses of the bridge.

Dynamic analysis

In addition to the static analysis, the dynamic behaviour of this bridge was also studied. Because the dynamic behaviour was not the first concern of this work, the fundamental frequencies were obtained only for the case when the length of cable 1 was equal to design length (Case 1) and when the length of cable 1 was 2.5cm longer than the design length (Case 2). Since the test value of the latter (3.15Hz) was less than that of the former (3. 5Hz), it can be concluded that the stiffness of the girder is enhanced by the increase in the length of cable 1, which is in complete agreement with the static analysis.

To understand the dynamic properties more profoundly, an analytical modal analysis was performed using a three-dimensional FEM model to determine the characteristics of the three-dimensional vibration modes shapes and frequencies. Table.1 gives the first ten frequencies and vertical, transverse and torsional mode shapes obtained from the analytical modal analysis (only the global mode shapes are included in the table). It was found that the mode shapes generated by these two cases were nearly the same, and so they are illustrated only once as shown in column 4 of Table.1. The descriptions of the mode shapes given in column 4 summarise the dominant type of vibration. It might be supposed that the dynamic behaviour in these two cases is similar since they both exhibit the same mode shapes, but actually the frequencies are very different confirming the difference in the girder stiffnesses. As can be seen, almost all the frequencies in column 3 when the length of cable 1 is longer than its design length are larger than those in column 2. As such, it can again be concluded that increasing the length of cable 1 results in an increase in the overall stiffness of the structure. This analysis shows not only a good correlation between the dynamic results obtained numerically and those obtained experimentally, but also a good correlation between the static and dynamic results.

Conclusion

In this paper the experimental and numerical behaviour of a self-anchored cable-net suspension bridge has been compared. Good agreement was achieved between the numerical solutions and the experimental results. This confirms that an appropriate numerical analysis during the design stage is essential to predict the ultimate behaviour of such a structure and that such an analysis can be used to investigate the effect of manufacturing errors on this predicted behaviour. The variation in the overall stiffness of the structure as a result of manufacturing errors in the cables was also discussed. It was identified that the longer the length of cable 1, the larger the stiffness of the structure, while the shorter the length of cable 2, the smaller the stiffness of the structure.

Table.1
Vibration modes from FE model (only the global mode shapes are included in this table)

Mode no.	Frequncy (Hz)		Mode shapes	Remark
	Cast 1	Cast 2		
1	3.044	3.150		vetical bending of girder
2	4.147	4.228		vetical bending of girder
3	9.015	9.011		vetical bending of girder
4	12.072	12.075		lateral bending of girder (coupled with the torsional of the girder)
5	15.822	15.881		vetical bending of girder
6	18.274	18.357		bending of tower
7	18.956	18.944		bending of tower
8	24.708	24.742		vetical bending of girder
9	35.497	35.510		vetical bending of girder
10	41.345	41.359		bending of tower

Cast 1, the length of cable 1 is equal to design length
Cast 2, the length of cable 1 is 2.5cm longer than design length

References

Xiu-Dao Mei.   Study on model experiment of the No.1 bridge Xinghai Bay, Dalian City. Dalian University of Technology. 2006.
Niels J. Gimsing.   Cable supported bridges. Technical University of Denmark, Lyngby, Denmark. 1998
F.W. David, H. Nolle. Experimental Modelling in Engineering. 1982.

Theme six:

The future of condition monitoring, assessment and maintenance

Novel testing facility to evaluate the structural reliability on large bridge elements subjected to tension

B. Köberl, Vienna University of Technology, Vienna, Austria
J. Kollegger, Vienna University of Technology, Vienna, Austria

Abstract

Fatigue is a leading cause of failure in mechanical components and structures that are subjected to repeated loads. The testing of structural components e. g. stay cables for stay cable bridges under cyclic loading constitutes one of the most important fields of experimental mechanics. The testing of specimens with 2 to 20 million load cycles is only feasible for small specimens but not for large structural bridge elements. Usually conventional servo-hydraulic testing machines are used so the testing time and energy consumption increase dramatically in regard to specimen size. Here we show a completely new approach to the testing of large structural elements by taking advantage of the resonance effect. This new approach affords the opportunity to carry out fatigue tests on large specimens under conditions close to the loading condition of a structure during its lifetime, e.g. low stress amplitudes and many millions of cycles. In numerous areas of structural and mechanical engineering there is a lack of such data simply because tests on large components are very time and energy consuming and thus, in the majority of cases, too expensive. More and detailed data on the fatigue strength of structural components will make a contribution to safer designs and simultaneously to more economic structures.

Introduction

The term "fatigue" does not seem to fit into precise technical terminology. However, there is a certain analogy between fatigue in the common sense of the word, and fatigue in metals. Under cycling external loading a complete fracture may occur, although no obvious damage can be recognized throughout the majority of the loading cycles. The magnitude of the applied forces may be so small that a single application does not result in any detectable damage at all [1].

In the middle of the nineteenth century failures were observed in railway axels and in bridge components that were subjected to cycling loading. The term "fatigue" was first applied 1839 by Procelet, an English Engineer [2]. The most important work aimed at characterizing fatigue as a material property was done by August Wöhler in the period from 1856 to 1870 [3]. Because of the strong contributions to the field, plots of the stress amplitude (for a given mean stress) versus the cycles to complete failure are called "Wöhler" or S-N curves.

Bridge design, construction and maintenance 2007, Thomas Telford, London

Fatigue is generally subdivided into two categories:

- Low cycle fatigue
- High cycle fatigue

In Low cycle fatigue the stress level can exceed the yield strength, with the result that the number of cycles N_f to failure is relatively low ($<10^3$). High cycle fatigue can occur when the stress range is lower than the yield strength, and failure may require 10^3 to 10^6 cycles. For some material, e.g. steel, there is a distinct stress limit σ_{fat} below which the specimen or component will have essentially infinite fatigue life. This is known as the fatigue limit or endurance limit and thus becomes an important stress level for the design preventing fatigue failure [4]. In Figure 1 a schematic drawing of a Wöhler curve can be seen. New research shows that fatigue failure for steel may occur also after this limit. C. Bathias and P. C. Paris give in [5] a review of this new research and show that fatigue failure in the giga cycle range is possible, for steel too.

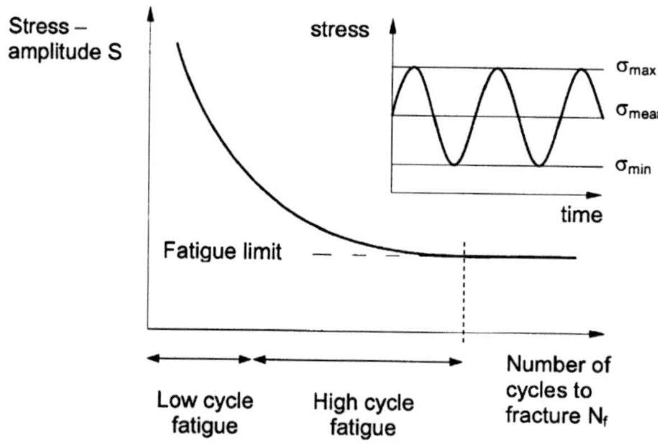

Figure 1: Schematic drawing of a Wöhler curve

Components in service are generally subjected to variable amplitude loading rather than to a constant amplitude loading used to generate S/N curves. The Palmgren Miner –Law allows a approximate prediction of the fatigue life time using a linear damage summation rule [6]. Using S/N- curves together with the Palmgren Miner law is still the most important tool for engineers to prevent fatigue failure. An also very common approach, mostly used in mechanical engineering and aeronautic engineering, to prevent fatigue failure is the Paris law. The Paris law combines fatigue with fracture mechanics and thus leads to a more realistic approach to describe fatigue failure.

Fatigue tests on large structural elements subjected to tension

In the course of the design process for real structures subjected to cycling loading the fatigue behaviour of the construction material as well as for the construction details, e.g. a welding joint, has to be considered. But there are cases where the behaviour of structural components under fatigue loading during the service life of the structures has to be considered, too. For example a cable stayed bridge, where the fatigue resistance of the stay cables is checked experimentally. Current design philosophies are based on fatigue tests under fixed amplitudes of cyclic stress [7][8]. The photograph of an anchorage for a stay cable anchorage for 37 strands is displayed in Figure 2 (right). To offer a better view of the fixing of the individual strands with wedges, a quarter of the anchorage was removed. Inside the anchorage the cable is fixed with wedges in the anchor body resulting in complicated three-dimensional stress states.

During the design process of a cable stayed bridge, tests with 2 million load cycles are usually carried out at an upper stress limit of 45 % of the guaranteed breaking strength of the strands and with a stress amplitude of 200 N/mm^2. The test specimen consists of the cable and two anchorages. The stress range of 200 N/mm^2 and the required 2 million load cycles were chosen in testing guidelines based on practical reasons. The straight stay cable specimen is shortened in comparison to the actual length in the bridge to about 3 to 5 m for practical purposes.

Conventionally the fatigue tests are carried out by means of servo hydraulic controlled jacks; a schematic setup for such a facility can be seen in Figure 2 on the left. In a conventional servo-hydraulic fatigue test, oil is pumped in and out of the hydraulic jack in such a way that the cable is subjected to a sinusoidal stress change. A test with a typical test frequency of one cycle per second (1 Hertz) takes 23 days and requires a large amount of energy for the hydraulic pump which operates with 200 to 400 litres per minute during the servo-hydraulic test operation. In fact this test method is rather cost-intensive for contractors and laboratories Fatigue tests, with more than 2 million load cycles, using the common servo hydraulic facilities are usually too expensive for these large components.

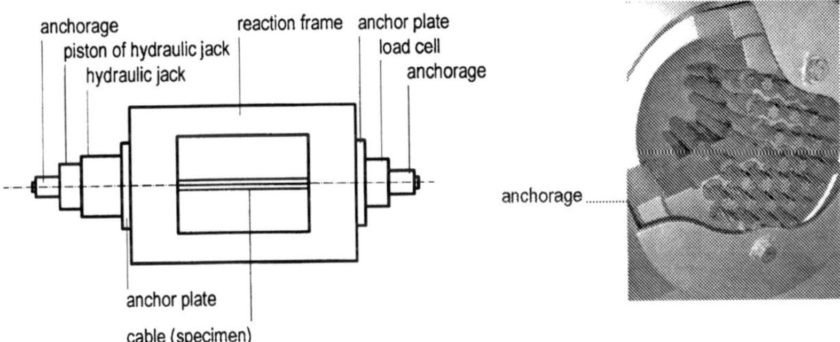

Figure 2: Schematic conventional testing setup (left), Anchorage for a stay cable with 37 strands (right)

Development of a novel fatigue testing facility

By taking advantage of the resonance effect a new technology for high cycle fatigue tests has been developed at the Institute for Structural Engineering at Vienna University of Technology. Testing frequencies from 20 to 50 Hz for large structural elements will be possible in the future. So this new approach affords the opportunity to carry out fatigue tests on large specimens, like stay cables or tendons for reinforced concrete under conditions close to the loading condition of a real structure during its lifetime - low stress amplitudes and many millions of load cycles [9].

A dynamic fatigue test for 2 million load cycles can be carried out, depending on the testing frequency in 27,7 hours to 11,1 hours. The energy requirement is just a fractional amount compared to conventional testing setups, because operating a hydraulic aggregate is no longer necessary. A survey of the Austrian patent office approved the capability of the development. The Vienna University of Technology applied for a patent in April 2005, which was granted in August 2006 [10].

Effectiveness of the new approach

In Figure 3 the schematic testing setup and the resulting model of a SDOF-System is shown. The specimen (e. g. a stay cable or tendon) and an auxiliary cable (also a stay cable or tendon) are tensioned by means of a static hydraulic jack. This static jack applies the mean force for the following fatigue test. After tensioning the specimen and the auxiliary cable the piston of the hydraulic jack is locked by a ring nut and the hydraulic pressure can be released. An unbalanced vibration generator, which is placed at a coupling unit, applies a sinusoidal load in axial direction of the cables. On the one hand the coupling unit is used to connect the auxiliary cable and the specimen and to carry the vibration generator and on the other hand it delivers the main part of the necessary mass.

If the testing frame is assumed to be ridged and the mass of the specimen and the auxiliary cable is neglected the eigenvalue (ω) and the natural frequency (f) of the system can be determined easily by using the following equations for a damped SDOF-System [11].

$$\omega = \sqrt{\frac{K}{M}} \cdot \sqrt{(1 - \xi^2)} \tag{1}$$

$$f = \frac{\omega}{2 \cdot \pi} \tag{2}$$

$$K = K_{Auxil.} + K_{Specimen} \tag{3}$$

The eigenvalue and the natural frequency respectively are depending on the stiffness of the specimen and the auxiliary cable as well as on the mass of the coupling unit. Variation of the stiffness of the auxiliary cable and the mass of the coupling unit can adjust the eigenvalue of the testing setup. If the rational frequency of the unbalanced vibration generator, which can be adjusted by a frequency converter, is equal to the natural frequency f of the setup the applied load is multiplied due to resonance. The dynamic magnification factor (V_{Dyn}) depends on the damping (ξ) of the testing setup. In order to achieve a high dynamic magnification factor damping has to be low. For example, if ξ is 0,05 % the dynamic magnification factor is equal to 100. In case of resonance the vibration generator has to apply only 1 kN to obtain a force of 100 kN in the specimen.

$$V_{Dyn} = \frac{1}{2 \cdot \xi \cdot (1 - \xi^2)} \tag{4}$$

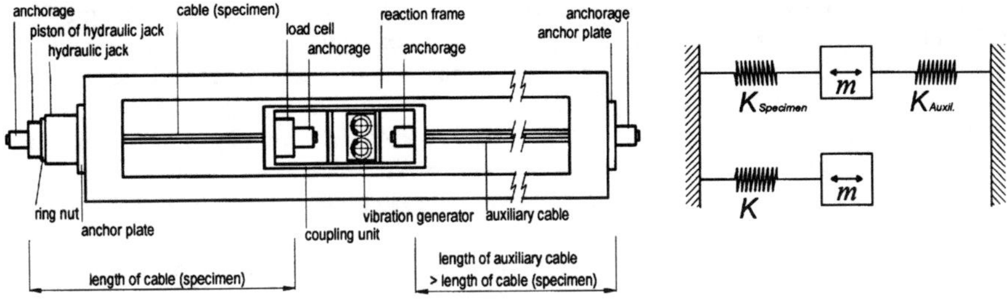

Figure 3: Schematic sketch of the novel testing setup

Preliminary tests

Preliminary tests, carried out at the Institute for Structural Engineering at Vienna University of Technology, demonstrated that the testing method can be used in practice. A vertical and a horizontal setup, similar to the sketch in Figure 3, were realized. For the specimen and the auxiliary cable tension rods made of steel with a diameter from 6 to 12 mm were used. The static loading was applied with a hydraulic jack and the cycling loading by an unbalanced vibration generator. The measurement equipment consisted of a load cell to measure the force in the specimen, strain gauges at the rods to measure the strain of the rods and three acceleration sensors, applied at the coupling unit, to measure the acceleration and therewith the displacement of the coupling unit.

Attaching additional mass at the coupling unit and different auxiliary cables made it possible to change the natural frequency of the testing arrangement. The rotational frequency of the unbalanced vibration generator was controlled by a frequency converter, thus each eigenvalue could be adjusted. Results of the test program can be found in the master thesis of Maier carried out at the Institute for Structural Engineering [12].

Construction of a testing facility with a load capacity of 20.000kN

Due to the promising results of the preliminary tests and the support of Vienna University of Technology a novel testing facility for large specimens was built in the laboratory of the Institute for Structural Engineering from January to May 2006. The testing unit is dimensioned for a static tensile load up to 20.000 kN, an upper load for fatigue tests up to 20.000 kN and a vibration range up to 2.500 kN. Due to practical reasons a horizontal set-up was chosen.

Design and numerical simulations of the testing setup

The u-shaped reaction frame is 16 m long, 2.8 m wide and 2 m high with a total weight of 150.00 kg. The reaction frame is made of reinforced high strength concrete and is post-tensioned in three directions in order to remain free of cracks. This is very important to achieve low damping and therewith a high dynamic magnification factor.

The laboratory of the Institute is located in downtown Vienna, so it was very important to avoid any negative effects to surrounding buildings. In order to achieve vibration isolation the reaction frame is mounted on spring bearings in the basement. To compute the reaction forces as well as the expected eigenfrequencies and the influence of damping during a fatigue test, dynamic Finite Element simulations were carried out. To verify the results analytical simulations on a two degree of freedom system were also carried out. In Figure 4 the finite element model and the model for the analytical simulation can be seen. The arrangement is for a fatigue test on a specimen with 55 strands. The stiffness of the bearing is k_1=26,2 kN/mm, k_2=459 kN/mm is the overall stiffness of the specimen and the auxiliary cable, m_1 is the mass of the reaction frame, m_2 is the mass of the coupling unit. F_A is the transmitted force in the foundation and r is the displacement of the coupling unit due to harmonic excitation. Both methods lead to almost the same results, the difference of the natural frequencies was quite small (numerical: f_1=2,06Hz, f_2=35,9Hz; analytical: f_1=1,95Hz, f_2=35,3Hz). The chosen stiffness of the spring bearing for the frame made it possible that only 0.3 percent of the free mass force F, which occurred due to the movement of the coupling unit, were transmitted into the ground. Thus the dynamic forces for the surrounding buildings are small enough and no negative effects should be expected.

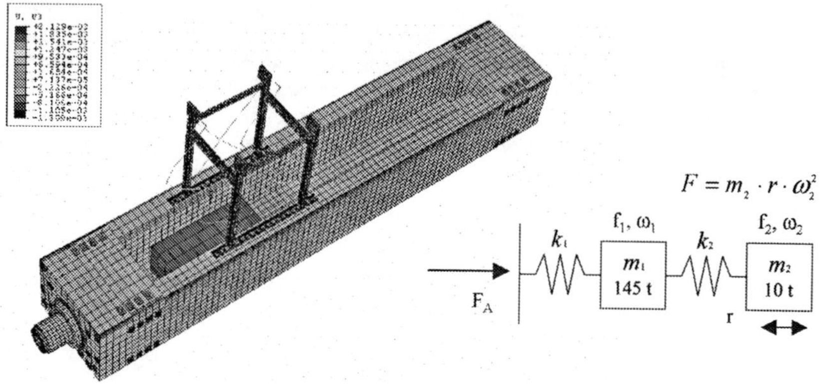

Figure 4: Finite Element and analytical model of the testing setup

Construction of the reaction frame and of the first testing setup

Due to practical reasons the reaction frame was built on a framework on the ground floor of the laboratory and was then lowered down into the basement. On the left side of Figure 5 a view from below the testing frame and the spring bearings for the frame can be seen. This picture was taken before the lowering process started. In Figure 5 (right) the reaction frame can be seen during the lowering process.

Later on a coupling unit, to connect both cables and to carry the vibration generator and a steel frame, to support the coupling unit, were provided. Both elements were made of steel. In November 2006 first tests were carried out with a specimen consisting of 9 strands and an auxiliary cable consisting of 12 strands (ST 1570/1770). According to ETAG 013 the specimen was tensioned with 65 % of the guaranteed tensile stress (1.553 kN). Tensioning the specimen and the auxiliary cable was done by a hydraulic jack with a capacity of 20.000 kN. After achieving the mean force of 1.553 kN the jack was locked by means of a ring nut. The following pictures in Figure 6 show the testing frame including the steel frame to support the coupling unit on the left side and on the right side the hydraulic jack including the anchor plates.

Figure 5: View from below the reaction frame (left), reaction frame during lowering on the right

Figure 6: Testing setup for the first experiments (left), hydraulic jack for tensioning the specimen – 20.000 kN load capacity (right)

RESULTS OF THE FIRST EXPERIMENTS

First experiments with the test set-up on a cable specimen with 9 strands and an auxiliary cable consisting of 12 strands can be seen in Figure 6. These tests were carried out in November and December 2006. The displacement of the coupling unit was measured by acceleration sensors as well as by displacement transducers. The force in the specimen was measured by a load cell.

The first resonant frequency of the system was determined by sweeping the frequency range of the vibration generator from 14 Hertz to 30 Hertz. The frequency response curve (Fig. 7 top left) is obtained by plotting the vibration response at each frequency step. The maximum forces in the stay cable specimen during each cycle are presented in the bottom of Fig. 7 showing a time period of 800 seconds. During the test procedure a steady state vibration response could be achieved after about 300 seconds. After turning off the vibration generator it took more than 225 seconds for the acceleration to cease, an occurrence that can be attributed to the low damping within the system. The axial force in the stay cable specimen based on the recorded measurements of the load cell and calculated from the measured acceleration and displacements are compared in Fig. 7 (top right). The small deviation of the negative force calculated from the measured displacement in comparison to the two other measurements can be explained by the fact that the tip of the displacement transducer briefly lost contact with the coupling unit at the point of highest acceleration in a cycle. The cyclic force in the specimen being tested was equal to 84.5 kN, which is 1987 times larger than the force exerted by the vibration generator (0.0425 kN at 20.5 Hertz).

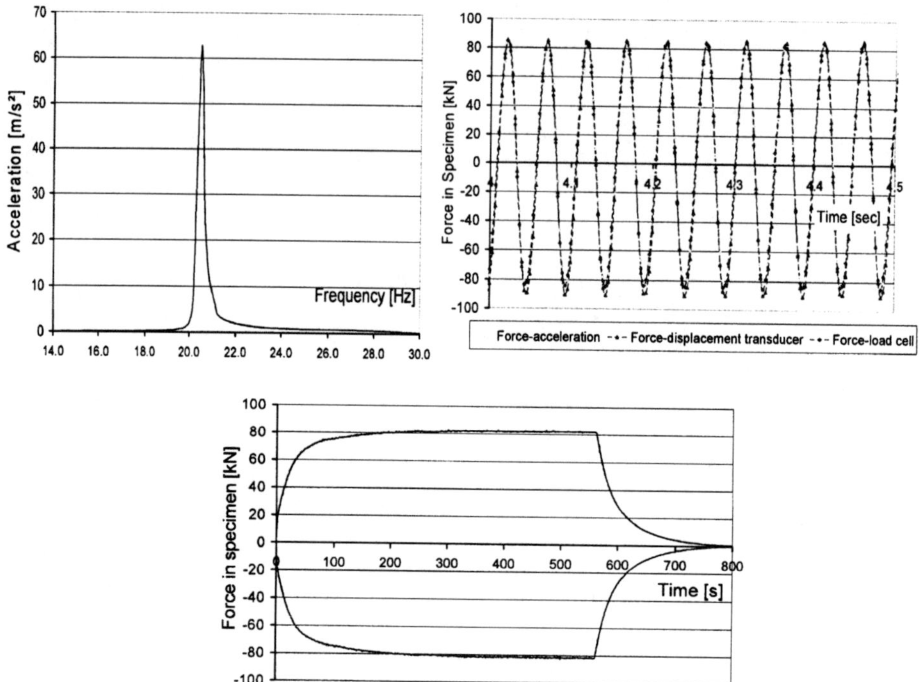

Figure 7: Frequency sweep showing the first eigenfrequency at 20.5 Hertz (top left), Cyclic force in a stay cable specimen during 10 load cycles (top right), Maximum forces in the stay cable specimen during a test lasting 800 seconds

Conclusions - Outlook

Depending on the specimen, the testing of large components can be carried out 20 to 100 times faster than with the conventional servo-hydraulic testing procedure. At the same time, energy requirements drop to a fraction (1/1000 or less) of the energy used in a conventional test. This new approach offers the possibility to carry out dynamic fatigue tests more economically in the future. Furthermore it affords the opportunity to carry out fatigue tests on large specimens under conditions close to the loading condition of a structure during its lifetime, e.g. low stress amplitudes and many millions of cycles. In numerous areas of structural and mechanical engineering there is a lack of such data simply because tests on large components are very time and energy consuming and thus, in the majority of cases, too expensive.

The principle of applying a static force by means of an auxiliary cable and testing at resonance is also applicable in loading situations other than the one described above, e.g. a beam subjected to a bending moment and tested for a cyclic bending moment.

References

[1] M. Klesnil, P. Lukaš; *Fatigue of Metallic Materials*; ELSEVIER; New York 1992

[2] A. J. McEvily; *Metal Failures – Mechanisms, Analysis, Prevention*; John Wiley & Sons, Inc.; New York 2002

[3]A. Wöhler; *Über die Festigkeitsversuche mit Eisen und Stahl*; *Zeitschrift für Bauwesen* **20**, 73-106; 1870

[4] A. Bedford, M. Liechti; *Mechanics of Materials*; Prentice Hall Inc.; New Jersey 2000

[5] C. Bathias, P. C. Paris; *Gigacycle Fatigue in Mechanical Practice*; Marcel Dekker; New Nork 2005

[6] A. Palmgren; *Die Lebensdauer von Kugellagern*; Zeitschrift des Vereins Deutscher Ingenieure 58, 339-341; 1924

[7] Fêdêration Internationale du Bêton (fib); *Acceptance of stay cable systems using prestressing steels*, fib 1ˢᵗ edition; 2005

[8] Post Tensioning Institute (pti) Publication; *Recommendations for Stay Cable Design, Testing and Installation*, pti 4ᵗʰ edition; 2001

[9]] S. Suresh; *Fatigue of Materials - Cambridge Solid State Science Series* (ed. Clarke, D. R., Suresh, S., Ward, M. I.); Cambridge University Press; Cambridge 1992

[10] J. Kollegger, B. Köberl, H. Pardatscher, M. Vill; *Verfahren zur Durchführung von Dauerschwingversuchen an einem Prüfkörper sowie eine Vorrichtung zur Durchführung des Verfahrens*; Austrian Patent AT 501 168 B1; 2006

[11] R. W. Clough, J. Penzien; *Dynamic of Structures*, McGraw-Hill, Inc.; 1993

[12] Ch. Maier; *Entwicklung einer Prüfvorrichtung für Dauerschwingversuche mit einer Oberlast bis zu 20.000kN*; Master Thesis – Vienna University of Technology; Vienna 2005

Improved safe-life prediction of existing bridge structures

Professor J. Menčík, Ing CSc, University of Pardubice, Czech Republic
Ing L. Beran, PhD, University of Pardubice, Pardubice, Czech Republic
Ing B. Culek, PhD, University of Pardubice, Pardubice, Czech Republic

Abstract

Due to high costs of new bridges, ways have been sought for more accurate assessment of the safety and residual life of existing bridges so that they can be operated longer, but without unacceptable risk. The paper informs about several prospective methods for reliability assessment of existing railway (steel) bridges: 1) evaluation of safety against failure due to overloading, based on a new definition of traffic load rating factor, 2) determination of loads and stresses using railway information systems and computer simulation, 3) estimation of fatigue life using Weibull curve and traffic monitoring, 4) fully probabilistic estimation of remaining life using three-parameter rain-flow analysis and Monte Carlo simulation, 5) estimate of safe life of a bridge structure with cracks using fracture mechanics approach, and 6) better evaluation of inspection results using fuzzy logic and other advanced methods.

Introduction

Bridges are usually in operation several decades. Due to corrosion, fatigue and other damaging processes, their condition gradually gets worse and the safety decreases. High costs of reconstructions necessitate keeping them in operation as long as possible. Therefore, ways have been sought for more accurate assessment of the actual safety and probable residual life of existing bridges, in order that they could be operated as long as possible, but without unacceptable risk. Today, several efficient tools can be used for this purpose: better methods of structural analysis, better knowledge about fatigue and fracture processes, new approaches to reliability and safety assessment based on probabilistic and simulation methods possible due to computer technology. For existing bridges, more accurate information can also be obtained about the actual state and about operation conditions, especially loading.

This paper informs about new methods developed at the Jan Perner Transport Faculty of the University of Pardubice in various research projects (e.g. Menčík et al., 2003, 2005), namely:
- evaluation of safety of railway bridges against failure due to overloading,
- determination of loads and stresses in bridge structures using railway information systems and computer simulation,
- estimation of remaining fatigue life using traffic monitoring, Woehler curve and damage accumulation law,
- fully probabilistic estimation of remaining fatigue life using three-parameter rain-flow analysis and Monte Carlo simulation,
- estimate of safe life of a bridge structure with cracks, using fracture mechanics approach,
- better evaluation of inspection results using fuzzy logic and other advanced methods.

Bridge design, construction and maintenance 2007, Thomas Telford, London

Evaluation of safety against failure due to overloading

The safety of a construction against overloading can be characterised by the factor of safety:

$$Z = R / S > 1 \quad ; \tag{1}$$

R is the resistance of the structure, and S is the effect of all loads. The codes for dimensioning of steel structures and for assessment of their safety use the LRFD approach, which calculates the design values of R and S as products of their characteristic values and the load and resistance factors. In bridges, the total load effect S is caused by the dead load, long lasting load and live load. However, only the live load due to vehicles varies and can be the cause for bridge overloading. Therefore, Šertler et al. (1994) proposed to transform the relation (1) by subtracting the effects of all loads except the traffic load, (ΣS_i), from the resistance R of the structure, and to work with so-called Standard Traffic Load Rating Factor, defined as the ratio of the net load carrying capacity and the load caused by the standard train UIC-71:

$$Z_{UIC} = (R - \Sigma S_i) / S_{UIC} \quad , \tag{2}$$

This factor characterises the safety against overloading failure by this train. This approach has been implemented into the codes for steel bridges of Czech Railways. It can be used for a structure as a whole or for its components, where it can also be expressed in terms of stresses.

Better information about actual safety of a particular bridge or its part can be obtained by relating its actual load-carrying capacity ($R - \Sigma S_i$) to the loads caused by the actual train, instead of the standard UIC train (Beran & Šertler, 2001). Such information is very important, e.g., when planning a route for an extraordinarily heavy train. For example, the actual factor of safety for the most loaded component, expressed by means of stresses, is:

$$Z_P = \left(f_y \varphi - \sigma_g \right) / \left(\sigma_P \delta \right) \quad , \tag{3}$$

where f_y is its yield stress (measured in the investigated part), φ is the value characterising the scatter of geometric parameters of the cross section, σ_g is the stress caused in the component by the own weight of the structure and the railway bed (obtained, e.g., by the finite element analysis), σ_p is the stress caused by the passage of the assumed heaviest train, and δ is the dynamic coefficient. The reciprocal value of the factor of safety, $1/Z_p$, expresses the degree of the utilisation of the load carrying capacity of the component. The histogram (or distribution) of Z_p or $1/Z_p$, constructed for the trains going over the bridge during some period, can characterise the degree of utilisation of its load-carrying capacity (Beran, 2002, 2004).

Determination of loads and stresses in bridges using traffic information systems and computer simulation

The information about actual loads and stresses in important parts of a steel bridge can be obtained by measurement or by computer simulation utilising the railway (or traffic) information systems. Direct measurement, which uses strain gauges fixed to the structure, can yield relatively accurate values. However, problems exist with long term monitoring lasting months or years: the sensors must have very high reliability and long life, and must also be protected sufficiently against weather and mechanical damage, including intentional one.

An alternative approach is based on computer simulation. The stresses in various parts of a structure can be calculated with sufficient accuracy using the finite element method, provided the loads are well known. Today, basic information about loading of a railway bridge can also be obtained from rail information systems. Railway companies store the data about all trains and their movement in the railway network. These data can be used to extract the relevant information about the individual trains going over a particular bridge: the types of the locomotives and the passenger and freight cars, and the weights of transported goods. After supplementing them with the weights of individual cars and their lengths and axle distances, it is possible to create the virtual load scheme for the pertinent train. Similarly, monitoring of movement of trucks (done, e.g., for toll purposes) could be used for the simulation-based stress analysis in road bridges.

A special procedure for the simulation-based approach was developed by one author of this paper (L.B.), consisting of the following steps. First, the finite element model of the bridge is created. Then, the influence line for internal forces and stresses at the investigated point is created by static analysis. Finally, the train passage is simulated by moving the load model along this influence line. For this purpose, a computer program was created, able to calculate the time course of stresses and to select the characteristic stress values specified according to the purpose of the analysis, such as the stress ranges for rain-flow sorting in fatigue analysis, or maxima values classification for the assessment of safety against overloading.

The proposed method was verified by comparing the calculated stresses with those measured by strain gauges. The comparison was done for two steel railway bridges: a truss bridge and a plate-girder bridge, both over the Elbe river. The strain analysis was performed using a finite element code IDA Nexis, with truss and plate elements; the train load data were taken from the information system CEVIS of Czech Railways. The stresses were measured using strain gauges glued at various points of each structure (main girders, cross beams and stingers). The data were processed using a dynamic amplifier system. The traffic was monitored for 24 hours, with about 150 train passages over each bridge. Fig. 1 shows a part of the FEM model of a truss bridge, and Fig. 2 shows the measured and calculated time course of stresses. One can see very good agreement between the measurement and model. The agreement was good also in other tested cases. For the common train velocities, the quasistatic model was sufficient, though dynamic effects could also be considered. More details can be found in Beran (2002), and especially in Beran (2004).

The results are promising and indicate that this method could be used for the evaluation of bridge safety and for the assessment of fatigue damage accumulation and of the remaining lifetime, as it will be shown in the next section.

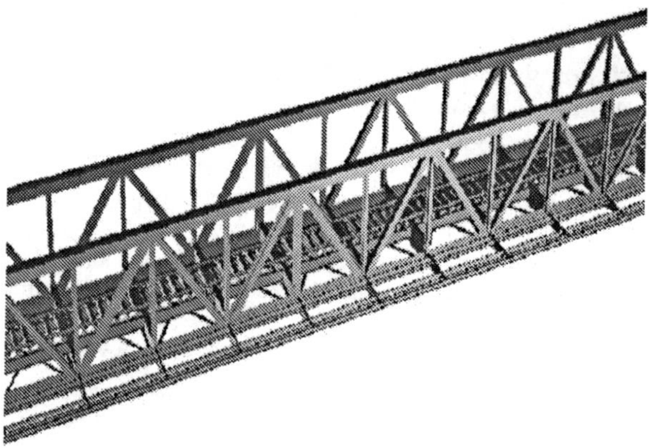

Figure 1. A part of the finite element model of a bridge (Beran, 2004).

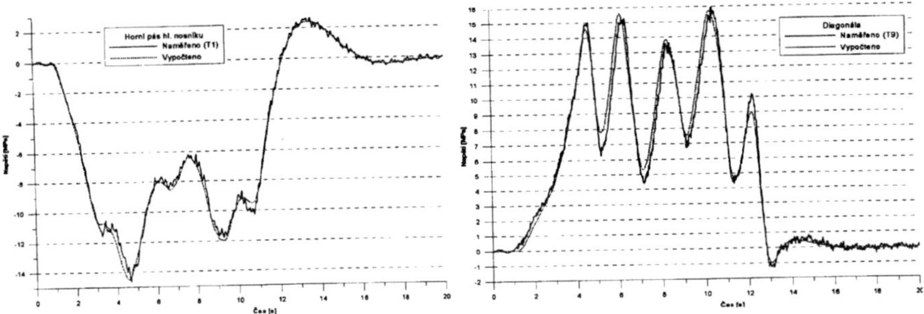

Figure 2. Time course of stresses caused by a passenger train – measured and calculated (Beran, 2004). Left graph – main girder, right graph – diagonal truss.

Residual fatigue life prediction based on Woehler curve

Woehler curve relates the stress range $\Delta\sigma$ and the number of cycles to failure N_f as

$$N_f = A\,\Delta\sigma^{-m} \ , \tag{4}$$

with constants A, m. The loading of bridges is random, and the concept of damage accumulation must be used. The most common, Palmgren-Miner linear rule, defines the relative damage D as

$$D = \Sigma(N_i/N_{fi}) \ , \tag{5}$$

where N_i = number of loading cycles for i-th stress range, and N_{fi} = number of cycles to failure for this stress range; Σ means summation for all $i = 1$ to n stress ranges. (Also other expressions for the fatigue curve and damage accumulation can be used.) The failure is expected if $D = 1$. Expressing N_{fi} from the Woehler curve, Eq. (5) can be rewritten as

$$D = \Sigma(N_i\,\Delta\sigma_i^{m})/A \ . \tag{6}$$

The fatigue effects of randomly varying stresses can be evaluated by sorting the stress ranges $\Delta\sigma$ between individual reversals and counting them using the rain-flow method. As also the mean value of stress, σ_m, plays a role, usually two-parameter rain-flow method is used, with $\Delta\sigma_i$ and σ_{mi} as parameters. The mean stresses consist of the mean stresses of individual train passages and the stresses caused by the dead weight of the bridge (obtained, e.g., by the FEM analysis). Equation (6) corresponds to symmetrical loading cycles. The effect of nonzero mean stresses is usually considered by replacing $\Delta\sigma$ by the so-called equivalent (symmetrical) stress range, calculated from both $\Delta\sigma_i$ and σ_{mi}.

For loading, consisting of various loading blocks (corresponding to individual train passages), the remaining life can be expressed as a fraction of the total possible life:

$$D_r = 1 - \Sigma D_j \ , \tag{7}$$

where ΣD_j (for $j = 1 \ldots z$ loading blocks) corresponds to the loading up to this instant. $D_r = 0$ corresponds to the total exhaustion of the residual life. For safety reasons, the bridge must be taken out of operation and repaired or replaced well before D_r drops to zero, for example for $D_r = D_{r,a} = 0.2$ (here, subscript a means alert).

If the damage in the individual loading blocks has random character, it is only possible to determine the failure probability, corresponding to some time t of operation, i.e. the probability of $D_r(t) \leq 0$. The number of cycles to failure is also a random quantity, even under constant stress amplitude, due to the scatter of material properties, represented by the „constants" A and m in the Woehler curve (4). When the damage is calculated in semiprobabilistic manner, a suitable quantile of A is usually worked with, e.g. 5%, while the exponent m is assumed constant. Fully probabilistic approach and simulation technique Monte Carlo allows one to respect the random variability of both A and m (Marek et al., 1996).

The remaining life of a railway bridge can be expressed by the number of passing trains. For this purpose, the best unit of damage is that caused by one train, D_1, calculated from (6) for the stress ranges and numbers of loading cycles corresponding to the pertinent train. The remaining number of passages till the alert state is then obtained as

$$N_r = D_{r,a} / D_1 \quad .$$

$$(8)$$

Various trains cause different damage. In the simplest approach, the damages by individual trains, calculated from the data obtained by bridge monitoring for 24 hours or more, can be used to construct a histogram of damages D_j. Inserting the average damage for D_1 into Eq. (8) gives the average number of trains till the limit state. The upper confidence limit $D_{1,U}$ gives the lower (conservative) confidence limit of the train number, $N_{r,L}$. More information can be obtained using the Monte Carlo simulation, based on the probability distribution of D_j and D_r. In this way, a histogram of N_r can be created, allowing one to calculate the number of trains corresponding to various probabilities of survival or vice versa.

Prediction of residual fatigue life by fully probabilistic simulation method

This method, developed by one author (B.C.) of this paper, uses the Monte Carlo technique and consists of the principal steps:
− obtaining the load response of the structure (and the fatigue curve of the material),
− processing of the measured load (and stress) data,
− creation of a new record of stresses by Monte Carlo simulation,
− determination of partial (cumulative) damages,
− calculation of the remaining lifetime.

The input data on loading are obtained by monitoring-measurement on the actual construction. However, their processing is based on three-parameter rain flow method: in addition to stress ranges and mean values, the individual loading cycles are sorted also with respect to their time and duration. This makes possible to consider the time order of various loading cycles in the Monte Carlo simulations, or even to modify it if more variability or changes in the load pattern should be considered.

The analytical part of the procedure begins with the standard two-parameter rain-flow (RF) decomposition of measured data, resulting in 2-parameter RF matrix. Simultaneously, the times of individual stress amplitudes are recorded, which represent the additional parameter in the resultant three-parameter rain-flow matrix. Then, the simulation part of the procedure follows, consisting of the following steps:
− generation of a new RF matrix,
− generation of new stresses, asigning them the corresponding times and ordering them,
− modification of this record in order to remove the errors caused by the simulation.

This makes possible the generation of stresses with a random time course. A special computer program was developed for this purpose (Culek, 2002a,b). The simulated record is not identical with the original record obtained by strain gauge measurement, but corresponds to it fully in the stochastic sense. The very good agreement between the measured and simulated data is illustrated in Fig. 3.

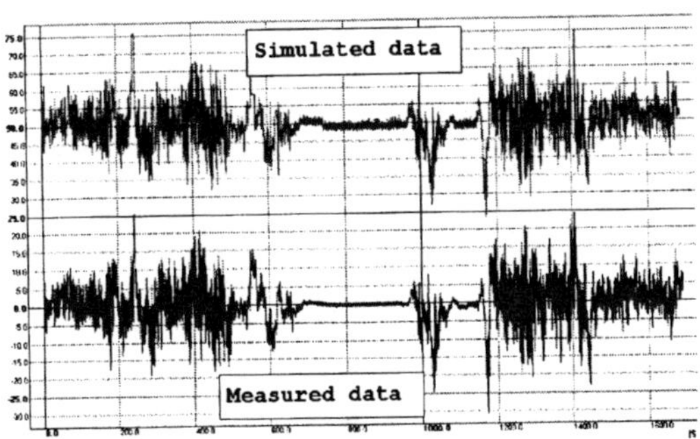

Figure 3. Stresses in the main girder: measured and simulated course. Horizontal axis – time, vertical axis – stress (Culek, 2002a,b).

The cumulative damage is calculated using a suitable law, e.g. Palmgren-Miner. However, also the fatigue curve is generated (Monte Carlo) in each simulation step, in order to consider the scatter of data in the original fatigue tests. Figure 4 shows a histogram of partial damages of a bridge component. Then, it is possible to calculate the statistical characteristics of the fatigue process and to model the development of failure probability with time.

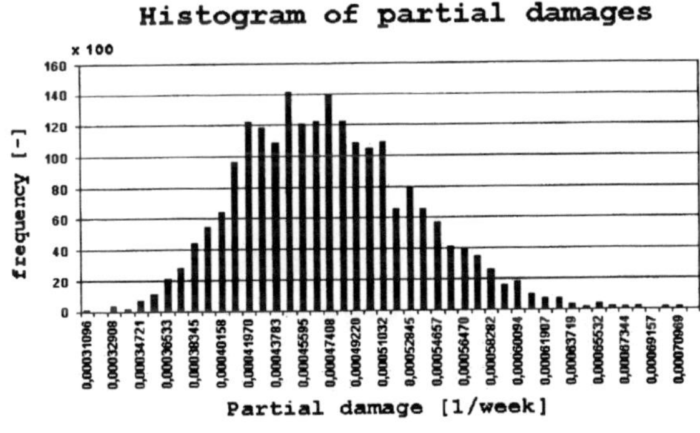

Figure 4. Distribution of partial damages in a component (Culek, 2002 a,b)

The method allows one to modify the original time course of load response, in order to respect the changes in the load pattern during longer time compared to that used for stress measurement. Moreover, it also enables the consideration of other long-term factors, such as the temperature variations during the year, or the change of properties of the structure due to aging. This is made possible by so-called influence factors, based on experimental data. More details can be found in Culek (2002a, b) and Culek & Culek (2005).

The method was tested on several steel constructions (bridges as well as bogies of railway cars), and the results show reasonable agreement between the predicted and actually observed times to failure. As short-term monitoring cannot sufficiently reflect all changes during the very long life of a bridge, e.g. introduction of new types of trains or changes in the volumes of transported goods, the repetition of monitoring and recalibration from time to time is necessary. Very promissive seems to be the method described in the previous section, based on the permanent data-taking from railway information systems.

Simple safe-life assessment of a bridge structure with cracks

If cracks were found in load carrying parts, the question is how long the bridge may be allowed in operation. This can be estimated using fracture mechanics. The base is the Paris-Erdogan law for subcritical crack growth,

$$v = da/dN = B \, \Delta K_I{}^m \quad ; \qquad (9)$$

da/dN is increment of crack length a per loading cycle, ΔK_I is the range of stress intensity factor in a loading cycle, and B, m are constants for given material and environment. Expressing ΔK_I as

$$\Delta K_I = \Delta\sigma \, Y(a) \, a^{1/2} \quad , \qquad (10)$$

where $\Delta\sigma$ is the stress range in a loading cycle and $Y(a)$ is a factor characterising the crack shape and size and the stress distribution, one can separate the variables a and N, and arrive at the expression for the number of cycles corresponding to the crack growth from the „initial" crack length a_0 to the length a:

$$N = \int_{a_0}^{a} \frac{da}{B \, \Delta\sigma^m \, Y^m(a) \, a^{m/2}} \quad . \qquad (11)$$

Replacing a by critical crack length, the number of cycles to failure can be determined. Usually, the critical length corresponds to the onset of fast crack growth after the stress intensity factor has reached the critical value K_{IC} (ultimate limit state). With respect to many factors of random character in Eq. (11), it is safer to work with the usability limit state, corresponding to a less dangerous value $K_I{}^* = qK_{IC}$, where q is a constant (for example, $q \leq 0.8$). The corresponding crack length is

$$a^* = [qK_{IC}/\sigma_{max}Y(a^*)]^2 \quad , \qquad (12)$$

where σ_{max} is the maximum stress during train passages.

Another reason for the use of smaller value a^* than critical is that the prediction of crack behaviour in a large interval of lengths is very difficult. A long crack deformes significantly the main stress field in the component, so that its path often becomes curved. Such behaviour must be modeled step by step. The situation during a moderate crack enlargement is easier to analyse, as constant crack length may be assumed. Also other simplifications are possible. For constant stress range ($\Delta\sigma$ = const), the number of cycles until the limit crack length a^* is proportional to $\Delta\sigma^{-m}$. (For varying stress range, equivalent constant stress range can be introduced.) Equation (11) can then be rewritten as

$$N = A \, \Delta\sigma^{-m} \quad , \qquad \text{with} \qquad A = \int_{a_0}^{a^*} \frac{da}{B \, [Y(a)]^m \, a^{m/2}} \quad . \qquad (13)$$

The constant A can be calculated using the constants B, m from Eq. (9), the length a_i of the detected crack, and the limit length a^*. With proper expression for the geometric factor $Y(a)$, given, for example, in Murakami (1987), the integral must be obtained in a numerical way. If, for simplicity, the calibration factor Y is assumed constant (e.g. the largest value $Y(a^*)$), expression (13) can be integrated directly.

Equation (13) has the same form as the Woehler curve (4); only the constants A and m have different values. This means that the concept of linear accumulation of damage is applicable also for crack growth processes with small changes of crack length. This allows one to use Eqs. (5) – (8) and all procedures described above, including the use of railway information systems and computer simulation. The main difference is that the stress ranges $\Delta\sigma$ are determined not by the rain-flow method, but by the methods usual in fracture mechanics; cf. Stephens et al. (2001). Strictly spoken, the linear accumulation concept is not correct for crack growth processes, as the crack velocity increases with its length. However, for crack growth within some interval, the linear summation of partial damages may be used as long as the total damage $\Sigma N_i \Delta\sigma_i^m$ does not exceed the value corresponding to the upper-limit crack length a^*.

It is possible to determine the damage in each loading cycle, but it is more convenient to work with damages caused by individual trains and to determine the approximate number of trains till the crack attains the allowable limit length a^*. Using the data from traffic monitoring, the distribution of damages corresponding to some traffic regime can be obtained. The concept of residual life D_r (Eqs. 7 and 8) enables one to determine the average number of expected train passages, or – using the Monte Carlo simulation – to calculate the number of trains corresponding to various probabilities of reaching the limit state.

Use of advanced methods for better evaluation of inspections

Besides fatigue, there are also other causes of bridge deterioration. For a more accurate lifetime assessment of structures with very long life, inspections and diagnostics are indispensable. However, detailed inspections are expensive, and must be substantiated. A suitable instant for an inspection or repair can be obtained by measuring the deterioration characteristics in some time intervals, fitting them by a suitable curve (usually of power-law type), and extrapolating to certain „alert" state. The fitting can be improved by assigning weights to the individual values, higher to the recent and lower to older ones. The random uncertainties can be accounted for by the use of probabilistic simulation methods such as Monte Carlo.

Planning of inspections is more efficient by ranking the parts of a particular bridge according to the failure consequences, probability of failure initiation, the remaining time to failure, and the risk that some defects remain unnoticed. Here, the approach common at the Failure Modes and Effects Analysis (FMEA) can be used: For each potential failure mode, weights are assigned to the three criteria: 1) failure consequences, 2) probability of occurence, and 3) early detectability. Then, so-called Risk Priority Number (RPN) is calculated as the product of all criteria for the particular failure mode. RPN makes possible easy ranking of the components, and creation of so-called safety maps of the structure. Critical components are then inspected more often („risk-based inspections").

In some cases, the information content can be increased and predictions improved by the use of Bayesian methods, enabling combination of information from various sources. Additional information can be obtained from repeated tests, long-term monitoring, or from experience with similar structures or components. The use of Bayesian methods is common for some methods of nondestructive testing (e.g. ultrasonic detection of cracks and improvement of information about their probability distribution). These methods also make possible the use of Weibull distribution shape parameter β from long term testing also for a new (small) group of experiments, provided the failure mode in the investigated case is the same as in the large series of previous tests.

In bridge inspections, only part of obtained information or records has quantitative character, while some information is vague or „fuzzy" („the girders are very rusty", „the condition of central bridge span is relatively good", "there is intensive water seeping near the second pillar", etc.). Information of fuzzy character appears in cases where: 1) no means for exact measurement exist, 2) verbal characterisation is common, 3) the obtaining of accurate quantitative information would be too expensive. Sometimes, the fuzzy information is quite sufficient: also when dealing with problems of everyday life, such as car driving, one works not with accurate values, but with verbal notions, such as „far, near", „fast, slow" etc.

The necessity of working with vague quantities has led to the creation of rules and procedures, called generally „fuzzy methods". They enable one to work with linguistic variables as well as numerical quantities, which can be „fuzzy" or „sharp", defined by one number. Fuzzy methods make possible the use of logic operations (IF, AND, OR, THEN...) or mathematical operations (SUM, PRODUCT...), and allow combination of various types of variables. After making the operations needed, the resultant (fuzzy) quantity is transformed by so-called defuzzification to a sharp value, characterizing, for example, the condition of a bridge structure („the damage degree is 5.5") and allowing the decision about repair.

Today, commercial software exists for these methods (e.g. Fuzzy Logic Toolbox in MATLAB, or specialised SW). The overal condition assessment of a structure consists of:

1) definition of parameters and criteria used in the assessment (for example, condition of the concrete plate, condition of steel reinforcement, condition of weather moulding, behaviour during train passage, etc.),

2) definition of various degrees of deterioration (or of characteristic response) for the individual criteria (i.e. definition of membership functions and transformation matrix),

3) assigning the attributes to the individual criteria according to the actual state (= creation of the transformation matrix), and the calculation of the resultant criterion.

The main problem is thus an adequate quantification of various criteria and formulating the membership functions. This can be done in cooperation with experts. The resultant reliability characteristics serve for the decision about repairs, or, in a bridge network, for their prioritising. Also this research has been in progress at the Jan Perner Transport Faculty, cf. Rudolf (2006) and Menčík et al. (2006).

Acknowledgment

The work was supported by the Grant Agency of Czech Republic, research project No. GAČR 103/05/2066.

References

Beran, L. & Šertler, H. (2001): Simulation of operational reliability of existing bridges. *Proc. 6th Conf. „Railway Bridges"*. SUDOP & Czech Railways, Prague, pp. 41 – 46.

Beran, L. (2002): The reliability assessment of an existing bridge. *Proc. 4th Int. PhD Symposium in Civil Engineering*. Sept. 19 – 21, Technical University Munich, pp. 23 – 31.

Beran, L. (2004): *Reliability and load carrying capacity of existing steel railway bridge structures*. PhD thesis. University of Pardubice, Jan Perner Transport Faculty, Pardubice.

Culek, B. (2002a): *Evaluation of fatigue life of steel constructions under complex loading.* PhD thesis. University of Pardubice, Jan Perner Transport Faculty, Pardubice.

Culek, B. (2002b): Method of three-parametric rain flow and its use for probability assessment of service life of bridge steel construction. *Int. Conf. "Traffic Effects on Structures and Environment (TESE 02)"*. Rajecké Teplice, Sept. 24 – 26, University of Žilina, pp. 81 – 86.

Culek, B. jr., Culek, B. (2005): Response of traffic and non-traffic loads of the steel railway bridges from the lifetime point of view. *22nd Danubia-Adria Symposium on Experimental Methods in Solid Mechanics*. Monticelli Terme, Sept. 28 – Oct. 1. University of Parma, pp. 250 – 251.

Marek, P., Guštar, M., Anagnos, T. (1996): *Simulation-Based Reliability Assessment for Structural Engineers*. CRC Press, Boca Raton, FL.

Menčík, J. et al. (2003): *Influence of defects and imperfections on safety and fatigue life of transport structures and bridges.* Final report of GAČR Project No. 103/01/0243. University of Pardubice & Grant Agency of Czech Republic.

Menčík, J. et al. (2005, 2006): *Determination of load carrying capacity and litetime of bridge structures.* Reports of GAČR Project No. 103/05/2066. University of Pardubice & Grant Agency of Czech Republic.

Menčík, J., Beran, L., Culek, B., Rudolf, P. (2006): Modern tools for bridge safety assessment and maintenance optimisation. *4th Int. Sci. Conf. „Challenges in Transport and Communication"*. Pardubice, Sept. 14 – 15. University of Pardubice, pp. 1099 – 1104.

Murakami, Y. (editor) 1987. *Stress intensity factors handbook.* New York: Pergamon Press.

Rudolf, P. (2006): Principles of hybrid method for condition evaluation of of bridge objects using soft computing methods for railway bridge management systems. *4th Int. Sci. Conf. „Challenges in Transport and Communication"*. Pardubice, Sept. 14 – 15. University of Pardubice, pp. 1129 – 1134.

Šertler, H., Vičan, J., Slavík, J. (1994): Reliability of existing bridge structures. *Proc. Conf. „New Requirements for Structures and their Reliability"*. Czech Technical University, Prague, pp. 65 – 70.

Stephens, R. I., Fatemi, A., Stephens, R. R., Fuchs, H. O. (2001): *Metal fatigue in engineering.* 2nd Edition. John Wiley & Sons, New York.

TWO CASE STUDIES OF THE MANAGEMENT OF BRIDGES DIAGNOSED WITH DELAYED ETTRINGITE FORMATION

D E Wimpenny, Halcrow Group Ltd, Swindon, UK
P S White, Halcrow Group Ltd, Swindon, UK
M A Eden, Geomaterials Research Services Ltd, Basildon, UK

Abstract

Delayed Ettringite Formation (DEF) is a relatively rare reaction in which sulphates within hardened concrete form ettringite leading to expansion and cracking. A significant factor in the development of the reaction is exposure of the concrete during curing to high temperatures, through steam curing, or from the heat generated by hydration of the cement.

The paper provides two case studies of the management of bridges diagnosed as having DEF: a UK railway bridge with precast steam-cured members and a highway structure in Malaysia with thick sections formed from in-situ concrete.

The railway bridge carries a mainline over an unclassified road and was constructed in 1969. The principal members are prestressed concrete deck beams bearing on precast cill beams over the abutments and precast reinforced concrete parapet beams. Cracking to the parapets and deck were first reported 16 years after construction. An assessment of the effect of progressive damage and deterioration was undertaken in 2004. A literature search identified a large number of references on DEF but little guidance on management of affected structures. A programme of testing, crack movement monitoring and durability modelling, followed by repairs, periodic condition surveys and acoustic monitoring was developed.

The highway structure, comprising a 1.5km viaduct carrying a six lane carriageway, was opened in 2002. The structure has split decks formed from externally prestressed precast concrete box girders resting on T-shaped piers. Cracking to several piers and one abutment was reported six months after opening. A series of investigations were commenced involving crack mapping and monitoring, testing of cores and detailed finite element analyses. The temperature of the concrete was calculated to have possibly exceeded 80°C during construction suggesting a risk of DEF. Monitoring of the structure indicated ongoing crack opening and laboratory testing confirmed expansive DEF. Remedial works were developed using waterproofing and strengthening.

Introduction

Ettringite or calcium sulfoaluminate hydrate is produced as a normal reaction product of cement hydration in immature concrete. DEF is a relatively rare phenomenon in which ettringite is developed later in the life of the structure leading to expansion and cracking.

Though there is uncertainty over the precise mechanism of deterioration, there appear to be a number of commonly reported risk factors associated with DEF, principally:

o   High curing temperature (typically 65°C or more)
o   Exposure to moisture in service
o   High sulphate and alkali content in the concrete mix
o   Presence of pre-existing cracks (eg plastic or early age thermal cracks).

The importance of high curing temperatures has meant that the occurrence of DEF has been associated with steam-cured precast elements. However, the phonomenon can also occur in insitu concrete when high temperatures are allowed to occur during hydration in thicker sections and where placing temperatures are elevated due to the ambient conditions (Quillin, 2001, Grantham et al, 1999, Dunster and Clayton, 2002). Expansion is believed to follow an 'S' shaped progression with time with damage generally becoming evident once the expansion gets to 0.5mm/m within 5 to 25 years of construction.

In recent years, the higher fineness of cement and the removal of sulphate limits in specifications may have acted to increase the risk of DEF, although this is offset by a wider appreciation of the risk of DEF and greater use of secondary cementitious materials.

There are relatively few documented cases of DEF occurring in bridge structures and little advice on managing structures with DEF (Merrill, 1998). This may be due to problems in correctly diagnosing the cause of deterioration. In particular, alkali silica reaction (ASR) and DEF have some common features and review of a case of deterioration to seven prestressed bridges attributed to ASR (Tordoff, 1990) suggests DEF contributed to the expansive cracking. Similar findings have been reported by others (Hobbs, 1999). Correct diagnosis is important as DEF can have a greater impact on compressive and bond strength than ASR.

This paper describes two case studies of investigating and managing bridges with DEF: a UK railway bridge with precast steam-cured members and a highway structure in Malaysia with thick sections formed from in-situ concrete.

UK Railway Bridge

Introduction
The structure carries four tracks of a mainline rail route over an unclassified road and was constructed in 1969. The deck has a span of approximately 6m, width of 17m and headroom of 4.9m. The principal members are 21 prestressed concrete deck beams bearing on precast cill beams over the abutments and precast reinforced concrete parapet beams (Figure 1). The parapet beams are approximately 1110mm deep, 910mm wide and 6m long reinforced with 28mm and 19mm diameter main bars and 10mm links. The deck beams are approximately 430mm deep, 760mm wide and 7.9m long. They are prestressed longitudinally by seventy-six 7mm wire tendons and are post-tensioned transversely by four 12 mm strands (Figure 2). The longitudinal tendons are partly unbonded, being run in 11mm diameter PVC ducts. The transverse wires are assumed to be grouted within PVC ducts which run through 50mm diameter metal sheathing cast into the units.

Cracking in a 'marbling' pattern on one of the parapet beams was first reported 16 years after construction. Longitudinal cracking was subsequently reported to the deck beams. Cores were

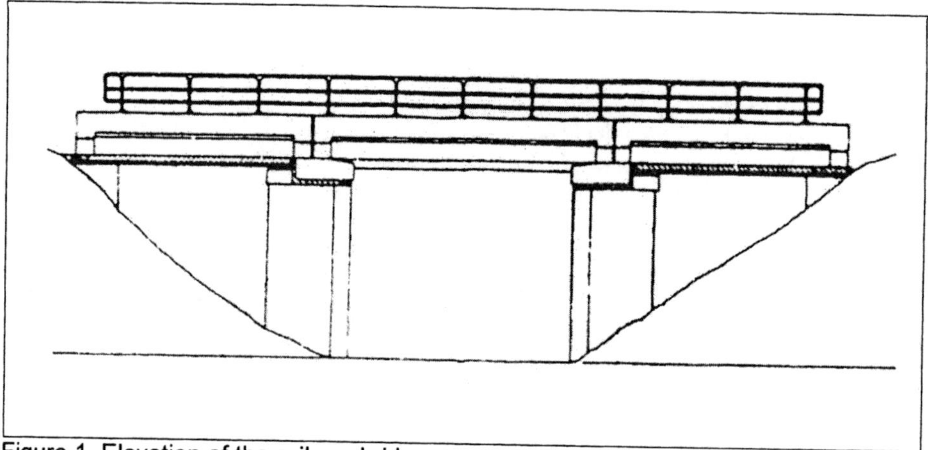

Figure 1. Elevation of the railway bridge

taken from the top of the deck beams in 1998 which suggested DEF as the cause with ASR less likely. At that time the cracking to the parapet beams was up to 4mm in width (Figure 2) and that to the deck beams was described as hairline but in a few cases up to 1.5mm in width.

Figure 2. Cracking to parapet beam

Unfortunately, no samples were taken from the parapet beams and few construction records were available. All the beams are understood to have been steam cured and all are exposed to water; the parapets are directly exposed to rain and the deck beams are exposed to water leaking though the deck due to a failed waterproof membrane.

DEF management strategy

The management strategy was divided into three stages as indicated in Table 1 below.

Table 1. DEF management strategy for UK railway bridge

Action	Purpose
Stage 1- Condition survey	
Hammer tapping, covermeter, half-cell potential, carbonation and breakouts	Determine the stability of the and extent of corrosion to reinforcement and prestressing
Cores from deck, parapet and cill beams for petrographic examination	Confirm DEF diagnosis and the characteristics of the concrete
Dust samples from deck, parapet and cill beams for sulfate and chloride testing	Confirm level of internal sulfates and any exposure to de-icing salt spray
Stage 2 – Expansion monitoring	
Cores from deck, parapet and cill beams for expansion testing (BCA, 1992)	Establish risk of further expansion for different member types
Monitor crack movement over 12 months	Determine progressive expansion and degree of movement for selecting repair products
Stage 3 – Repair and Monitor	
Improve deck drainage and apply waterproofing	Reduce water ponding and penetration to deck
Seal cracks >0.2mm width with sealant And apply elastomeric coating	Reduce moisture ingress and deterioration through steel corrosion and freeze-thaw
Visual and tapping survey every 10 years	Confirm stability of concrete
Long-term acoustic monitoring of prestressing	Provide early warning of wire failure

The objective of the first two stages is to confirm the underlying deterioration mechanism, the effect of any cracking on reinforcement and prestressing corrosion and the likely development of expansion in the future. This is important in determining the long-term prognosis for the structure and the suitability of any repair treatment. The expansion testing in Stage 2 provides an indication of future worst case expansion. The test method involves subjecting 70mm diameter cores to storage in water at 20°C or 38°C and monitoring the length change over time. This method has been previously used successfully for a DEF affected structure (Grantham et al, 1999) and is combined with remotely monitoring the movement across cracks on the structure. The chloride and carbonation depth data from site together with the observed corrosion at breakouts provides an input into numerical models to predict future deterioration.

The repairs in Stage 3 are aimed at excluding moisture and preventing the structure from becoming unsafe due to spalling of concrete and corrosion of the reinforcement and prestressing. Crack sealing and surface treatments have to consider the possibility of ongoing reaction and further expansion. In particular crack sealing may require a low elastic modulus sealant applied in a chase to allow further reaction products to form in the unfilled crack.

There was specific concern over corrosion of the prestressing in the deck beams and acoustic monitoring of wire failure using the Soundprint System (CBDG, 2002) was proposed. However, the client chose replacement of the deck beams during several track possessions.

Malaysian Highway Structure

Introduction

The highway structure, comprising a 1.5km viaduct carrying a six lane carriageway, was opened in 2002. The structure has split decks formed from externally prestressed precast concrete box girders resting on T-shaped piers, comprising an 18.7m long crosshead tapering from 3.5m to 2m in depth, and a 3.6m wide octagonal column (Figure 3). The abutments are also T-shaped similar to the piers, but for the most part buried. Further details of the structure are given elsewhere (Buckby et al, 2006).

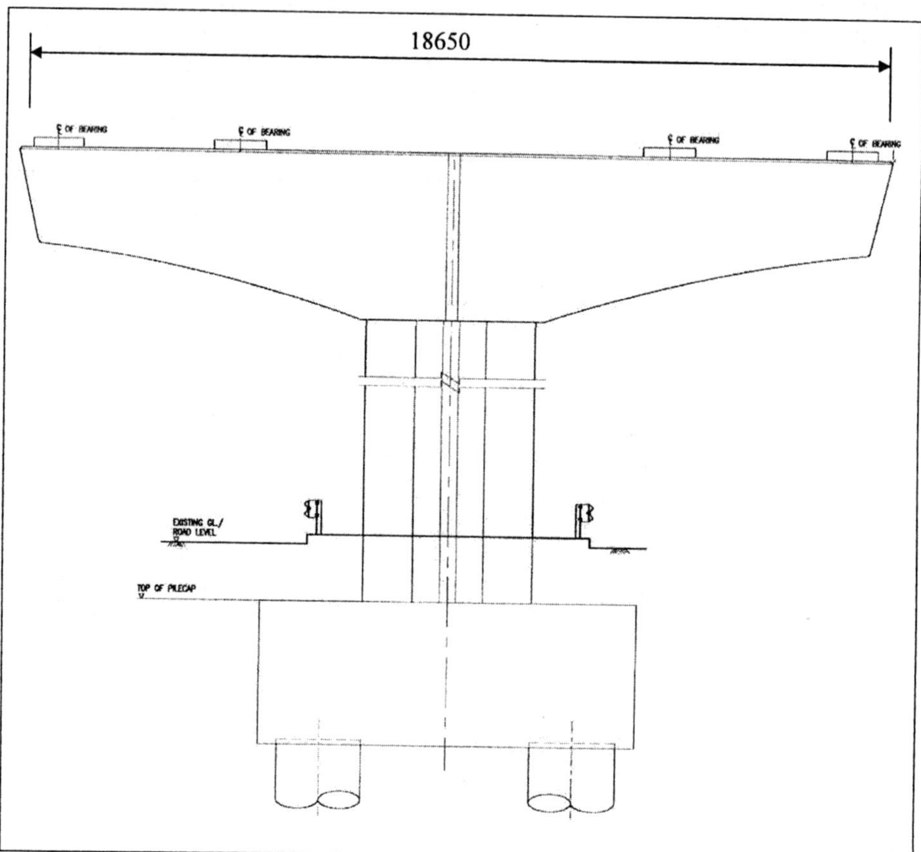

Figure 3. Elevation of T-shaped piers

Cracking to several piers and one abutment was reported six months after opening the viaduct. A detailed survey of the crack mapping was undertaken in 2004, by which time some of the most severely cracked crossheads had cracks several millimetres wide. Some of the cracking could be attributed to inadequate reinforcement. However, the variability, severity and presence of a lateral displacement across some cracks suggested other mechanisms, such as an expansion reaction, may be contributing to the cracking. A thermal analysis using published guidance indicated values in excess of 80°C could have been reached during initial curing leading to a risk of DEF.

DEF Management Strategy

The management strategy has three stages as indicated in Table 2 below. Partial closure of the bridge meant there was considerable pressure to resolve the problem and this meant several of the stages occurred in parallel.

DEF was unexpected and a number of steps were taken to confirm the finding. Potentiometers and strain gauges were installed to measure crack movement and response of the reinforcement for selected crossheads. These were remotely interrogated via a GSM modem. This indicated progressive expansion which did not coincide with traffic loading but showed a higher rate in the rainy season. Cores were taken for petrographic examination which found ettringite in cracks and rims around aggregate particles characteristic of DEF. The reaction products were confirmed using scanning electron microscopy (SEM) and microprobe analysis. Cores were subject to expansion testing using the BCA and Duggan test methods (BCA, 1992 and Grabowski et al, 1992). These confirmed damaging expansion levels of up to 0.2% at 20 days, compared to a safe limit of 0.05%. The primary cause of the expansion in the cores subjected to BCA expansion testing was confirmed as being DEF using petrographic techniques and electron microprobe analysis. (Eden, White and Wimpenny, 2007).

Table 2. DEF management strategy for Malaysian highway structure

Action	Purpose
Stage 1- Confirmation of diagnosis	
Crack movement and reinforcement strain monitored	Confirm progressive expansion and localised reinforcement yielding unrelated to live loading
Cores from crosshead for petrographic examination and SEM work	Confirm DEF present from distribution and nature of reaction products
Cores from crosshead for Duggan expansion testing	Confirm potential for DEF based on expansion levels
Cores from crosshead for BCA expansion testing	Confirm occurence of DEF based on expansion levels and petrographic examination of the specimens after expansion testing
Stage 2 – Numerical Modelling	
Finite element analysis of temperature within section during construction	Confirm the likely zone of concrete at risk of DEF
Spreadsheet model developed to assess critical sections of the crosshead	Predict likely structural impact of DEF and determine the strengthening requirements
Non-linear Finite element analysis of stresses and damage based on predicted expansion	Confirm the likely structural impact of DEF and the strengthening requirements in more detail
Stage 3 – Repair	
Grout existing cracks with an epoxy resin based grout	Seal the existing cracks and restore the tensile strength of the concrete
Apply waterproof coating to crossheads	Exclude water ingress from rain and ponding water
Strengthening Scheme	Prestressing and plate bonding to address present and possible future loss of strength

A 2D finite element analysis was undertaken to predict the temperature in the crosshead during construction (Figure 4). No concrete temperature records were available, except at placing (approximately 32°C) and assumptions were made on worst case and average ambient conditions, including solar gain. The worst case and average case analyses indicated temperature values in excess of 80°C and 70°C respectively for over 100 hours at the core of the section. The temperature values from this analysis were used to predict DEF expansion at different locations in the cross-head which were in turn used within the structural analysis.

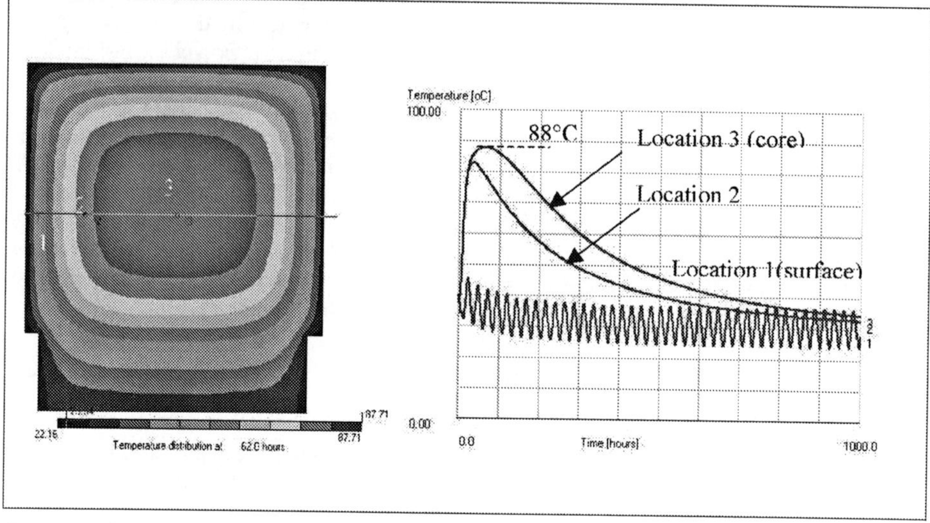

Figure 4. Section through crosshead showing predicted temperature development

The significant structural effects of ongoing DEF expansions were found to be:
o   Cracking around the perimeter of the section
o   loss of ultimate moment capacity due to cracking of the soffit
o   failure of laps in main tension and in shear links.

The issue of lap failure was a particular concern as two of the four layers of main flexural reinforcement in the top of the crosshead were lapped at the centre of the crosshead with laps that were not full strength laps.

In order to assess the strength of DEF affected crossheads and identify suitable strengthening works, a spreadsheet model was developed for analysing critical sections of the crossheads from first principles. A 3D non-linear finite element analysis was also undertaken.

The spreadsheet gave similar results to conventional section analysis software but in addition was able model the variation of DEF expansion across the cross-section, the effect of the stresses due to DEF expansion on the concrete and reinforcement and the loss of stiffness and strength of the concrete section due to ongoing DEF. The spreadsheet model included options for modelling the tensile strength of the concrete, the stress-strain response of lapped bars and the effect of yielding of bars and strengthening by plate bonding and external prestressing.

The spreadsheet and non-linear FEA models were used to predict the damage to the structure for different DEF expansion levels. There was particular concern about ongoing expansions leading to yielding of the reinforcement, failure of laps, and potentially prejudicing the strengthening work at higher levels of expansion. At 0.1% DEF expansion it was found that there was significant cracking around the top and sides of the section but that the soffit remained intact and the effect on the ultimate moment capacity was negligible. However, as the modelled expansion increased the moment capacity reduced due to cracking in the soffit and lap failure necessitating strengthening work to reinstate the moment capacity at the root of the cantilever. The non-linear finite element analysis produced a pattern of cracks in the crossheads over the column very similar to that observed on site. Simulating the effect of increasing expansion lead to cracking developing further down the vertical face of the crosshead until it extends into the soffit.

The crossheads were subject to frequent wetting due to the longitudinal gap between the split decks, leaking drainage, and wind driven rain. The rapid development of the reaction may have been promoted by the high ambient temperatures and humidity in Malaysia. The waterproofing of the structure was regarded as essential to control future expansion and specifications were developed for a waterproof coating, which included requirements for:

a)   tolerance to damp substrate and high relatively humidity during application
b)   ability to transmit vapour transmission not less than $3g/m^2/day$ out of the concrete
c)   transparency to permit cracks to be viewed and monitored
d)   a minimum elongation to failure of 100% and crack bridging of up to 2mm width.

The strengthening scheme developed for a typical crosshead is shown in Figure 5. The longitudinal prestressing is required to strengthen the crosshead at the main lap regions

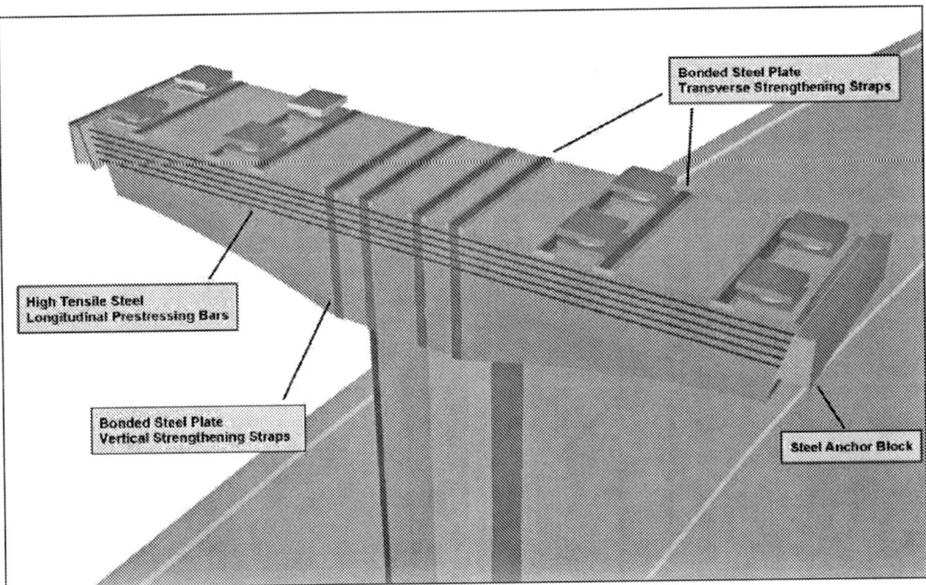

Figure 5. Typical crosshead strengthening scheme

primarily by providing a relieving moment, such that if the two top layers of bars fail at the laps the external prestress in combination with the remaining reinforcement would be sufficient to develop the design moment. The transverse bonded steel plate straps are required to strengthen the crosshead due to deficiencies in the transverse reinforcement. These straps also provide reinforcement to control ongoing cracking from the limited ongoing residual DEF expansion expected after waterproofing. The vertical bonded steel plate straps are required to control ongoing cracking in the weakened crosshead-column connection region where the existing reinforcement has yielded and lost its stiffness at wide cracks.

Conclusions

DEF can occur in both steam cured and insitu concrete leading to serious cracking. There are few published cases relating to bridges, which may be due the deterioration having been diagnosed as ASR. High curing temperatures during initial curing, and exposure to water in service are critical risk factors, and it has been found that the structural effects of DEF can be significantly different from the effects of ASR.

The two case histories described are very different. One relates to a 1969 railway structure in the UK formed from steam-cured precast beams and the other to a Malaysian highway structure constructed in 2004 with insitu concrete. Although the DEF management strategies have been tailored to each of the particular structures, there are common elements:

o   Removal of samples to confirm the cause of cracking
o   Monitoring of crack width movement and expansion testing of cores to assess future expansion and movement for selection of the repair products
o   Measures to exclude moisture in the form of coatings and sealing cracks.
o   Strengthening or replacement of the affected elements

A number of techniques were applied to assist in the investigation and management of DEF, including SEM to confirm the distribution of ettringite in the concrete, monitoring of crack movement and modelling of temperature, stresses and deterioration. Long-term monitoring of movement and, in some instances, acoustic emissions may provide a means of confirming the effectiveness of any remedial measures and provide early warning of ongoing deterioration.

It is considered important for the assessment and management of DEF affected structures that the occurence of DEF is correctly identified, modelled and mitigated.

References

BCA, 1992. The diagnosis of alkalis-silica reaction, report of a working party, British Cement Association Publication 45.042, Crowthorne.

Buckby R J, White P S, Mills CA and Quillin K, 2006. Severe Cracking In Insitu Concrete Substructures Due To Delayed Ettringite Formation (DEF) : Diagnosis, Assessment And Remediation, Hong Kong International Conference on Bridge Engineering, November 2006.

CBDG, 2002. Guide to testing and monitoring the durability of concrete structures, Concrete Bridge Development Group.

Dunster A and Clayton N, 2002. Avoiding deterioration of cement-based building materials and components, Lessons from case studies:5, Building Research Establishment, CRC Ltd.

Eden MA, White PS, Wimpenny DE, 2007. A laboratory investigation of concrete with suspected delayed ettringite formation – a case study from a bridge in Malaysia. 11th Euroseminar on Microscopy Applied to Building Materials, 5-9th June 2007, Porto, Portugal.

Grabowski E, Czarnecki B, Gillot J E, Duggan C R and Scott J F, 1992. Rapid Test of Concrete Expansivity due to internal sulfate attack., ACI Materials Journal, September/October 1992, pp 469-480.

Grantham M G, Gray M J and Eden M A, 1999. Delayed ettringite formation in foundation bases, A case study, Proceedings of Structural Faults and Repairs, Edinburgh University Press.

Hobbs D W, Expansion and Cracking in Concrete Associated with Delayed Ettringite Formation, Ettringite: The Sometime Host of Destruction, ACI SP-177, Erlin B (Edt) .pp 159-181.

Quillin K, 2001. Delayed ettringite formation: in-situ concrete, BRE Information paper IP11/01, CRC Ltd.

Lawrence B L, Myers J J and Carrasquillo R L, 1999. Premature concrete deterioration in Texas Department of Transportation Precast Elements, Ettringite: The Sometime Host of Destruction, ACI SP-177, Erlin B (Edt) .pp 141-158.

Merill D B, 1998. Durability of concrete in Highway Facilities, Proceedings of Conference, University of Houston, www//gem1/cive.uh.edu.

Tordoff M A, 1990. Assessment of Pre-stressed concrete bridges suffering from alkali-silica reaction, Cement and concrete composites, Vol 12, pp 203-210.

A Prototype of Bridge Monitoring System under Ubiquitous Environment

Mu-wook Pyeon, Konkuk University, Department of Civil Eng. Seoul, Korea
Jee-hee Koo, Ubiquitous Land Implementation Research Department, Korea Institute of Construction Technology, Korea
Yang-dam Eo, Agency for Defense Development, Daejeon, Korea
Jae-sun Park, Konkuk University, Department of Civil Eng. Seoul, Korea

Abstract

As facilities reach their life-time, the importance of monitoring for safe use of them is more increasing. For replacement of the existing manual measuring, the introduction of a ubiquitous technology is highly needed. The ubiquitous technology refers to a new paradigm of intellectualizing physical space and organically integrating its objects through the means of ubiquitous computing and network. The forthcoming ubiquitous revolution is expected to serve a significant role for civil engineering field. In this study, for the practical use of the system that manages u-bridge, Authors proposed a system scenario and then, based on it, constructed a model system consisting of sensor network, relay system, and service system. Authors suggested how to connect the system with UFID and the bridge management system. The proposed U-bridge monitoring system was tested for the verification and the practical operation. Overall, this study clarified the list of requirements and basic technologies to implement intelligent facility monitoring system in the ubiquitous computing environment

1. Introduction

In Korea, large scale bridges like the Youngjong Grand Bridge are equipped with a measuring-monitoring system for effective management. Generally, in operation of measuring-monitoring system, breakdown of sensors or on-site systems could be considered as the greatest problem, but sensors have a lifecycle of 3~5 years and even in case of troubles of them, the locations of sensors or device are easily detected so it is possible to take immediate measures, but when networks such as network lines or transponders have problems, it is hard to catch the error-occurred-locations, and takes much time and cost to replace them. Therefore, studies for switching the telecommunications part, which is the most important in operation of measuring-monitoring system.

On the other hand, ubiquitous computing is a technology that watches, tracks, and optimizes conditions and positioning of facilities by combining sensor with wireless network.[1] The ubiquitous technology appearing as a new paradigm of computerization designates the

Bridge design, construction and maintenance 2007, Thomas Telford, London

technology that intellectualizes physical space and integrates its objects organically on the basis of ubiquitous computing and network. If we allow facility monitoring to have ubiquitous wireless network and transmission and real-time measuring technologies, it could be possible for us to not only monitor a lot of facilities at one control tower but also reduce unnecessary costs and work force.

In this study authors also suggested ways to utilize sensor network and general directions of ubiquitous facility monitoring. Moreover, we suggested a series of scenarios for bridge monitoring system under ubiquitous environment to apply ubiquitous technology. In addition to that, attention is given to the effective interconnection of bridge monitoring information and bridge management systems and how, through a ubiquitous environment, to connect the UFID (Unique Feature IDentifier) authorized by Korea with a GIS-based bridge management system (BMS) operated by the Ministry of Construction and Transportation. In addition, authors constructed u-bridge(ubiquitous bridge) monitoring model system, which could be a prototype of facility management system in ubiquitous environment that is expected to be demanded tremendously

2. Overview of bridge monitoring in ubiquitous environment

The formation of ubiquitous environment starts from the point that invisible special-functioned computers are introduced into the environment and things and they become intelligent by themselves. The computers in things recognize the configurations of their surroundings and also have functions of sensing, monitoring, and tracking changes of things and environment even from a long distance. The computers, without intervention from human beings, mutually telecommunicate and provide useful information in real-time, or could make decisions in place of men. Additionally, things themselves could take actions or measures for human beings.[2] To realize a space-science in bridges, which represent civil construction, means securing ubiquitous technologies of new concept and the next direction of monitoring to go.

The services using ubiquitous computing are classified into five steps of communication service, information-offer service, situation notification service, action guidance service and intelligent action service.[3] The service level becomes increasingly higher and required technologies become complicated as service goes up from communication to intelligent action.

Therefore, our aim, the ubiquitous-based bridge monitoring technology level, is defined five levels in Table 1.

Table 1. Five steps of u-bridge monitoring technology development

Classification	Items of technology development
Communication service	- Step where a tag containing information of things is attached to bridge construction - To attach a measuring tag to major sites of bridge or roads
Information-offer service	- Step where sensors send/receive information of situations passively - Managers understand the current situations by using tag or sensor which is attached to bridge
Situation notification service	- Step where sensors become intelligent and meet situations actively -Step where in case of that possible dangers take place, sensors send warning message to managers
Action guidance service	- Step where active sensors are attached to all bridge structures

Intelligent action service	- Step, where bridges become themselves intelligent and limit the traffic of bridges
	- Step where the whole nation and the world are connected to the ubiquitous system
	- Step where central server controls men and vehicles which are passing bridges along with bridges and optimizes bridge conditions

In this study, along with the information-offer and situation notification services that transmit data measured from sensors to server by wireless telecommunication, we tried to construct u-bridge management system where the action guidance service works.

3. UFID and BMS in U-bridge Monitoring System

3.1 Technology Trend of UFID

UFID (Unique Feature IDentifier), currently under construction by the Ministry of Construction and Transportation in Korea, is used as a common key to manage, search, and apply topography to judge locations. It is the only electronic identifier distributed to the entire nation's topography[4]. UFID is commonly used as a reference to topography for different organizations and is also a basis of the LBS(Location Based Services). A systematic UFID application and its management and system are therefore indispensable for the management of national land in the near future.

The features of UFID are shown in Table 2.

Table 2.   Items and Main Contents of UFID

Items	Main contents	Length
Version code	code to identify UFID, versions of UFID	Two digits
Topographic feature	System made by the National Geographical Information Institute, used in the Map version 2.0	four digits
Agency code	Organization to manage topography	Six digits
Serial Code	Serial No. of the same topography within a region (having the same location information)	three digits
Location data	1" x 1" Unit Lattice Identifier, ex: 36425173829 is located on latitude 3642'51", longitude 12738'29"	eleven digits
Elevation data	In the case of buildings, the ground and underground expressed as 01-90 and 91-99, respectively. Others that get additional document at attribute information are given attribute flags In the case of roads, expressed as 01, 02 by the numbers of layers of overpasses, and 91, 92 in the case of tunnels or underground roads	two digits
Attribute Flag	Where one particular attribute is found in the same topography, it is used to refer to other attribute DBs	one digit
Error Check	Code to identify transmission errors of ID	one digit

3.2 UFID in Ubiquitous Environment

In the near future, all information of items in the national territory can be managed in a ubiquitous environment, and, in the course of this, location information is the most basic and important information. In some cases, there might be a great deal of uncertain information for locations. To apply a ubiquitous environment efficiently, this kind of basic location information management system is needed and should be necessarily considered in the facility monitoring under this environment. In this study, a measure to connect UFID to the location management system of ubiquitous-based bridge monitoring is studied.

1) The Expression of Location Information

The current standard of location information used by UFID is the expression of a representative location of a unit target. The precision required under a ubiquitous environment, however, is higher. A RFID(Radio Frequency IDentification) or USN (Ubiquitous Sensor Network) is located on the unit structure or on all the inside elements of the target. Particularly in the case of the facility monitoring process, location information from the unit measurement sensor is very important for the behavior analysis of the bridge and the process requires coordinate information at centimeter level. Therefore, an additional standard for location information that expresses each part of the bridge is required.

2) The Expression of Detailed Attributes

Like the location standard, UFID itself has a limitation on expressing detailed attributes for the ubiquitous bridge facility monitoring. Although the classification of the facility itself and the managing organizations can be expressed, UFID needs a much more detailed approach. In particular, the bridge is comprised of various materials while road facilities are above deck. Different sensors are set to monitor these things. The database structure that expresses the detailed attributes should be designed appropriately.

3) Linkage to the Telecommunications Environment

Various wired and wireless telecommunications play a great role in the ubiquitous environment, where telecommunications modules united with sensors will transmit information from USN (Ubiquitous Sensor Network).[5] In the progress of a ubiquitous system, each ubiquitous device in the facility will get identification numbers based on telecommunications protocol in the short term; all facilities will be given telecommunications addresses like TCP/IP, telephone numbers or identifiers, in the long-term. The most proper identifier for this is IPv6, which could be an important means to transmit the status information of facilities, including location information, to users or systems and will be necessary for meeting the situation requirements, as well.

3.3 Connection UFID to BMS

1) BMS DB Table Analysis for the Connection with UFID

The DB table is analyzed for the connection of UFID and BMS. BMS is the GIS-based bridge management system (BMS) operated by the Ministry of Construction and Transportation.[6] The analysis is classified into location information, materials attributes, quantitative information, maintenance information and management information items. This is a fundamental process for connecting UFID with detailed locations or materials attributes of bridges, as shown in Table 3.

Table 3. BMS Detail Input Item

Classifications	Detail input item	Related item				
		Location Inf.	Materials attributes	Quantit-ative Inf.	Mainten-ance Inf.	Mgnt. Inf.
Common materials	4 items, incl. bridge No.					O
Basic materials	21 items, incl. usage	O	O	O		O
Bridge specifications	8 items, incl. lines	O		O		

General specifications	31 items, incl. date of start of work	O	O	O	O	O
Cross materials	3 items, incl. cross kinds and states					O
Structural materials (Superstructure)	21 items, incl. span-length	O	O	O		
Structural materials (Substructure)	17 items, incl. site No	O	O	O		
Inspection materials	38 items, incl. inspection date	O				O
Load Bearing Capacity evaluation	11 items, incl. evaluation date		O	O		
Repair Records	13 items, incl. preparation date	O		O	O	O

2) Location Expression in BMS

As mentioned above, the location expression of sensors or materials of a bridge cannot be fully satisfied using the existing UFID only, which leads to a new method to express exact locations of sensors and materials.

By analyzing the currently-used input items of BMS, the item related to the location on the bridge is drawn out. The contents are shown below.

o Bridge Length : Calculates the length between bridge spans of the abutment of the bridge along with the center of the bridge and then rounds this number off down to two decimal places in m units

o Bridge Width : Defines bridge width as the right angular width to the bridge axle from the superstructure

o Bridge Height : Defines bridge height as the longest perpendicular-direction length from the deck to the ground surface

o Water depth : In the case of bridges over a river or the sea, the maximum water depth of the bridge is calculates. Water depth here means at the time of inspection.

o Number of Spans : Defines the number of spans as the space between one bridge pier and its neighboring bridge pier. The total number of spans are input.

o Maximum Span Length : Calculates the maximum span length. Span length means between the supports of spans but for the convenience of calculation in the case of a single bridge, between the platform of the bottom, and in the case of a continuous bridge, between the platform and the center of the bridge.

o Span Length   : Inputs the span length in charge

o Column No. : Inputs the number of columns into the data of the substructure. Column No. starts with 1 in order and is entered in three digit-natural numbers.

3.4 Measures to Connect BMS and UFID

The existing BMS and UFID separately manage the same bridge with different systems and identifiers. In this situation, the integration of these identifiers makes the information of different systems shared with each other. For this, the linkage of BMS and UFID is proposed as follows:

1) The relation table (or field) with UFID is added to the bridge management system DB and the basic geographical information management system of the UFID server. E.g., "Bridge No. 0000 = UFID XXX"

2) The common 'Bridge No.' item of the BMS DB detail input items is replaced by UFID.

4. Scenarios for U-bridge Monitoring System

4.1 Main application technologies of bridge monitoring in ubiquitous environment

1) Sensors
In the existing bridge measuring and monitoring system, various sensors are needed according to types of bridges. For example, cable-stayed girder bridges require nine types of sensors including geodimeter, cable tension gauge, displacement gauge, and PCM/ILM bridges need five types of sensors including reaction gauge and accelerometer. In this study, we tried to confirm a possibility to construct the system that supports ubiquitous environment for the action guidance service and wireless remote area data transmission and sensor network within wireless local area environment rather than monitoring of structural engineering bridges considering accuracy of data. Therefore, through customizing accelerometer among the existing sensors, we tried to construct an embedded module made suitable for ubiquitous environment and provide actual wireless monitoring environment in this study.

2) Local Area Sensor Network
In this study, we select and apply a wireless local area telecommunication technology for the communications among various sensors in bridges. WPAN (Wireless Personal Area Networks) is effective in transmitting information among small number of users within a range of short distance. Besides, WPAN is emerging as a solution that is inexpensive and has good power efficiency along with its easy embodiment in various devices (PDAs, notebook computers, headsets, bar-code readers, sensors, PCS phones, etc)
Currently commercialized products for WPAN are IEEE 802.15.1 Bluetooth and IEEE 802.15.4 Zigbee. Bluetooth, as a stable technology, is already commercially used, while Zigbee still under development stage.[7] Zigbee has an advantage over Bluetooth in terms of power efficiency ,easy network configuration, and the range of telecommunication so that we built a sensor network by applying both Bluetooth and Zigbee.

3) Wireless Remote Area Network
In facility monitoring technology used currently, wire network is most applied to between sensor and data logger, or data logger and main system. When you construct facility monitoring system with wire network and optical cable isn't set where facilities are, then you

cannot nearly construct system. From that, we attempt to solve that problem by constructing remote transmission with wireless. Field Hub, an on-site system part for the inside of a bridge structure, performs collection, transmission and process of data. That is, a local area network-module to receive data that is collected by the sensor part and another module to send this data to remote area should be built in.   Now, main methods of wireless remote area network are CDMA, satellite, ADSL, PSDN and we constructed wireless remote area network by making practical application of CDMA network.

4) Server System

The purpose of server system construction lies in processing a bunch of data collected from lots of servers. Main server system part, Field Box processes and decodes various data. Regardless of sensor network or sensor types, in order to process data, server system should introduce standardized mode like SensorML. And the system retrieves, inquires and analyzes the collected data on the web. For this, we developed data telecommunication and middleware for web service by using SOAP (Simple Object Access Protocol), a XML-based protocol.

4.2 Scenarios for u-bridge monitoring system construction

The design of u-bridge monitoring system should fulfill these basic elements as following below.

① Relay system (Field Hub) should be linked with signal lights, crossing gates at a bridge approach place to control them, making action guidance service available.

② A wireless local area telecommunication system should be constructed to maintain system efficiently, cost effectively and flexibly.

③ Ad-hoc networking should be embodied for systematic network construction between sensors and telecommunication and also be intelligent enough for rational measuring.

④ Sensors that perform on-site measurements should be small-size system with portability, considered that measurement is easily made in ubiquitous environment anytime, anywhere.

⑤ XML(Extensible Markup Language)-based system should be implemented, where final server(Field Box), which receives measured data and processes massive data using Distributed Computing and, data format could be easily integrated in heterogeneous platforms making data transmission easier.

⑥ Final server should process collected data and perform exact structure interpretation and find information for the degree of bridge-aging and its maintenance and sense dangerous situations of a bridge, based on analyzed data. In case of emergency states, it should form a proper judgment according to danger level immediately.

5. A Prototype of U-bridge Monitoring System

5.1 Overview of the whole system

A comprehensive system that we want to implement by making use of the application technologies that are required for u-bridge facility monitoring is mainly comprised of a sensor group to perform measurement, terminal& relay system to control sensors and server group to receive and store and process data along with structure analysis.<Fig. 1>.

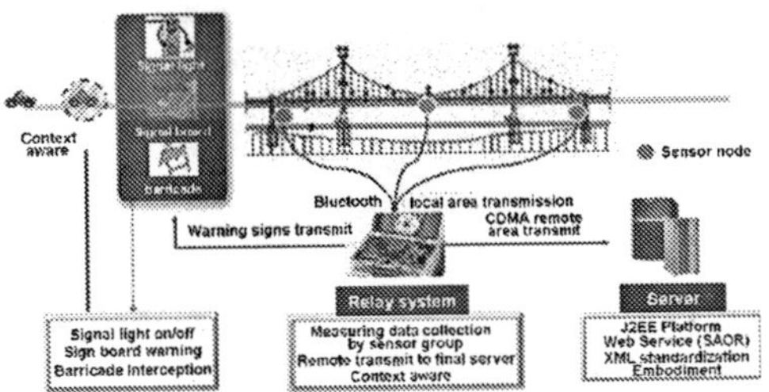

Fig. 1.   The whole composition diagram of u-bridge monitoring system

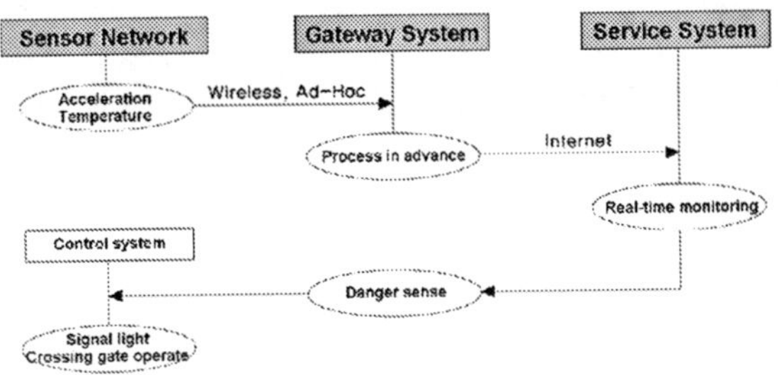

Fig. 2.   Flow chart of ubiquitous facility monitoring information

Fig. 2 shows a sequence diagram by system requirements, along with the associative relation between unit systems and data flow.

To embody a model system, we worked for sensor network building, relay system production and server system implementation in order. After completing them, we made a model bridge and did field tests of it. We made service architecture based on web service technology for the service-oriented architecture. Regarding measuring data, we introduced XML to it and had it programmed to be self-descriptive and decodable. Table 4 is shown below main hardware, software, network-embodiment form and transmission data-type of the constructed model system.

Table 4. Outline of model system

Classification	Sensor network	Relay system	Service system
Role	Data collection	Data process& relay	Data storage& service
Hardware	Nano24E	PC	Server Computer
Software	NanoOS+Qplusn	Linux + Java Gateway Application	Linux + Java JBoss+Axis, MySQL

			Fieldbox Middleware
Network	Zigbee Ad-hoc	Wire/Wireless LAN	Internet
Data form	XML	XML, SOAP	HTML, XML, SOAP

5.2 Model bridge test

For the data transmission tests of the constructed sensor network, we made a   bridge model of cable-stayed girder bridge type and did the test in laboratory. Fig. 3 shows a model bridge for the test.

Fig. 3.   Model bridge made for test

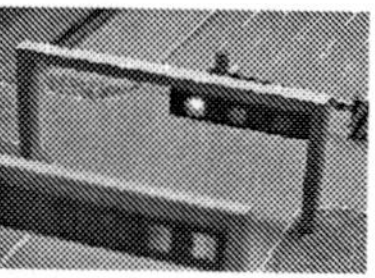

(a) Control device   (b) Sensor module   (c) Excitation device   (d) Traffic control

Fig. 4. Model bridge test

After producing sensor network, authors conducted laboratory tests using this model before application to actual bridge and this model was made connecting with the real-time monitoring system of web GIS system. The model bridge was equipped with a sign board for warning message, signal lights and a crossing gate to cope with emergency situations. Fig. 4(a) shows the control device of the model bridge.

Fig. 4(b) shows sensor module attached to the model bridge for the test. By using Bluetooth and Zigbee telecommunication, the sensor module transmits data to on-site system (Field Hub), and then the on-site system examines whether or not the data from the sensor are right with the help of signal-based decoding method, and then determines it will operate the control system. Fig. 4(c) shows a test of data transmission when adding load more than its threshold value to the bridge, Fig. 4(d) shows that the control device is operating for this.

4. Conclusion

Ubiquitous technology is still much under discussion. As a case study of applying the technology to the construction sector, authors prepared a system scenario necessary for the practical use of u-bridge management system and constructed a model system consisting of sensor network, relay system and service system based on that scenarios related with BMS and UFID. A model bridge test was conducted for the prototype of ubiquitous facility management system.

This study clarified the list of requirements and basic technologies to implement intelligent facility monitoring system in the ubiquitous computing environment.

Authors expect that the range for application will be more extended in the near future. Particularly, in the construction sector, the ubiquitous technology is expected to present a large amount of benefits, which can make the current on-site measuring instruments intelligent. More studies about application of ubiquitous computing to various construction sectors should be continued. For this, it will be   essential for construction and IT industries to have a collaborative relationship.

References

1. Ubiquitous mobile computing, Alexander Joseph Huber, Jinhan publishing Co., 2003

2. Embedded S/W platform technology development for u-Korea, Kim Heung-Nam, u-Korea Forum foundation commemoration seminar materials, 2003

3. Technical Components for Ubiquitous Computing of Construction Processes, Mu-Wook Pyeon, Dong-Ho Ha , Byoung-Kil Lee, Jin-Nyoung Lee, PRIMA2005, 2005

4. Application Technology of Unique Feature Identifier,   the Korean Society of   Open GIS, 2002

5. Jim Garret, Advanced Infrastructure System, http://www.ices.cmu.edu/ais.html, 2004

6. Facility information integrated management system, Korea Infrastructure Safety& Technology Corporation, 2002.1

7. Pervasive Computing Handbook, Uwe Hansmann et al., Jinhan publishing Co., 2003

Theme seven:

Measure and monitoring of bridge
deformations and deflection

Monitoring of highway bridges in areas under mining exploitation influence

Dr Eng M. Salamak, Silesian University of Technology, Gliwice, Poland
Dr Hab Eng J. Weseli, Silesian University of Technology, Gliwice, Poland
Dr Eng A. Radziecki, Silesian University of Technology, Gliwice, Poland

Abstract

The construction of a highway in the Silesian urban agglomeration, in the operation area of a number of coal mines, is a specific task. This is mainly because of the need to take into account in the design of the highway, bridges and all of accompanying facilities the necessity of protection against strong ground surface deformation. The safe use of highway bridges necessitates constant monitoring of their condition. Although monitoring is usually performed during the use of the bridge, it sometimes has to be implemented at an earlier stage. The situation that occurred during the construction of the Silesian segment of the A4 motorway may be defined as "interaction of highway construction with mining operations". Monitoring was performed in accordance with a programme developed earlier. That programme included indication of the type of measurements to be taken at each individual structure with the marking and arrangement of control points. The scope of observation was gradually expanded as construction operations proceeded.

Introduction

The Upper Silesian Agglomeration is a cluster of neighbouring cities forming a strip about 70 km wide with highly concentrated industrial and business activities. The Agglomeration is inhabited by ca. 10% of the population of Poland, and the average population density is over 1900 persons/km^2. It is also a large transportation centre - due to both geographical location as well as the volume of transport. It is an important component of the European transport network (Figure 1). The predominant traffic direction is east-west (Berlin-Cracow-Lvov) along the existing A4 motorway and the local thoroughfare (DTS). The total traffic flow along this direction may in the near future reach 150 thousand vehicles per day. The other traffic direction, north-south, comprises mainly the A1 motorway under construction, part of the transeuropean north-south route, which links Scandinavia with the South of Europe.

All these transport routes run across areas of intensive underground mining activities. There are nearly 40 coal mines in the region. Total output of these mines is close to 100 million tonnes of coal per year. One of the effects of these activities are ground surface deformations which cause damage to the infrastructure.

This calls for the taking into account of protection against strong effects of ground surface deformation in the design of civil engineering structures and for continuous monitoring of the condition of the latter. That is one of the factors, which contribute to the high cost of motorway construction here, amounting to nearly 15 million Euro for 1 km, which is about three times as high as in other regions of Poland.

Bridge design, construction and maintenance 2007, Thomas Telford, London

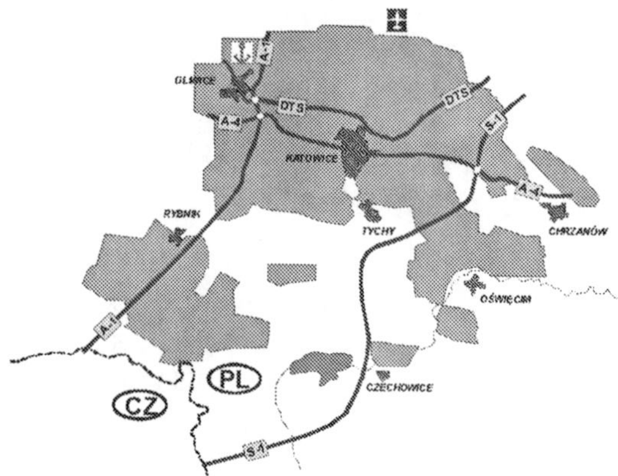

Figure 1. Main transport corridors in the Silesian region against the background of mining areas

Linear structures in areas of mining operations

The construction of structures in an area where numerous coal mines operate is specific. This is mainly because of the need to take into account in the design of structures and facilities the necessity of protection against strong ground surface deformation. This applies particularly to structures of linear configuration that spread over large distances. Such structures, in the first place, include transport ways.

Much work has been done in Poland in the field of research and practical applications in the area of protection of buildings. Numerous regulations and guidelines concern the design and protection of buildings. Mining classification, which is based on the intensity of existing or expected deformations, has even been developed for technological and legal purposes. The parameters of this classification (cf., for instance, [4]) include:
- horizontal strain in subsoil (ε),
- radius of curvature of previously flat surface under the structure (R),
- inclination of previously horizontal surface (T).

In general, the first of these parameters is the weakening indicator of soil, the increasing of soil lateral pressure and the degree of change in the structure foundations geometry. The second parameter illustrates the effect of non-uniform subsidence of supports and associated strong bending in structural components. The third parameter is not in fact a real measure of threats for the load-bearing capacity of structures; it mainly determines the deterioration of the performance of the structure (discomfort and, possibly, damaging effect of failure to maintain vertical and horizontal alignment of structural components). However, the effects of mining deformations on linear structures are more extensive. Even the largest buildings (office buildings, sports arenas, temples) are not subject to effects specific for transport lines and accompanying bridges.

First of all it is important to know the absolute value of ground subsidence (w), which is in fact needless for close buildings. The main parameters important for the design of transport lines and accompanying facilities (drainage facilities in particular) are height relations. Changes in these relations caused by ground surface deformation affect the geometry of the transport line, and consequently the conditions of traffic. Proper drainage of road pavement and substrate may also be affected, or even impossible. Standards concerning the vertical relations in such structures are very strict. For instance, the Polish acceptance requirements for motorways specify the following permissible deviations of actual elevations from design specifications: substrate layer 2 cm, base layer 1 cm, wearing course 1 cm. In continuous bridges of tensioned structure the permissible difference in the subsidence of adjacent supports usually cannot exceed 1 cm.

In linear structures the horizontal strain ε can (in addition to effects on buildings mentioned above) additionally affect the stability of embankment slopes, or even cause change of the shape of embankment, the effect of which is that the road may have to be closed.

The role of the third parameter, inclination T, increases in the case of facilities that accompany transport lines. A bridge, for instance, is not a compact structure (a building), but usually consists of three solid bodies that are to some extent independent and the mutual displacement of which, caused by translation and rotation, are of fundamental importance in the evaluation of ground surface deformation effect on the equivalent stress of structure components.

Linear structure erection phase

Recently another important aspect of linear structures has appeared. It is sometimes difficult to avoid substantial time shifts in the completion of individual segments of a linear structure during its construction. In science and provisions of law regulations concerning the protection of ground surface in mining areas, if looked upon closely, one principal assumption is made: that the structures on the ground surface have a definite structural form. This form can only be changed in the course of damages caused by ground movement, whereas ground surface deformation can be "steady" or "transient". If damage causes substantial change in the form of the structure (including deterioration of usability or increased risk of failure), the proper form is restored by default in the phase of steady deformation.

Recent experience and research work, particularly that gained and carried out during the construction of motorways (large linear structures) [1] indicate that the above assumption may not always be acceptable. In circumstances of working against the clock and maximising profits, it is impossible to avoid carrying on an extensive construction without the occurrence of transient ground surface deformations on some segments of the route. It has therefore become advisable to formally determine a separate process within the "lifetime" of the structure and develop technical and administrative aspects of procedures. This process has been given the name "interaction of motorway construction with mining operations" [8].

Here the interaction of motorway construction with mining operations will mean the coincidence in time of two processes: contemplation and undertaking of mining operations and design and construction of the structure during and within the reach of the mining operations, wherein the type and erection technology of the structure require restrictions to be placed on the method, intensity or even the range of mining operations under the structure.

This clearly indicates that there is a need to apply forecasting and determination of effects directly from displacement of points in space. The information provided by the three components (x,y,z) or (u,v,w) of displacement of individual points of surface being deformed is more valuable than that contained in traditional parameters ε, R i T.

Trends in the design of bridges

Bridges are built in the course of transport routes. It is therefore required that safety conditions be fulfilled for both the structure itself as well as for the traffic of transportation means over the structure. In an area that is subject to deformations, both these conditions may deteriorate. In such cases they have to be restored to the initial state. The methods, means and processes applied to put this into effect are usually mutually contradictory. For this reason two trends can be distinguished:

- protection of the structure: providing low sensitivity of the structure to deformation effects [9] and, consequently, minimisation of activities of restoring the proper condition,
- protection of the traffic: providing low level of perception of the existing deformations by traffic participants and, consequently, minimisation of activities in the area of structure surface.

Until recently it was the first of the two trends mentioned above that clearly prevailed or even completely dominated. The second of them gains on importance as fast traffic lines (motorways and express railway lines) are built. The differences between these trends may be illustrated by the following two examples.

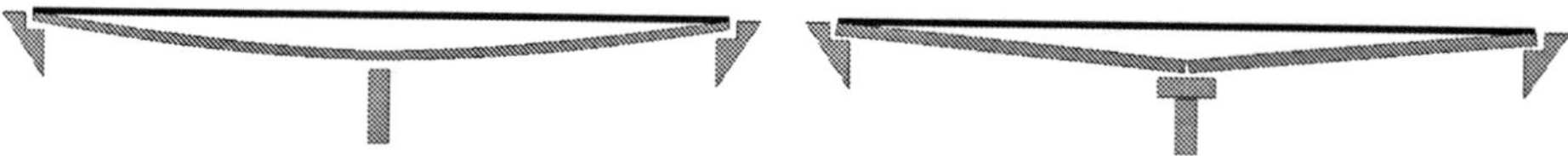

Figure 2. Bending lines of continuous structure (left) and freely supported structure (right)

A double-span beam structure may be designed as a system of two freely supported spans (Figure 2) or as a double-span continuous beam. The former is, at least in theory, totally insensitive to large differences in subsidence of supports, and therefore it does not require any repairs of structure. However, the sharp deflection of the driving path over the middle support introduces discomfort to the fast-moving users or even poses a hazard. In order to eliminate this, the surface must be repaired (smoothed out). On the contrary, the bending line of a continuous beam formed due to subsidence of the support remains smooth and does not affect the comfort of driving. It may, however, generate dangerous moments of forces. Thus the structure requires repair (rectification of support position).

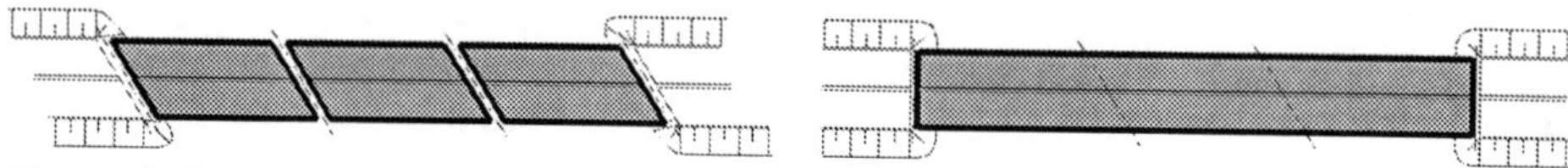

Figure 3. Examples of expansion joints in oblique bridges

Oblique expansion joints in motorway bridges, which are disagreeable and dangerous for the traffic, can also be eliminated in a similar manner (Figure 3). Such joints appear as a result of following the first trend. A bridge like this, even with closely spaced expansion joints, is not completely free from the effects of non-uniform subsidence of supports, and the mutual displacement of spans generates acute flaws in the smoothness of the surface due to sharp oblique deflections. Intermediate gaps should be eliminated in a motorway, and the terminal ones should be perpendicular to the road axis. This improves the comfort of driving, even when there is no ground deformation, but it requires that particular care be taken of the structure.

On the A4 motorway in Silesia both freely supported and continuous structures have been successfully applied (Figure 4).

Monitoring

Regardless of the design being aimed at better protection of the structure or of the traffic, the system requires to be monitored. The difference will consist only in the manner, programme and scope of observation. Both traditional methods of measurement based on inclinometers [6] as well as advanced space technology techniques, such as InSAR satellite radar interferometry [3], can be applied. In general, the objective of operating a monitoring system is to provide safety of use in a broad sense. The system, therefore, in addition to reception (measuring) devices, must comprise state analysers (processors) and protective or corrective devices (manipulators). Depending on the type of anticipated ground deformations, and particularly on the anticipated rate of deformation, such a system may function in a more or less automatic manner. In case of slow continuous deformations, traditional land-surveying methods may prove to be sufficient to determine the condition of the structure and, for instance, the adjustment of the positions of bearings (correction) may be affected by means of jacks of any type. The protection of traffic can also be based on occasional changes of road signs, setting up of detours and similar actions of routine road maintenance. If, however, sudden discontinuous deformations (cave-ins, steps, landslides) are expected, then the condition should be monitored automatically and analysed in real time by an adequately programmed computer which can operate permanent hydraulic cylinders (in the case of, for instance, correction of span elevation) or activate traffic protection device (signalling devices, barricades, etc.).

A new issue in the field of correction is the protection of structures sensitive to the displacement of supports (statically indeterminate). In bridge WD-21, for instance, shown in Figure 4, of critical effect, particularly in structures made of pre-stressed concrete, are the changes in the distribution of vertical reactions to ground deformations. The subsidence of supports and correction of the elevation of bearings is necessary here not only for the sake of traffic safety, but also for the reliability of the structure. An elegant solution would be the use of bearings capable of controlling the reaction and adapting it automatically to the determined permissible values.

Monitoring should be performed throughout the lifetime of the structure - from the start of its construction and during the entire period of its use. During the construction phase it is often necessary to use temporary observation points, which are then transferred to new components of the structure as construction proceeds, and are eventually arranged in their final locations (when construction is completed). This procedure enables the detection of any initial stresses caused by ground movement and occurring in the structure when it is commissioned.

An example of monitoring performed during construction

In this section, selected results of the monitoring process performed during the erection of bridges along a 10-kilometre stretch of the A4 motorway near the city of Katowice is presented. This section of the motorway was constructed in two stages differing in the level of protection against the effects of mining activities.

The monitoring process was applied to all facilities (bridges, footbridges, shallow tunnels and culverts), although not all of them were located within areas of mining activities. Effort was made to observe these facilities from the very start; that is from the moment of concreting foundations. Control measurements were made periodically, usually every two months. In the case of local intensive ground movements, frequency of measurements was increased, even up to one reading every two weeks.

Measurement results formed basis for running analyses of the condition of individual facilities, for drawing conclusions and making recommendations, if necessary. These were included in reports that documented in a continuous manner the changes in spatial arrangement of the structures being erected. The monitoring process was performed mainly for the needs of construction operations. However, it also provided a complete set of observation points at the facilities and a record of the state at the moment of commissioning of the motorway.

Figure 4. Motorway bridges: WD-21 freely supported (left) and WA-12 continuous (right)

Examples of the effects of mining deformations on two bridges being erected are presented below (Figure 4). One of them, overbridge WD-21, was erected during the first stage of construction. The other, underbridge WA-12, was erected in the subsequent stage. There was some difference in design specifications related to the protection of bridges against the effects of mining activities. During the first stage it was decided that bridges (and the road itself) would be protected against mining damages of the 2nd class. In the next stage the bridges, unlike the road and other structures, were to be protected against mining damages of the more severe 3rd class. Three bridges constructed during the first stage had a structure of multi-span continuous slab. Support structures of two of these bridges (including WD-21) were made of reinforced concrete, that of the third one was made of pre-stressed concrete. After the requirements were raised for the subsequent stage, the designers not only adopted higher parameters of bearing and expansion joints movement, but they also abandoned the use of continuous spans and introduced expansion joints over each support (WA-12).

Monitoring was performed on the basis of a "Measurement schedule" developed beforehand. The schedule specified the type of measurements and designation and location of control points in each structure. In general, the following data were recorded:
– vertical displacement of points on support structures,
– horizontal displacement of corner points of the structures,
– inclination of supports relative to two perpendicular vertical planes,
– changes in gaps in expansion joints and, if necessary, movements on selected bearings.

Example of an arrangement of all measurement points, along with designation and numbering thereof on a structure, is shown in Figure 5 and Figure 6. Vertical displacement measurement points for support structures were arranged on upper surfaces of spans (on walkways and median strips) within the axes of all supports. Numbers assigned to these points are preceded with letter "N". Coordinates in the horizontal plane were additionally specified for the corner points. Shifting of the locations of these points was measured with the use of GPS.

During construction, changes in elevation were recorded from the moment of completion of foundations. When elevation measurements were made on final benchmarks on the support structure, the previous elevations (of transient points) were brought conventionally to the level of final points. This way the record of support subsidence is maintained from the start of the construction.

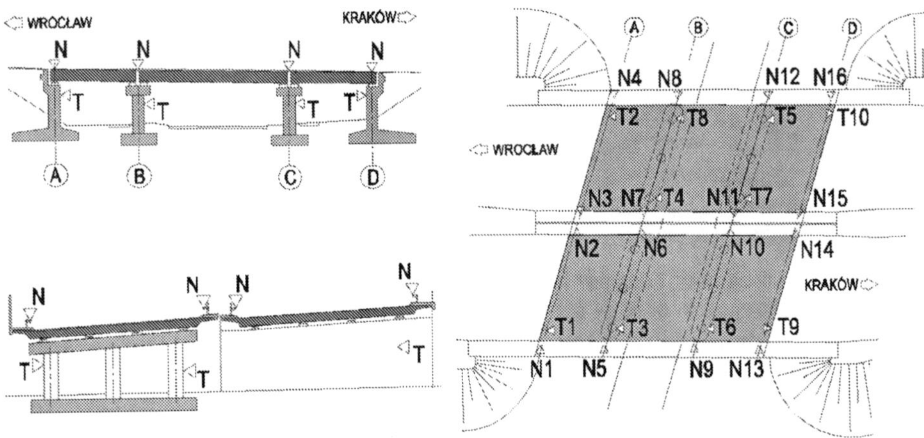

Figure 5. Points of subsidence and inclination measurement of supports on bridge WA-12

Numbers assigned to points used for measurement of vertical alignment of supports are preceded with the letter "T". Each time the inclination of supports in two mutually perpendicular planes is measured. In the case of expansion joint measurements, numbers assigned to points are preceded with "Du" when joints perpendicular to structure axis are measured, and "Dv" when longitudinal joints are measured.

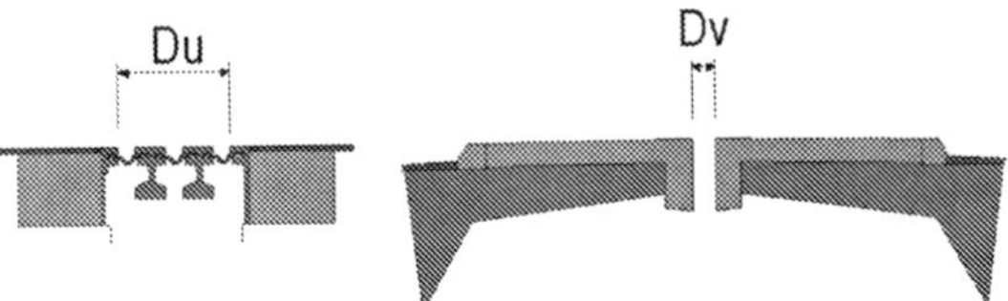

Figure 6. Measurement of displacement on transverse and longitudinal expansion joints

From the beginning of the motorway construction the mines tended to move their working faces away from the motorway strips. Therefore, in most cases, the effects of mining on the structures being built faded with time. The structure of bridge WD-21 subsided during one half of year by −72 to −40 mm. Subsidence increments in subsequent measurement cycles are illustrated in Figure 7. The graph on the left refers to points over supports along the longitudinal axis of the bridge, whereas the graph on the right shows increase in subsidence of two adjacent supports between which the differences were largest.

The maximum difference in subsidence between two adjacent supports reached the value of Δ=16 mm. Differences between other supports are distinctly lower and did not exceed 10 mm. In addition, as subsiding proceeded, there was a tendency to level out the differences.

Inclination of columns did not exceed 1.0%, and the measured movements in expansion joints followed the changes in ambient temperature.

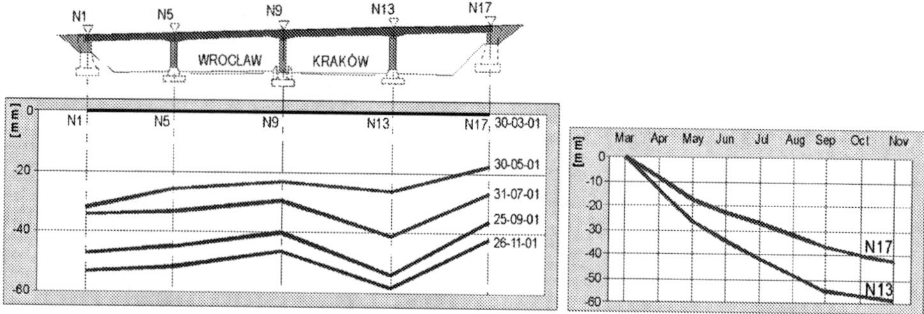

Figure 7. Subsidence of supports of bridge WD-21: in subsequent measurement cycles (left), changes in time of subsidence of supports N13 and N17 (right)

The conditions of construction of the next, western section of the motorway were much more difficult. Strong impact of mining activities was observed in three areas. Bridge WA-12 is situated in one of these areas. The bridge was erected in circumstances of continuous subsidence and other displacements caused by coal mining activities. The effects of ground surface deformation in the near vicinity of the site, with subsidence as large as 2.0 m, required the alteration of the design (e.g. the motorway gradeline) and, consequently, of the scope and schedule of work.

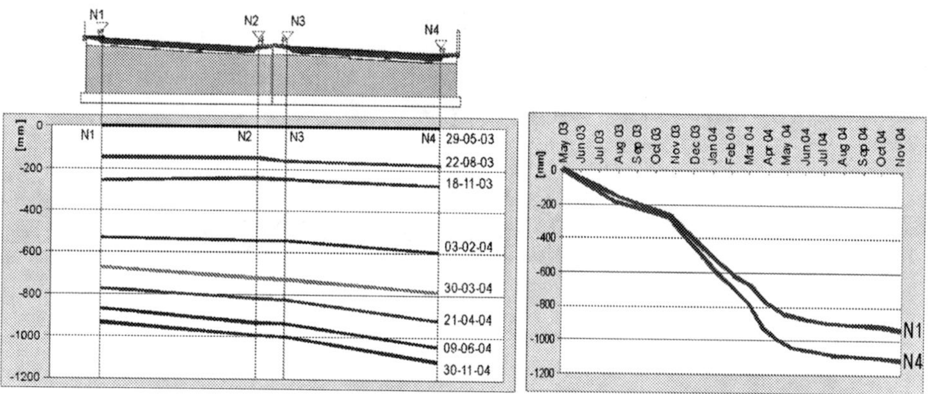

Figure 8. Subsidence of abutment of bridge WA-12: in subsequent measurement cycles (left), changes in time of subsidence of points N1 and N4 (right)

The largest observed subsidence of a complete support of bridge WA-12 exceeded 1110 mm. Figure 8 shows the subsidence of the west abutment of this bridge recorded within 1.5 year after erection. As in the previous figure, the left hand graph illustrates changes in subsequent measurement cycles, the graph on the right shows subsidence increments of two extreme points of the abutment. Subsidence value is more than one metre, in addition there is significant difference in subsidence along the width of the abatement. In this case this resulted in the change in transverse inclination by more than 0.5%.

During the final period of construction favourable conditions were created for finishing all work. This was mainly the result of stopping mining operations with strong impact on ground surface that were planned before and the result of the working face being moved away from the WA-12 bridge construction area. Nevertheless the resultant subsidence values were larger than anticipated.

Summary

Interacting processes should be separated out as a separate branch from the issues of protection of civil engineering structures on mining areas. A special procedure of planning coal extraction and road construction design should be implemented as part of the interacting processes. In such cases it is advisable to set up a coordinating and supervising body provided with legal powers and superior to the parties to this process [2]. Such a supervising body should start acting at the design stage. An indispensable component of such a design should be a joint schedule of construction works and mining operations, which would form the basis for drawing up a special type of forecast of ground deformation. Bridge design must provide for modification of span elevation.

The special type of forecast mentioned above should include isolines of three coordinates of displacement of three surface points at defined moments of time. It should also clearly define specifications for coal mining operations as basis for inspections by the supervising body. The forecast may additionally specify tolerances of the predicted values. Predicted spatial displacement values are necessary and sufficient, whereas classification of mining areas should be used for guidance only, as mining area classification is not adequate for the assessment of linear structures.

The monitoring of sixteen bridges erected along a 10.6 km long stretch of the A4 motorway between the cities of Ruda Ślaska and Katowice carried out in the years 2001 – 2004 enabled running assessment of changes in the position of support structures and, when necessary, taking corrective measures in due time. A complete database of observation points in all the structures was set up and initial measurements were taken (prior to the commissioning of the motorway) for further observation.

Taking into account the production plans of the mines and the still observed displacement of the erected structures, the monitoring process (required by Polish regulations) that has been commenced, must absolutely be continued. It should form part of a system of safe use of the motorway. In view of the specific features of the climate and the area subject to ground surface deformations, it is necessary to utilise state-of-the-art teledetection methods [7]. Traditional radar interferometry is difficult to apply under conditions that prevail in the region of Silesia due to the sensitivity of this technique to environmental impact. High hopes are placed on a modified PSInSAR (Persistent Scaterrer Interferometry Synthetic Aperture Radar) method, which utilises objects that strongly reflect radar waves [5]. The experience gained during the construction of the A4 motorway and the knowledge acquired will be useful for the construction of the Silesian section of the A1 motorway, which has just begun.

References

[1]   Motorways in mining areas [in Polish], GIG, Konferencje 25, Katowice 1998
[2]   Mining operations and ground surface protection – Experience gained from mines in the Walbrzych region [in Polish], A. Kowalskiego (Edit.), GIG, Katowice 2000
[3]   Ge L., Chang H.-C., Rizos C., Monitoring Ground Subsidence due to Underground Mining Using Integrated Space Geodetic Techniques, ACARP Report C11029, 2004

[4]   Kłosek K., Mining area deformation and motorways [in Polish], Magazyn Autostrady, 11/2006, pp. 20-25

[5]   Marinkovic P., Recursive Persistent Scatterer Interferometry, Proc. of Advances in SAR Interferometry from ENVISAT and ERS missions, Rome, 28.11-02.12.2005, 2005

[6]   O'Connor K., Clark R., Whitlatch D., Dowding C., Real Time Monitoring of Subsidence Along I-70 in Washington, Pennsylvania, Transportation Research Record No. 1772, Soil Mechanics 2001, pp. 32-39

[7]   Perski Z., Jura D., ERS SAR Interferometry for Land Subsidence Detection in Coal Mining Areas, Earth Observation Quarterly, No. 63, 1999, pp. 25-29

[8]   Weseli J., On the interactive construction and mining [in Polish], Prace Naukowe GIG, Konferencje 41, Problemy ochrony terenów górniczych, Katowice 2002, pp. 413-417

[9]   Weseli J., On the rules of aligning bridge bearings in order to ensure freedom of expansion [in Polish], Inzynieria i Budownictwo, 10/2004, pp. 538-540

Deflections and Frequency Responses of the Forth Road Bridge Measure by GPS

C J Brown, School of Engineering and Design, Brunel University, Uxbridge, UK.
G W Roberts, IESSG, University of Nottingham, Nottingham, UK.
C Atkins, IESSG, University of Nottingham, Nottingham, UK.
X Meng, IESSG, University of Nottingham, Nottingham, UK.
B Colford, Forth Estuary Transport Authority, South Queensferry, West Lothian, UK.

Abstract
Using kinematic GPS to monitor the deflections of large suspension bridges is an area of joint research at the University of Nottingham and Brunel University. The technique to enable the magnitude of both the deflections and frequencies to be measured is well established. Many novel techniques have been developed which tackle the issues of multipath on such structures as well as tropospheric issues and integration of the GPS with pseudolites and accelerometers to enhance the satellite signals.
The following paper reports some elements of experiments carried out using the Forth Road Bridge in Scotland in February 2005. 46 hours of data were collected on a continuous basis. Two GPS reference stations were located on the south side of the north-south orientated bridge. A further 5 GPS receivers were placed upon the bridge's deck, situated in an optimum layout, and two GPS receivers were placed on top of the southern support towers. The receivers were Leica survey grade SR530 and 510 receivers, and data was gathered at a rate of 10Hz.

INTRODUCTION

The following paper details the ongoing research in deflection monitoring and deformation monitoring of structures, notably bridges. The use of kinematic GPS is being used for this and the work has been ongoing for about a decade at the University of Nottingham and in collaboration with Brunel University [1], [2], [3]. It is possible to measure 3D deformations of discrete points upon the bridge at rates of up to 100Hz using GPS alone. It is also possible to measure sub-centimetre precisions over the short baselines used for this work. Typically the baseline lengths from the reference to rover receivers are approximately 1km or so.

Initial trials were carried out on the Humber Bridge and since then a number of trials have been carried out on other bridges including the Wilford Suspension Bridge in Nottingham, the Millennium Bridge in London and more recently the Forth Road Bridge in Scotland. During the trials, survey grade dual frequency GPS receivers are used. However, more recently

Bridge design, construction and maintenance 2007, Thomas Telford, London

research has also been carried out investigating the use of single frequency survey grade receivers (code and carrier) [4], [5]. Further to this, trials have been carried out investigating the use of cheap hand held GPS receivers as it is now possible to output the carrier phase from these receivers [6].

The work itself investigates the use of kinematic GPS as well as comparing this to the modelling of such structures. The overall aim is to be able to use a fixed number of GPS receivers located upon the bridge at pre-determined discrete locations and comparing the movements at these points with the FEM. Once agreement between the real data and the FEM established, then it is then possible to use the FEM to model how the remainder of the bridge moves based upon this real data.

THE FORTH ROAD BRIDGE GPS TRIALS

The Forth Road Bridge has an overall length of 2.5 km and a main span length of 1,005m. It was opened in 1964. Traffic has steadily increased over this bridge, from 4 million vehicles in 1964 to over 23 million in 2002. In addition, the heaviest commercial vehicles weighed 24 tonnes; the current limit is 44 tonnes. When the bridge was opened, it brought to an end an 800 year history of ferry-boat service across the river at Queensferry in Scotland.

The data gathering trials were conducted over a nearly continuous 46 hour period from 11am on the 8 February 2005 to 9am on the 10 February 2005. For the whole period, 7 GPS receivers were located upon the bridge, as illustrated in Figure 1, and two reference GPS receivers were located on the viewing platform adjacent to the FETA (Forth Estuary Transport Authority) building, Figure 2. The GPS receivers gathered data at a rate of 10Hz. In addition, an Aplanix INS was located adjacent to point E [7]. The layout of the GPS antennas meant that a GPS antenna was located at each of the east side mid span, 1/4 span, 3/4 span and 3/8 span as well as the west side mid span and on top of the two southern towers. A selection of Leica SR530, SR510 and GX1230 surveying GPS receivers were used in conjunction with lightweight and choke ring GPS antennas.

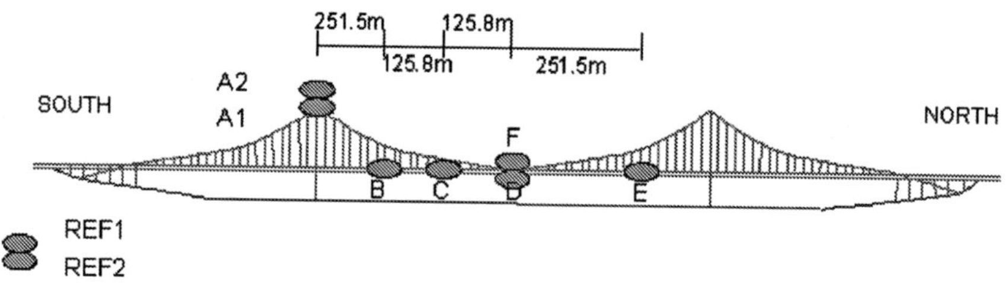

Figure 1. Schematic of the Forth Road Bridge, and GPS receivers.

Figure 2. Two GPS reference receivers located adjacent to the Forth road bridge (left). A GPS choke ring antenna attached to the bridge (right).

During the second night, two 40 tonne lorries were hired by FETA, accurately weighed and used as a control loading of the bridge. These trials were carried out a couple of hours after the high winds experienced subsided slightly, and during these specific trials the bridge was closed off to other traffic. The trials were carried out in the early hours of the morning, when the traffic flow over was at a minimum, and only closed whilst the control lorries passed over the bridge ad re-opened whilst they turned around before subsequent crossings. The lorries started the trials at the North end of the bridge, and the manoeuvres were as follows.

- 1 lorry ran from North to South
- 1 lorry ran from South to midspan on west side, stopped then the other lorry moved north to south
- 1 lorry moved from north to south and stopped at midspan, other moved south to north
- 1 lorry moved from south to north, and then both moved side by side north to south

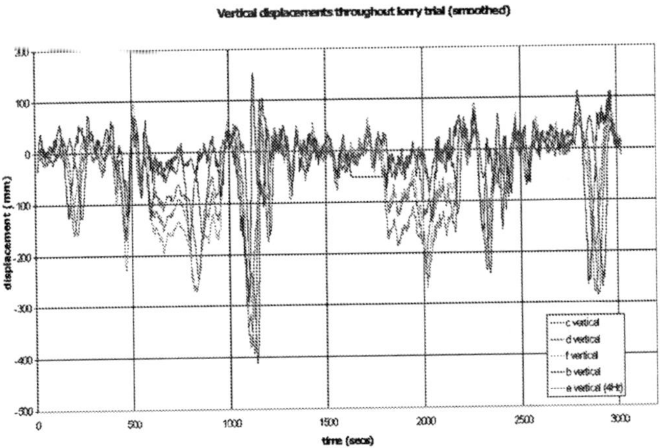

Figure 3. Height Deflections of the Bridge During the lorry trials.

During these trials, the lorries travelled at 20 mph. Figure 3 illustrates the overall movements experienced by the bridge in the height component for the whole trials. The results show that the bridge deflected by up to a maximum of 400mm. This is not due to the combined 80tonne loading, but happens when the bridge is opened to the waiting bunch of vehicles following the controlled manoeuvres, but represents a wave of traffic as it crosses the bridge.

Figure 4 illustrates the final manoeuvre whereby the two lorries travelled from North to South whilst located side by side at 20mph. The graph also shows the physical location of the lorries at any time e.g. Midspan, North Tower etc

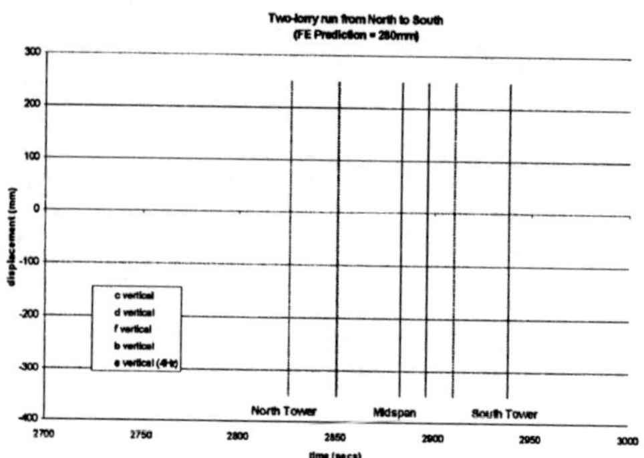

Figure 4. Height deflections with the two 40tonne lorries passing over the Bridge side by side.

Three main phenomena are evident in Figure 4. Firstly the deflections are offset from each other. Secondly, the GPS receivers located at sites D and F, midspan, deflect by different magnitudes, even though they start off at the same height. This is due to the torsional movement of the bridge. The lorries, travelling on the left hand side of the carriageway from North to South, were in fact travelling on the East side of the bridge. Hence the eastern side (site D) deflects more than the Western side (site F). Thirdly, the reader should note that the bridge consists of three separate spans, each connected through a cable which passes over the top of the towers. As the lorries pass over the Northern side span, the load pushes this smaller span down, which in turn pulls the hanger cables down and the suspension cable which they are attached to. This then results in the suspension cable pulling up on the main span. This is evident in Figure 4 at around 2,800s. The lorries pass into the main span, and their passage over the measured positions are shown in Figure 4. As the lorries pass into the southerly side span, upward movement of the main span – described above – is observed.

NATURAL FREQUENCIES

Matlab's Fast Fourier Transform (FFT) algorithm was used to apply spectral analyses to the data gathered by each receiver so that the natural frequencies of the bridge deck and towers could then be identified. As stated above, the 'north-south' movement of the tops of the towers corresponds with vertical movements of the bridge deck. Evidence for this can be seen in the Figures below. Referring back to Figure 1, Figures 5 and 6 show the vertical deflections of the receivers placed on the mid-span of the deck (receivers D and F) at 23:00 – 23:01 on 8th February. Figures 7 and 8 show the corresponding 'north-south' movement of the tops of both towers (receivers A1 and A2). It appears that when the height of the deck increases at both receivers on the deck, both receivers at the tops of the towers move 'south'. Hence it would be expected that a natural frequency identified on the bridge deck through spectral analysis would correspond with a frequency identified on the towers.

Two time periods were chosen in which to analyse the natural frequencies of the bridge: a quiet period at 02:00 – 03:30 hours on 9th February, and a period with heavy rush-hour traffic loading at 07:30 – 09:00 hours on the same day. The weather conditions for both periods, including both wind speed and direction, were very similar. A spectral analysis was applied to each session for each receiver. At this time, considering the formula frequency = $\sqrt{(}$ stiffness / mass $)$ [8], it was thought that GPS may be able to detect slight decreases in natural frequency in the rush-hour period compared with the quiet period due to the increase in traffic loading.

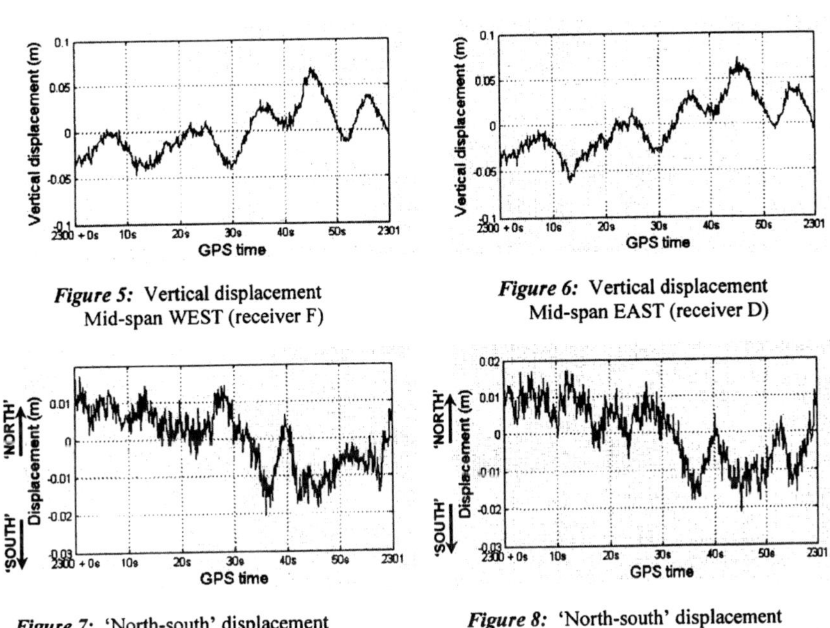

Figure 5: Vertical displacement
Mid-span WEST (receiver F)

Figure 6: Vertical displacement
Mid-span EAST (receiver D)

Figure 7: 'North-south' displacement
Tower WEST (receiver A2)

Figure 8: 'North-south' displacement
Tower EAST (receiver A1)

As well as seeking the vertical and lateral frequencies of the bridge deck, the torsional frequency of the deck was also analysed. This was achieved by subtracting sample readings gathered by receiver F (mid-span, western side) from the corresponding samples gathered by receiver D on the eastern side. The resulting movement is shown below in Figure 7 which shows a very clear sinusoidal wave. This example is taken from the period 07:45 – 08:00 on 9th February. Figure 8 shows the corresponding spectral analysis, illustrating a clear torsional frequency of 0.27Hz.

In the same way the 'north-south' torsional movement of the towers was analysed by subtracting the data from the east tower with the data from the west tower. The resulting movement for 07:30 – 07:35 hours on 9th February is plotted in Figure 9. Again a sinusoidal wave is evident. The spectral analysis for all torsional movement between 07:30 – 09:00 on

9th of February is shown in Figure 10. The three peaks shown correspond to the frequencies 0.106Hz, 0.181Hz and 0.268Hz (see Table 2 below).

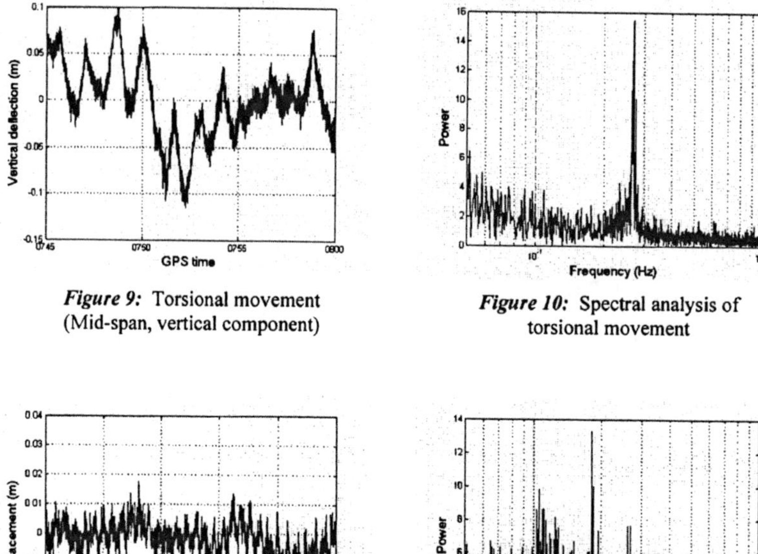

Figure 9: Torsional movement
(Mid-span, vertical component)

Figure 10: Spectral analysis of
torsional movement

Figure 11: Torsional movement
(Towers, 'north-south' component)

Figure 12: Spectral analysis of
torsional movement

Results from the spectral analyses are summarised in Table 1. All frequencies are given to four decimal places.

The first natural frequencies of the bridge deck are approximately 0.10Hz vertically and 0.07Hz laterally – i.e. the lateral mode gives the lowest natural frequency. These frequencies are easily identifiable at all receivers on the bridge deck. Also, the frequency of the vertical movement of the deck (0.10Hz) corresponds with the frequency of the 'north-south' movement of the top of the towers, as predicted at the start of this section.

Table 1: Frequencies (Hz) detected through spectral analyses – 'quiet' and 'busy' period

		02:00-03:30 (quiet period)	07:30-09:00 (busy period)
Mid span E	Vertical movement	0.1041	0.1026
	Lateral movement	0.0679	0.0668
Mid span W	Vertical movement	0.1041	0.1026
	Lateral movement	0.0679	0.0668
Mid span E & W	Torsional movement	0.2699	0.2684
3/8 span	Vertical movement	0.1040	0.1027
	Lateral movement	0.0679	0.0668
1/4 span	Vertical movement	0.1041	0.1027
	Lateral movement	0.0673	0.0668

The torsional movement of the bridge deck has a frequency of approximately 0.27Hz, and this naturally corresponds to secondary frequencies identified in vertical and lateral movement at other locations on the deck since these are components of torsional movement. This frequency is detected in the movement (both 'north-south' and torsional) of the towers during the 07:30 – 09:00 period, but was perhaps too weak to be detected in the early morning period of 02:00 – 03:30 due to the lack of traffic to excite the bridge.

Table 1 also shows that the frequencies identified in the quiet period were also identified in the busy 'rush-hour' period, almost without exception. The Table also suggests that the natural frequencies identified during the rush-hour period are very slightly lower than in the quiet period, and this is the reason why the frequencies in Tables 1 are stated to four decimal places. For example, the first vertical modal frequency of the bridge deck was consistently identified as 0.104Hz in the quiet period and 0.103Hz in the busy period. The first modal frequency of lateral movement is identified as 0.068Hz in the quiet period (except at 1/4-span) but as 0.067Hz in the busy period. Other examples, and some exceptions, can also be identified. Such small changes in frequency could be explained by the lack of frequency resolution in the spectral analysis, as discussed above, but the general lowering of frequencies between the two sessions could perhaps be due to the higher traffic loading during the rush-hour. Natural frequency is a function of both the mass and stiffness. Natural frequency determination of complex structures is challenging, but in essence for a suspension bridge it is a function of $\sqrt{(\text{stiffness} / \text{mass})}$, [8] so, in the case of bridge monitoring, an increase in mass would yield a relatively small reduction in natural frequency. The small changes in frequency reflect the fact that the total mass of extra traffic during the rush-hour, which may amount to a few hundred tonnes, is relatively insignificant compared to the mass of the bridge itself, which is approximately 39,000 tonnes (Forthbridges.org.uk, 2006).

CONCLUSIONS

The trials have shown that it is indeed feasible to use GPS on such structures to measure the magnitude and frequencies of the bridge's deflections in 3-D. This is possible at a rate of up to 10Hz, and all the results are synchronised to each other. Although the trials were carried out in a post processing manner, it is possible to have carried out these trials in real time.

The results have been compared to a FEM of the bridges, and this could well be the basis of future bridge monitoring whereby real GPS data from specific points on a bridge are used in conjunction with FEM, or similar model, to assess the behaviour of a bridge. If the structure deteriorates over time or if any specific mishaps occur then these actions may well be picked up through the model and GPS data.

REFERENCES

1. Ashkenazi, V., Dodson, A. H., Moore, T., and G. W. Roberts (1996) Real Time OTF GPS Monitoring of the Humber Bridge, Surveying World, May/June 1996, Vol. 4, Issue 4, ISSN 0927-7900, pp 26-28.
2. Ashkenazi V. and G. W. Roberts (1997) Experimental Monitoring the Humber Bridge with GPS. In: Proc. Institution of Civil Engineers; Civil Engineering, Nov 1997, vol 120, Issue 4., pp. 177-182. ISSN 0965 089 X.
3. Brown, C. J., Karuna, R., Ashkenazi, V., Roberts, G. W. and R. Evans (1999) Monitoring of Structures using GPS, In: Proc Institution of Civil Engineers, Structures, February 1999, pp 97 - 105, ISSN 0965 092X.
4. Roberts, G. W., Meng, X., Cosser, E. and A. H. Dodson (2004) The Use of Single Frequency GPS to Measure the Deformations and Deflections of Structures. In: Proc of the FIG Working Week, May 2004, Athens.
5. Cosser, E., Roberts, G. W., Meng, X., and A. H. Dodson (2003) The Comparison of Single Frequency and Dual Frequency GPS for Bridge Deflection and Vibration Monitoring. In: Proc of the Deformation Measurements and Analysis, 11th International Symposium on Deformation Measurements, International Federation of Surveyors (FIG), May 2003, Commission 6 - Engineering Surveys, Working Group 6.1, Santorini, Greece.
6. Cosser, E., Hill, C.J., Roberts, G. W., Meng, X., Moore, T. and A. H. Dodson (2004a) Bridge Monitoring with Garmin Handheld Receivers. In: Proc of the 1st FIG International Symposium on Engineering Surveys for Construction Works and Structural Engineering, June 2004, Nottingham, UK.
7. Hide, C; Blake, S; Meng, X; Roberts, G W; Moore, T; Park, D; An Investigation in the use of GPS and INS Sensors for Structural Health Monitoring, Proc ION GNSS 2005, 13-16 September, 2005, Long Beach, California.
8. Ghali, A., Neville, A. M., Brown, T.G., 2003. Structural analysis: a unified classical and matrix approach, 5th ed. London : Spon Press, 2003.

Using GPS to Measure the Deflections and Frequency responses of the London Millennium Bridge

G W Roberts, The University of Nottingham, UK
C J Brown, Brunel University, UK
X Meng, The University of Nottingham, UK
P R B Dallard, Arup, UK

ABSTRACT

The following paper was initially published in the ICE Bridge Engineering Journal[1] and reproduced for the conference with their kind and full permission. Previous work by the authors on the Humber Bridge has proved that it is possible to use carrier phase kinematic GPS (Global Positioning System) to give the absolute 3D position of receivers fixed on such a structure. Through gathering the GPS positional data at a rate of up to 10 Hz, both the displacement and the movement of the structure can be calculated.

Trials conducted on the London Millennium Footbridge in 2000 monitored movements using three GPS receivers. GPS data were obtained over a total period exceeding 11 hours and were post-processed.

The London Millennium Bridge is well documented; it runs approximately north-south from St Paul's on the north bank towards the Tate Modern on the south bank of the Thames. After initial processing, the results gave plausible vertical and lateral (sideways) displacements, but also described implausible longitudinal movements along the axis of the bridge. DSP (Digital Signal Processing) techniques applied to the data are reported.

Due to the satellite constellation in the UK, a large void exists from the zenith to the horizon in a northerly direction. This lack of GPS satellites results in poor satellite geometry in the North-South direction, and hence poor North-South precision, and so this geometric phenomenon explains the unsatisfactory element of the Millennium Bridge results. More importantly a technique to overcome the issues is described. A software based GPS data simulator, developed at the IESSG, was then used to re-create the actual Millennium Bridge data. Subsequently pseudolite data was simulated, and showed that if a pseudolite had been used on the day of the field trials more reliable results would have been obtained.

Notwithstanding this, the lateral vibration frequencies obtained from the GPS data agree extremely well with those found in the modal surveys carried out by Arup, even when displacement amplitudes are small.

Bridge design, construction and maintenance 2007, Thomas Telford, London

INTRODUCTION

Kinematic GPS (Global Positioning System), real time and post processed, is now widely used in engineering applications, such as in landslide observations, construction plant control and structural deformation monitoring[2-7]. In these applications, three-dimensional positioning precision in a local coordinate system is required. Some of these applications have specific precision requirements in the vertical direction, for example when analyzing land subsidence, while others may be concerned with the north and east direction, such as when measuring the structural deformations caused by wind loading. It is important to understand the achievable positioning precision with the current GPS constellation and potential augmentation techniques for such engineering applications.

The London Millennium Footbridge[8] links St Paul's with the Tate Modern Gallery. The footbridge is a shallow suspension bridge with cables below the deck, allowing pedestrians unobstructed views of the surrounding area. The bridge has spans of 81m (north), 144m (centre) and 108m (south). The opening of the bridge on 10 June 2000 attracted large crowds of people. It was estimated that up to 2,000 people were on the bridge at any one time. Unexpected lateral vibrations of the bridge occurred.

Arup undertook extensive research into the cause of the vibration and showed that it was due to a phenomenon that has become known as Synchronous Lateral Excitation. People generate small lateral forces when they walk. Within a crowd of pedestrians walking on a fixed surface, these lateral forces are essentially uncorrelated. On a surface that is subject to lateral vibration, like a bridge deck, pedestrians modify the way they walk such that they generate a correlated lateral force acting to further excite the vibration. Arup were able to derive a model for this type of pedestrian excitation and use it successfully to design a retrofit damping system. As part of their studies, Arup carried out modal surveys of the bridge to obtain natural frequencies and mode shapes. This data confirmed the accuracy of the finite element models originally used in the design of the bridge. Today, the Bridge is fully operational and is a major tourist attraction.

While the bridge was closed for assessment, and prior to installing the damping system, the authors were able to conduct trials upon the bridge in order to assess whether GPS could effectively monitor the deflections of such a structure over the longer term. The bridge remained closed to pedestrians, but on one of the days there was a significant (blustery) wind load acting.

This paper details the trials and some of the outcomes. Many of the outcomes again confirm the accuracy and precision of the technique and its potential and ease of use for application to long-term structural deformation monitoring. However, one element of the data analysis suggested an unrealistic deformation along the axis of the bridge (North to South direction). Many checks have subsequently been made that demonstrate plausible explanations for the data, and illustrate that the GPS constellation results in noise that could be interpreted as inaccurate measurement.

KINEMATIC GPS

GPS is an US owned and operated satellite based positioning system. Initially conceived in the 1960s and developed into the 1970s, the first satellite was launched in 1978. It was designed as a precise timing and positioning system. Consisting of more than 24 satellites, orbiting the Earth at approximately 22,000 km, the original idea was for a system capable of providing 3D coordinates anywhere in the world where the sky is seen. Different accuracies are available depending on the type of equipment and processing software used. Basic handheld GPS receivers allow accuracies of approximately 10m to be achieved, whilst survey grade GPS receivers allow sub cm accuracies to be achieved.[9-11].

The code used for stand-alone positioning is transmitted via a carrier wave; Kinematic GPS uses this carrier wave to define distances. Using the precise carrier wave one receiver is positioned relative to another - the reference receiver typically being located over a survey station with a known coordinate. Today, resolution of a millimetre or better can be achieved by using survey grade GPS receivers. Kinematic GPS can be carried out in either a real-time manner - where low powered UHF (Ultra High Frequency) data links or GSM (Global System for Mobiles) links are typically used to transmit the data from one receiver to the other, or post processed - where the data is gathered at both reference and rover and post processed. In either case specialised processing software is required. It is thus possible to achieve relative 3D positions with excellent precision, currently at a capture rate of up to 100 Hz. The distance from the reference to rover (baseline) is typically anything up to a few kilometres. Such a distance would be defined as a relatively short baseline in surveying terms but enables measurements on large-scale structures to be easily achieved.

Various error sources exist in the GPS, as the signals from the satellites are simply low powered radio signals. Errors, such as troposphere and ionosphere refraction of the signals, clock errors in both the GPS receivers and satellites, as well as ephemeris errors (GPS satellite location data) are differenced away using kinematic GPS because of the relative nature of the positioning and the fact that the error is the same on both ends of the baseline. However, these errors become de-correlated as the baseline length increases. This is why, in good practice, the reference receiver is placed as close to the rover as possible. One remaining error source, however, is multipath. This is not the same at both ends of the baseline, and is caused through the GPS satellite signal reflecting off adjacent structures, and the GPS signal therefore reaching the antenna via more than one route. This leads to noisy results.

EXPERIMENTAL METHOD

The viability trials were conducted on the 22nd, 23rd and 24th November 2000. The reference station and the three rover stations each consisted of Leica AT504 choke ring antennas connected to Leica SR530 dual frequency GPS receivers. Choke ring antennas (a GPS antenna surrounded by concentric rings of metal) were used to absorb multipath signals.

The antennas were attached at pre-defined locations on the bridge handrail using rigid clamps. The reference receiver was fixed in a stationary location in an adjacent building (St Paul's School). The locations meant that the roving antennas had an unobstructed view of much of the sky. The GPS data was gathered at all the receivers at a rate of 10Hz and stored locally on memory cards. The reference station's coordinates were established through processing all the data gathered relative to Ordnance Survey active stations located around London.

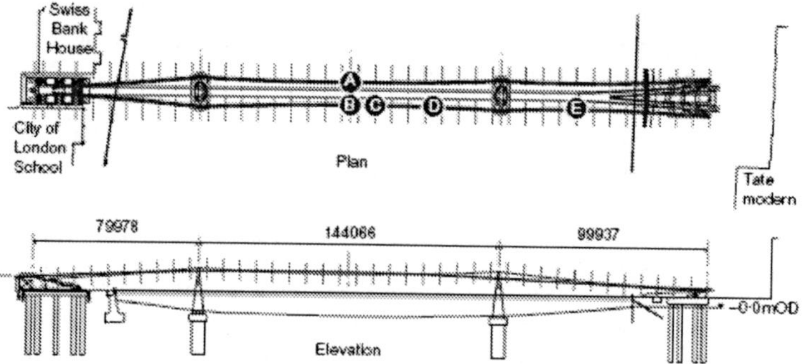

Figure 1. The locations of the bridge's monitoring stations

A Kistler tri-axial accelerometer was also used, co-located at one of the bridge GPS positions, gathering data at a rate of 200 Hz. The other roving GPS receivers were located at two of the four prescribed locations upon the bridge. The locations and sessions are illustrated in Figure 1. All the data were gathered and periodically downloaded for post processing.

During the trials there was sporadic pedestrian movement on the bridge; only construction workers and the researchers were allowed on the bridge at this time, but no significant loading (other than wind) was applied to the bridge during the trials.

In all, 11 hours of GPS data were gathered at a rate of 10Hz, and 4 hours of accelerometer data at a rate of 200Hz. This resulted in nearly 2 million and 3 million data points respectively. The GPS post processing was carried out using both Leica's SKI-PRO and the IESSG's Kinpos. Figures 2 and 3 illustrate the reference GPS receiver and choke ring antennas and one of the bridge monitoring stations respectively.

Figure 2. Setting up the reference GPS Receiver (Leica SR530)

Figure 3. An integrated device for housing a GPS antenna and a Kistler accelerometer

Initial results

The GPS data were post-processed, and the results converted from GPS coordinates into "bridge coordinates". The orientation of the bridge is known in WGS84 (World Geodetic System), derived from the coordinates of two GPS stations along the length of the bridge. WGS84 is an international reference system, defined in such a way as to "best fit" the world as a whole[12]. In descriptions below, the lateral direction is the horizontal normal to the

Bridge's span and the longitudinal runs along its length. The bridge runs approximately North-South.

A datum for the deflections is required. In processing the data for engineering purposes, a datum for long term deformation monitoring would be adopted as well as for measuring short term deflections. As there was no significant live load the datum was taken as the mean deflection for all the observations. Figures 4, 5 and 6 illustrate a typical data set of the initial results collected from the west central span on 24 November. A zero datum is used for convenience but is an arbitrary reference point for the results. This was achieved by taking the average of all the data points for the moving bridge. The same datum is used throughout this paper. These results are taken over approximately a 36 minute period, resulting in nearly 21,727 data points for each of the three orthogonal coordinate directions. This is about 5% of the available data.

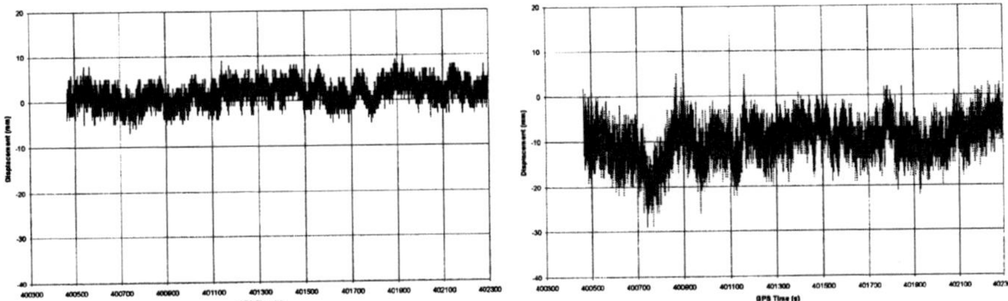

Figure 4. Unfiltered lateral deflections   Figure 5. Unfiltered longitudinal deflections

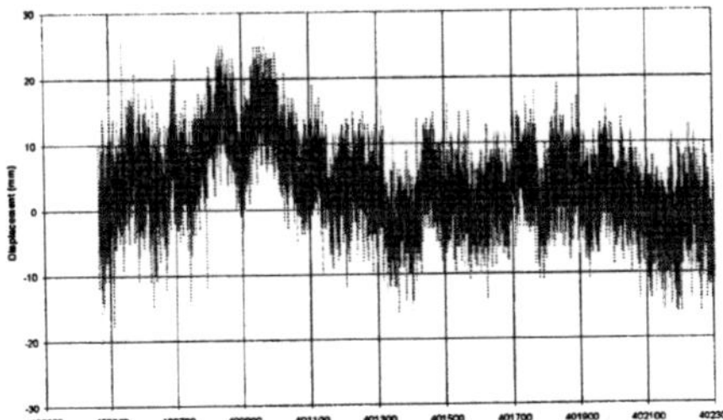

Figure 6. Unfiltered vertical deflections

Signal Processing

There are many formal digital signal processing techniques available. In this work a hybrid approach has been taken to process the data. After processing through systems specific to GPS, the basic data produced consists of Cartesian coordinate data relative to the bridge axis as described above. Significant software advances have meant that most of this data is "clean", and has very few "erroneous" points, cycle slips, or periods in which data recording was stopped for some reason. Nevertheless, the data as shown in Figures 4-6 is somewhat difficult to interpret for engineering purposes. Particularly where the structural deformation is small, the high noise/signal ratio means that some form of filtering is essential. There are many possibilities in the engineer's armoury, ranging from a simple moving average to Butterworth filters or similar[13, 14]. In Figures 7a and 7b the outcomes from a moving average are presented and show that this simple approach acts to clarify the results considerably. However, to obtain principal frequencies from the data, FFT (Fast Fourier Transforms) and

PSD (Power Spectral Density) (convolution) analysis is more useful; further results using these techniques are presented below. In addition, when the detailed analysis of the longitudinal movements were carried out, a wavelet analysis was completed in order to identify the component parts of the signal.

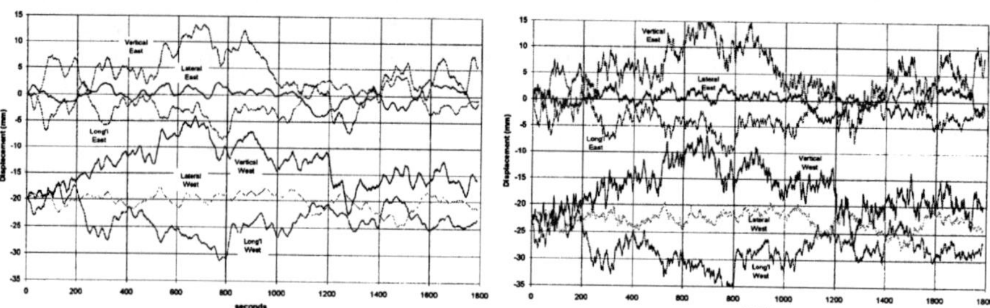

Figure 7a. A 20s filter applied to the two Mid-span GPS receivers' deflections

Figure 7b. A 5s filter applied to the two mid-span GPS receivers' deflections

RESULTS

There is, as expected, good agreement between the two GPS receivers located on opposite sides of the Bridge's midspan centre (Figure 7). Closer inspection (Figure 8) shows that there is significant synchronous variation in the vertical displacements at these two locations as a result of torsion movement along the bridge axis, and any frequency of this movement can be more readily found from these derived results.

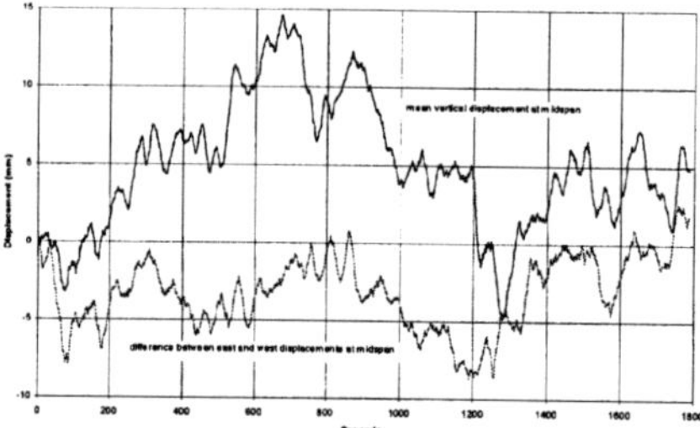

Figure 8. Vertical movements – mean and difference of the two mid-span receivers

Figures 9a and 9b illustrate the lateral dynamics of the middle span and the southern span respectively. The power spectrum analysis carried out on the results for the main span show that a primary frequency value of 0.52 Hz and a secondary value of 0.97 Hz were obtained using the GPS results. The southern span's results show primary values of 0.75 Hz. These frequencies agree well with those found in the modal surveys of the bridge. The actual frequencies are known to vary slightly with temperature (which affects the cable tension) and excitation amplitude (which determines whether joint stiction will be overcome). In the modal surveys the first lateral mode of the centre span occurred between 0.49 and 0.51Hz, the second mode between 0.95 and 0.98Hz and the first mode of the south span 0.78 to 0.82Hz. It should also be noted that the amplitude of displacements is small. The range of Figure 9a is 20mm, and of Figure 9b is 4mm. The accuracy of these results suggests that the technique can be applied to structures with vibrations of quite small amplitude.

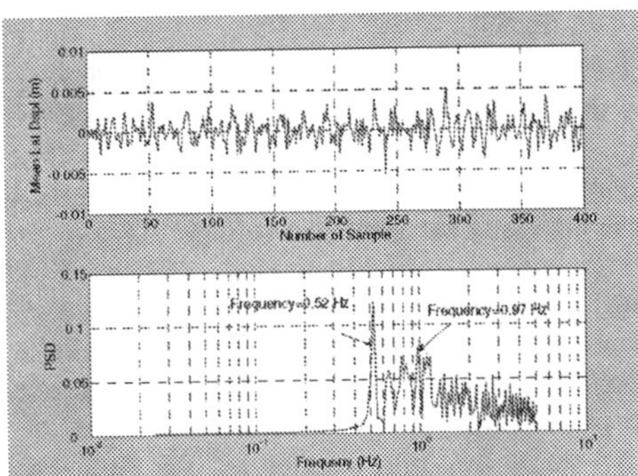 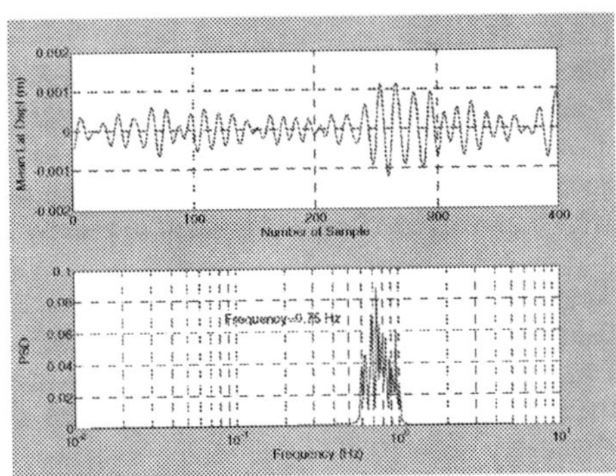

Figure 9a. Mean lateral movements and power spectral density (PSD) of the lateral dynamics on the middle span, modal survey results 0.49-0.52 and 0.95-0.98 Hz

Figure 9b. Mean lateral movements and power spectral density of the lateral dynamics on the south span, modal survey results 0.78-0.82 Hz

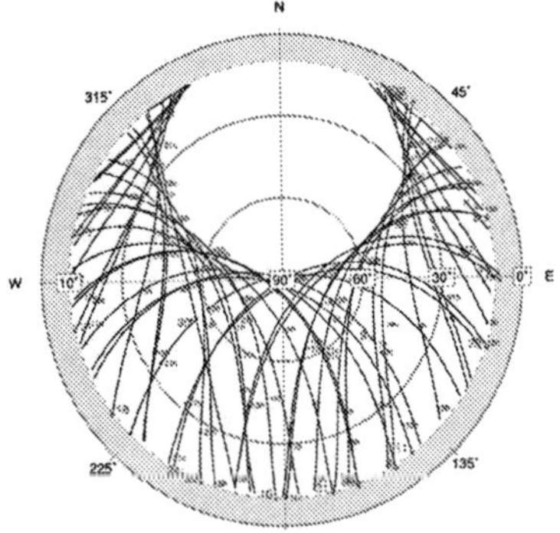

Figure 10. A 24h sky plot of GPS satellites over the Millennium Bridge

However, data also suggested that the longitudinal deflections of the bridge were in fact larger than those in the lateral direction. This was thought to be unbelievable for such a bridge. When detected, this caused some concern and was inexplicable at that time. The authors considered whether this was a measurement or interpretation problem; clearly further investigation was required as the results in the lateral direction were so good, but poor in the longitudinal. It was thought likely to be the result of noise in the GPS data. Multipath causing significant distortion of the signals is a common problem[15, 16] and was one of the first possibilities examined, but an initial assessment using an adaptive filtering technique showed this to be very unlikely. However, a 24-hour sky plot for the Millennium Bridge (Figure 10) shows there is never a GPS satellite due north of the survey site. This will have implications in the northerly Dilution of Precision (DOP). A GPS Measurement of DOP is an indication of expected accuracy. The East-West component has the highest quality, followed by North-South, and then the vertical component. Upon further analysis, it was concluded that the geometry of the GPS satellites would also introduce noise into the GPS coordinates in the North-South component. The Millennium Bridge is orientated approximately in a North-South direction, and this is thought to be why apparent displacements in the longitudinal component are larger than in the lateral direction.

Pseudolites

The trials illustrated a possible problem with GPS for such precise deflection monitoring. Research has been conducted into the integration of pseudolite data in order to enhance GPS[17,18]. A pseudolite is basically a pseudo-satellite; a ground based GPS data source. The pseudolites can be programmed to transmit a modulated code on the L1 carrier wave. Certain GPS receivers can be programmed to receive these code and carrier phase values.

The IESSG and the University of New South Wales have developed a programme of collaborative trials for deformation and deflection monitoring. Figure 11 shows a pseudolite being used during such trials. Early results indicate that the use of pseudolites does improve the GPS positions, and can also be used as a positioning technique in its own right[19,20]. However, research into pseudolite data would not allow any improvement of the GPS data that already been gathered. Instead, simulated pseudolite data were produced to provide additional results in order to see whether or not an improvement would have been feasible if pseudolites had been used during the trials.

The IESSG has developed a GPS data simulator that will output RINEX (Receiver Independent Exchange Format) GPS data for a receiver at a given location and time. The simulator was used to re-create the original bridge data, the results of which compared well with the real bridge data. The simulated longitudinal results were larger than the lateral results, showing the simulator to be able to create the GPS data correctly. Figure 11 illustrates the simulated results. The next step was to introduce code and carrier data from a stationary satellite located at a pre defined location within the simulation i.e. a pseudolite. All the data produced from the simulator, in RINEX format, were then processed using Kinpos.

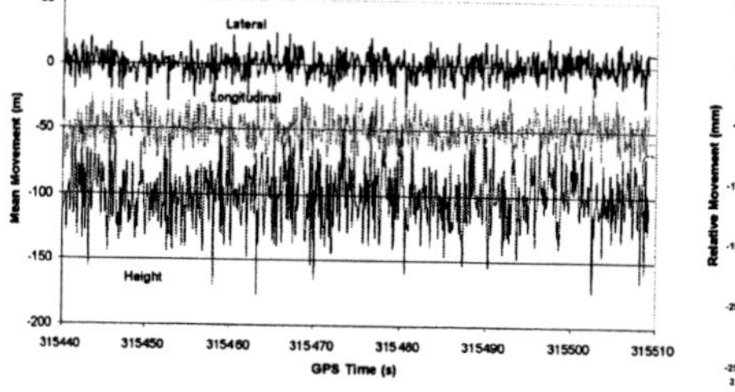

Figure 11. The GPS only simulated data

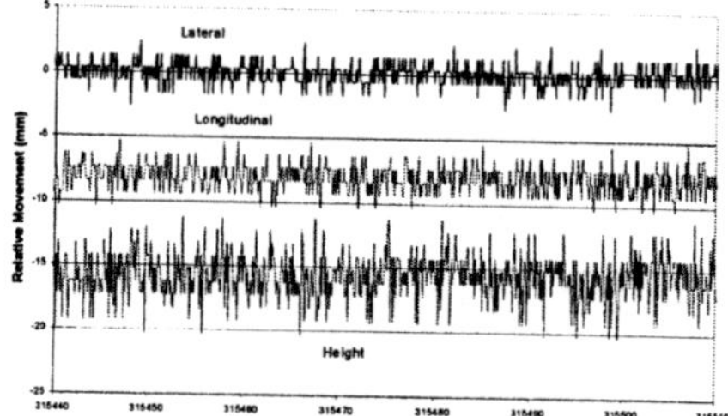

Figure 12. Simulated GPS results with one pseudolite

At this stage no weighting was used for the pseudolites within the processing. The simulation was used purely as a means to evaluate whether pseudolites could be a feasible way to improve such a scenario. Figures 12 illustrates the Bridge data with the introduction of one pseudolite, compared to Figure 11 which shows the results without any pseudolites. These show that the introduction of one pseudolite dramatically improves the results in both the North direction as well as in the vertical. However, the simulator has placed the pseudolites in the optimum position, which in reality may not be physically possible. Therefore, notwithstanding the difficulties described above, the careful subsequent investigation leads the authors to now have increased confidence in the data set for lateral and vertical movements.

Due to the restrictions placed on transmitting at the GPS frequencies, the researchers are now concentrating such research on using a similar technology called Locatalites. These are similar but transmit data on Wi-Fi frequencies.

DISCUSSION

The use of GPS in the monitoring of structures is not a "mature" procedure, and great care in both the experimental and data processing phases is required. Since the experimental work was carried out there have been many advances in GPS, and it is one of these – the development of pseudolites – that has finally enabled the authors to report not only the positive and negative elements of the original tests, but also a potential solution when satellite geometry is poor.

The positive aspects include the amount of data that can be obtained and that the base station does not have to be co-located with the measurement station (i.e. may be separated by some kilometres).

Difficulties include multipath, and the need to be careful with satellite geometry problems. Notwithstanding these issues, the results demonstrate good agreement between predicted and measured lateral and vertical deformations.

For the lateral deflections, the agreement in frequency appears very good. The effects of torsion on the structure are also clearly identified, and by selecting positions carefully the torsion measurement is readily amplified.

CONCLUSIONS

This paper has demonstrated the use of GPS as a means of continuous monitoring of structures, along with the importance of continuing research into such techniques. It has once again shown that good agreement between predicted and measured deflections and frequency of movement can be obtained.

The development of pseudolites may be an important underpinning technology that would help GPS measurement in other applications.

ACKNOWLEDGEMENTS

The authors are grateful to Arup for allowing access to the Bridge. We would like to thank St Paul's School for allowing us to place a reference station on their site. In addition, we would acknowledge The University of Nottingham for funding the trials and Brunel University for contributing to the project. In particular we would like to thank colleagues Dr Andy Evans and Mr Theo Veneboer (formerly at The University of Nottingham) for their help with the experimental work, and Dr Francisco Andrade (formerly at Brunel University) for his assistance in carrying out the wavelet analysis. The authors would also like to thanks and acknowledge Prof Chris Rizos and Dr Joel Barnes of the University of New South Wales for their assistance with pseudolites.

REFERENCES

1. Roberts, G. W., Meng, X., Brown, C. J., Dallard, P., "GPS Measurements on the London Millennium Bridge". Proceedings of the Institution of Civil Engineers Bridge Engineering, Vol. 159, issue BE4, pp 153-161, December 2006

2. Lovse, J.W., Teskey W.F., Lachapelle G. and. Cannon M.E, "Dynamic Deformation Monitoring of a Tall Structure Using GPS Technology", Journal of Surveying Engineering, ASCE, Vol. 121, No. 1, pp. 35-40, 1995

3. Ogundipe, O., Roberts G.W. and Dodson A.H., "Construction Plant Control and Guidance - The Development Continues". Civil Engineering Surveyor GIS Supplement, pp. 4 - 9, 2002

4. Ding, X.L., Chen Y.Q., Huang D.F., Zhu J.J., Tsakiri M. and Stewart M. "Slope monitoring using GPS: a multi-antenna approach", GPS World, No. 3, Vol. 11, pp. 52-55, 2000

5. Roberts, G W, Dodson A H and Ashkenazi V., "Experimental Plant Guidance and Control by Kinematic GPS", Proc ICE (Civil Engineering), No 138, pp 19 - 25, 2000

6. Brown, C.J., Karuna R., Ashkenazi V., Roberts G.W. and Evans R.A. "Monitoring of Structures using GPS", Proc. ICE (Structures and Buildings), No. 134, No.1, 1999

7. Cheng, P, Shi, W.J. and Zheng W., "Large Structure Health Dynamic Monitoring Using GPS Technology" GPS Earthquake FIG XXII International Congress, Washington, D.C. USA, April 19-26, 2002

8. Dallard P., Fitzpatrick A.J., Flint A., Le Bourva S., Low A, Ridsdill Smith R.M. and Willford M., "The London Millennium Footbridge", The Structural Engineer, Vol.79, No. 22, 20 Nov 2001. (Available on the Arup website www.arup.com/millenniumbridge/).

9. Roberts, G.W., Real Time On-The-Fly Kinematic GPS. PhD Thesis. The University of Nottingham, UK. 1997

10. Roberts, G W, Dodson A H, Ashkenazi V, Brown C J, Karuna R and Evans RA., "The Use of Kinematic GPS and Finite Element Modelling for the Deformation Measurements of the Humber Bridge", Proc GNSS 99, 3rd European Symposium on Global Navigation Satellite Systems, pp 230-235, Genoa, Italy, October 1999.

11. Ashkenazi, V., Dodson A.H., Roberts G.W., Brown C.J. and Evans R.A., "The Use of Kinematic GPS and GLONASS to Monitor the Deflection of the Humber Bridge under High Loading", Proc GNSS 98, The 2nd European Symposium on Global Navigation Satellite Systems, pp (VIII-O-06) 1 - 6, Toulouse, France, October 1998.

12. Moore, T, Chen, W, Ashkenazi V and Dumville M., "WGS84 and GNSS for UK Marine Navigation", Proc ION GPS 2000, 13th International Technical Meeting of the Satellite Division of the Institute of Navigation, Salt Lake City, USA, September 2000.

13. P.A. Lynn, W. Fuerst, B. Thomas, "Introductory Digital Signal Processing with Computer Applications", John Wiley, 1997.

14. Antoniou, A. Digital Filters: Analysis, Design, and Applications. 2nd ed. New York, NY: McGraw-Hill, 1993.

15. Roberts G.W., Meng X. and Dodson A.H., "Using Adaptive Filtering to Detect Multipath and Cycle Slips in GPS/Accelerometer Bridge Deflection Monitoring Data", Proc XXII International Congress of the FIG, TS6.2 Engineering Surveys for Construction Works and Structural Engineering II, Washington DC, USA, April 19 - 26 2002

16. Teferle, F N., Bingley R M, Dodson A H, Apostolidis P and Staton G, "RF Interference and Multipath Effects at Continuous GPS Installations for Long-term Monitoring of Tide Gauges in UK Harbours", in Proceedings of the 16th Technical Meeting of the Satellite Division of the Institute of Navigation, ION GPS/GNSS, Portland, Oregon, 9-12 September 2003

17. Meng, X., Roberts G.W., Dodson A.H., Cosser E., Barnes J., and Rizos C. Impact of GPS satellite and pseudolite geometry on structural deformation monitoring: Analytical and empirical studies. Journal of Geodesy, 77, pp. 809-822, 2004.

18. Lee, H.K., Wang J., and Rizos C. "An integer ambiguity resolution procedure for GPS/Pseudolite/INS integration", Journal of Geodesy, 79, pp.242-255, 2005

19. Cosser, E., Meng, X., Roberts, G.W., Dodson A.H., Barnes, J. and Rizos, C., "Precise engineering applications of pseudolite augmented GNSS", 1st FIG Int. Symposium on Engineering Surveys for Construction Works & Structural Engineering, Nottingham, U.K., 28 June - 1 July, Paper TS10.1, 2004

20. Dodson, A.H., and Roberts, G.W., "The monitoring of bridge movements using GPS and pseudolites". 11th Int'l Symposium on Deformation Measurements, Santorini, Greece, 25-28 May, pp. 563-572. 2003

The Potential of a Ground Based Transceiver (*LocataLite*) Network for Structural Monitoring of Bridges

Dr J. B. Barnes, Prof. C. Rizos, Mr A. Pahwa, Mr N. Politi, School of Surveying and Spatial Information Systems, University of New South Wales, Sydney, Australia
Mr J. van Cranenbroeck, Leica Geosystems AG – Surveying and Engineering Division, Heerbrugg, Switzerland

Abstract

Locata technology is becoming part of Leica Geosystems solution for the structural monitoring applications such as bridges. This paper assesses the performance of the *Locata* technology using a test *Locata* network (*LocataNet*) established at the University of New South Wales. Using this network a long term static tests and a simulated deformation movement test, with GPS as a comparison, were conducted. This paper described the *LocataNet* established at UNSW and presents the results and analysis of the tests conducted. Overall the paper demonstrates the suitability of *Locata* for structural deformation monitoring type applications (such as bridges) where there is reduced or unavailable satellite coverage.

Introduction

Ideally the movements of man-made engineering structures should be monitored on a continuous basis and with high accuracy in order that departures from the expected movements of a structure can be detected quickly and necessary action taken. In the past few years the Global Positioning System (GPS) has been applied to monitoring the structural deformation of bridges, dams and buildings (Roberts et al., 2004), by permanently installing GPS receivers at key locations on the engineering structure so as to provide cm-level positioning information on a 24/7 basis. However, the major problem with such GPS receiver installations is that, the accuracy, availability, reliability and integrity of position solutions is very dependent on the number and geometric distribution of the available satellites. This means that the precision of positioning solutions will vary by typically up to 3 times during the day in Sydney, Australia (from an analysis of PDOP values). The large variation in positioning precision obtained with GPS is undesirable for a continuous deformation monitoring system. More-over, the accuracy of the height component is typically 2-3 times worse than for the horizontal (because of the geometrical distribution of the satellite constellation and the poorer quality of data at low elevation angles). This situation becomes worse when the line-of-sight to GPS satellites becomes obstructed, as on a bridge, and there may be insufficient GPS satellites for positioning.

Another limitation of the GPS technology for precise (cm-level) real-time continuous positioning is the requirement for differential corrections or measurements from a single reference station or Continuously Operating Reference Station (CORS) Network. Acceptable performance from RTK GPS in structural deformation monitoring type applications is therefore heavily dependent on the reliability of the wireless data link used, and on a

Bridge design, construction and maintenance 2007, Thomas Telford, London

relatively unobstructed sky-view, where there are at least five satellites with good geometry available. To address these significant limitations of the GPS *Locata* has developed a novel positioning technology.

Locata Positioning Technology

Locata's solution to "difficult" Global Navigation Satellite Systems (GNSS) environments is to deploy a network of terrestrially-based transceivers (*LocataLites*) that transmit positioning signals. These transceivers form a positioning network called a *LocataNet* that can operate in combination with GPS (such as in urban environments) or entirely independent of GPS (for indoor applications). One special property of the *LocataNet* is that it is time-synchronous, potentially allowing single point positioning (no differential corrections and data links required) with cm-level accuracy.

In the current system design the *LocataLites* transmit their own proprietary signal structure in the 2.4GHz ISM band (license free). This ensures complete interoperability with GPS and allows enormous flexibility due to complete control over both the signal transmitter and the receiver. Details of the current system design have been detailed previously in Barnes *et al.* 2005.

On the 19th July 2006 Leica Geosystems announced publicly on their website the signing of a co-operation agreement between Leica Geosystems and *Locata* Corporation for the distribution and support of *Locata* technology in two key market areas, namely:

- open cast mining – for machine automation and mine monitoring operations, and
- structural deformation monitoring – for structures such as bridges, dams and buildings.

In addition, Leica Geosystems will develop the first integrated GPS/*Locata* receiver.

As a first step in assessing the suitability of the *Locata* technology for deformation monitoring applications the University of New South Wales has conducted tests to assess the *Locata* network stability and the level of movement that can be detected by the system. The remainder of this paper describes the *LocataNet* established at UNSW and some of the tests conducted.

LocataNet installation used for the trial

To test and evaluate the performance of *Locata* technology for the purpose of structural deformation monitoring, a small semi-permanent *Locata* network (*LocataNet*) was set up at the University of New South Wales (UNSW) from early January to mid February 2007. The term *LocataNet* describes a network of *LocataLites* (at least four *Locata* transceivers) that transmit the positioning signals (in the 2.4GHz ISM band). Typically a *LocataNet* is deployed around the area where the *Locata* positioning signals are required. Once a *LocataNet* is established a *Locata* receiver (or rover) can determine its position independently of other positioning technologies (GNSS etc).

The *LocataNet* established at UNSW is illustrated in Figure 1. It consists of 10 *LocataLites* situated on top of three buildings. The *Locata* receiver antenna was situated on the roof of the Electrical Engineering building (Elec. Eng in Figure 1), and the distance from the *Locata* receiver antenna to *LocataLites* ranged from approximately 5 to 80 metres. The *Locata* receiver's omni-directional antenna was mounted on a tripod, and the *Locata* receiver was located in an office below via a 30m low-loss coaxial antenna cable. Each *LocataLite* (LL)

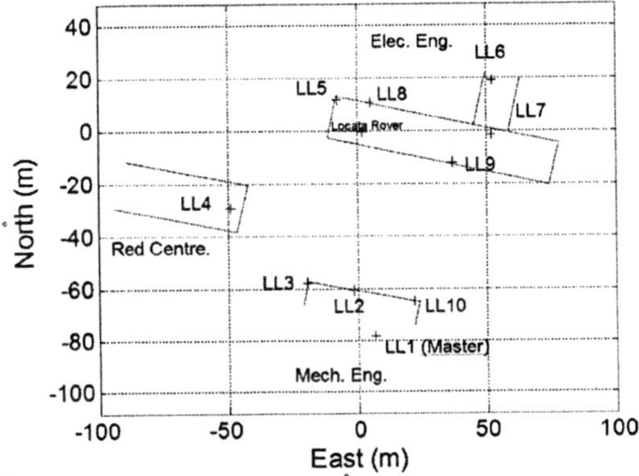

Figure 1 *LocataNet* of 10 *LocataLites* established on the roof-tops at UNSW.

was assigned consecutive PRN codes (except LL8), starting from the "Master" in a clockwise direction. In operation, the "Slave" *LocataLites* 2-10 time-synchronise to the "Master" *LocataLite* 1. A *Locata* receiver using these positioning signals can compute a carrier-phase single point position with cm-level accuracy (without requiring a differential reference receiver and data links).

Each of the *LocataLite* sites consists of three main components: a pole with three antennas attached, a *LocataLite*, and a power source. For a fully operational *LocataLite* utilising spatial diversity, two transmitting antennas and one receiving antenna are required. In the UNSW setup, directional patch antennas with beam width of 70 degrees were used for both transmission and reception. The transmitting antennas were positioned towards the rover antenna work area and attached to a vertical pole with a separation of approximately 75cm; the receiving antenna was directed towards the "Master" *LocataLite* and mounted just below the top of the transmitting antenna (see Figure 2).

The *LocataLites* were enclosed in customised weatherproof boxes, allowing for external connections to the antennas, data communication ports and power sources. The external interface is then wired to the *LocataLite* inside, as shown in Figure 2.

With the exception of the "Master" *LocataLite*, which operated on a mains power source, the *LocataLite* locations were powered by 12V/55AH batteries, which allowed a continuous run

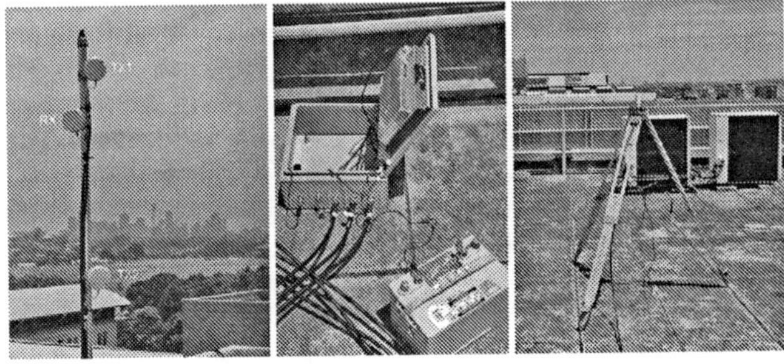

Figure 2 *LocataLite* (5) antenna setup (left) and weatherproof enclosure (middle). *Locata* receiver antenna setup (right).

time of over 24 hours per battery. Y-splitters were connected to the power cables, which enabled the connection of a replacement battery in parallel to the exhausted one before disconnecting the latter, thus providing uninterrupted power to the *LocataLites*.

The coordinates of the transmitting antennas were surveyed using a combination of carrier-phase differential GPS (using Leica System 500 processed using Leica Geo-Office) and a reflectorless total station.

During early January to mid February the *LocataNet* was in continuous operation for several days at a time, without any network failure. A number of static tests of several hours in length were conducted during this time, and the next section gives a description and results from a typical static test.

Long term static test

In deformation monitoring applications (such as bridges), the monitored structures are generally relatively static and it is any deviation from this state that requires early detection. The long term stability of a positioning solution is therefore critical for deformation monitoring applications. For the purposes of this test, the network setup described in the previous section was used and Figure 2 shows the setup of the *Locata* receiver antenna on the tripod.

The *Locata* receiver in the office was connected to a laptop computer via two serial ports. After powering up the receiver the *LocataLite* signals are acquired and tracked within 10s of seconds. For a single point carrier-phase solution the receiver currently requires initialising at a known point to resolve the carrier-phase ambiguities. When *LocataLites* transmit on a second frequency in the 2.4GHz ISM band (expected in the next 9-12 months) the *Locata* receiver will be able to resolve ambiguities On-The-Fly. The coordinates of the *Locata* receiver were surveyed using differential GPS, at the same time as the *LocataNet* survey was conducted. The receiver was initialised via a command through the laptop and then the receiver output single point carrier-phase solutions at a 1Hz rate in the NMEA format, which was logged and visually displayed. In addition to this the real-time position solution, raw data (containing pseudorange and carrier-phase) were logged. Data in this particular test were collected for approximately 13.5 hours. Due to the fact that the elevation angle to the *LocataLites* from the receiver location are all less than 8 degrees, the geometry in the vertical is very poor. The following results will therefore concentrate on the horizontal component.

Results and Analysis

Figures 4 and 5 show the horizontal scatter plot of the position error (with respect to the true position surveyed using GPS) and the individual East and North positioning error components. The mean position error in both East and North are less than 1mm and the standard deviation in East and North was 2.1 and 1.5mm respectively. The slightly larger standard deviation in the East component is due to the fact that the dilution of precision in the East-West (0.543) component is slightly worse than the geometry in the North-South direction (0.530). Visually from Figures 3 and 4 it is clear that the overall precision and stability of the position solution is very good over the 13.5 hour period with no evident long term drifts. However, there are approximately 7 position solutions (out of ~48600) that could be considered as outliers, and the largest with a maximum error of 2 cm in the North component. *LocataLites* internally monitor their time synchronisation integrity. If the time synchronisation is not within specification the *LocataLite* takes steps to ensure the *Locata* receiver "sees" the signal as "unhealthy". However, in this particular *LocataNet* the distances from the *LocataLites* to the rover are very short (5-80 metres). At this distance it may have been possible for a rover to occasionally track "unhealthy" signals. On a *LocataNet* with distances of several hundred metres to the rover this is not likely to be an issue. In addition, these are "single events", so they could be easily removed using a filter or using data snooping techniques.

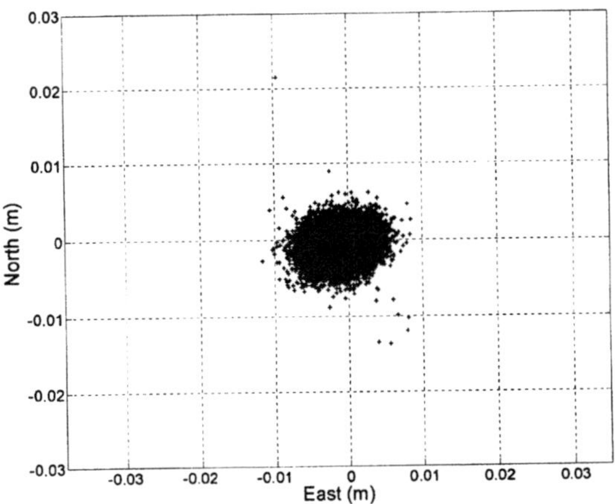

Figure 3 Horizontal error scatter plot for long term (13.5 hour) static positioning test.

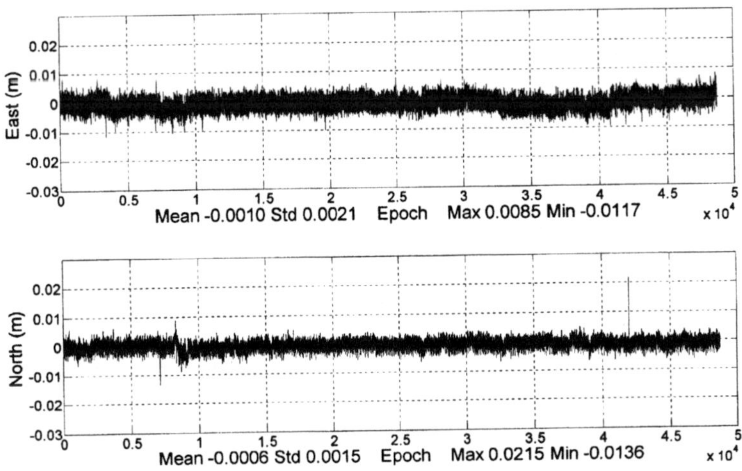

Figure 4 East and North error for long term (13.5 hour) static positioning test.

Simulated deformation movement test

In some structural deformation monitoring applications (such as for bridges) the positioning technology used must be able to detect centimetre to millimetre level movements. The purpose of this test was to establish if the accuracy of the *Locata* technology allowed one centimetre level movements of the *Locata* receiver's antenna to be detected.

For the purpose of this test, the rover antenna was required to move accurately over a small distance in a pre-defined pattern. The process needed to be automated and repetitive in order to test the system over a long-term period. To satisfy these requirements a HP XY plotter table was used. Both a *Locata* receiver antenna and a Leica GPS system 500 AT502 antenna were mounted to the printing-head of the plotter (as shown in Figure 5). The use of such a plotter enabled control of the device using a serial port connection to a laptop. The plotter supports the HPGL graphic language and thus, by creating appropriate computer scripts, it allowed the automation, repetition and accuracy of movement which was required.

The plotter, with the antennas attached, was placed on a levelled table on the roof of the Electrical Engineering building near the *Locata* rover antenna used in the static test. This location had a clear line-of-sight to all surrounding *LocataLites*. The coordinate of the *Locata* receiver antenna at the centre of the plotter table was surveyed using a reflectorless total station. In addition the plotter table was orientated so that the X and Y axes were as closely aligned with true North/South and East/West as possible.

It was decided to make this test more "challenging" by only using five of the *LocataLite* locations and thereby making the network geometry worse (and more "real world"). The five *LocataLites* used were LL1, LL4, LL5, LL7 and LL8. Conducting the test in a similar way to the static test, the *Locata* receiver was first initialised at the know point and the receiver then output positions at a 1Hz rate. After one minute, both antennas were moved 1 cm in the West direction. After one minute of static data collection, the antennas were moved a further 1 cm to the West. This procedure was repeated until the antenna was 12 cm to the West of the initial position. The antenna was then moved 1cm to the East repeatedly until the antenna was a full 12 cm East of the initial position. The antenna was then moved by 1cm steps in the West direction again until the antenna was back at the initial start location. The procedure described above was then repeated giving a total of 149 static points (each with 1 minute of data), with the entire test taking approximately 2.5 hours to run.

The GPS receiver data was post-processed using Leica Geo Office relative to an MC500 Leica GPS reference station with a AT504 choke ring antenna, located approximately 55 metres from the test area.

Figure 5 HP XY plotter table with *Locata* and Leica AT502 antennas.

Results and Analysis

Figures 6 to 8 show epoch-by-epoch position solutions from *Locata* and GPS for the horizontal trajectory and in East/North components. Visually from the figures the *Locata* solution is more stable and repeatable than the GPS solution. The *Locata* position solution has consistent positioning geometry with a HDOP of 0.64 with 5 *LocataLites*. In comparison the GPS HDOP varies from 1.5 to 4.1 with 5 to 9 available satellites. The section of poorer GPS geometry can easily be seen in the middle section of the data for the North component. For the *Locata* North time series there is a repetitive pattern of movement in the North direction (as the antenna moves East-West), with a maximum deviation of about 2.5 mm. There are two possible explanations for the repetitive movement in the North-South direction. First, the error could be due to the actual movement of the plotter head. The second possible reason is multipath error. In an RF-based terrestrial positioning system the multipath error at a particular position in the network will have a similar multipath error if the same position is reoccupied. This is assuming the transmitter locations and local factors (buildings etc) do not change. The repetitive nature of the error signature in this particular test suggests that it may be possible to reduce the multipath error in a relatively static environment through calibration, although further investigations would be required to verify this.

The mean static position of each location was computed (from each 1 minute of static data) for the *Locata* and GPS solutions. These are plotted in Figure 9 for the East and North components. Figure 10 shows the first 24 mean static points for the East component. In addition the East and North standard deviation of each static point for *Locata* and GPS is shown in Figure 11. For *Locata* the largest standard deviation in the East and North coordinate components was 3.2 mm and 1.2 mm respectively, with the smaller North component being due to better geometry (lower DOP). For GPS the largest standard deviation in the East and North coordinate components was 4.0 mm and 5.3 mm respectively, which are correlated with the section of worse satellite geometry. The distance 'travelled' with each 1 cm step was computed based on the mean position values, and the error computed, assuming a 'true' step value of 1cm. Figure 12 shows the error in the distance moved with a maximum error of 2.7 mm for *Locata* and 7.2 mm for GPS. This indicates that a 1 cm move can easily be detected using *Locata*, but for GPS cannot always easily be detected due to the varying satellite geometry. In addition the *Locata* solution can be improved by positioning the *LocataLites* in a more optimal network configuration. This was demonstrated in Barnes *et al.* 2007 with 10 *LocataLites* in the UNSW network, which gave a maximum horizontal error of 1.3 mm for a 1 cm antenna move.

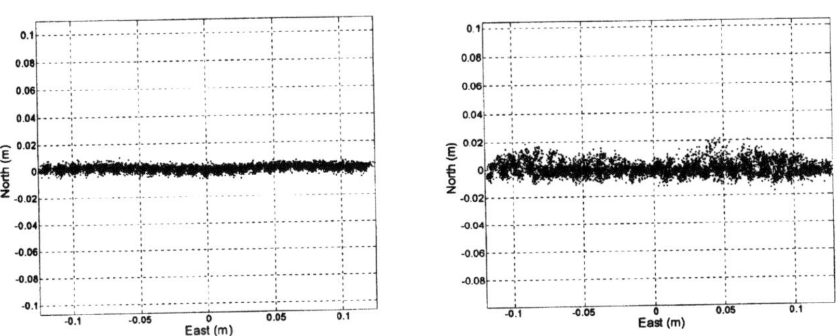

Figure 6 Horizontal trajectory: *Locata* (left), GPS (right)

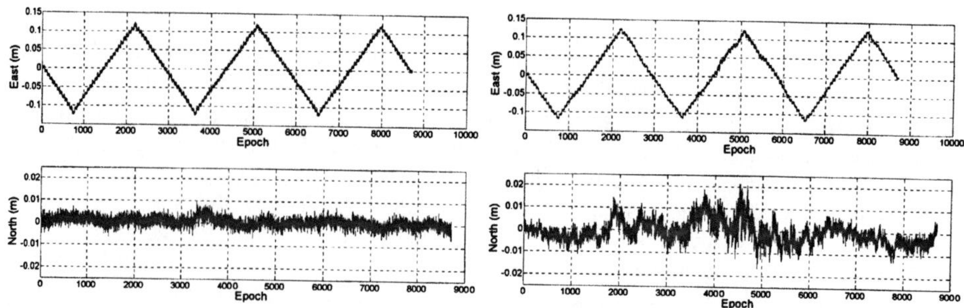

Figure 7 East and North time series: *Locata* (left), GPS (right)

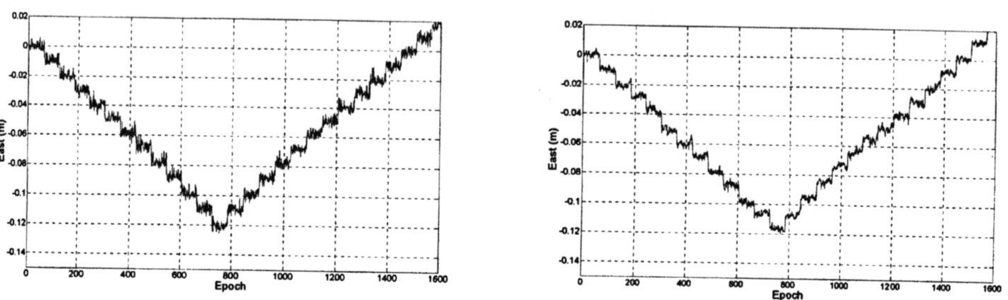

Figure 8 East time series, 1st 1600 epochs: *Locata* (left), GPS (right)

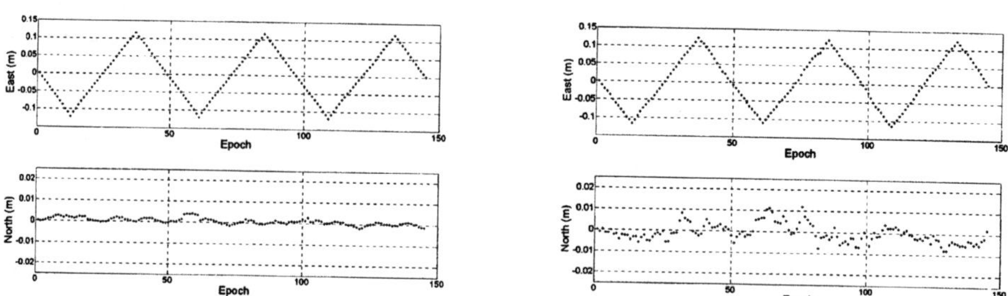

Figure 9 Mean static East and North time series: *Locata* (left), GPS (right)

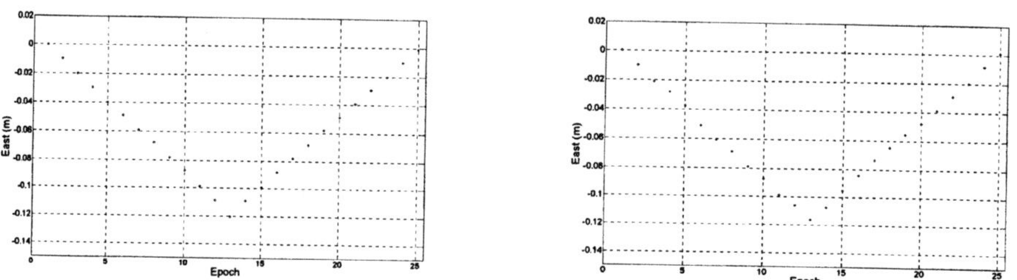

Figure 10 Mean static East 1st 24 moves: *Locata* (left), GPS (right)

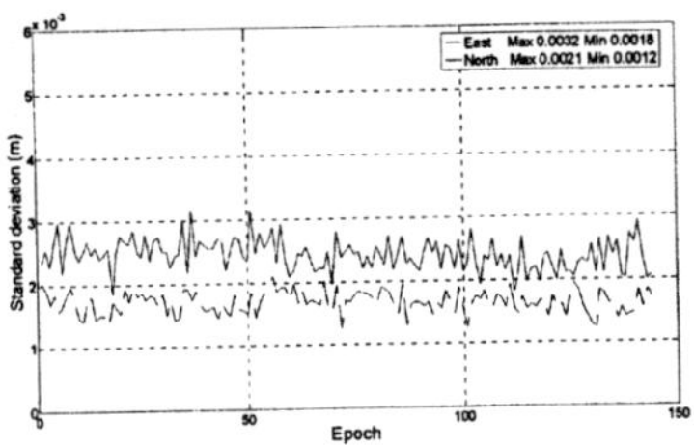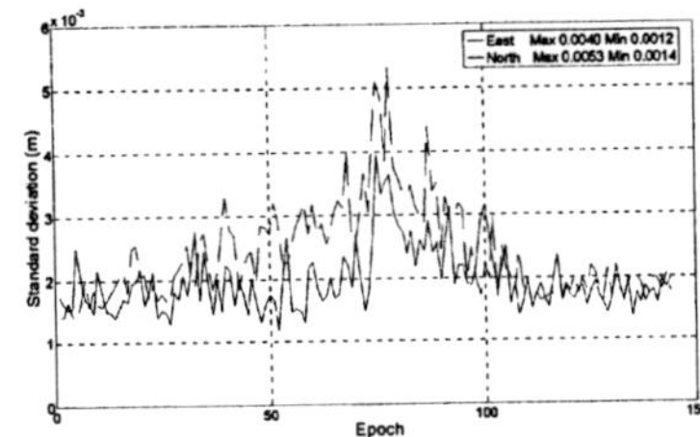

Figure 11 Standard deviation of static East and North: *Locata* (left), GPS (right)

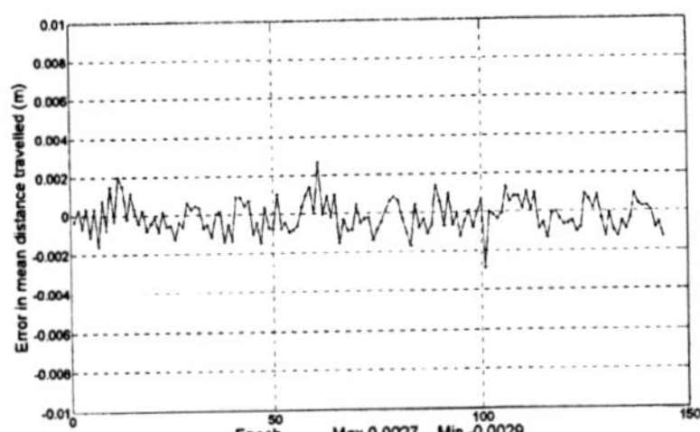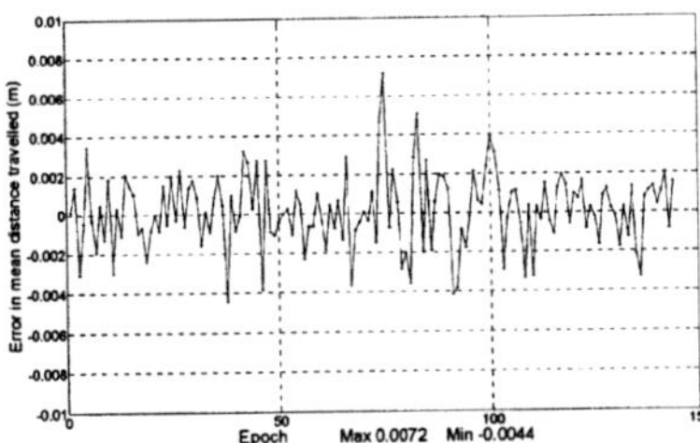

Figure 12 Error in distance travelled for each 1cm move (computed from mean position values): *Locata* (left), GPS (right)

CONCLUSIONS

In this paper a *LocataNet* was successfully established at the University of New South Wales for assessing the suitability of *Locata* technology for structural deformation monitoring applications. Using this network a long term static test and a simulated deformation movement test were conducted. The static test over approximately 13.5 hours verified the long term stability of the *LocataNet*. The resulting position standard deviation of the test was approximately 2 mm, and there were no evident long term drifts. The position solutions in this test were computed on an epoch-by-epoch basis with no filtering or smoothing, once a second. For structural deformation monitoring applications it is likely that a combined epoch solution or smoothed solution would be more appropriate. Therefore work is now focused on methods to combine several epochs of data to generate solutions with higher precision and better integrity.

In the simulated deformation movement test a *Locata* receiver antenna and GPS antenna were repeatedly moved by 1 cm steps and static data was collected for one minute after each move (149 static points in total). For the GPS solution the maximum error in distance moved computed from the mean static positions was 7.2 mm, and indicates that a 1cm movement of the antenna cannot always be detected due to the varying satellite geometry. However for *Locata* the maximum error in distance moved was 2.7 mm, and suggests that *Locata* technology can easily detect movements of 1cm. For Locata this result can be improved with a more optimal *LocataNet* design.

In the tests conducted for both *Locata* and GPS the atmospheric effects are insignificant due to the size of the *LocataNet* and the close proximity of the reference station in the case of GPS. Work is now under investigation to remove multipath error via calibration and improve the positioning results further. In addition tests will now focus on larger *LocataNet* installations (where tropospheric effects are greater), and at real structural deformation monitoring sites (such as bridges and dams).

Overall the tests conducted have demonstrated that *Locata* technology has the potential to meet the expected requirements for structural deformation monitoring type applications (such as bridges) where there is reduced or unavailable satellite coverage. The *Locata* technology is very soon ready for trial investigations to begin at real structural monitoring test sites (bridges, dams etc).

References

BARNES, J., CRANENBROECK, J.van, RIZOS, C., PAHWA, A., & POLITI, N., 2007. Long term performance analysis of a new ground-transceiver positioning network (LocataNet) for structural deformation monitoring applications.*FIG Working Week, Strategic Integration of Surveying Services*, Hong Kong, 13-17 May

BARNES, J., RIZOS, C., KANLI, M., PAHWA, A., SMALL, D., VOIGT, G., GAMBALE, N., & LAMANCE, J., 2005. High accuracy positioning using *Locata's* next generation technology. 18th Int. Tech. Meeting of the Satellite Division of the U.S. Institute of Navigation, Long Beach, California, 13-16 September, 2049-2056.

ROBERTS, G.W., COSSER, E., MENG, X., & DODSON, A.H., 2004. Monitoring the deflections of suspension bridges using 100Hz GPS receivers, *17th Int. Technical Meeting of the Sat Div of the Institute of Navigation*, Portland, Oregon, September.

Innovation and Experience in GNSS Bridge Real Time 3D- Monitoring System

Joël van Cranenbroeck, Business Development Manager
GNSS Networks and Geodetic Monitoring
Leica Geosystems AG, Heerbrugg, Switzerland

Wu Xinghua, GNSS Reference Station Business Manager
Leica Geosystems Greater China, Beijing, PR China

Vincent Lui, Business Manager
Leica Geosystems Hong-Kong, PR China

Transportation Authorities are continually challenged to provide and maintain safe and efficient highway networks. Not only are bridges an integral part of these networks, they also represent a multibillion-dollar investment. To meet this challenge and safeguard this investment, transportation authorities need to understand completely the condition and behavior of the bridge structures, so that the bridges can remain open to traffic, be resistant to the elements, and be undaunted by the millions of loading cycles per year – all with minimal maintenance expense.

Realistically, the high cost of maintenance – often exacerbated by the budget-driven policies of bridge owners – frequently lead to the deferment of routine bridge repairs and reservation measures. These policies can contribute to an occasional bridge failure, which is completely unacceptable and forces more costly actions.

To manage bridges effectively, more needs to be done to access the day-to-day and long-term condition and behaviour of in-service bridges, so that preventive measures can be taken, and deterioration rates can be better understood.

Owing to the advantages of high accuracy, all-weather conditions and no requirements of inter-visibility between measuring points, GNSS, the acronym standing for Global Navigation Satellite System and including the US Global Positioning System, GLONASS its Russian equivalent and the future European Galileo and the Chinese BEIDOU (Big Dipper) is playing a more and more important role in high precision positioning missions in structure/construction health monitoring.

A properly configured GNSS measurement system can meet most of the possible static and dynamic measurement needs in such applications for absolute positioning and relative displacement. In other words the required precision and accuracy can be approached with an architecture of the GNSS single/dual (L1 or L1/L2) frequency carrier phase, data sampling rate, communication between GNSS receivers and control data centre and the method of data processing.

Bridge design, construction and maintenance 2007, Thomas Telford, London

GNSS for Structural Health Monitoring

GNSS is a very interesting tool for monitoring because it has a number of distinct advantages over terrestrial positioning technologies. GNSS is able to measure at high rates with low latency, operate in all weather conditions, has synchronized measurements, does not require line of sight to ground marks/targets, can measure over long baselines, has low maintenance and a long service life and can provide timing for other sensors, such as accelerometers.

These unique characteristics make GNSS particularly interesting for monitoring large structures such as long bridges, dams, high rise buildings but also for seismic and land slide applications and for the provision of control for other instruments, such as robotic total stations, in unstable areas.

Each point to be measured must have an antenna, a receiver, ground mark, power, communications and, possibly, protection against lightning and vandalism or theft.

The reference stations' receivers from where the baseline to the monitoring receivers will be processed must be installed in a stable place as all the results will refer to this. The minimum required is only one but to ensure an independent control on the solution and to have also an internal control on the relative stability of their location it is suggested to set-up at least two.

Leica Geosystems has developed dedicated GNSS receivers like the Leica GMX901 (see Figure 1) and the Leica GMX902 (see Figure 2) for monitoring applications complemented by GNSS antennas that have the capacity to mitigate and reduce the multipath effects induced by the structure itself in many cases.

The Leica GNSS Monitoring Receivers

Figure 1 Single Frequency GMX901   Figure 2 Dual Frequency GMX902

The Leica GPS AT504 GG Choke Ring Antenna (see Figure 3) provides the state of art in GNSS signals tracking even in multi-path environment and radio jamming and the Leica AX1202 GG Geodetic antenna (see Figure 4) fits any GNSS monitoring project until the location can be guaranteed free of multi-path effects. During the design phase of a GNSS monitoring project, the choice of the proper antenna will be one of the most important topics.

Leica GNSS Antennas

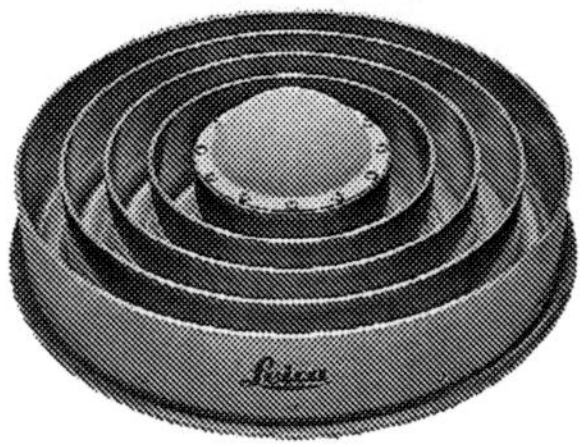

Figure 3 AT504 GG Choke Ring Antenna

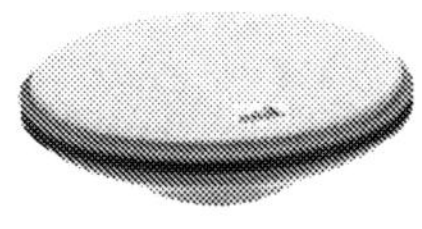

Figure 4 AX1202 GG Geodetic Antenna

In monitoring applications, accuracy is of paramount importance, so only ambiguity-fixed positions are of interest. A highly reliable ambiguity resolution strategy is needed to prevent wrong fixes, which will be detected immediately by the monitoring system as an apparent movement.

The processing kernel that has been developed for the bridge GNSS monitoring solution is based on that used in Leica Geosystem's high-end RTK GNSS sensors and the LGO (Leica GeoOffice) post-processing software.

Real Time and Post-Processing modes

The kernel, which is integrated into the Leica GNSS Spider reference station and GNSS monitoring software, is able to process single and dual frequency data from GPS and/or GLONASS in real time and post processing. Three ambiguity resolution techniques have been implemented and successfully tested on many projects: Kinematic On the Fly (OTF), Initialisation on a Known Mark and the new Leica innovative Quasi-Static approach. The OTF technique allows full dynamics of the rover antenna suitable for use in, for example, formula one racing. The quasi-static approach uses an assumption of lower dynamics such as would be experienced in most monitoring applications like the one on the Yeong Jong Bridge in Korea.

The traditional approach to real time GNSS monitoring is to deploy RTK enabled receivers to the field, which receive corrections from a nearby reference station; self computing their positions. This distributed processing approach has some distinct disadvantages:

Two communications lines are required per point that is measured (one to receive the corrections and one to transmit the resulting coordinates,

- Only one baseline can be computed per point,
- Single frequency RTK is not supported,
- Post processing is not possible, and
- Archiving of the raw observations is not possible.

In the decentralized approach used by Leica's GNSS Spider Software, only a single communication channel is required to send the raw observations to the monitoring server.

Multiple baselines may be computed for each point using different reference stations or processing parameters. Single frequency RTK is supported, as is post processing and archiving of both raw data and real time 3d positions. In the case of unreliable communications, it is also possible to log directly in the memory of the GNSS and then download the data periodically for post processing, rather than relying on having a permanent open communication channel. In that case the GNSS receivers used must have local storage capacity on flash card memory.

The Leica GNSS Spider software is dual-purpose software. It offers comprehensive GNSS reference station capabilities for the configuration and control of GNSS sensors, archiving of data and dissemination of correction data for single-base and network RTK positioning.

In addition to the reference station capabilities, GNSS Spider has advanced GNSS baseline processing capabilities for monitoring applications. The marriage of reference station and GNSS monitoring features produces a flexible and powerful application with sophisticated communications, processing, data management and security functionality.

GNSS Spider may be combined with Leica GNSS QC coordinate analysis software as well as with any third party monitoring and analysing software for integration with other geo-technical sensors and to leverage the GNSS QC advanced limit checks, messaging and analysis features. The integration is easily made by streaming out the results in real time as well through TCP/IP ports, serial interface or Modem. All the results can be stored in text files as well for further analysis investigations.

The baseline processing in Leica GNSS Spider is divided into two parts: real time processing and post processing. The Leica GNSS Spider also has the capacity to re-process complete observation files in RINEX (Receive Independent Exchange) format in those two modes. It's particularly interesting during the design phase whereby receivers can be placed temporarily to collect 24hour or 48 hour data and then processed at a later stage. The advantage of this is that performance of the system can be analysed prior to permanently fixing cabling for power and communications as well as estimating any errors associated with multipath.

Real Time Monitoring With GNSS Spider

The real time processing kernel is based on that used in the Leica GNSS RTK rover, but has been modified for monitoring applications. The Leica Smart Check technology, which is an evolution of the repeated search process, is used to continuously re-verify the ambiguity fix to ensure the highest reliability. With this improved kernel GNSS Spider is able to compute RTK-fixed positions from both single and dual frequency data at extremely high reliability.

Three ambiguity resolution techniques are available: Kinematic on the fly (also known as OTF or While Moving initialisation), Initialisation on Known Marker (IOKM) and Quasi-Static Initialisation (QSI). The OTF ambiguity resolution allows for full receiver dynamics during the initialisation at the cost of reliability, especially for single frequency processing. The IOKM ambiguity resolution assumes strictly limited receiver dynamics (which is not practical for monitoring) but has much higher reliability. The QSI technique is combination of the previous two techniques – it allows for the antenna to be in motion during the initialisation but not to the same extent as OTF initialisation.

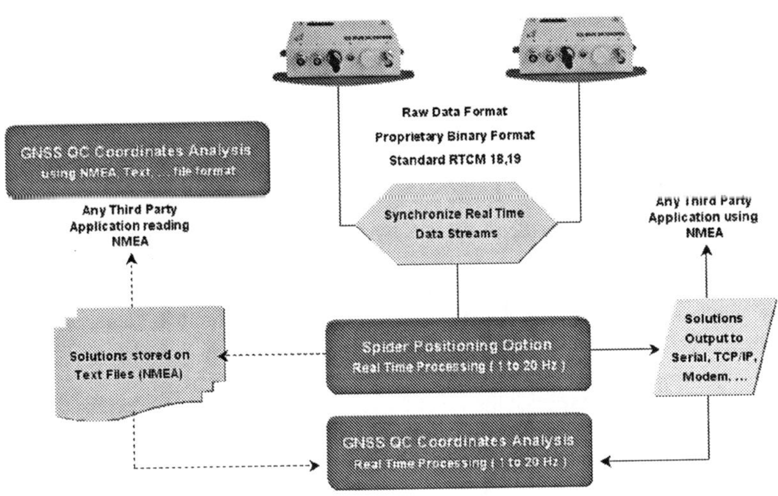

Post Processing Monitoring With GNSS Spider

The post-processing kernel used in GNSS Spider is based on that used in LGO. Like with the real time processing, a repeated search process is used to ensure highly reliable ambiguity resolution.

In addition, the initialisation on a float marker is used to further improve the reliability. Post-processing intervals of between 1 minute and 24 hours are possible for dual frequency data and between 10 minutes and 24 hours for single frequency data.

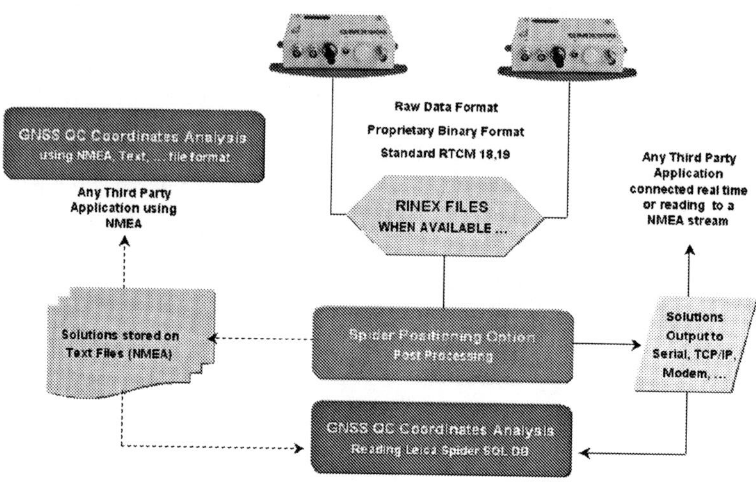

Reducing Inherent GNSS result noise by using Low Pass Band Filtering

Due to the nature of the different sources that affects the solution of GNSS measurements like atmospheric delays, the orbital errors and the multi-path effects in some extend, the results are generally noisy and will not reflect at first look the full potential of the solution.

Therefore it is necessary to reduce the noise by using digital signal processing filtering techniques. Leica GNSS Spider has that capacity and various tests have demonstrated that up to 30% to 45% of the noise are effectively reduced.

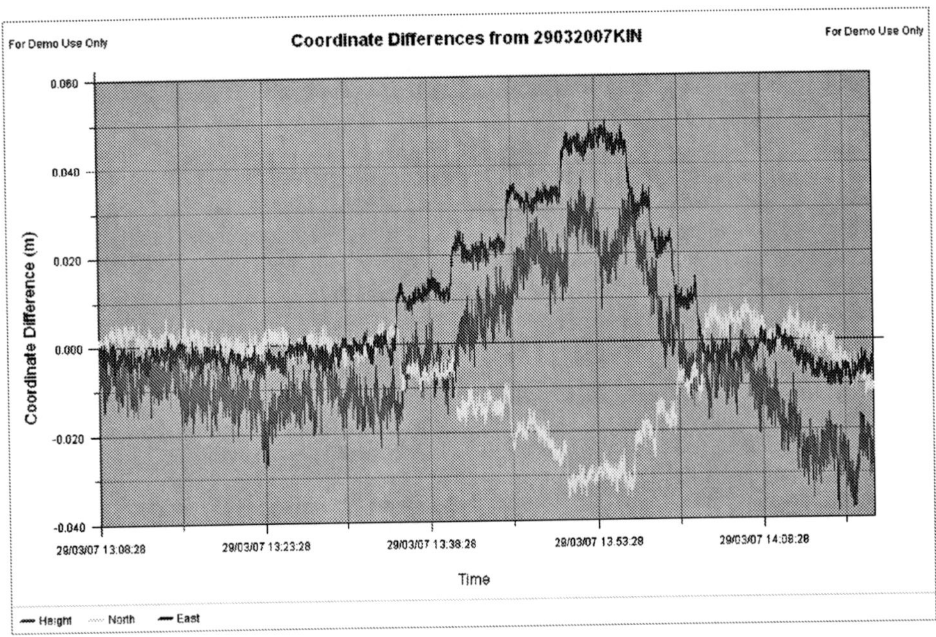

Real Time results unfiltered and resulting from a maximum displacement of 4 cm.

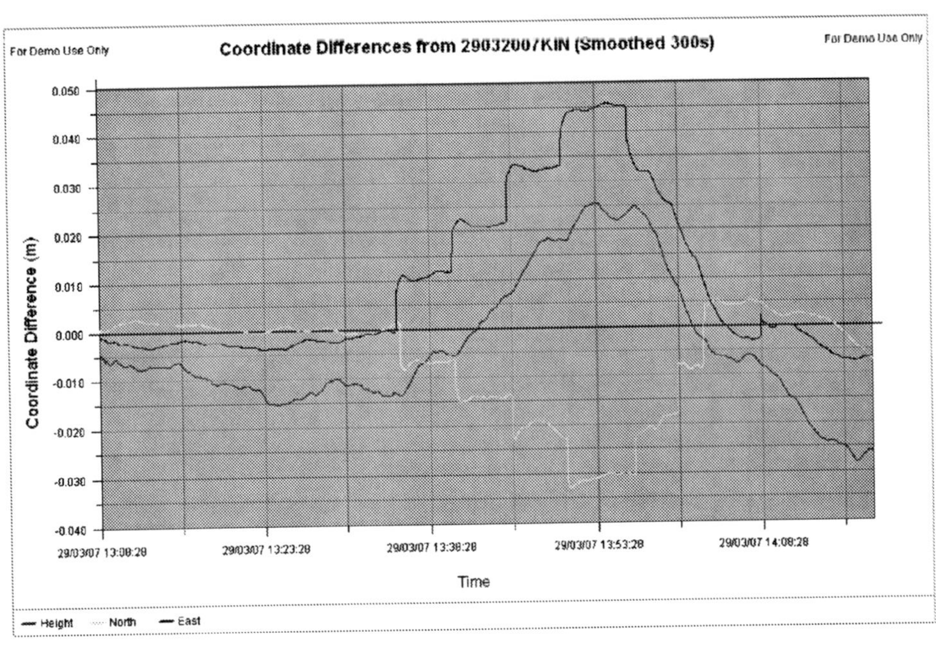

Real Time results filtered and resulting from a maximum displacement of 4 cm.

Implementation of a GNSS Bridge Monitoring solution

A typical GPS Bridge Monitoring Solution proposal is based on a dedicated <u>monitoring design</u>, which must address the following questions:

- How many monitoring station must be deployed on the infrastructure, where they should be located and which kind of support (mast, fixture) is allowed?
- Where will be the location of the GNSS Reference station(s). How many do we need? These stations play a very important role in the monitoring point determination and must be located on a very stable area…
- Do we need the results in real time or/and in post processing?
- What is the infrastructure project size? (3 km over 300 meters e.g.)
- Do we need to convert the results into a local datum? Which local datum? Who will deliver the parameters?
- How can we power up safely the GNSS receivers? Lightning protection? Do we need to interface the GNSS receivers with other sensors like tilt-meters, meteorological stations etc.?
- Where we will the control centre hosting the Leica SPIDER Server be located? To which application(s) must we interface the output results? Do we have to provide a PC computer with peripheral(s)?
- A technical map must be provided with the location of the reference station(s), the monitoring stations and the control centre.
- What environmental conditions surround the GPS monitoring stations?
- What is the required accuracy for X-Y coordinates? (Horizontal displacements)
- What is the required accuracy for Z-coordinate? (Vertical displacements)
- Required measurement frequency? (20Hz, 5Hz, 1Hz, slow motion, static)
- Expected movement's entity? (mm/hour, cm/day, dm/year, etc.)
- Planned/preferred communication link between measuring stations and Control Centre?
- Object extension: which is the maximum distance between two monitoring stations?
- Sky visibility. Presence of obstacle (structures, cables.) that could restrict the satellites visibility?
- Is there a budget already allocated for this monitoring project?
- Planned time-schedule: Monitoring system installation? Measurement start?
- How many GNSS sensors and their GNSS antenna's, including various accessories like cables, power supplies, uninterruptible powering system, lightning protection, mast and adapters must be provided?
- Does the receiver have to log the raw data or/and just to stream? How many ports? Does it have to synchronize other monitoring equipment on GPS time (PPS)?
- What will be the solution for the communication interfaces and lines?
- Which modules of Leica GNSS Spider software must be delivered?
- What kind of analysis software will be used and which interface should be delivered?

<u>Type of services to provide :</u>
- Feasibility study, support for the monitoring project design, location of the monitoring stations, processing scheme, simulations, multi-path analysis…
- Comprehensive and clear quotation including delivery terms…
- Installation, tests, training, support and commissioning…
- Maintenance contract including upgrade proposals for both hardware and software…

Conclusion

The ability to monitor real-time data from a remote location is a critical issue due to the divergence often found between the location of structures and the location of the people charged with monitoring those structures.

Leica Geosystems can provide GNSS monitoring systems that gather the data and feed it to a remote location via the Internet to give you 24/7 monitoring capability from any location.

Whether the movement occurs over a period of seconds, minutes, hours, days, weeks or months, the system is able to track the movement. Parameters can be set such that any movement outside a designated range can automatically notify the responsible people. This timely information gives the operators time to take an appropriate response and avoid any critical failures.

For further information please contact:

Leica Geosystems AG Switzerland
Joel van Cranenbroeck, Business Development Manager for Geodetic Monitoring
Heinrich-Wild-Strasse
CH-9435 Heerbrugg
Phone/Fax: +32 (0)81/41 26 52
Mobile: +32 474/98 61 93
Joel.vancranenbroeck@leica-geosystems.com

www.leica-geosystems.com

Closing plenary session:

From the past to the future

RC arch bridges – lessons from the past and future developments

Prof Dr Jure Radic, Faculty of Civil Engineering University of Zagreb and Civil Engineering Institute of Croatia, Zagreb, Croatia
Jelena Bleiziffer, MSc, Faculty of Civil Engineering University of Zagreb, Croatia
Igor Gukov, MSc, Polytechnic of Zagreb, Croatia
Damir Tkalcic, MSc, Civil Engineering Institute of Croatia, Zagreb, Croatia

Abstract

A State-of-the-art in reinforced concrete arch bridges has been presented with discussion on design, construction, performance in service and subsequently performed repairs on Adriatic RC arch bridges. Lessons from the past that should be incorporated in the design of future bridges have been pointed out, as well as requirements for future research in order to increase the competitiveness of RC arch bridges and to ensure the preservation of existing structures.

Introduction

During the whole of construction history, the arch has been a resistant structure par excellence, and still is today. Its strength is due to its shape and in its geometry converted into a resistant mechanism where compression predominates (Troyano, 2003).

Figure 1. Krk Bridge – the world's largest conventional reinforced concrete arch bridge is located in Croatia

Concrete arches have been favoured for large spans up to the middle of the twentieth century. Then, other structural systems were developed for large spans, which made arches non-competitive, primarily due to large construction costs. In the past, heavy falsework was needed for the arch construction, which was often more demanding than the arch itself, creating problems in the supply of materials, construction, foundations and stability. Even today, when arches are usually erected using the cantilever procedure thus avoiding the need for scaffolding, expensive auxiliary equipment is required during the construction process, because of the significant differences in internal forces during the construction stages and at the final stage, once the arch is completed.

In today's technology, reinforced concrete arches are generally considered well suited for spans ranging from 200 to 400 m, which coincides with the span range of steel girder or truss bridges and cable-stayed bridges. Arch bridges have considerable advantages because of their inherently interesting form, strong visual appeal and attractiveness. But, for an arch span to be competitive economically over other bridge types, the right site with good soil to provide adequate foundation condition is needed. Such surroundings are an important structural requirement to resist the thrust at the arch springing, but also provide an appropriate backdrop for exceptionally, aesthetically successful structures. The most preferable sites for an arch structure are river canyons or valleys with high bluffs. When the arch is placed below the deck, such topographic features provide visual and structural resistance to both horizontal and vertical reactions at the springing points.

Such favourable topographic conditions can be easily found along the Croatian Adriatic coast. Development of national infrastructure networks required the bridging of numerous rivers and bays in this region. Furthermore, since Croatia is a land of many islands, the sea-strait crossings providing fixed road links to the mainland are of utmost importance.

Two periods of extensive bridge construction in the Croatian coastal area can be distinguished. The first took place during 1960's and was initiated by the growth of tourist-trade in the region at the time. The latter, still in progress, started immediately after the establishment of Croatian's sovereignty and is associated with the extension of the principal road network which should integrate the most distant Croatian territories and provide connection to European traffic corridors. In both cases reinforced concrete arch bridges were favoured on many projects in most striking surroundings precisely because of their attractive shape. It should be noted that at several sites along the coast, steel arch bridges were constructed. Among these the most notable are the old Maslenica bridge and Morine bridge. Although these proved to be structurally and aesthetically pleasing structures, construction in reinforced concrete is preferred in Croatia. It allows for utilization of local materials and is thus cheaper. Furthermore, structural solutions comprising reinforced concrete arches have been a long tradition in Croatia.

State-of-the-art: Adriatic RC arches

Some of the most striking concrete bridges in Croatia are located at the Adriatic coast. Adriatic Arch Bridges are world-renowned not only because of their large spans, but also due to the introduction and subsequent improvements introduced in construction of reinforced concrete arches using the suspended cantilever technique. Seven large RC arches have been erected in the Adriatic region over the past four decades: Sibenik Bridge (1966), Pag Bridge (1968), Krk Bridge – two arch spans (1980), Maslenica Bridge (1996), Skradin Bridge (2005) and Cetina Bridge (2007). The Sibenik Bridge is the first concrete arch in the world that was erected entirely by the cantilever method (Figure 2). Arch segments were concreted on a scaffolding platform, 27 m long. The arch cantilevers were supported by stays anchored into strong abutments.

Figure 2. Sibenik Bridge (left) and Krk Bridge (right) construction

The stays were indigenously designed. They comprised steel profiles, whose load carrying capacity was considerably increased by combining them with prestressing tendons. The steel profiles stiffness also reduced arch displacements during construction. After a segment was completed the scaffolding was moved forward by a floating crane. The Pag Bridge was constructed much in the same way, but without auxiliary steel pylons atop the arch springings, and with backstays anchored directly into the ground.

Further development of the cantilever erection method for reinforced concrete arches was achieved with the construction of the Krk Bridges from 1976 to 1980. An innovative method in which concrete spandrel columns, temporary steel tension tie top chords and diagonals were combined to form a truss cantilever, was devised to erect the two arches of the Krk Bridge, the 390-m arch span still being the largest conventional reinforced concrete arch span in the world (Figure 2). The three-cell box arch was constructed in two stages. First, the centre box cell was constructed of precast panels for the top and bottom plates and two interior webs, and then these were joined by cast-in-place concrete. After the hydraulic jacks in the crown of the centre cell of the arch were activated, the temporary steel ties were removed and the two outside arch cells were constructed, using the centre arch rib for support.

These great engineering achievements were followed with three more reinforced concrete arches constructed in the past decade: Maslenica Bridge, Skradin Bridge and Cetina Bridge (Figure 3).

The 200-m span Maslenica Bridge was completed in 1997. The superstructure comprises a grillage of eight simply-supported precast prestressed girders made continuous over the middle supports by utilizing non-prestressed reinforcement and interconnected by a 25-cm thick concrete deck plate cast-in-site, with cross girders provided only at the supports (Candrlic et al., 1999).

The arch of the Skradin Bridge spans 204 m with a rise of 52 m and is of double-cell box cross-section. 22.56-m wide superstructure is composed of steel girders and a reinforced concrete deck-plate (Radic et al., 2003b).

The construction of a reinforced concrete arch bridge across the Cetina River commenced in the year 2005 and the bridge will be opened to traffic in 2007. The beautiful river canyon is an environmentally protected area, and an arch structure was selected as it harmoniously blends into the landscape.

Figure 3: Contemporary Adriatic RC arch bridges:
Maslenica Bridge (left), Skradin Bridge (center) and Cetina Bridge (right)

The arch span of 140 m with a rise of 21.5 m is fixed and of a single-cell cross-section with constant external dimensions of 2.5 x 8.0 m. In order to improve the side view of the arch, the lateral surfaces are curved. The 10-span continuous bridge superstructure consists of precast prestressed concrete girders, a cast-in-situ deck plate and cross-girders at the supports only. The typical superstructure span is 21.60m.

Performance in service: lessons from the past

The large RC arch bridges described are located in the Croatian Adriatic coastal area – a very aggressive environment (Radic et al., 2003a) in terms of frequent strong winds spraying sea-salt on the exposed concrete surfaces, thus speeding up bridge deterioration due to chloride corrosion of the reinforcement. It should be noted that the Adriatic Sea is a rather warm sea, with a salinity of 3.5-3.8 % of water mass. Additionally, high average annual temperatures, high moisture contents and occasional winter temperatures below 0°C accelerate chloride penetration and reinforcement corrosion. Over the years many deficiencies and the advanced stages of deterioration processes have been identified on older Adriatic Bridges.

Chloride attack due to maritime exposure, followed by cracking, delamination, splitting and peeling off of concrete is identified as the major deterioration mechanism (Figures 4 & 5). One of the causes for rapid structural degradation was also an underestimation of the role of maintenance in the past and poor planning of maintenance operations, mainly due to a lack of funding for regular maintenance activities (Radic et al., 2005a). Even in aggressive environments, some structural damage might have been avoided altogether or at least diminished, had there not been design errors and poor detailing in particular. Substantial damage called for many expensive and complex repairs to be carried out on older bridges.

Figure 4. Sibenik (left), Pag (center) and Krk (right) Bridges:
Deterioration of bridge superstructure

Figure 5. Sibenik (left), Pag (center) and Krk (right) Bridges: Deterioration of columns

Figure 6. Repair works on Pag Bridge

Figure 7. Repair works on Krk Bridge

Repair works on the Pag Bridge started after only a decade of its service, but did not prove efficient in terms of stopping the corrosion process. Major reconstruction started in 1991 with the repair of the arch and was finally finished in 1999 when the original concrete superstructure was dismantled and replaced by a completely new structure in steel. Columns were repaired by encasement in steel and concrete.

The repair works on the Krk Bridge started several years after its completion focusing on the superstructure supports. In the 1990s the works were broadened to include the repair of columns at the smaller arch. Different repair techniques had to be devised for spandrel and approach columns.

The smaller arch has been recently repaired by removal of the contaminated concrete, its subsequent replacement with shotcrete and the addition of a protective coating (Figure 7). The repair of the larger arch presents major challenge to engineers. The arch is actually supported on submerged inclined struts. After many years of research and testing of various corrosion protection systems, cathodic protection was selected for this part of the structure (Beslac et al., 2007). All of these repair works are not only expensive, but technically demanding tasks and very difficult to perform.

Sibenik Bridge is somewhat less exposed to an aggressive maritime environment and only minor rehabilitation work has been performed so far. The bridge was thoroughly inspected last year, and the repair methodology is presently being discussed.

Future developments

Currently, research is being conducted aimed at further development in three areas:
- maintenance and management of existing bridges
- design exploring the span limits for RC arch bridges
- cost reduction to improve the competitiveness

Maintenance and management of existing bridges

The short overview of the in-service performance of Adriatic arch bridges given in the previous section clearly indicates the lack of a consistent maintenance policy that would provide for efficient and effective management of these bridges. Croatia is currently placing large efforts to develop an asset management system for Croatian motorways, which would address all motorway structures: bridges, tunnels, pavements, drainage, geotechnical structures, other roadway components (safety barriers, wind barriers, sound barriers etc.) and buildings (roadside service facilities, maintenance and traffic control centers etc.). In the course of this work it was established that major structures should be hand-picked from the entire bridge stock and treated separately. This conclusion came mainly from the experience with the in-service performance and maintenance work carried out on large reinforced concrete arch bridges, set in the extremely aggressive environment of the Croatian Adriatic coastline.

This experience of the in-service performance of older large reinforced concrete arch bridges was carefully considered while designing the more recent Maslenica and Skradin Bridges, and the long-term performance of the structure has already been analysed at the planning stage. Structural members and concrete cover were significantly increased in size compared to the first generation of Adriatic arch bridges to achieve a robust and durable structure. High quality concrete was used with a water-cement ratio of less than 0.4. Portland cement with a 20% slag addition was utilized. All these measures led to an increased initial cost, but were justified with a lower probability of costly and complex repairs in the future, which were experienced on the older bridges.

Additionally, Maslenica and Skradin Bridges were equipped with a range of sensors for the long-term control of stresses, strains and corrosion progress with the intention to closely monitoring both structural performance and durability related performance in order to facilitate the future maintenance activities by triggering timely adjustments and interventions (Figure 8). Unfortunately, the Maslenica Bridge monitoring project was stopped due to lack of funds, but the monitoring system installed on Skradin Bridge, consisting of a smaller number of gauges, is set up and running.

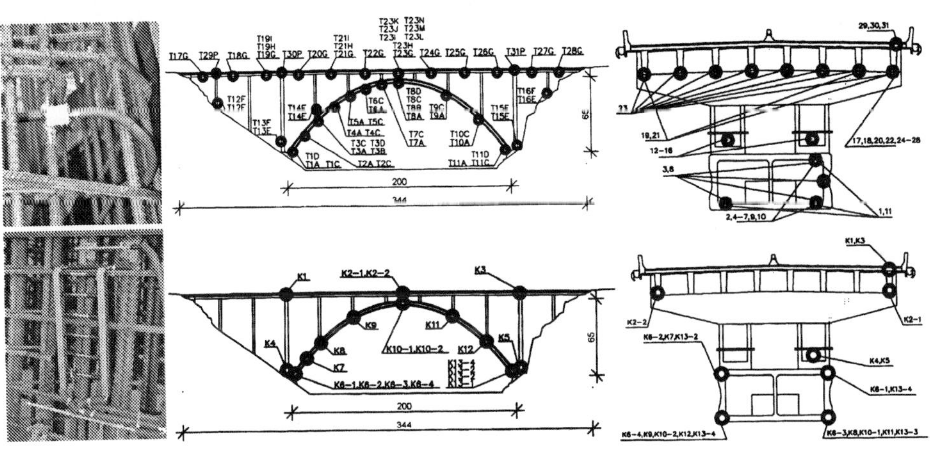

Figure 8. Location of strain-gauges (top) and corrosion sensors (bottom) installed on the Maslenica Bridge

The lessons learned from the in-service performance and monitoring of large Adriatic bridges has taught us that major structures need to be treated differently to the rest of the bridge stock. These bridges are often complex structures which include out-of-ordinary structural solutions. Thus, the technical feasibility of the repair, and not the cost itself, is generally the governing issue in management decisions. It is impossible to decide on a repair methodology by simply following standard repair procedures.

Major bridges are generally regarded as "irreplaceable" so the management of such bridges should incline towards developing an efficient and effective preventive maintenance program, and not so much towards minimum cost.

Additionally, inspections of major bridges are more exhausting and require well-experienced personnel to identify the damage and determine its cause and consequences. Bridge management system modules should consequently be adaptable to enable the collection of a broader set of data. This is especially important if a structural health monitoring system is installed on a structure.

These premises form the basis for developing a management system for the large Adriatic arch bridges, which will be adapted to the specific traffic, technical and economic requirements of each bridge. The intention is to develop bridge-specific maintenance strategies and programmes, as well as having dedicated bridge engineers for each of the bridges. The program would be tailored to individual structural aspects and the present structural condition. Additionally, as some of these bridges are equipped with a range of sensors and structural health monitoring data should be used for structural condition evaluation and deterioration prediction, thus triggering timely and appropriate maintenance and rehabilitation work.

Exploring the limits of RC arch bridges

Very large span arch bridges may be constructed if the dead weight is reduced by adopting a much lighter composite superstructure instead of the concrete one. An example of such a structure is the Skradin Bridge described earlier in more detail. The bridge structure combines a reinforced concrete arch and a superstructure composed of steel and concrete. Instead of a conventional prestressed concrete superstructure made of precast girders and a cast in place slab, it consists of steel girders and reinforced concrete deck-plates (Figure 9). This enabled the crosshead beams on the columns to be omitted and the size of the arch cross-sections to be reduced, resulting in 35% less mass in the Skradin Bridge when compared to the Maslenica Bridge. This smart combination of the two materials reduces the consumption of resources and shows the one possible way of achieving future record-breaking concrete arch spans (Radic et al., 2005b).

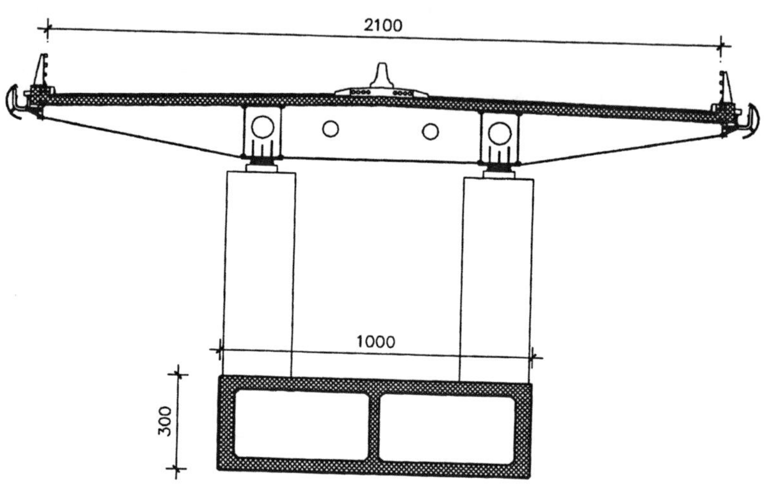

Figure 9. Composite superstructure of the Skradin Bridge

Reactive Powder Concrete, a Portland-cement based material with a compressive strength ranging from 200 to 800 MPa and of high ductility, may be utilised for large engineering structures in the future (Candrlic et al., 2001). A study of an arch bridge for a Bakar straits crossing, spanning 432 m, has been performed. The bridge is completely composed of precast segments with external cables and will be made of Reactive Powder Concrete with a strength of 200 MPa (RPC 200). Webs and flanges of a box-type section will be 12 to 20 cm thick. The construction of the arch is designed to be free cantilevering while the superstructure may be constructed either by free cantilevering or by incremental launching from both sides of the bridge. Both methods provide support to the segments by stays radiating from auxiliary steel pylons. Reactive Powder Concrete represents a big step in the enhancement of durability. At the same time it allows for new shapes and reduced thickness of cross-sections of precast segments. Both the arch and the superstructure are designed as triple cell boxes of aerodynamic shape with very thin webs and flanges. Further, this research was extended to investigate the structural aspects and construction technology of arch spans ranging from 500 to 1000 m, designed using similar principles. This study shows a number of new possibilities, but, to realise such bridges, further research is needed.

Construction methods to improve competitiveness

To extend the competitiveness of RC arch bridges, further research is needed to develop construction methods which would further reduce the cost. The construction of contemporary Adriatic arch bridges, using state-of-the-art techniques led to conclusion that it is still the cost of auxiliary equipment (cranes, temporary stays and pylons, anchors etc.) needed to erect the arch that prevails over the cost of structural materials. For instance, the cost of auxiliary equipment to erect Skradin Bridge was 60% of the total cost. A solution might be to develop such a structural system for RC arch bridges that integrates both the arch and the bridge superstructure in a single load-carrying structural system during the construction stages. Currently a research is underway to develop solutions in concrete and steel as well as to explore the possibility of combining the two materials (both in the arch and superstructure).

Conclusions

State-of-the-art in reinforced concrete arch bridges has been presented with discussion of the design, construction, in-service performance and subsequent repairs of Adriatic RC arch bridges, pointing out the lessons learnt from the past that should be incorporated into the design of future bridges.

Arch bridges have a main advantage over other bridge types in aesthetic terms, due to their inherently interesting form, but further research is required to increase the competitiveness of RC arch bridges in the future, as well as to ensure preservation of existing bridges.

The economy of RC arch bridges may be improved mainly in terms of reducing the construction costs by improving construction techniques. Larger spans may be achieved by using high performance concrete, as well as through the intelligent combination of steel and concrete. Maintenance and management of large RC arch bridges poses significant challenges to their owners, as the repair works are costly and require outstanding engineering efforts.

References

Beslac, J., Tkalcic, D. & Barisic, E. (2007) Present state of Krk Bridge repair and protection, *fib Symposium Concrete Structures – Stimulators of Development, Dubrovnik, Croatia, 2007.*

Candrlic, V., Bleiziffer, J. & Mandic, A. (2001) Bakar Bridge Designed in Reactive Powder Concrete, *Third international arch bridges conference, Paris, France, 2001*: 695-700.

Candrlic, V., Radic, J. & Savor, Z. (1999) Design and Construction of the Maslenica Highway Bridge. In *Proceedings of fib Symposium 1999, Vol. 2*: 551-556. Prague, Czech Republic.

Radic, J., Bleiziffer, J. & Tkalcic, D. (2005) Maintaining Safety and Serviceability of Concrete Bridges in Croatia, *Journal of Bridge Structures*, 1/3, pp. 327-344.

Radic, J., Savor, Z. & Gukov, I. (2005) New contribution to concrete arch bridges construction, In *Sixth International Bridge Engineering Conference Proceedings*.

Radic, J., Savor, Z. & Puz, G. (2003) Extreme Wind and Salt Influences on Adriatic Bridges, *Structural Engineering International*, 13/3, pp. 242-245.

Radic, J., Savor, Z. & Puz, G. (2003) Sixth Large Reinforced Concrete Arch on the Adriatic Coast. In R.K. Dhir, M.D. Newland & M.J. McCarthy (eds), *Proceedings of the International Symposium «Role of Concrete Bridges in Sustainable Development»*: 257-266. London: Thomas Telford Publishing.

Troyano, L.F. (2003) "Bridge Engineering – A Global Perspective, Thomas Telford, London, UK, 2003.

A Thousand-meter span Cable-stayed Bridge across Yangtze River - Sutong Bridge

You Qingzhong   Jiangsu Provincial Sutong Bridge Construction Commanding Department, Nantong, China
Zhang Xiongwen   Jiangsu Provincial Sutong Bridge Construction Commanding Department, Nantong, China

Abstract

Sutong Bridge is a cable-stayed bridge across Yangtze River with a main span of 1088m. The bridge site conditions can be characterized as an expansive river, with fast flowing tides and deep currents, thick Quaternary period sediments, complicated weather and also heavy traffic. Furthermore, the riverbed here is easily subject to scour. This article outlines the technical challenges encountered in both the design and construction stages such as scour protection, lowering the steel cofferdams, construction platform establishment, stay cable fabrication, closure and the corresponding solutions.

Project outline

Geographical location

A highway from Shenyang city, a heavy industry base in northeast China, through Shanghai, Suzhou and Hangzhou in Yangtze River Delta, to Haikou city, a famous scenic spot, is being built in the coastal area of the eastern part of China. The Sutong Bridge is an important project in the route taking it across the Yangtze River (Figure 1). The Sutong Bridge is located in the Yangtze River Delta, connecting the two cities of Nantong and Suzhou. It will further strengthen the Yangtze River Delta's role in promoting economic development throughout China. The project is very important for expediting cultural integration between different regions in and it is a project with many technical challenges.

Figure 1   Location of Sutong bridge

Construction condition and design standards

Meteorological conditions
In general, the bridge area is characterized by abundant rain with a highest and lowest

Bridge design, construction and maintenance 2007, Thomas Telford, London

temperature of +42°C and -13°C respectively. There are 179 days in a year with wind speeds higher than a strong breeze (identified as Class VI), over 120 rainy days and 31 foggy days on average in a year. Moreover, it is threatened by some adverse weather conditions, such as rainstorms and tornados, etc.

Hydrographic conditions

The bridge location is mainly subject to flood and runoff volumes, with a flood season from May to October and low water period from November to April. The flood peak normally occurs during June to August. The width of the Yangtze River at the bridge site is approximately 6 km. The river features deep water and high current velocity, characterized as a medium tidal effect reach. This semi-diurnal tide has an average cycle period of about 13 hours, i.e. two high and low tides per day. The highest tides are mainly due to storm tides. Tidal height could exceed 7m when a maximum astronomical tide coincides with a typhoon. Water depth adjacent to the mains pier is approximately 30m with a typical velocity of over 2.0m/s, a vertical line average flow velocity of 3.68m/s, and a max water flow velocity of over 4m/s. The tidal range is between 2m and 4m, and the wave height is between 1.0 and 3.0 m.

Geological conditions

The bridge location has an overburden formed in the Quaternary period of over 270m of mainly silty sands and silty clays, together with some thin layers of sands and gravels. The riverbed consists mainly of clay and silty sands and is subject to scour. There is a V-shaped deep groove (Figure 2) at the bridge site. The water area below the -10m isobath has a width of around 1800m and the water area below the -20m isobath has a width of around 1100m. The greatest water depth exceeds 30 m.

Fig. 2   Bridge site riverbed Shape

Navigational conditions

The bridge site has complicated navigational conditions with heavy traffic, featuring many different kinds of ships. On average, there are more than 3300 vessels passing through the bridge site every day and with a peak volume of more than 6000 vessels, among which 81% are common cargo ships and 18% are river steamers or sea-going pleasure craft. More than 400 are oil tankers, liquefied gas carriers, chemical carriers, dangerous cargo ships and large or medium sized vessels. The peak time for navigating the site normally occurs 1 hour before or after the turn of the tide.

Technical standard

The Sutong Bridge is designed as 6 traffic lanes in dual directions and the design traffic speed is 100 km/h. The design life is 100 years. The bridge has a standard width of 34m, a longitudinal slope ≤ 3% and a horizontal slope = 2%. The bridge is designed to resist standard seismic intensity VI (the seismic standard with 1000 and 2500 year return periods). The reference wind speed at 10 m height with a 100 year return period is 38.9m/s. The bridge has a navigational clearance of 891m × 62m, satisfying the navigational need of the 50,000 tonnage container and 48,000 tonnage bulk cargo fleets. The ship impact resistance of the main bridge piers is designed to be 13,000 tons transversely and 6,700 tons longitudinally.

Bridge structure

General layout

That part of the Sutong Bridge which is over water has a total length of 8146m, consisting of a north approach bridge, the main bridge, a special fairway bridge and a south approach bridge. The north and south approach bridges have a total length of 3485 and 1650m respectively with continuous, post-tensioned concrete box girders with typical spans of 30m and 50m. The special fairway bridge has a total length of 923m, consisting of a continuous rigid box girder with a span configuration of 140m + 268m + 140m, and a prefabricated cantilever concrete girder bridge with a span of 75m. The main bridge of the Sutong crossing is a seven span double-pylon, double cable-plane, steel box girder, cable-stayed bridge with a span arrangement of 100 + 100 + 300 + 1088 + 300 + 100 + 100 = 2088m (Figure 3).

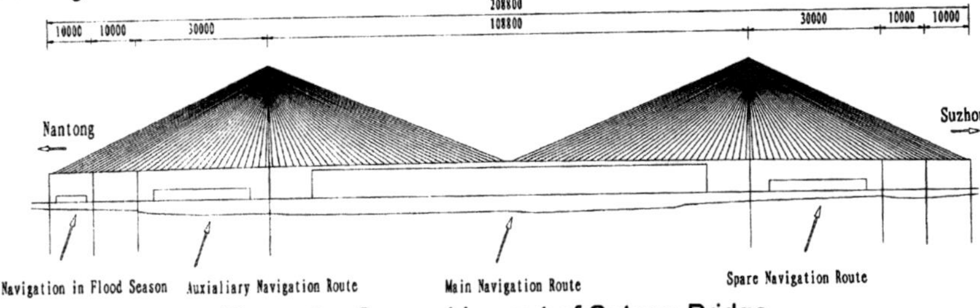

Figure 3   General Layout of Sutong Bridge

Structural system

The main bridge of the Sutong Bridge adopts a semi-floating structural system. A combined system of rigid position restriction with a rated stroke of ±750mm and a dynamic damper was adopted for the pylon-deck connection in the longitudinal direction. Wind resistant bearings are adopted to restrict the fixed pylon-deck connection in the transverse direction. Longitudinal sliding bearings are installed between the deck and the transitional piers and between the deck and the auxiliary piers to restrict relative transverse movement.

Pylon foundations

The main pylon foundations are both supported by 131 bored piles with diameters of 2.85m and/or 2.5m (Figure 4). The piles are arranged in an interlocking triangular pattern and are designed as friction type piles with a length of 117 m. The pile cap is shaped as a dumbbell. The overall dimension of the pile cap at the bottom of each pylon leg is 51.35m × 48.1m and the thickness varies from 5m at the edge to a maximum of 13.324m.

The bridge deck

The bridge deck is a slender steel box girder which was designed to have a good aerodynamic performance. The total width of the girder is 41.0m and the depth at the girder centre-line is 4.0m (Figure 5). Stays are anchored to anchor boxes in the girders. Top flange stiffening comprises 300mm deep U-troughs and solid plate transverse bulkheads are used. Two longitudinal diaphragms are formed as truss structures except at the vertical supports, ballast areas and pylon areas where solid plate is used.

Stay cables

Parallel wires were selected for the Sutong Bridge stay cables; the diameter of the high-strength zinc-coated wires being φ 7mm. There are a total of 272 cables for the whole bridge and the longest one is 577m. Typical spacing of the stay anchorages on the deck girder is 16 m and 2 m on the pylons.

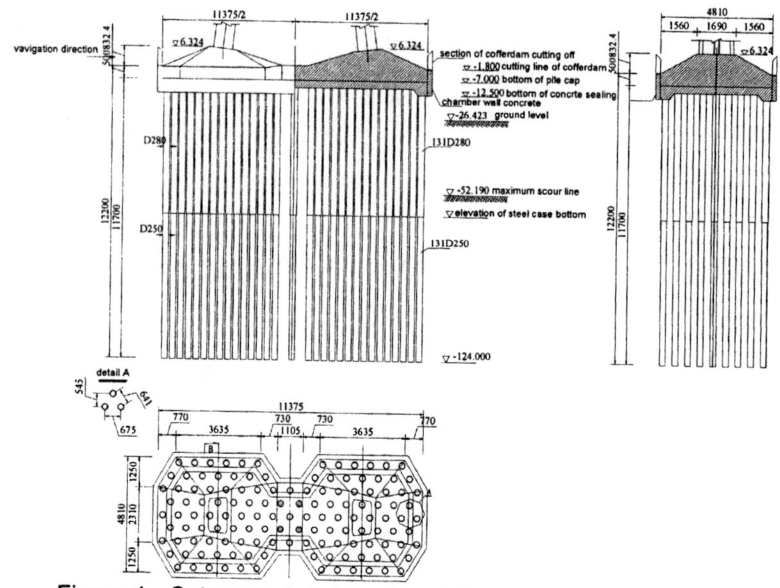

Figure 4   Sutong main bridge foundation structural diagram

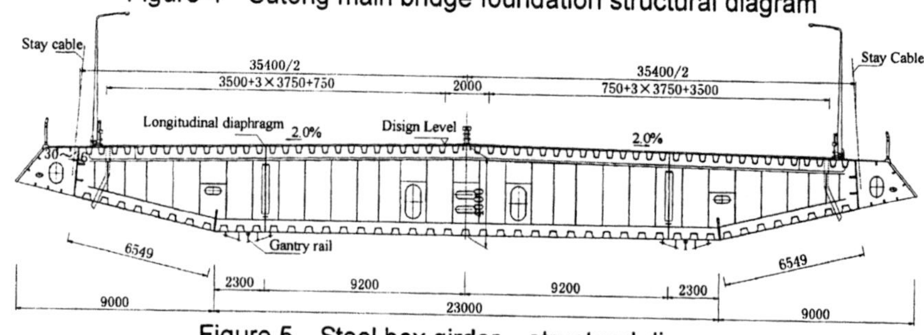

Figure 5   Steel box girder – structural diagram

Pylons

The pylons are inverted Y-shaped concrete pylons. Their total height is 300.4m, with the upper pylon legs being 91.4m, the middle pylon legs being 155.8m and the lower pylon legs being 53.2m. The cross section of the pylon legs is hollow apart from the bottom section where it is solid. There is a cross-beam at an elevation of 64.3m under the deck girder. Stay cables are anchored to an anchor box in the pylons. There are in total 30 anchor boxes over a height of 73.6m. The height of each standard section is between 2.3m and 2.9m.

Key techniques

Group pile foundations

Large pile cap steel cofferdam
Ship impact was a critical technical issue that had to be solved during the construction of the

pile groups for the foundations of the Sutong Bridge. A ship impact analysis showed that the pile group foundations had sufficient ship impact capacity. Since it is critical to avoid structural damage induced by ship impact a steel cofferdam filled with concrete was used for this ship impact protection. The steel cofferdams used to prevent local ship collision with the foundations were then also used to provide a dry environment for pile cap construction (Figure 6).

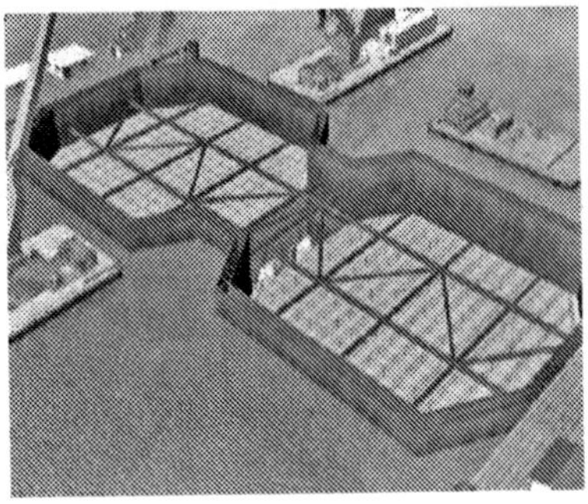

Figure 6 Sketch of steel cofferdam

The main challenge with the construction of the steel cofferdam was to lower it into position in one. As such, the difficulty can be described by the following five aspects:

- Large size of steel cofferdam, with length, width and height of 117.35m, 51.7m and 16.9m respectively;
- Heavy structure, approximately 5000 tons;
- Numerous lifting points, possible uneven load distribution;
- Large lowering distance, approximately 14m, during which collision between the bottom plate and steel casing might happen;
- Steel cofferdam is vulnerable to sway after lowering into the water due to current and wave effect.

In order to solve the five problems mentioned above, the following measures were taken:

- Choose appropriate time for the lowering operation. In order to avoid a sudden change in the cofferdam's internal force due to buoyancy as soon as the cofferdam enters into water, lower the cofferdam close to water surface during slack tide, stop for a while and then continue lowering until the bottom plate of the cofferdam is immersed in water during the flood tide. Between 1 and 13m of water was injected into cofferdam wall depending on whether there was a slack tide or flood tide after the cofferdam was lowered into the water so as to maintain the head difference between the inside and outside of the wall. Moreover, it was ensured that the strands were always under tensile stress before the cofferdam was lowered to its design elevation;
- The large number of lifting points. The cofferdam was synchronously lowered from 12 lifting points using 40 hydraulic jacks balanced using a computer control system. The lowering operation was stopped for adjustment in there was more than 1cm height difference;
- Conflict between the cofferdam and the steel casing was avoided by monitoring the position of the steel casing during the lowering operation and taking effective measures to avoid this conflict.
- Displacement limits were established and a guide device was used at both inside and outside of the cofferdam so as to avoid sway due to current impact. Vertical positioning of the cofferdam was achieved by means of 32 brackets on the peripheral steel casings. On completion of the vertical positioning, the plan position of the cofferdam was adjusted during a slack tide after which it was fixed. Full advantage was taken of the full force of the flow during positioning. The cofferdam wall was connected to the outside steel pipe piles as well as inside the steel casings as a whole so as to increase the integrity and stability of the cofferdam after it had been lowered into its design elevation.

The lowering operation typically lasted for 10 hours and was deemed to have been successful once it had been positioned within a synchronization tolerance of 1cm and a plan accuracy of 1cm.

Scour protection

Pile foundations must be sufficiently embedded into the soil however, local scour can lead to a reduction of this depth. Experimentation has shown that the north pylon of the Sutong Bridge was expected to experience the largest scour depth. Scour depth due to water flow with a return period of 20 years can reach 21.5m and scour depth due to water flow with a return period of 300 years can reach 27.2m. Moreover, the max scour may occur due to just one period of high water flow velocity. In order to satisfy ship impact and seismic requirements, the bottom elevation of the steel casing must reach -62.0m, taking riverbed scour into account. However, there is one thin layer of pebbles between elevations -52.0m and -54.0m, which it is very difficult to penetrate with the steel casing. In order to ensure the safety of the foundations, they were designed to accommodate scour protection. A scheme with a combination of pre-production and permanent protection was adopted for the Sutong Bridge so as to avoid regular maintenance during the service period.

The concept of the permanent scour protection was:
- Protect foundation riverbed by dumping sandbags, graded stones and armour stones;
- Divide the riverbed into a central area, a permanent protection area and a falling apron.

Different scour protection structures based on different water flow properties were adopted in each of these areas to ensure the long service life of the scour protection.

The provision of the scour protection can be divided into four steps. The first step is to dump sandbags for protection of riverbed bottom and avoid scour. The thickness of sandbag layer for the central area is 2.0m and 1.0m in both the permanent protection area and falling apron. The sandbags were dumped before the establishment of the construction platform and drilling of the bored piles. The second step was to place the steel casings and install the construction platform. The third step was to dump a layer of graded stones with a thickness of 1.0m, which followed the drilling for the steel casing. The fourth step was to dump armour stones so as to form the permanent protection structure. The thickness of the armour stone layer for the central and permanent protection areas was 1.5m, on the inside of the falling apron was 1.89m and on the outside of falling apron was 3.15m (Figure 7).

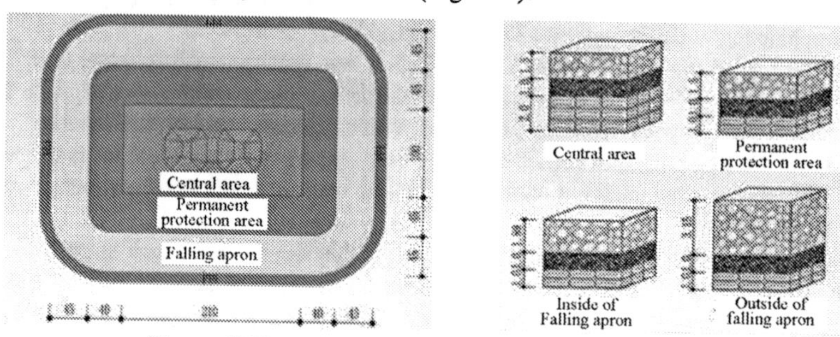

Figure 7 Foundation scour protection structure

Permanent scour protection works started in July 2003 and were completed in May 2004. There was a total of 300,000 m^3 of sandbags, 260,000m^3 of graded stones and 530,000 m^3 of armour stones dumped into the river. Monitoring over the last three years indicates that the structure in the protection area is stable. Scour is within the tolerance specified by the scour experiment, although local scour has occurred in the falling apron. In general, the riverbed is stable, which is in reasonable agreement with expectations. Implementation of the permanent scour protection has guaranteed the depth of embedment of the foundations into soil, therefore improving their ship impact and seismic resistance. Consequently it was verified that bottom elevation of the steel casing could be raised to -52.0m after carrying out the permanent scour protection, which successfully solved the difficulty of drilling the steel casing into the pebble layer.

Set-up of construction platform

The establishment of the bored pile construction platform was a key technical challenge during the construction of the piled foundations. In particular, there were two difficulties:

- It had to be constructed in about 30m depth of water around the main piers, with an average current velocity above 2.0m/s and a peak velocity of over 4.0m/s. In these conditions and because the direction of the water flow was very changeable, a way of accurately locating the position in which to drill the steel casing was expected to be very difficult.
- The safety of the bored pile construction platform was threatened by the significant scour that could occur during construction.

It was initially proposed to adopt a construction scheme using φ1.4m steel pipe piles as supporting piles. However, subsequent test piles showed that it was difficult to ensure the safety of the construction platform by means of these due to the deep water and high flow velocity and it was also very difficult to ensure the accuracy of the steel casing placing due to the tidal effect. Therefore, this method was optimised as follows:

- Set up construction platform in October to avoid the flooding season.
- Carry out the riverbed pre-protection before the setting up of the construction platform so as to avoid increased scour and a reduced embedment of the steel pipe piles and steel casings into soil.
- Use φ2.5m steel pipe piles to set up an initial platform upstream of the pile cap and then install a specially made cantilever guide framework on this initial platform to guide the positioning of the steel casing and to ensure the accuracy of the steel casing drilling.
- Drill the steel casing using two interlocked APE 400 vibration hammers.
- Integrate the steel casings that have already been drilled, move the guide framework and place the remaining steel casings from upstream to downstream so as to form a new construction platform (Figure 8).

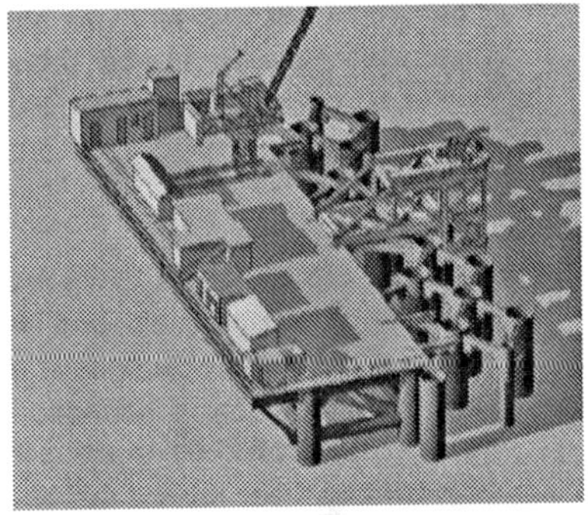

Figure 8   Construction platform establishment scheme

This method solved the problem of the establishment of a bored pile construction platform and the accuracy of the steel casing drilling was within a planar deviation of 5cm and a verticality of bored piles of less than 1/200 during actual construction.

300m High pylons

Two alternatives, i.e. steel pylons or concrete pylons, were proposed for comparison to determine the type of pylons for the Sutong Bridge. Steel pylons would be light, which is favourable for dimensional control; and they can be fabricated in a workshop and assembled on site, which helps to reduce construction period. However, steel pylons have a relatively low structural damping, poor wind resistance capability and high construction costs. As a result, concrete pylons were finally selected.

Figure 9   Anchor box

The main requirements for the design of steel anchorage box in the pylons are discussed as follows:

- Ensure structural reliability and durability
- Ensure convenient construction, inspection and maintenance (Figure 9).

The method chosen allows cracks in concrete pylon wall, but the width of these cracks shall be controlled so as to be less than 0.2mm. To achieve this, an appropriate quantity and spacing of common rebar is used in the design. Full-scale model tests have shown that the width of crack induced by normal loads is about 0.14mm, and the width of this crack induced by 1.7 times normal load is 0.17mm. Mathematic calculations and site monitoring indicate that the stress in the concrete at the bottom of the anchor box under normal service loads is relatively low and overstressing will not happen. It proves that the design of the steel anchorage box in the pylons is reasonable.

Pylon geometry control was the main technical challenge during the construction of the 300m high pylons. In order to ensure the structural geometry of the bridge, in the reference condition, the target geometry was required to satisfy the following tolerance requirements:

- Pylon leg verticality $\leq$ 1/3000; Axial line deviation of pylon legs $\leq$ 30mm;
- Elevational deviation at the pylon tops $\leq$ 10mm;
- Axial line deviation of pylon legs at the pile cap $\leq$ 10mm;
- Pylon leg cross section dimension deviation $\leq$ 20mm.

The main difficulties during construction were as follows:

- The pylons are high and flexible structures. They are sensitive to environmental factors, therefore compensation for these factors is needed.
- It was necessary to carry out 24-hour surveys since setting out could only really be done at night, which was not efficient in relation to the effective construction time that could be achieved.

To overcome these challenges, two main measures were taken:

- Establish a temperature and wind compensation system. 6 temperature observation points and 3 wind observation points were established along the pylons so as to realise real time monitoring of wind and temperature, establish a correlation between the pylon's deformation, temperature and the wind loading, and compensate for the influence of these effects on the displacement;
- Investigate and develop a prism tracking method. The principle of this method was to accurately measure the position of the finished section and make temperature and wind compensation adjustments.

Based on the position of the finished section, the position of the tracking prism was measured to establish a relative relationship. In the daytime, the setting out points could then be determined by reference to the prism. This method had the following advantages:

- It was possible to carry out some construction setting out work in the daytime, which would otherwise have to be carried out at night. Setting out and measurement in the daytime doesn't necessarily need temperature and wind compensation, which can significantly improve the efficiency of the activity. By using this method the cycle for the construction of one segment construction was reduced to five days;
- A reduction in the elevational deviation by transferring the relative positions of neighbouring segments from one to another.

Pylon construction for the Sutong Bridge was completed in September 2006 with the construction tolerance and verticality of pylons being controlled within 10mm and 1/40000 respectively.

Superstructure

Erection of steel box girder

According to the location of the boxes and the construction conditions, different erection methods were adopted. The large girder sections in the side span and auxiliary span, as well as the girder sections in pylon area were erected by a large floating crane (Figure 10). Standard sections for side span and main span as well as closure sections were installed by two deck cranes (Figure 11). The deck cranes were made in China. The maximum lifting weight and height were 450t and 80m respectively. For convenience, temporary supports were installed adjacent to both the auxiliary piers and the transitional piers. Temporary piers were also provided in the side spans 197m from the pylons. In order to ensure structural stability during construction, temporary fixings were used between the tower and deck temporary to provide vertical, longitudinal and horizontal restraint to the steel box girder. Hydraulic jacks were used for closure and this was achieved for the side spans in January 2007 and for the main span in June 2007.

Figure 10　Side span erection

Figure 11　Main span erection

To deal with the wind resistance issue during construction, wind tunnel tests were undertaken by the Southeast Communication University. A test model with a scale of 1:125, was used to model the tower crane, the trolley for cable erection, the deck crane, the cable stressing platform, safety guardrail, inspection gangway, baffler and the temporary structure. Wind tunnel tests showed that when there was vortex induced vibration in the steel box girder, the resonant wind speed for the minimum torsion frequency was more than 15m/s, Therefore, it was concluded that vortex induced vibration would not affect construction. It is a fact that buffeting to a different extent happens at each construction stage, however, buffeting will not lead to resonance of the stay cables as the acceleration at a 15m/s wind speed is less than that at 30m/s. The stress in the steel box girder is still within tolerance at the maximum cantilever stage and during the balanced cantilever stage. That is to say, structural safety or stability was deemed to be acceptable.

Fabrication, erection and damping of the long cables

The high strength galvanized parallel wires used for the stay cables were developed and produced by the Baosteel Group, PRC. The material properties of these wires were as follows: Tensile strength≥1770Mpa; Yield strength≥1450Mpa; Elastic modulus $(1.95{\sim}2.10) \times 10^5$ MPa. Specially made B87MnQL rod was selected. The stay cable has multiple corrosion protection systems to ensure an appropriate service life (Figure 12). Galvanizing coated the steel wire and the outer surface of the strands was wrapped with PVF and a double layer of high-density polyethylene applied by hot extrusion. Chill cast anchorage sockets were selected. The chill casting filling consists of steel balls, epoxy resin, hardener, a diluting agent toughening agents and filling materials. The anchorage was protected by hot galvanizing. In order to ensure fabrication accuracy of the long cables, a gauge wire was selected from specially made steel wires for measurement of the Elastic modulus of the strands. The gauge wire was specially made on a bench with a fabrication accuracy of over 1/23000.

Based on different lengths and weights of cable, two different erection procedures were selected. For cables 1~20#, in which the maximum force was 160t and the maximum lifting weight was 25t, tensioning from the pylon end was selected. For cables 21~34#, the pulling force and pulling distance were relatively large. The anchorage pulling force for cable 34# even reached 430t and thus when pulling from the pylon end it was difficult to ensure a symmetrical pull. To overcome this a tensioning scheme from the deck end was adopted. The Second Navigational Engineering Bureau of China Port undertook the cable erection.

The aim was to control the amplitude of stay vibration to within 1/1700 of the stay length. A combined method of stay dampers and aerodynamic measures was applied to mitigate vibration (Figure 13). Vortex induced vibration was controlled by means of dampers, which provided 3% additional damping force in the stays. Dimples on the stays and aerodynamic measures were also installed to mitigate the vibration induced by wind and rain. A structural wind resistance study showed that no vortex induced vibration would happen on the completed cable-stayed bridge and there were no grounds for believing that vibration and internal linear resonance would be a problem. As such, auxiliary stays were not needed, but for the sake of safety, an embedded structure for auxiliary stays and an increased bearing capacity of outer stays was still provided. In the case of the maximum balanced cantilever, the frequency of the stays was widely distributed, which resulted in internal linear resonance and parameter vibration. Temporary cable ropes were installed to mitigate this vibration. They

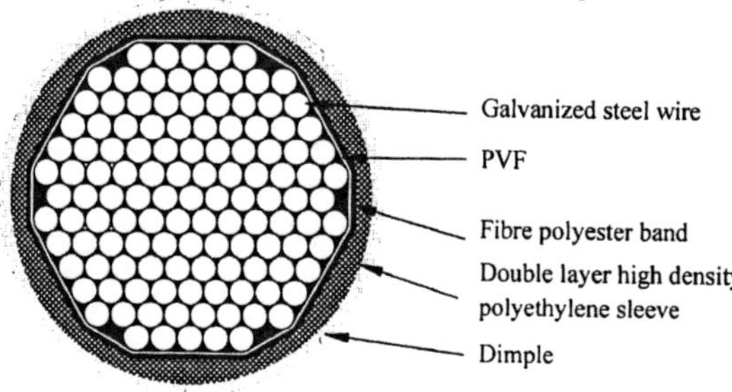

Figure 12   Structure of stayed cable

Figure 13   Stayed cable with dimple

were used to fix the stay and steel box girder after erection of each stay from 1# to 15#. For stay cables 16~34#, the stay was connected to the steel box girder and two neighbouring stays by a rigging screw so as to temporarily mitigate vibration.

Conclusion

The Sutong Bridge is the longest cable-stayed bridge in the world. For such a large-scale project, its construction conditions were very complicated. It provided multiple technical challenges including deep-water foundations, 300m steel concrete composite pylons, a 540m cantilever and extra-long stay cables. Construction of the bridge has shown:

a) It is feasible to adopt piled foundations for a large bridge in a fast flowing tidal area. There were three enormous challenges during the design and construction of the foundations. They were riverbed scour, assuring pile verticality assurance and pile cap construction. Riverbed scour during the construction and service stages was solved by a combined method of temporary scour protection and permanent scour protection. Pile verticality was assured by use of a specific positioning guide frame, while lifting the steel cofferdam as one solved the difficulties associated with pile cap construction.

b) It is also feasible to adopt steel/concrete composite pylons under complicated load conditions for extra-large cable-stayed bridge pylons. Appropriate design of the pylon anchorage area can efficiently deal with the load transfer problems and improve structural durability. The use of a monitoring prism is an effective way to deal with the necessary setting out under complicated construction conditions.

c) It is also feasible to use semi-floating structural system for a cable-stayed bridge with a span of over a thousand meters. Large floating cranes and deck cranes were used to lift the large, wide box girder segments.

d) Extra-long stay cables provide a major technical challenge requiring high-strength wire, fabrication accuracy, a long life span, and appropriate damping measures. It has been proved that high-strength wire and stay cable developed and fabricated in China can fulfil these requirements. Both dimples on the PE sheath and dampers on the cables can control the vibration of extra-long parallel wire cables. Also temporary cables can effectively prevent cable vibration during construction.

References

[1] You Qingzhong, Wu Shouchang, Li Zhen, Sutong Bridge Project and Key Techniques. The 10th East Asia Pacific Conference on Structure Engineering and Construction, Aug., 2006.
[2] Anton Petersen, Ejgil Veje, You Qingzhong, Dong Xuewu. Sutong Bridge- A Cable Stayed Bridge with A World Record Free Span of 1088 m. KSSC Journal, Mar., 2005.
[3] Zhang Xigang, Peng Deyun. Sutong Bridge, China- A New Record for Cable stayed Bridge. The 10th East Asia Pacific Conference on Structure Engineering and Construction, Aug., 2006.
[4] Robert B. Bittner, Zhang Xigang, Ole Juul Jensen, Ole Rud Hansen. Design and Construction of the Sutong Bridge Foundation. American Ociean Engineering Society, Oct., 2005.
[5] C. Truelsen, Wu Shouchang, O. Juul Jensen, Gao Zhengrong. Design of Scour Protection For Sutong Bridge, P. R. China. ICSE2006, May, 2006.

Key Technologies in the Tower Structure of the Sutong Cable-stayed Bridge

Zhang Hong   Second Navigational Engineering Bureau CCCG, Wuhan ,China
Zhang Xigang China Highway Planning and Design Institute Consultants, INC. CCCG, Beijing,China
Xiao Wenfu   Second Navigational Engineering Bureau CCCG, Wuhan ,China
Dai Jie   JiangSu Provincial Communication Planning and Design Institute, Nanjing, China

Abstract

Sutong Bridge is the world's longest cable-stayed bridge. It measures 2 088 m in total length and has a 1088 m length center span. With a height of 300.4m, the pylon of Su-tong Cable-stayed Bridge is the tallest amongst similar bridges in today's world. Topics on the technologies used in the construction of the pylon, including construction methods, geometry control, and wind-resistant studies during construction, are discussed in this paper.

Description

108km away from the mouth of the estuary of the Changjiang River, the Sutong Bridge connects Suzhou with Nantong, and is one of national key projects that has drawn attention worldwide. The project comprises the north interlink, a river-crossing bridge and the south interlink, with 6 carriageways for traffic from two directions. The river-crossing bridge is composed of a north approach, the main bridge, a special fairway bridge and a south approach, with a total length of 8146m. The main bridge is designed as a cable-stayed bridge with a main span of 1088m; the longest in the world today. The configuration of spans of the main bridge is 100+100+300+1088+300+100+100, amounting to a total length of 2088m. See figure 1 for the main bridge configuration.

Figure 1   Configuration of main bridge

The pylon of the main bridge is an inverse Y, with a height of 300.4m.   It consists of a lower, middle and upper column, with a lower crossbeam between the lower legs. Box sections are used for the pylon legs and cross beams, and a steel-concrete composite structure for the anchorage zone in the upper pylon column. The pylon structure is illustrated in Figure 2, and the upper pylon column in Figure 3.

Bridge design, construction and maintenance 2007, Thomas Telford, London

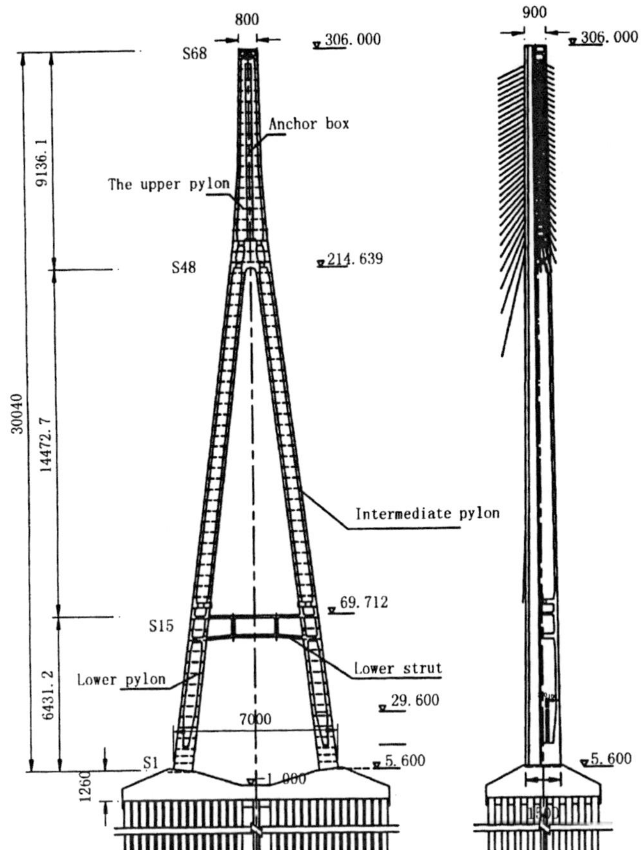

Figure 2 Structural configuration

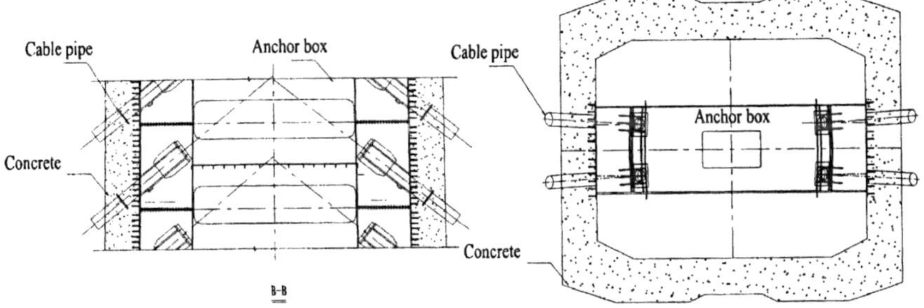

Figure 3 Structural details of upper pylon

Construction features and problems

The pylon height greatly exceeds that seen in ordinary structural designs, so the pylon needed to be constructed with high precision. Because the pylons sit in the main navigation channel of the Changjiang River, where the hydrological conditions are difficult to define and there are 177 out of 365 days windy days when the wind velocity exceeds 22m/s, there were many difficulties which had to be faced during construction. These difficulties can be summarized as follows:

1) Installation of pylon anchor box: It was going to be difficult to position the segmental anchor box, especially when the demand on installation precision was so high, ambient conditions so unfavorable, and erection had to be carried out at extreme height.

2) Casting of the high performance concrete for the construction of the pylons.

3) Geometric control: During construction the pylons would be sensitive to the effects of temperature difference due to sunlight and the action of the wind.

4) The analysis of the wind-induced oscillations and a study of the relative wind-resistant measures that might be implemented.

Key construction technologies

Erection of the first anchor box section

The first anchor box section with a weight 45.82 tons, is a base segment for subsequent ones, and is connected directly to the concrete pedestal. Installation was achieved using the following procedures:

1) Cast the pedestal concrete after adjusting the bolts and reaction columns embedded at the four corners of the concrete pedestal, leaving 8cm thickness of concrete to be cast later.

2) Install steel bearing plates, and adjust their level precisely with adjusting bolts.

3) Adjust precisely the planar position of the first anchor box segment with a guiding device and jacking system.

4) Use cement mortar grout to fill up the holes of the anchor bolts.

5) Cast the remainder of the concrete in the concrete pedestal.

6) Tighten the anchor bolts after the required concrete strength has been achieved.

Concrete work in anchorage zone

The anchorage zone of the pylon is designed as a steel-concrete composite structure. Therefore, the steel anchor box has to work together with the concrete to take the loads, and the vertical component of the tension in the cables has to be transferred to the concrete pylon leg through shear keys set between the side of the anchor box and the concrete casing around it. During construction, tension stress may occur due to the effects of temperature, shrinkage and creep of the concrete, and the restraints provided by the shear keys. Considerable tension stress may also be present in the casing concrete during the erection of cable stays and during the service life of the bridge.   In consideration of structural durability, the width of concrete cracks has been limited to less than 0.2mm by the designers. To achieve this goal, fibers were added into the concrete mixture. A fiber reinforced concrete was therefore developed so that it could also be pumped to enable the construction of the high pylon.

Geometry control

To achieve the target geometry after completion of the bridge, the necessity for geometry control during construction is evident. As they are extremely high and flexible structures, the pylons are sensitive to environmental effects due to their relatively small rigidity. Sensitivity analysis on the factors that might influence their behaviour was carried out for the pylon construction. Based on the analysis results, it was evident that factors such as temperature, wind actions, shrinkage and creep of concrete would have significant effect on the control of the pylon construction. To control the effects, two different construction control methods were used for the lower & middle pylon legs and the upper pylon column due to their different structure type and different construction methods.

Allowable deviation for pylon construction control

After completion of the pylon, its target geometry must satisfy the following allowable deviation criteria under base conditions:

1) Inclination deviation less than 1/3000, deviation of axis of pylons less than 30mm.

2) Deviation of pylon top elevation less than 10mm.

3) Deviation of axis of pylon at the connection place between pile-cap and pylon less than 10mm.

4) Deviation of sectional dimensions less than 20mm

Construction control method for the Middle and lower pylon legs

Normally, as-built surveys and construction setting-out are carried out at night between 24.00hrs on the previous day to 6:00hrs on the next day, so that the effects of temperature and wind may be avoided. Because the pylon is so high, a temperature difference of 2°C in the direction along the bridge axis can cause a movement of 33mm at the top of the middle pylon leg, which would be greater than allowed for in the design criteria. Hence the environmental effects caused by temperature and wind must be allowed for.

A tracing prism method was introduced for controlling construction. A tracing prism was installed on the top of the completed section. By finding out the theoretical as-built and as-set positions of the tracing prism, the real position of a surveying mark could be determined. This method works well only when the theoretical position of the tracing prism in base conditions (temperature 20°C, no wind) has been determined. More details of the method are illustrated in Figure 4.

Upper pylon column

The upper pylon column is designed as a steel-concrete composite structure. During its construction, an anchor box was first erected, and the concrete was cast later on. Therefore, the geometry control of the upper pylon column construction depends on the erection control of the anchor boxes. To control the geometry of the anchor boxes, the erection of the first anchor box segment is critical. Upon the anchor box being installed and connected to the bolts, it had to be aligned with the assembly geometry. Thus the position and inclination of the anchor box depends on the geometry of the first anchor box segment. As long as the geometry of the anchor box exceeds the design criteria, it must be corrected.

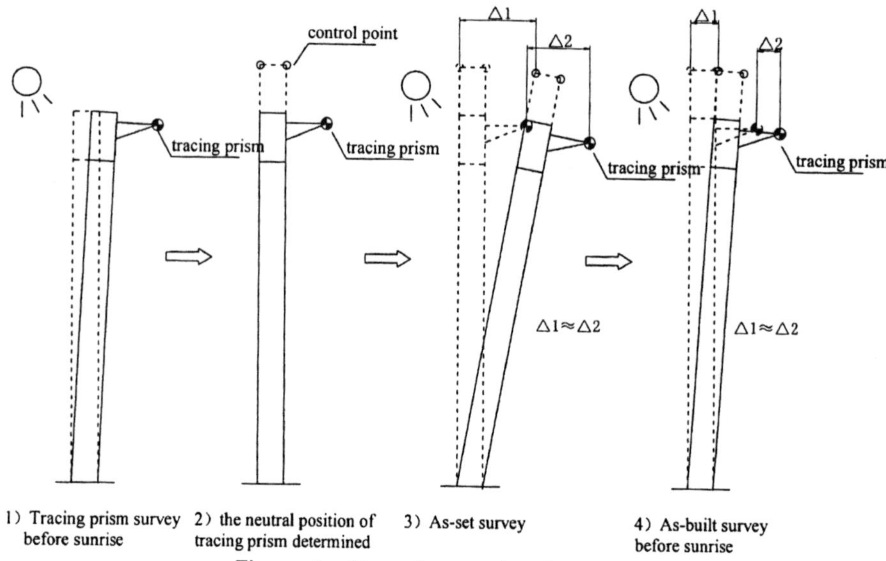

1）Tracing prism survey 2）the neutral position of 3）As-set survey 4）As-built survey
before sunrise   tracing prism determined   before sunrise

Figure 4　Steps for construction control

The inclination deviation of the first anchor box segment shall be less than 1/3000. The centerline of the steel anchor box shall not deviate by more than +/-5mm from that of the concrete section. The steel anchor box shall be installed on four separated horizontal bearing plates, and the out-of-plane deviation shall be less than 0.42mm.

Combining the inclination deviation of the first anchor box segment, accumulated fabrication errors, and the out-of-plane deviation of the first anchor box segment, the deviation at the top of the pylon could be up to 52mm, which would exceed the design criteria and would therefore need be corrected. To do this, 7 shim plates with a nominal thickness of 12mm were placed at the matching positions.　If the inclination of an anchor box deviates from the design criteria, it can be adjusted by changing the thickness of the shim plates. Care was needed when making such adjustments, because both horizontal and vertical connection plates had to be well connected.

Correction for temperature and wind actions

The target geometry and ideal geometry for construction are referred to a temperature of 20° C with no heat gradient, and no wind action. Therefore, each measurement value from the tracing-prism and as-built survey needs to be corrected to suit the practical values after taking the effects of wind and temperature actions into account in the following way:

1. Input the data collected from thermocouples installed on the pylon top to enable them to be analysed by building a mathematical model, so that the theoretical position of the tracing prism can be determined.

2. Conduct as-set and as-built surveys according to the variation in the position of the tracing prism.

Results of geometry control for the middle and lower pylon leg.
Generally, the linear errors caused by the uniform temperature changes inside the pylon structure are compensated above a height of 80m, and those caused by local temperature changes are compensated above a height of 170m. The geometry control is satisfactory and the deviations of the axis of pylon are mostly within 10mm.

Wind tunnel test and analysis of oscillation suppression

Wind tunnel test
The target of the aero-elastic wind tunnel test was to assess the safety risk due to wind actions, and the effects of wind actions on the construction equipment/plant and on the operational conditions for personnel during the construction stages. In addition, a system to suppress wind induced oscillations during construction was proposed based on the test results.

According to the erection method of the pylons, 3 representative construction stages (or cases) were chosen for the test. The pylon was exposed to a uniform fluid field and boundary turbulence field (5.5%, 11%) for testing.

Case 1: full height, without construction equipment/plant

Case 2: full height, with construction equipment/plant

Case 3: middle pylon leg in its maximum cantilevering state (up to EL.212.4m) with construction equipment/plant

For the model test, the geometry scale was 1:100, the wind velocity scale was 1:6, the damping ratios were 0.5%, 0.8% and 1.6%. The wind tunnel test may be summarized as follows:

1) In the condition of uniform flow, the phenomena of vortex-excited resonance was found in the 3 cases in a direction along the bridge axis, and the most adverse wind incident angles were 0°, 3°, and -1.5° for the 3 cases respectively. In the short-term, additional dampers were proposed to effectively suppress the oscillation induced by the wind.

2) The vortex-excited resonance decreases significantly when the flow velocity of the fluid field was raised.

3) The lateral oscillation was comparably much smaller than the oscillation in the direction along the bridge axis, and no remarkable vortex-excited resonance was found.

Analysis of the necessity of oscillation suppression for the pylon
There was no necessity for an oscillation suppression system for the pylons, if the following conditions wee satisfied:

1) Under working wind velocity (15m/s), the responding acceleration velocity is not bigger than 30cm/s^2.

2) Under the design wind velocity, the stresses in both the pylon and pylon crane must not exceed the allowable stresses.

Based on the wind tunnel test, a study was conducted to analyse the necessity for oscillation suppression during pylon construction, and the results indicated that although there would be no vortex-excited oscillation occurring during some of the construction stages under the design wind velocity, the maximum amplitude would be 137mm at the pylon top and 178mm at the top of pylon crane, and that the maximum acceleration velocities would be $0.11m/s^2$ and $0.14m/s^2$ respectively at the top of the pylon and the pylon crane. During some of the construction stages, the maximum amplitude would be 47mm and 129mm and the maximum acceleration velocities would be $0.41m/s^2$ and $1.05m/s^2$ respectively at the top of the pylon and pylon crane, but the wind velocity in the analysis was greater than the working wind velocity (15m/s). The conclusion was that no oscillation suppression system was needed for the pylon structure.

Conclusion

Multiple advanced construction technologies were applied during the tower construction of the Sutong bridge. Significant challenges associated with the height of the pylons and their loacation have been overcome as has been described in this paper. Thanks to the endeavours of construction team the two towers of the Sutong bridge were completed on Oct 4th 2006.

References

[1]SATOU Yoshiyuki:Erection of "Tatara Bridge",IHI Engieering Review,Vol.36 No.2, June 2003,PP65-84

[2]Zhang Hong,Yasutsugu Yamasaki:Assessment Report on Tower Wind Tunnel Test Results,April 2006

Construction Geometry Control Of The 268m Span Prestressed Concrete Box Girder Bridge Over The 2nd Navigational Channel Of Sutong Bridge

ZHOU, Xinya, China Zhongtie Major Bridge Engineering Group Co. Ltd.
WONG, Carlos, Ove Arup & Partners Hong Kong Ltd.
WEN, Wusong, China Zhongtie Major Bridge Engineering Group Co. Ltd.
LUI, James, Ove Arup & Partners Hong Kong Ltd.

Abstract

Sutong Bridge, crossing Chang Jiang (Yangtze River) between Nantong and Changshu, comprises of a 1088m cable stayed bridge in the major navigation channel and a 268m prestressed concrete box girder bridge at the 2nd navigation channel. Together with the approach viaduct the length of the river section is about 8 km. Construction started in 2004. When completed, the main bridge will become the world's longest span cable stayed bridge whereas the 268m concrete box girder bridge will be amongst the world's top 5 longest span of its kind. Balanced cantilever construction method of insitu concreting on traveler formwork is adopted. This paper describes the plan, procedures and measures taken during the construction to control the bridge profile so that it will match with the design road alignment at the targeted year after the concrete having crept to its final value.

Keywords

Sutong Bridge, 268 m span, construction alignment control, concrete box girder, prestressed concrete bridge, balanced cantilever.

Introduction

Sutong Bridge, crossing Chang Jiang (Yangyze River) between Nantong and Suzhou, comprises of a 1088m cable stayed bridge in the major navigation channel and a 268m prestressed concrete box girder bridge at the 2nd navigation channel. Together with the approach viaduct the length of the river section is about 8 km. Construction started 2004. When completed, the main bridge will become the world's longest span cable stayed bridge whereas the 268m concrete box girder bridge will be amongst the world's top 5 longest span of its kind. Figure 7 shows the advanced stage of the bridge construction.

The span arrangement, see Figure 1, is 140m + 268m +140m. The bridge is divided into two box girders, namely Bridge A and Bridge B and each one carries one carriageway of 3 lanes plus a hard shoulder. The box girder varies in depth from 4.5m at the crown to 15m at the two piers following a parabolic curve; see Figure 2 of the section.

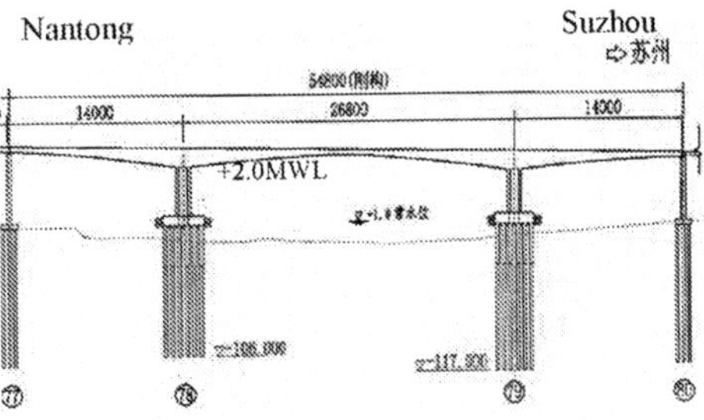

Figure 1 Span arrangement

Piers are made up with two hollow walled type rectangular columns 7.5m x 2.5m x 0.5m thick. They are cast with the box girder to form a rigid frame structure. The road profile contains a longitudinal gradient of 1.5% forcing the deck to follow. Transverse fall of 2% is used. Bearings are used at the end piers, which are also shared by the adjacent viaducts.

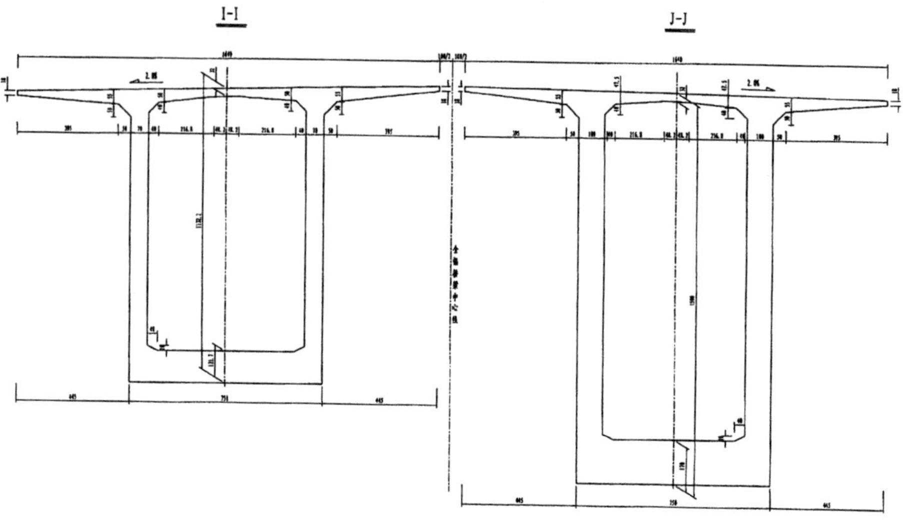

Figure 2 Cross Section

The deck is prestressed in three directions: longitudinal, transverse and vertical directions. Longitudinal prestressing consists of slab tendons, web tendons and stitch tendons for the closing segments. Transverse prestressing is for the top flange slab only. Vertical prestressing is by stressing bars in the webs. Construction method is by balanced cantilever cast insitu on traveler formwork for 31 segments.

The construction specification requires that the deck profile shall be matching with the design road profile at the 30th year after opening. It demands an accurate assessment of the concrete creep effects so that a pre-camber can be set on the formwork such that it compensates the time-dependent deflection at the 30th year. By the time of writing this paper in April 2007, the bridge is at it final stage of casting the closing segments at the side spans and at the mid-span.

Control Strategy

Control Target

The control target is to have the bridge structure profile to match with the design road profile at the 30th year since opening of the bridge, i.e., the concrete bridge will creep from its original pre-cambered profile to the road profile after 30 years, and this requires a reasonably accurate estimate of the creep effect and the corresponding pre-camber values together with a strict construction plan.

Target Control Tolerance

Global Vertical Alignment is within ±50mm.
Staging Vertical Alignment is within ±20mm
Section cross fall is within ±0.2%.

Control Plan
Control plan is executed in the following sequence:
1. Pre-camber prediction calculation
2. Construction carried out according to the pre-camber prediction
3. Site measurement after casting, and feed-back results to geometry control unit
4. Error analysis and review of parameters, and prepare specs for correction (if any)
5. Correction (when required)
6. Move on to next stage of construction or to compute the next stage pre-camber values based on the new parameters or new site measurement data.

Implementation of Control Plan
Pre-camber Prediction Calculation
The pre-camber prediction is based on the deflection calculation using space frame analysis. Typical results are shown in Figure 3. Three programs in the forefront of this application are used to cross check the results. They are the Midas, Ibdas and Tdv used by MBEC, COWI and Arup respectively. The basic principle of all the three programs in this particular application is that they all use space frame with rigid offsets catering for the varying C.G. of each element (or the segment) and for the application of the tendons, see Figure 4. Each program is believed to have some differences in the treatment of the cables but the fundamental principle is the same. The analysis follows the casting sequence. The major parameters used in the calculation are:

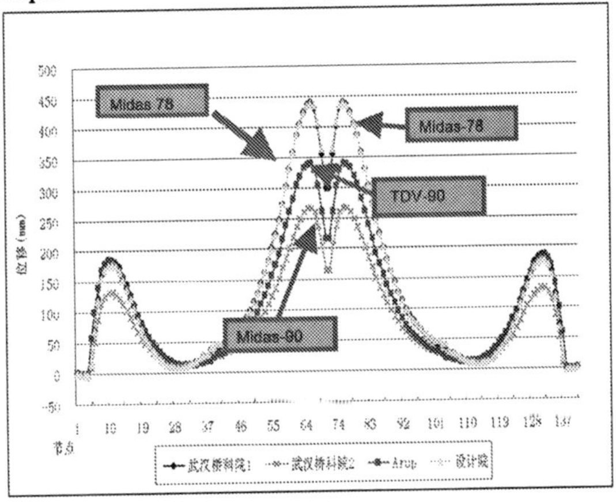

Figure 3 Comparison of using ceb78 and ceb90 in pre-camber calculation

Section Property
Gross section of the concrete outline without the contribution of reinforcement, but with tendons activated according to the construction sequence, i.e. the section with voided ducts, the section with un-bonded tendons and the section is grouted with bonded tendons.

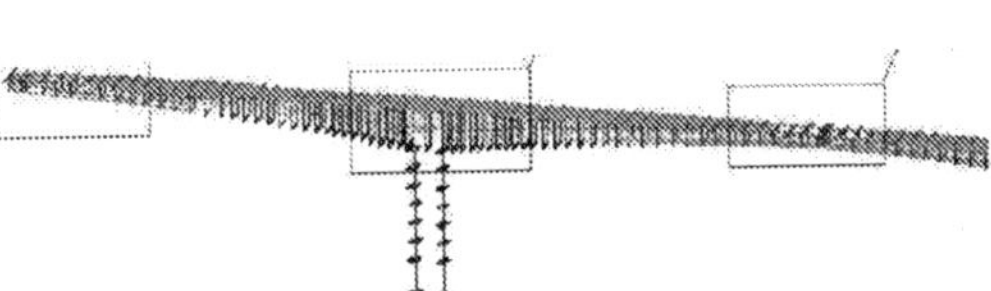

Figure 4 Space frame calculation

Concrete Property
The concrete grade is 60 with 20% PFA. The concrete property in the calculation follows the ceb-fip78 model of the development of Young's modulus, creep and shrinkage and the concrete strength.

Tendon Property
The use of uPVC ducts in China is not popular, hence a study on its friction coefficient and the performance under vacuum assisted pressured grout is carried out but before the test results are available typical values as recommended in JTG D62-2004 [ref. 1]are used, i.e. friction coefficient is taken as 0.15 and wobble coefficient is taken as 0.0015. Anchorage

draw-in is 6mm. The test results [ref. 2] recommend that the friction and wobble coefficient be taken as 0.16 and 0.001, which are not far outside the assumed range.

Foundation Stiffness
Pile cap measured 33.2x49.6x7.0m is supported on 48 nos. 2.8m diameter bored piles. A study to investigate the influence of the foundation stiffness upon the deflection concluded that the foundation could be approximated by a fixed support at the pile cap top level, thus simplify the modeling.

With the above assumed and measured parameters, the pre-camber values are calculated as shown in Figure 3, which shows the difference between using the ceb-fip90 and ceb-fip78. It is obvious that the pre-camber values as predicted by the latter are larger than the former one, since the ceb-fip78 model gives greater creep value.

Construction Control
Traveler Formwork
Traveler formwork is the Doka system, see Figure 5. The load-deflection characteristic curve is obtained by test on site of the formwork setup. The deflection is eliminated in the setting of the formwork according to the curve at the particular weight of the segment. The centre of the gravity is also determined in the test, which is taken as 1.0m from the face of the already cast segment on the airside.

Temporary Construction Load
Machine and equipment are all weighed and their existence in a segment construction period is recorded with location.

Figure 5 Traveler Formwork

Section Dimension Control
The formwork is also used to make trial casting (Figure 6) from which sectional dimension is measured and the dimensional error can be estimated. The error in dimension is used to predict the error range in weight, which is in the range of 1%. The influence of such error in weight upon the pre-camber value is then estimated, and is about 2mm.

Prestressing Control
While the stressing is done by force control, the extension is used to check the tolerance, which shall be within ±6% of the anticipated value. Typical average deviation is 4.5% to -1.0%. This tells the fact that the friction and wobble

Figure 6 Trial Casting of a Segment

coefficients in the calculation have been over-estimated. Later tests prove that this is the case.

Site Measurement after Casting

Survey
The box girder has large area exposed to sunlight and wind. To minimize the influence of temperature on the leveling survey, the work is carried out in the early morning before sunrise. This is also the calm period when the wind speed is generally slowing down. The bridge site is designated as the wind energy rich region; therefore it is not surprise to find out the wind speed at site is very high (wind speed exceeds 10m/s 60% of the time in a year). The strong wind could produce sufficient lift force acting on the cantilever deck causing measurement error. A relatively calm climate is therefore targeted for surveying. Survey is taken to the control points set in the segment covering the longitudinal profile and cross fall measurements.

Environmental Data
The environmental data such as wind speed and temperature are continuously recorded and monitored.

Strain/Stress Monitoring
Strain gauges are installed at the bridge piers and at the deck at the critical location. They are used to monitor the stress in the concrete for 2 purposes: to provide another set of data to cross-check with the calculation, and to give a warning sign should the stress in the concrete become too high.

Error Analysis
Error analysis is conducted after obtaining the site measurements and the following are identified in terms of pre-camber values:
- Surveying error including accidental, those due to wind load and system error: ±5mm
- Setting out error: ±5mm
- Concrete surface variation: ±5mm
- Traveler formwork deflection variation: ±2mm
- Segment weight deviation induced error: ±2mm
- Prestressing force deviation induced error: ±2mm

The maximum error when all the "misfortunes" happen at the same time would be about ±20mm, which is within the staging tolerance range.

Correction
The correction is done in several levels from the do-nothing Level 1 to the rectification work in Level 4 and they are:

Level	Action
1.	No correction if the error is less than the staging tolerance (~20mm).
2.	Half of the error is corrected in the remaining segments if the error is greater than the staging tolerance but less than the correctable tolerance (~30mm).
3.	Correct the target alignment if the error is greater than the correctable tolerance but is less than the global tolerance (~50mm).
4.	Carry out rectification work to rectify the error if it is greater than the global tolerance.

Correction review is set at the Segment #19 when sufficient data are available to indicate a need for the major correction of the pre-camber in the remaining segments. The major correction so far is the support settlement due to foundation yielding, which record reads 25mm on the Pier 79 and 20mm on Pier 78. The deflection is then distributed evenly in the remaining segments such that the Segment #31 will fully compensate the support settlement. Method and procedures for the level correction are shown in Figure 8. Other corrections carried out are for the revised Young's modulus of the concrete, and the increase of the top flange thickness by 25mm, as a result of design change.

Results of the Construction Geometry Control
Predicted Pre-camber
The predicted pre-camber is shown in Figure 9, which shows the three computer programs produce practically identical results in the main span but not in the side spans. The difference is due to the difference in interpretation of the closing procedures.

Measured Concrete Property
Experimental study [ref. 2] for the creep and shrinkage characteristic of the concrete used in this project has been carried out. The study compares different codified shrinkage and creep models with the tested values, as shown in Figure 10 for shrinkage and creep. The tests indicate that the shrinkage predicted by ceb-fip78 is less than the tested values in magnitude of 100 micron whereas the predicted creep factor expressed as the ratio of creep strain to elastic strain under the sustained load is about 2 times of the tested value. The creep factor calculated using ceb-fip78 is the highest amongst all the codified models, including ceb-fip90, which is about 57% of the ceb-fip78 model. From calculation, the 10-year mid-span deflection increment for ceb-fip90 and ceb-fip78 model is 39mm and 147mm respectively.

The average concrete cube strength exceeds 70 MPa; the tested Young's modulus 49 MPa is also higher than the codified value of 40 MPa. The increase of Young's modulus has direct effects on the pre-camber calculation, as such the pre-camber calculation is corrected midway in the construction phase.

Vertical Alignment
The vertical alignment difference between the measured levels and the computed levels (including pre-camber values) is shown in Figure 13 and 15 for the two cantilevers supported by the Pier 78 and Pier 79 plotted as the level difference of the predicted verses the measured data, after Segment #29 is cast. The maximum difference is 38mm at Segment #21. The curves show that many of the corrections are done when the difference reaches 20mm except for a few cases. The error analysis reveals that in most of the cases, the error came from the setting of the formwork. After casting of segment #31, it is expected that there will be level difference between the two cantilevers. As envisaged in the closing segment construction method (Figure 11 and 12), the two cantilevers are to be pushed apart 60mm then to secure the jacks and stabilize the two cantilevers. Clamper beams are installed between the two cantilevers and stressing bars are installed for strand jacking to bring the two cantilevers into leveling. Depending on the stiffness of the two cantilevers (one pier is taller than the other) the level difference may not be equally shared by the cantilevers. However, the imposed deflection at the cantilever tip (should be no more than 50mm) will not impose adverse stress in the structure. The stress will become lock-in stress after the closing segment is cast and cured and the removal of the clamping beams, and the jacks. It will be gradually dissipated under the creep action of the concrete.

The built level and the computed level comparison are shown in Figure 14 and 16, and the stress comparison between measured and computed values, at the maximum bending location of Segment #2, is shown in Figure 17. It shows that the measured levels are greater than the predicated levels echoing the fact that the actual creep is less than the predicated value. The error analysis for the stress difference between the two indicates that the error of stress difference could be 3 MPa, due to the uncertainty nature of the actual creep and the computational creep. It can be seen from the plots that the max. deviation is less than 2MPa, which is within the error tolerance range of ±3MPa. The max. stress is about -16MPa.

Discussions and Conclusions

The bridge has been successfully constructed up to the closing stage. The pre-cambered bridge deck surveyed after casting of the Segment #29 shows that the vertical geometry differs not more than 38mm from the calculated profile, and is less than the global tolerance (50mm). The stress difference also shows good correlation of the two. The casting of the remaining 2 segments is also expected to be within the limit. The careful preparation works such as the load test of the traveler formwork to gain the reference load-deflection curve; the trial casting; the skilful technicians who operate the traveler formwork; the dedicated survey teams and the dedicated geometry control unit are the key factors to success. The experience gained in this project could be used as a reference or a work example for similar bridge construction.

The adoption of the ceb-fip78 creep and shrinkage model however, may over-estimate the creep deflection, or the pre-camber values. The study carried out by the South East University [ref. 2] shows that concrete property is close to the ceb-fip90 model, which reports a creep value about half of that of ceb-fip78. One argument that supports the use of the ceb-fip78 model is that the traffic loads patterns in China differ substantially from the pattern of European countries because the traffic load in China tends to be constantly running over the bridge resemble a permanent load. A suggestion to take 1/3 of the live load as permanent load for the determination of creep deflection has been proposed by members of the expertise review panel [ref. 3]. The use of a greater creep factor would, to a certain degree, safeguard the bridge from creeping below the design road profile, which is always psychologically disturbance especially to the general public. Figure 7 shows the bridge in advanced stage.

ACKNOWLEDGEMENT
The authors sincerely thank the Sutong Bridge Construction Commanding Department for their supports and kind permission of publishing this paper. Sincere thanks are also extended to MBEC and Arup for their encouragement and supports on the writing of this paper.

Figure 7 Construction Progress on site sees the completion of 30 segment leaving

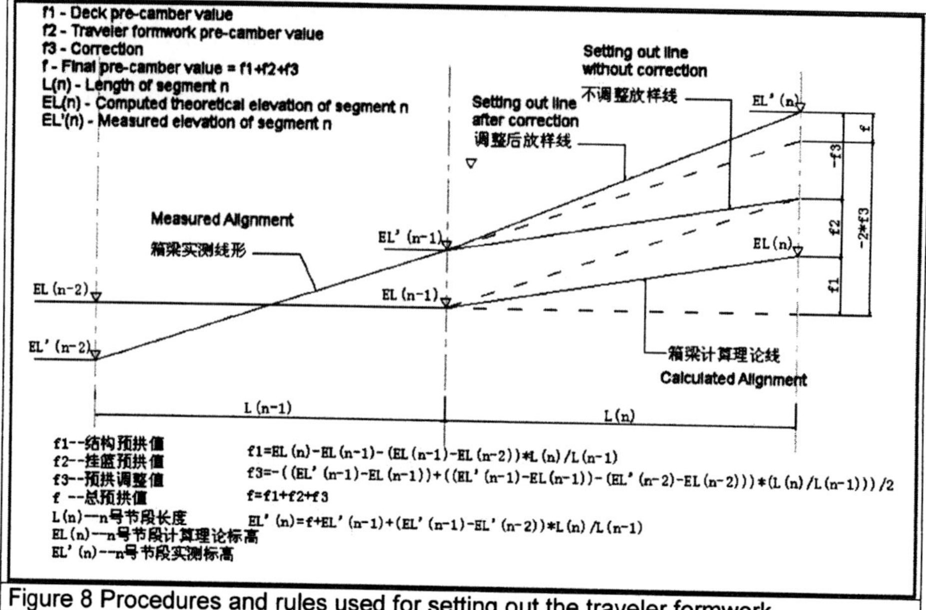

Figure 8 Procedures and rules used for setting out the traveler formwork

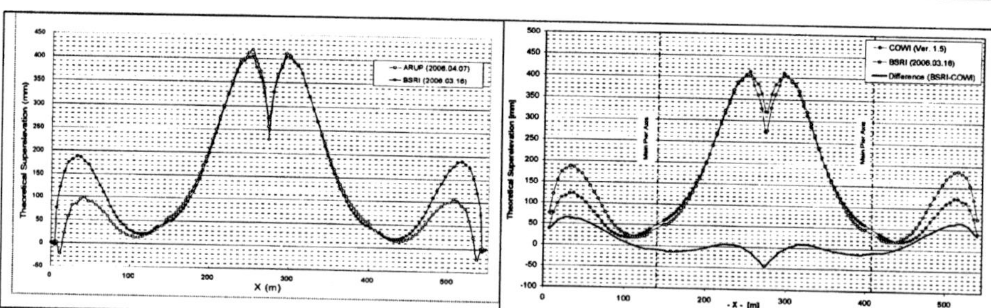

Figure 9 Comparison of pre-camber calculation between Arup-MBEC, and between COWI-MBEC.

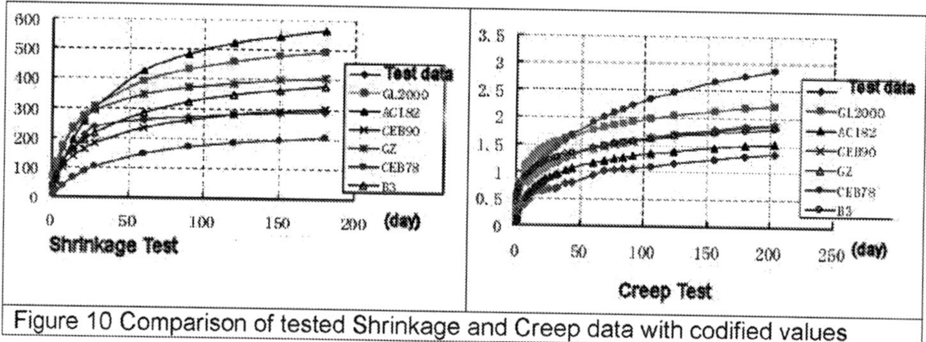

Figure 10 Comparison of tested Shrinkage and Creep data with codified values

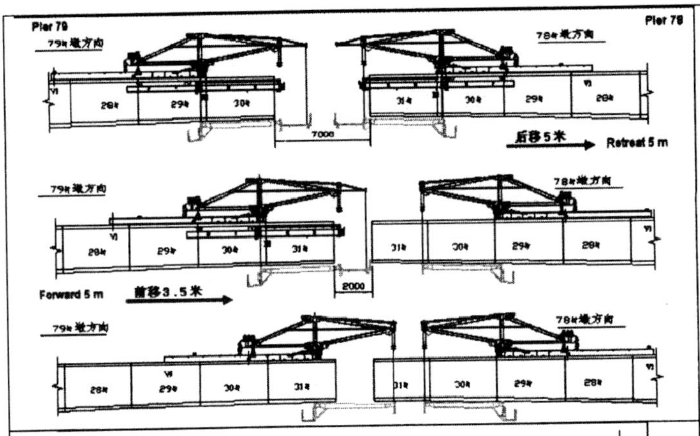

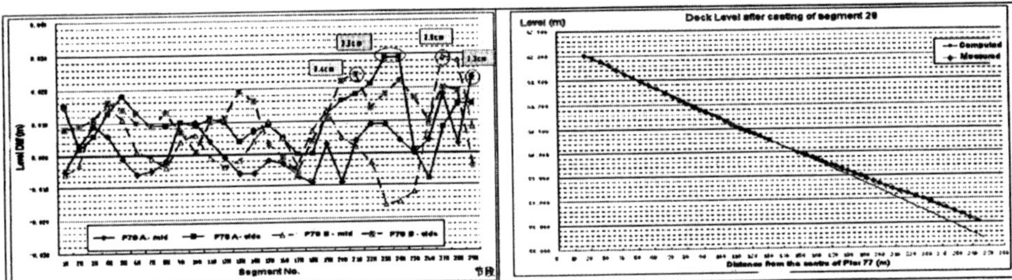

Figure 11 Construction sequence and arrangement for the Segment #31 before closing up.

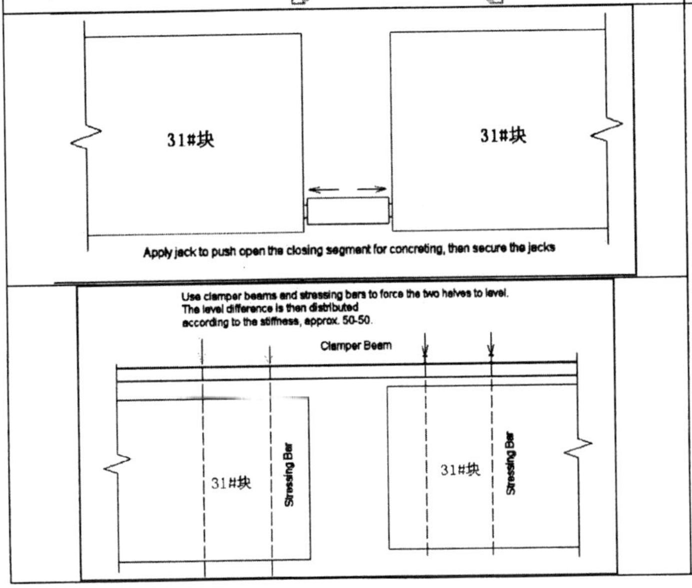

Figure 12 Construction method proposed for the closing segment at the mid span. Closing of side spans is similar. The side spans will be closed prior to the closing of the mid span.

Figure 13 Level difference along the alignment after Segment #29 is cast (plot for cantilever supported by Pier 79)

Figure 14 Predicted and measured level for cantilever supported by Pier 79, after casting Segment #29 (note the max. deviation is at the cantilever at the midspan.)

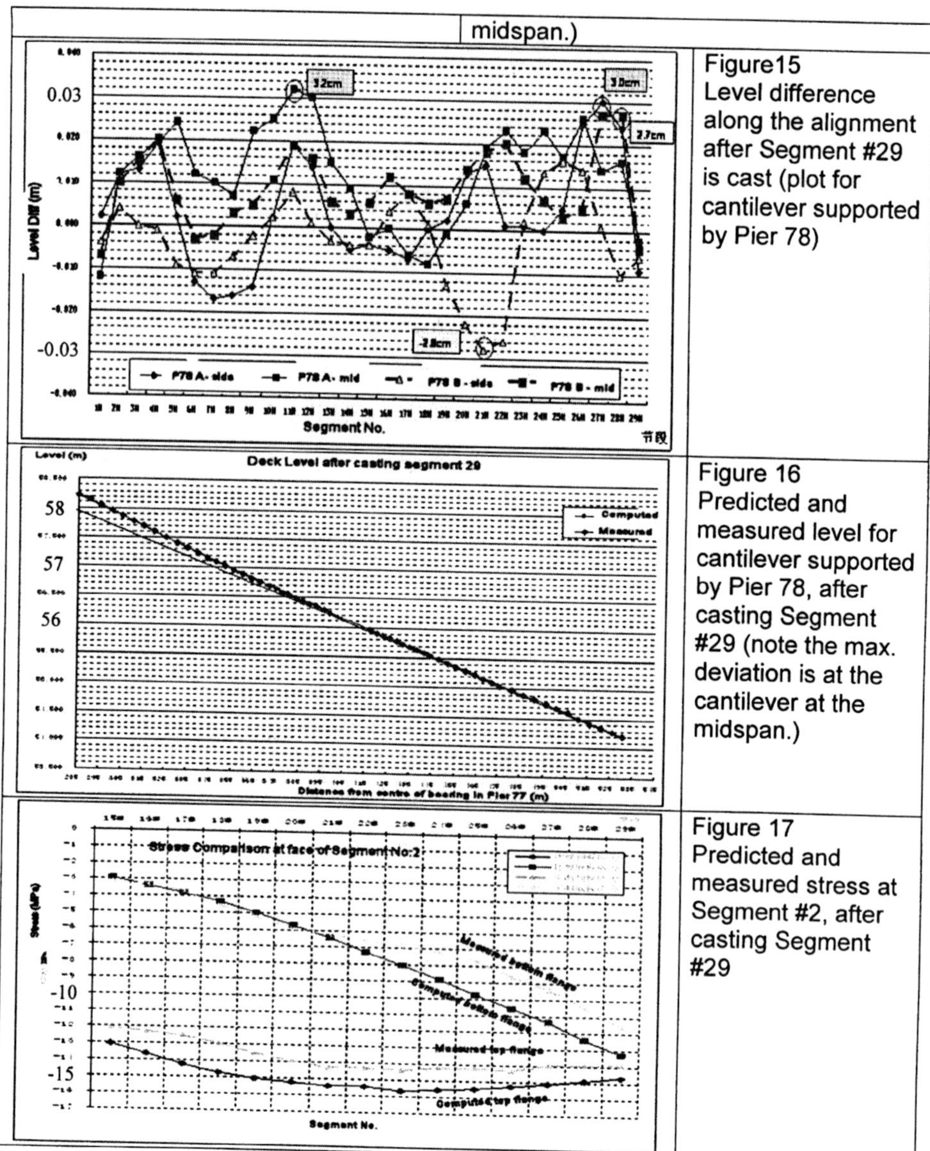

Figure15 Level difference along the alignment after Segment #29 is cast (plot for cantilever supported by Pier 78)

Figure 16 Predicted and measured level for cantilever supported by Pier 78, after casting Segment #29 (note the max. deviation is at the cantilever at the midspan.)

Figure 17 Predicted and measured stress at Segment #2, after casting Segment #29

REFERENCES

[1]   JTG D62-2004 "公路钢筋混凝土及预应力混凝土桥涵设计规范(Specification of design of highway reinforced and prestressed concrete bridges and culverts)", 中华人民共和国交通部 (Ministry of Communication, People's Republic of China, 2004.

[2]   "苏通大桥预应力施工工艺及关键参数试验研究总结报告(Final Report of Study of Prestressed Concrete Key Parameters in Application for Sutong Bridge", 江苏省苏通大桥建设指挥部、中港集团第二航务工程局、东南大学，(Sutong Bridge Construction Commanding Depatment, China Harbour Engineering Company (Group), SouthEast University). 2005.

[3]   Sutong Bridge Expertise Review Panel Meeting, Nantong, 2005.

DESIGN CHARACTERISTICS OF A NEW SELF-ANCHORED SUSPENSION BRIDGE WITH SPACE RETICULATE CABLE

Wang-Qian, PhD student of Civil Engineering, Bridge Science Research Institute, Dalian University of Technology, Dalian 116023, China
Zhang-Zhe,Professor of Civil Engineering, Bridge Science Research Institute, Dalian University of Technology, Dalian 116023, China.
Li-Wen Wu, Lecturer of Civil Engineering, Bridge Science Research Institute, Dalian University of Technology, Dalian 116023, China.

Abstract: In contrast to ordinary suspension bridges, the Dalian Xinghai-bay No.1 Bridge is a well-designed structure in which the cable force is divided into two parts. The cable system for the bridge was designed as a space reticulate structure. The large horizontal force in the dorsal cord is transmitted via a subterranean collar beam to the foundation under the bridge tower. Since this type of self-anchored suspension bridge is adopted for the first time, some innovative and unique problems were presented in the design procedure. The particular design concept for this bridge is reported in the paper. The finite-element method was used for comparing various solutions and analysing the stablity of the tower. Following this analysis, a steel pipe truss structure was selected. The main conclusion from the analysis shows that the stability of the tower satisfied the standard requirement. This paper is a major contribution in the development of self-anchored suspension bridges.

Project background

Dalian Xinghai-bay No.1 Bridge (Fig 1) is a well-designed self-anchored suspension bridge, with a main span of 134m. The net-like arrangement of cables looks like a casting fishing net which is intended to represent the work and happiness of local fishermen. [1]

The main beam is made of steel, with a width of 9m, a beam depth of 1.3m with a cross slope of 1%. The thickness of the top plate and the bottom plate for the main beam is 12mm. The longitudinal gradient of this bridge is 8% and the radius of the vertical curve is 800m. In addition, a series of cross slabs are set at 6m intervals along the deck. Two solid concrete

Bridge design, construction and maintenance 2007, Thomas Telford, London

cross girders have been adopted at the extreme ends of the main beam. The structural form of the tower is a hollow steel pipe truss, the longitudinal inclination angle of which is 70 degrees and the included angle between the two towers in the lateral direction is 9.08 degrees.

As illustrated in Fig 1, the towers are connected by two cross beams in the transverse direction. The top cross beam is used to anchor the cables, while the other one contributes to the stability of the structure.

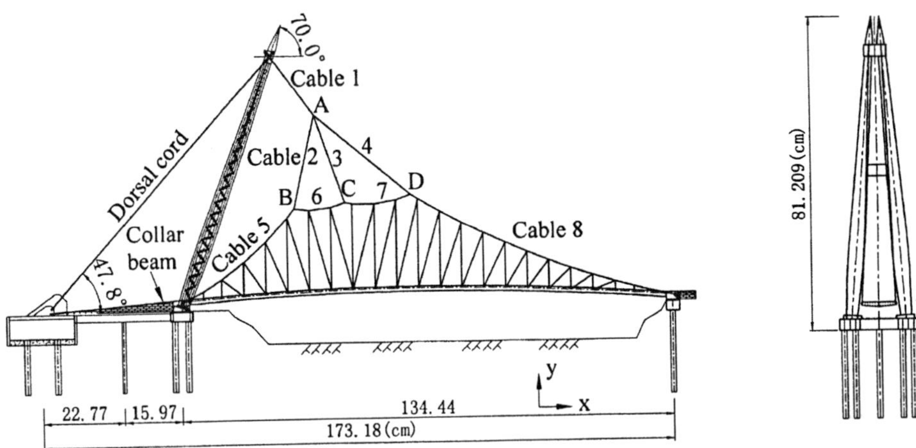

Fig1. Elevation view and side view of Xinghai-bay No.1 Bridge of Dalian

Nine different kinds of cables have been utilised in this bridge, as shown in Fig1. The distinctive design concept applied in this bridge is that the cable forces are divided into two sequential systems. The forces from the main beam are transmitted to Cables 2, 3 and 4 directly by suspenders, and subsequently these forces are transmitted to the tower by Cable1.

The large horizontal force in the dorsal cord is transmitted via a subterranean collar beam (Fig1.) to the foundations under the bridge tower. The main tower, dorsal cord and the collar beam form a trianglular loading system, which was the main force structure of this bridge.

Tensile force of suspenders

As a self-anchored suspension bridge, the stiffening girder flexes upwardly with the bridge deck by the action of the horizontal pressure of the main cables. Hence the stiffening girder was regarded as a circular arch during the analysis. An upwarp, which reduces the tension in the suspenders, corresponds to a tension. The tension in the suspenders in the built state is obtained by iterating the cable horizontal pressure and the tension of the suspenders. [1]

Geometry of the main Cable in the built state

The position of the control point

In addition to the structural behaviour, the aesthetics of the structure should also be considered. The underlying principles in the design process were as follows:

1) The control points A, B and C should be set in line.
2) The resultant forces of AB, AC and AD must converge at the anchorage on the top of the main tower.
3) As can be seen in Fig1, the position of A influences the geometry of the main cable and the distribution of the internal forces. Hence the coordinate of A should be well designed. One coordinate which can uniform the cable force is selected from the coordinates which satisfies conditions 1) and 2) above.

Based on these principles, the correct positions of A, B, C and D were determined according to an "inheritance algorithm" [2].

Geometry of the main cable

An ideal geometry of the main cable was determined by means of non-linear finite element analysis, taking into account aesthetics and structural behaviour.

According to the theory of load increment, the relationship between the displacement increment and the structural stiffness is presented as: $\{dF\} = [K]\{d\Delta\}$ (1)

$\{dF\}$ — load increment;

$[K]$ — structural stiffness

$\{d\Delta\}$ — displacement increment

Fixing the load increment, the integral stiffness increases with a reduction of $\{d\Delta\}$.

The procedure used in the analysis is illustrated in Fig 2.

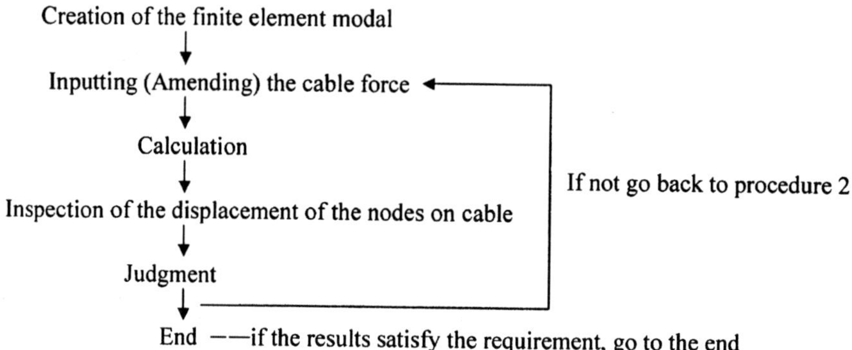

Fig2. Implementation procedure of the finite element analysis

Initially, the internal force (which is later amended) of the cable elements was inputted; then the structural finite element analysis was performed; thirdly the displacement of the nodes caused by both the internal force and external loads was inspected (the smaller the better). These three steps iterate until the displacement is considered small enough. Thus, a set of displacements of nodes and corresponding initial cable forces are obtained. After the deformation of the cables, caused by the action of the certain initial cable force and external load were obtained, the ultimate geometry of main cable was determined. [4]

In order to reduce the error caused by human error, the objective function and constraint condition are applied to inspect the displacement and ensure the precision of the results, which is the main principle of dynamic monitoring method.

Unstressed length of cable

Under normal temperature, the length of the structural parts for which stress/strain equals zero is called the unstressed length. The unstressed length of a cable is obtained by deducting various formations from the cable length under ultimate loadings. As shown in Fig1, the connections of the cables and the suspenders are complex, which requires that the ropegrip must be well analysed. In order to keep the cables smooth, some arc ropegrips are adopted, as illustrated in Fig 3.

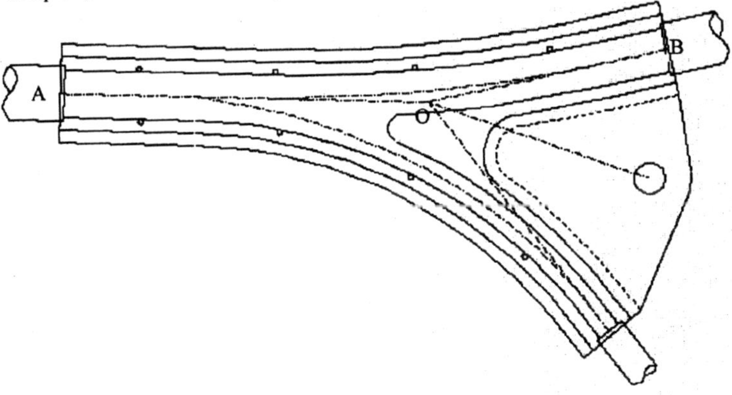

Fig 3. Ropegrip construction drawing

In the analytic process, $l_{OA}+l_{OB}$ is regarded as the approximate length between A and B. To eliminate the approximate error, the effect of the ropegrip should be considered by the

equation: $$\Delta = (l_{AO} + l_{BO} - l_{AB}) - \frac{F_A l_{AO} + F_B l_{BO}}{EA} \tag{2}$$

Δ— error caused by ropegrips

l_{OA}— the straight line length between O and A

l_{OB}— the straight line length between O and B

l_{AB} — the curve length between A and B

F_A — force of cable l_{OA}

F_B — force of cable l_{OB}

E — elastic-plastic modulus of cable

A — area of cable

By deducting the approximate error, the accurate unstressed length of a cable can be obtained.

Fabricating length of suspenders

After the erection of the main cables, the position of the deck is determined by the length of the suspenders linking the stiffening girder to the main cables. Hence, the fabricating length of the suspenders should be calculated exactly. The fabricating length is determined by:

$$L_i = L_i^{(1)} - \Delta L_i^{(1)} - \Delta L_i^{(2)} \tag{3}$$

L_i is the length of suspenders from the center of the cable to the anchorage in the stiffening girder, which is amended by $\Delta L_i^{(1)}$. The corrected length is calculated according to the external diameter of the cable, the inner diameter of the ropegrip and the distance between the ropegrip and the center of cable. Furthermore, the elastic stretching length $\Delta L_i^{(2)}$ should be

deducted, which can be expressed:
$$\Delta L_i^{(2)} = \frac{N_i \left(L_i^{(1)} - \Delta L_i^{(1)} \right)}{EA} \tag{4}$$

N_i — the tension of the suspender numbered i

E — elastic-plastic modulus of suspender

A — area of suspender

Scheme comparison of the tower

Two design elements were considerd in the Dalian Xinghai-bay No.1 Bridge which is located at a beauty spot: aesthetic function and economic considerations. Two alternative design schemes were considered for the tower. One was a spatial steel tube structure; the other was a concrete-filled steel tube. In order to choose the better one, a 3-D finite element analysis was performed to simulate the ultimate behaviour of the tower, as shown in Fig 4.

According to the analysis, the results showed that the spatial steel tube structure had more merits. Due to its significant self-weight, the tower made with a concrete-filled steel tube produced a large negative moment at the root and under traffic loading, the tube was subject to tensile stresses whereas the concrete was subject to compressive stresses, which means that the concrete and the steel tube do not act as a composite structure. In addition, the concrete-filled steel tube tower required larger foundations which increased the fabrication cost. Furthermore, there would be some difficulties during pouring the concrete into the tube.

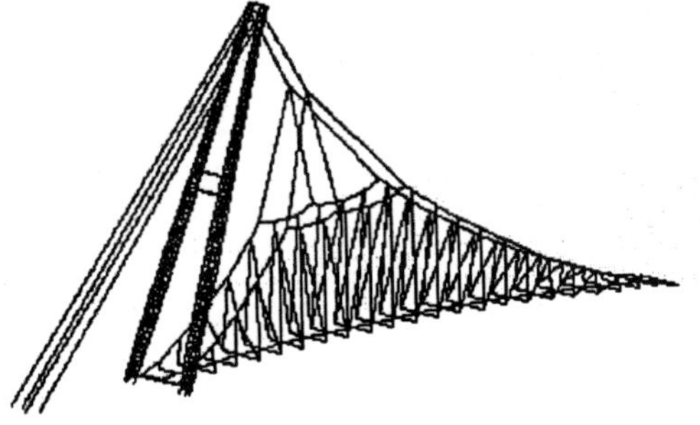

Fig 4. Finite element model of the main tower

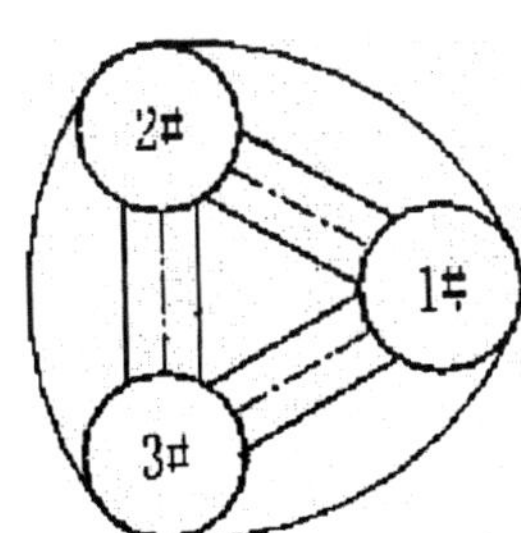

Fig 5 Segment of steel tower

Considering all these factors, the designer finally decided on the use of a spatial steel tube structure. The tower which looks like the letter "A" in the transverse direction is a spatial truss structure with both legs of the tower formed by three steel pipes (Fig 5). [3, 6]

Stability analysis of the main tower

Loading cases

Six kinds of possible load combinations were considered in order to verify the stability of the towers, as follows:

1) dead load
2) dead load+traffic load
3) dead load and wind load in the transverse direction
4) dead load+traffic load+temperature load
5) dead load+temperature load+ wind load in the transverse direction
6) dead load+traffic load+ wind load in the transverse direction+ temperature load

Stability analysis of linear elastic buckling of tower

It was assumed that the structure was always in the linear elastic state, which means that the internal force is in direct ratio to the external load. The finite element model, shown in Fig 4, was used to analyse the stability of the tower. From the analysis, the stability coefficient, which is the eigenvalue of the tower, and four buckling modalities (illustrated in Fig 6-Fig 9) for the six load cases were obtained.

Fig 6. First-order unstable modality.   Fig 7. Second-order unstable modality

Fig 8. Third-order unstable modality   Fig 9. Fourth-order unstable modality

As seen in Figs 6-9, the first three flectural modalities represent the state in which the tower is unstable in the plane X-Y (Fig 1), while the fourth modality represents lateral buckling. Since the tower inclines forward and the deadweight is large, flexion emerges initially in the plane X-Y, when the load reaches a critical value. The specific stability coefficient and the unstable modality are reported in Table 1.

Table 1. Stability coefficient of elastic flexion

Load case	1	2	3	4	5	6
Stability coefficient	7.0563	4.8836	7.0555	4.8792	7.0464	4.8692
Unstable modality			Unstable in plane X-Y			

In order to ensure the safety of the structure, the buckling limitation should be ensured to develop after the strength limitation. The norm prescribes that the linear elastic stability coefficient must be greater than 4. As can be seen in Table 1, the stability coefficients in six kinds of possible load cases all satisfy this requirement.

Geometric nonlinear analysis of tower

Due to possible erection errors, material defects and deviation of the point of application of the load, a linear elastic analysis may not be appropriate in practice. Hence, when analysing the stability of the tower, geometric nonlinearity should be taken into account. [5]

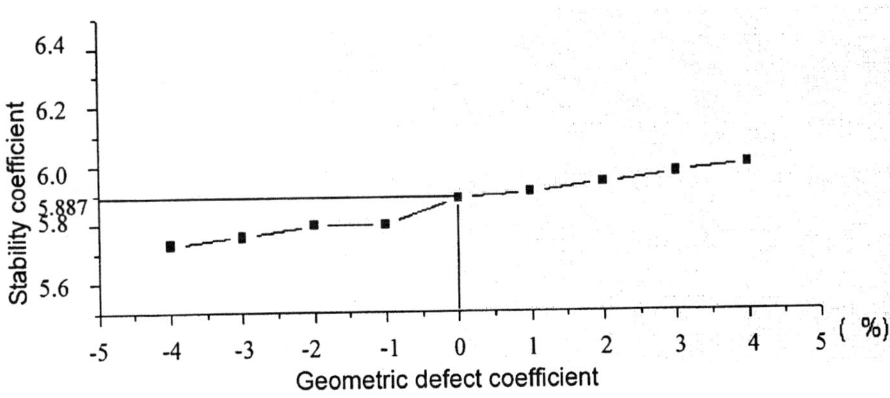

Fig 10. Stability coefficients of the tower with different initial geometric defect under dead load

The first –order flexion modality was selected as the basis of the initial defect. By multiplying constants, different kinds of initial defects including the displacement of the coordinates of the nodes on the tower were simulated. Fig 10 shows the stability coefficients of the tower with different initial geometric defects under dead load. (The negative values correspond to coordinates which are in the reverse direction to the flexion direction of the first-order unstable modality (Fig 6). The stability coefficient in Fig 10 changes from 5.713 to 6.003, which are close to the stability coefficient of 5.887 observed without any geometric defect. The main conclusion here is that the initial geometric defect only marginally affects the stability of the tower.

According to this conclusion, -1% of the first-order flexion modality was selected as the initial geometric defect coefficient. Using this value, an analysis of 6 kinds of load cases was performed. The flexion modalities are all for cases of buckling in the plane X-Y (Fig 11). The results for the stability coefficients are presented in Table 2.

Table2. Stability coefficient of tower considering the geometric nonlinearity

Load case	1	2	3	4	5	6
Stability coefficient	5.7950	3.5040	5.7943	3.4988	5.8935	3.4988

The damage to the structure caused by some forms of deficiency can be predicted by considering the construction details such as the node coordinates, etc. However, the buckling damage cannot be foreseen. Hence the buckling damage must be developed after any strength damage. In order to ensure this condition the linear elastic stability coefficient should be greater than 4 and the stability coefficient, considering the geometric nonlinearity, should be greater than 2, as prescribed by the norm. [7]

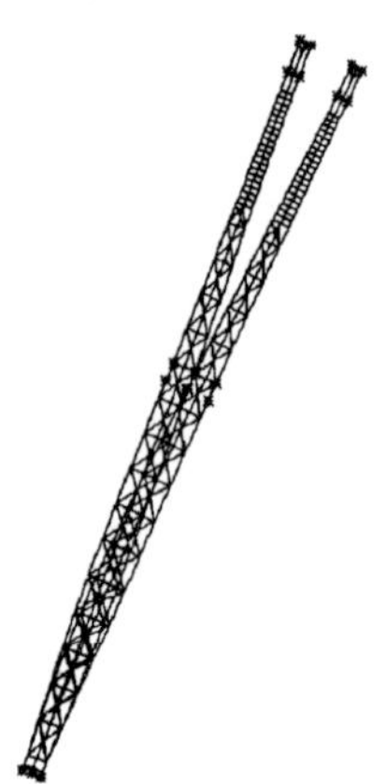

Fig 11. Flexion modality of tower considering the geometric nonlinearity under dead load

As seen in Table 1 and Table 2, no matter whether the geometric nonlinearity is considered or not, the stability coefficients of the tower satisfy this requirement, which indicates that the tower of this bridge is stable.

Conclusions

This paper reports upon the main design concept for the Dalian Xinghai-bay No.1 Bridge, which is a new self-anchored suspension bridge with space reticulate cables. This was the first time that this kind of special self-anchored suspension bridge was adopted.

The calculation principles for the tensile forces in the suspenders, the geometry of the main cable, the unstressed length of cable and the fabricating length of suspenders are described.

Two types of tower arrangements were considered – a spatial steel tube structure and a steel pipe truss structure. The latter one was selected due to its merits, as described in the text above.

The finite-element method was used for analysing the stability of the tower, by which the stability coefficient of the tower was obtained. The results show that the linear elastic stability coefficient is greater than 4 and the stability coefficient considering the geometric nonlinearity, is greater than 2 in six possible load cases. This fully satisfies the requirement of the norm. Hence it was concluded that the tower of this innovative bridge was stable.

References:
[1] Li WenWu, The establishment of main cable curve and inclined tower force analysis for Xinghai-bay footbridge. Masters Dissertation. Dalian University of Technology. 2006.6

[2] Li Min Qiang, Theory and application of the inheritance algorithm. Beijing: Scientific Publishing Company. 2002

[3] Zhe Zhang, Yonggang Tan, Lei Shi. A new searching approach on the calculation of the target configuration of cables for suspension bridges [A]. 2003-10-1 International Meeting of Shanghai

[4] Yonggang Tan, Zhe Zhang, Cailiang Huang, Analytical method for the configuration of the main cables of self-anchored suspension bridges[J]. Computers & Structures
[5] Qiu Wen ling, Nonlinear Analysis and Experimental Study of Self-anchored Suspension Bridge. Doctoral Dissertation. Dalian University of Technology. 2004.6

[6] Holland J.H, Adaptation in Natural and Artificial Systems [M]. The University of Michigan and MIT Press,Cambridge, MA(1989)
[7] Fan Li Chu, Bridge engineering. Beijing: China Communications Press. 2001

Optimization of bridge management strategy based on time-dependent reliability and pre-posterior decision analysis

Q. Qin, Professor, Tsinghua University, Beijing, China
J. X. Wang, Graduate, Tsinghua University, Beijing, China

Abstract: On the basis of results on the corrosion rate of tensile steel reinforcement induced by the chloride attack and concrete carbonation, moment resistance deterioration models of highway RC girder bridges due to the steel corrosion are developed with the pitting corrosion for the former and average corrosion for the later. Component and system time-dependent reliabilities of such bridge decks are developed for safety evaluation. Updated deterioration models based on inspection results are used for pre-posterior decision making to optimize bridge inspection/repair schedules which is able to consider if an inspection was worth. The optimization is for both monetary costs and owner's utilities of bridge inspection/repair. A computer program for such optimization based on Monte Carlo method and statistic regression is also developed with different bridge inspection/repair cost models. A 40 year old highway bridge in Beijing is taken as an example. The results show such optimization beneficial.

Flexure resistances of RC Bridges designed for a long working life in the atmospheric environment reduces due to steel corrosion, and reasonable bridge inspection/repair schedules are necessary. Optimum bridge inspection/repair schedules should be based on the time-dependent reliability. Following Qin & Li (2005), the paper updates corrosion models due to the chloride attack, adds steel corrosion models due to the concrete carbonation. The pre-posterior decision making are used for optimizing their inspection/repair schedules.

Steel corrosion due chloride attack

Chloride laden environments may induce steel corrosion of RC bridges. Chloride ions penetrate the concrete cover and steel begins to corrode as soon as the chloride around steel has the critical concentration. Fidjestoel & Tuutti (1998) studied the corrosion of two phases and effects of cement types, admixtures and environments on the corrosion, and indicated on the basis of abundant experimental results that the corrosion due to the chloride attack mainly is the pitting corrosion related to the macrocell effect rather than the general corrosion relating to the microcell one. The pitting corrosion affects the flexure resistance of RC girders more significantly. The studies assumed residual lifetime as the time until start of steel corrosion or the time when 20~30% of steel are corroded. Stewart & Rosowsky (1998) developed a deterioration model to calculate reliability for flexure failure of a RC continuous slab bridge due to steel corrosion induced by de-icing salt or atmospheric marine exposure and discussed effects of the cover depth of concrete on the reliability.

Fick's second law of diffusion gives the time for corrosion initiation of the jth steel bar is

$$t_{fl} = \frac{X^2}{4D_c}[\mathrm{erf}^{-1}(\frac{Cl_s - Cl_{cr}}{Cl_s - Cl_0})]^{-2} \tag{1}$$

where X= net thickness of the concrete cover; D_c= chloride ion diffusion coefficient; Cl_s= equilibrium chloride concentration on the concrete surface; Cl_{cr}= critical chloride concentration initiating corrosion; Cl_0= initial chloride concentration. The reducing diameter of the jth steel bar at time t, in a, is

$$D_j(t) = \begin{cases} D_{j0} & t \le t_{fl} \\ D_{j0} - 2\lambda_c(t - t_{fl}) & t_{fl} < t < t_{fl} + D_{j0}/2\lambda_c \\ 0 & t \ge t_{fl} + D_{j0}/2\lambda_c \end{cases} \tag{2}$$

where D_{j0}= initial diameter of the steel bar; λ_C= corrosion rate due to chloride attack.

Chlorides at a certain level of concentration cause local breakdown of the passive layer, leading to the localized and intense pitting corrosion (Fidjestoel & Tuutti 1998). The corrosion current density measured with nondestructive tests should be modified for the pitting corrosion. Stewart & Rosowsky (1998) assumed the corrosion rate for the pitting corrosion as

$$\lambda_C = C_R \lambda_{C0} = 0.01174 C_R i_{corr}, \quad \text{in mm/a} \tag{3}$$

where C_R= influence factor for pitting corrosion, a normal random variable with a mean of 3.0 and a coefficient of variation (cov) of 0.33, λ_{C0}= general corrosion rate (Fontana & Greene 1978), in mm/a, i_{corr}= average corrosion current density, in $\mu A/cm^2$.

Liu T. & Weyers (1998) indicated that the corrosion rate decreases rapidly for a short time period immediately after corrosion initiation, and becomes stable about one year later. Therefore, i_{corr} for the phase before concrete cover cracking induced by corrosion can be represented by the stable value

$$\ln 1.08 \, i_{corr} = 10.69 + 0.618 \ln 1.69 Cl - 3\,034/T - 0.000\,105 R_c \tag{4}$$

where Cl = chloride concentration on the steel surface, in kg/m^3; T = temperature at steel surface (°K); R_C= resistivity of concrete cover (Ω). Concrete cover cracking due to the steel corrosion accelerates the steel corrosion and is often taken as the limit state of durability and the serviceability limit state of concrete structures. Up to now, however, there are few results on steel corrosion after concrete cover cracking, and the paper assumes that concrete cover cracking does not change the corrosion rate.

Zhao & Jin (2004) developed an analytical model of λ_C at the time t_{cr} of the concrete cover cracking due to steel corrosion and discussed parameters influencing λ_C. They did not consider the amount of corrosion product to fill the porous zone around the steel/concrete interface and gave conservative results. Considering this, Liu YP. & Weyers (1998) determined the time to concrete cover cracking induced by steel corrosion. The analytical results were compared to experimental ones. Following the idea, Thoft-Christensen (2000) assumed the concerned parameters as normal variables and showed that t_{cr} from corrosion initiation fits a Weibull distribution and is

$$t_{cr} = \alpha W_{crit}^2 /(0.196 \, \pi \, i_{corr}), \quad \text{in a} \tag{5}$$

where α= ratio of the molecular weight of steel to one of the rust product; W_{crit}= limiting amount of corrosion products needed for concrete cover cracking in a unit length of steel bar

$$W_{crit} = \frac{\pi D \rho_{rust} \rho_{st}}{\rho_{st} - \alpha \rho_{rust}} \left[\frac{X f_t}{E_{ef}} \left(\frac{a^2 + b^2}{b^2 - a^2} + v_c \right) + d_0 \right] \tag{6}$$

where ρ_{rust}= density of corrosion products; ρ_{st}= density of steel; E_C= elastic modulus of concrete; $E_{\text{ef}} = E_C/(1+\phi_{\text{cr}})$ = effective elastic modules of concrete; ϕ_{cr} = creep coefficient; f_t= tensile strength of concrete; $a = (D + 2\ d_0)/2$; $b = X + a$; v_c= Poisson ratio of concrete; D = diameter of steel bar; d_0= thickness of the annular layer of concrete pores.

Steel corrosion due to concrete carbonation

Concrete carbonation reduces the pH-value in the concrete pore system and leads to steel corrosion when the iron is not in a passive state. Differently from the chloride attack, corrosion due to concrete carbonation induces the general corrosion. Niu (2003) developed practical models of the general corrosion of steel due to concrete carbonation. The models are validated by data measured in several dozen structures in different provinces in China.

The initiation time t_1 of steel corrosion due to concrete carbonation, in a, is

$$t_1 = \left[(X - x_0)/k\right]^2 \tag{7}$$

where X in mm; x_0= carbonation residue, in mm;

$$x_0 = 4.86\left(-RH^2 + 1.5RH - 0.45\right)(X - 5)(\ln f_{ck} - 2.30) \tag{8}$$

where f_{ck} = characteristic value of the compressive strength of concrete, in MPa; RH = humidity of the air, in %; k = carbonation factor,

$$k = 2.56 K_{mc} k_j k_{CO_2} k_p k_s \sqrt[4]{T}\,(1 - RH)\,RH\left(\frac{57.94}{f_c}\kappa_c - 0.76\right) \tag{9}$$

where f_c = cube compressive strength of concrete, in MPa; κ_c = bias for f_c; K_{mc} = random variable of uncertainty in the analytical models; k_j= factor for steel locations with values 1.4 at corner and 1.0 elsewhere; the factor for concentration of CO_2, k_{CO2}= 1.1~1.4; k_p = factor for effects of vibrating and curing concrete and form stripping at the casting field, with a recommended value 1.2; k_s = factor for the working stress in steel, with a value 1.1 for the tensile stress. The steel corrosion rate before cover cracking is

$$\lambda_{e1} = 46 k_{cr} k_{ce} e^{0.04T} X^{-1.36}(RH - 0.45)^{2/3} f_c^{-1.83} \tag{10}$$

where k_{cr} = factor for steel locations, 1.6 at corner, 1.0 elsewhere; k_{ce} = factor for local environment. The corrosion rate of steel after cover cracking is

$$\lambda_{e2} = \begin{cases} 2.5\lambda_{e1}, & \lambda_{e1} > 0.008 \\ 4.0\lambda_{e1} - 187.5\lambda_{e1}^2, & \lambda_{e1} \le 0.008 \end{cases} \tag{11}$$

The time to cover cracking due to steel corrosion from the time of corrosion initiation, in a, is

$$t_{\text{crk}} = k_{\text{crs}}[(0.008c/D) + 0.00055 f_c + 0.022]/\lambda_{e1} \tag{12}$$

where D in mm; k_{crs} = factor for steel locations with values 1.0 at corner and 1.35 elsewhere. The varying diameter of the jth steel bar is

$$D_j(t) = \begin{cases} D_{j0}, & t \le t_{j1} \\ D_{j0} - 2\lambda_{e1}(t - t_{j1}), & t_{j1} < t \le t_{\text{crk}} + t_{j1} \\ D_{j0} - 2\lambda_{e1}(t_{\text{crk}} - t_{j1}) - 2\lambda_{e2}(t - t_{\text{crk}}) \ge 0, & t_{\text{crk}} + t_{j1} < t \end{cases} \tag{13}$$

Time-dependent reliability of deteriorating RC girders

The flexure resistance of a RC girder with rectangular or T cross sections is

$$R_M(t) = \left(\sum_{j=1}^{m} \pi D_j^2(t)/4\right) f_y\left[h_0 - f_y \sum_{j=1}^{m} \pi D_j^2(t)\Big/8 f_c b\right] \tag{14}$$

where f_y = tensile strength of steel; h_0= distance from the girder top to the center of steel; f_c= compressive strength of concrete; b = width of rectangular cross section or flange of T cross section of the girder; m- number of tensile steel bars.

$R_M(t)$ in Eq. (14) is a non-stationary stochastic process. The paper calculates the resistance using Monte Carlo method and then derives equations of the means and coefficients of variance of the related variables because of complicated Eq. (14). According to Liu & Weyers (1998), the dead load random variable S_D fits a normal distribution with probability density denoted by p_{S_D} and has a bias $\kappa_{SD}=1.0148$ and a cov $\delta_{SD}=0.0431$.

On the basis of the data acquired by a long-term current traffic survey on the Beijing 110# highway and the results given by the K-S testing and the small size sample W^2 testing, all axel loads are filtered Poisson processes (GB/T 50283-1999) and the parent probability density of the 3rd axel load W_{G3W3} of a truck with 3 axels is

$$p_{G3W3}(x) = \frac{1}{\sqrt{2\pi}\,0.0779\,4x}\exp\left[-\frac{(\ln x - 5.9115)^2}{2\times 0.0779\,4^2}\right] \tag{15}$$

The time-dependent reliability of the ith girder without repair in a period $[0,T]$ is (Ellingwood & Mori 1991)

$$R_{Si}(T) = \exp\left\{-\lambda T\left[1 - \int_{-\infty}^{\infty} p_{S_D} P(T, s_D)ds_D \Big/ T\right]\right\} \tag{16}$$

$$P(T, s_D) = \int_0^T (\int_R P_{G3W3}(r - s_D)p_{R(t)}(r)dr)dt \tag{17}$$

where $p_{R(t)}$ = density of the flexure resistance; P_{G3W3} = distribution of W_{G3W3}; λ = Poisson's arrival rate of the traffic load. Assume that a bridge deck is a series system consisting of n independent T-girders, the time-dependent system reliability is

$$R_s(T) = \prod_{i=1}^{n} R_{Si}(T) \tag{18}$$

The authors developed a Fortran 90 computer program TDSR for the time-dependent reliability of deteriorating RC bridge decks in environments of chloride attack and carbonation.

Example: Baihe bridge, built in 1966, is a RC highway bridge with 6 span of 15m. Figure 1 shows its cross section and transverse locations of wheel loads. Figure 2 shows the cross section of a T-girder. All girders consist from concrete of grade 25 with reinforcement bars of $8\phi32$mm and $2\phi20$mm. The related parameters are shown in Table 1.

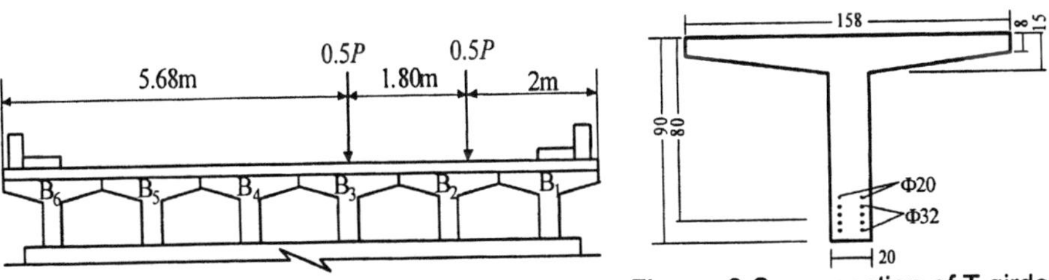

Figure 1. Deck of Baihe highway bridge　　　　Figure. 2 Cross section of T-girder

Considering effects of the cross linking members between T-girders, the study used finite element methods and calculated the midspan bending moments in girders, due to the deal load, a Normal variable with bias $\kappa_{SD}=1.0148$ and a cov 0.0431.

Case for chloride attack: t_{ji} in Eq. (1) fits a log-normal distribution, with a

Table 1. Rabdom variables in Eq. (14)

Variable	f_y	f_c	D_{i1}	D_{i2}	h_0	f_t	X
	MPa	MPa	mm	mm	mm	MPa	cm
Mean	303.3	27.8	32.0	20.0	809.9	1.95	3.05
cov	0.1014	0.1928	0.0250	0.0250	0.0229	0.15	0.050
Distr.	Norm.	Norm.	Norm.	Norm.	Norm.	Norm.	Norm.

Variable	D_c	Cl_s	Cl_{cr}	Cl_0	E_c	α
	cm^2/a	Kg/m^3	Kg/m^3	Kg/m^3	MPa	
Mean	0.234	39.2	11.03	0.735	2 850	0.573
cov	0.31	0.16	0.16	0.10	0.15	0.009
Distr.	Norm.	Norm.	Norm.	Norm.	Norm.	Unifo.

mean 19.9a, and cov 0.52. R_c=1500Ω and T= 285°K in Beijing. The regression analysis on the Monte Carlo results shows that i_{corr} fits a normal distribution, with a mean 1.01 and a cov 0.062. ρ_{rust}=3 600kg/m^3; φ_{cr}=2.0; ν_c=0.18; d_0= 12.5 μm and ρ_{st} = 7 850 kg/m^3. The times to concrete cover cracking, t_{cr32}, and t_{cr20} for steel bars of diameter 32mm and 20mm fit Weibull distributions with a mean 3.79a and a cov 0.23 for 32mm bars and mean 2.18a and a cov 0.22 for 20 mm bars, respectively.

The moment method of estimation and regression on the results of Eq. (14) from Monte Carlo methods shows that R_M fits a log-normal distribution. The mean and standard deviation of ln R_M can be described by functions of the time as

$$\mu_{\ln R_M}(t)=4\cdot10^{-7}t^3 - 8\cdot10^{-5}t^2 + 4\cdot10^{-5}t + 7.4265, \quad \text{regression coefficient } 0.999\,5 \qquad (20)$$

$$\sigma_{\ln R_M}(t) = -7\cdot10^{-9}t^3 + 2\cdot10^{-6}t^2 + 0.0001t + 0.1137, \quad \text{regression coefficient } 0.990\,6 \qquad (21)$$

Case for concrete carbonation: the characteristic value and bias of the cubic compressive strength of concrete are 17.5MPa and 1.5868, respectively. The model uncertainty in Eq. (9) fits a normal distribution with a mean 1.0980 and cov 0.0710, and k_{co2}=1.2. k_{ce}=3.5 (Nowak & Collins, 2000); RH =75% in Eq. (10). The monthly averaging values of RH are 44~77% in Beijing, a rather larger value is taken to consider the effect of alternatively dry and humid states.

The times of the corrosion initiation t_{Ic} for bars at corners and t_{Is} for bars near by sides fit log-normal distributions, with a mean 17.71a and a cov 0.18 for bars at corners and a mean 8.96a and a cov 0.18 for bars nearby sides, respectively. The times to concrete cover cracking t_{crC} at corners and t_{crE} near by sides fit log-normal distributions, with a mean 20.07a and a cov 0.14 at corners and a mean 31.81a and a cov 0.144 near by sides, respectively. Estimations of the moment method and regression analysis on results of Eq. (14) from Monte Carlo methods show that R_M fits a log-normal distribution and the time-dependent mean and standard deviation of ln R_M as follows

$$\mu_{\ln R_M}(t)= 2\cdot10^{-7}t^3 - 4\cdot10^{-5}t^2 + 0.0005t + 7.4216, \quad \text{regression coefficient } 0.999\,3 \qquad (22)$$

$$\sigma_{\ln R_M}(t) = 2\cdot10^{-9}t^3 - 2\cdot10^{-7}t^2 + 1\cdot10^{-6}t + 0.1115, \quad \text{regression coefficient } 0.061\,9 \qquad (23)$$

Its low value is because of an oscillating feature of $\sigma_{\ln R_M}$. However, the middle curve of the oscillating curve is very close to Eq. (23).

The largest bending moments in each T-girder at midspan are produced by W_{G3w3} when only the 3rd axel stays in a span. The transverse wheel locations are shown in Figure 1. The time-dependent reliabilities of single T-girders and complete deck are given by calculations of Eqs. (16) and (18) by using Monte Carlo methods. The cover cracking is an irreversible serviceability limit state. ISO 2394 recommends a target reliability index 1.5 for it, corresponding to a target reliability 0.066 81. The times t_{crk} to cover cracking corresponding to a fractile 0.066 81 and the median for both chloride attack and concrete carbonation are calculated for the example.

Figures 3 and 4 show time-dependent reliabilities $R_S(t)$'s and related time-dependent generalized reliability indexes $\beta_G(t)= \Phi^{-1}[R_S(t)]$, where $\Phi=$ the standardized Normal distribution, of the six girders and complete deck of Baihe bridge for the chloride attack. The two vertical lines in the two figures are the times to cover cracking of 10.10a and 22.08a corresponding to a fractile 0.066 81 and the median, respectively.

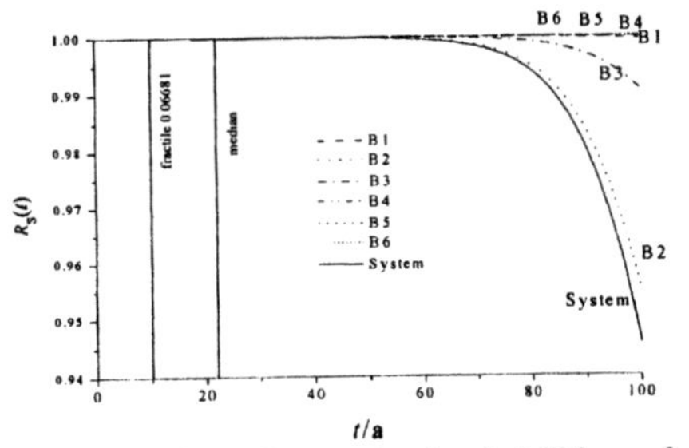

Figure 3. Time-dependent reliabilities of T-girders and deck system under chloride attack

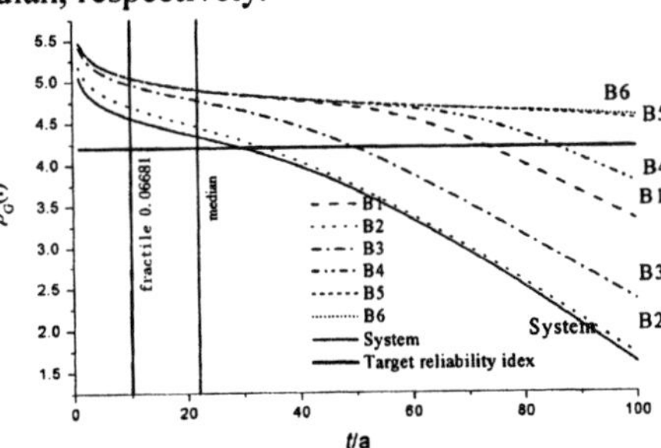

Figure 4. Time-dependent generalized reliability indexes of T-girders and deck system under chloride attack

Figures 5 and 6 show $\beta_G(t)$'s and related $R_S(t)$'s of the six girders and complete deck for the concrete carbonation. The two vertical lines in the two figures are the times to cover cracking of 24.42a and 29.03a corresponding to a fractile 0.066 81 and the median, respectively. The four figures show that $R_S(t)$'s based on stochastic processes have a value 1 at $t = 0$, for an initially safe bridge, a value different from the target reliability of the traditional structural reliability theory, and related $\beta_G(t)$'s have a value ∞ at $t = 0$, a value different from the target reliability index (Qin. & Yang 2005). The four figures also show t_{crk}'s corresponding to a fractile 0.066 81 and the median, respectively. GB/T 50283-1999 recommends a value of 4.2 for the target reliability index β_T of the ultimate limit state of ductile girders of the safety class 2. Figures 4 and 6 show that t_{crk}'s are much earlier than the times when the time-dependent reliability of girders reduces to the target reliability behind GB/T 50283- 1999. Concrete cover cracking influences bridge serviceability and durability, and t_{crk}'s is important for planning bridge maintenance. The CPU time for the example on Pentium III computer is less than 20″.

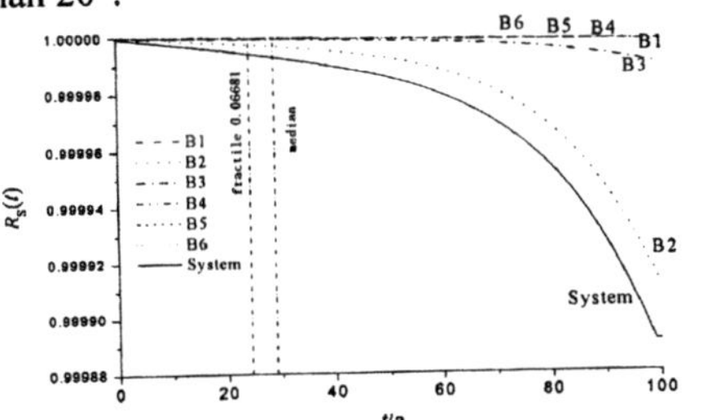

Figure 5. Time-dependent reliabilities of T-girders and deck system under concrete carbonation

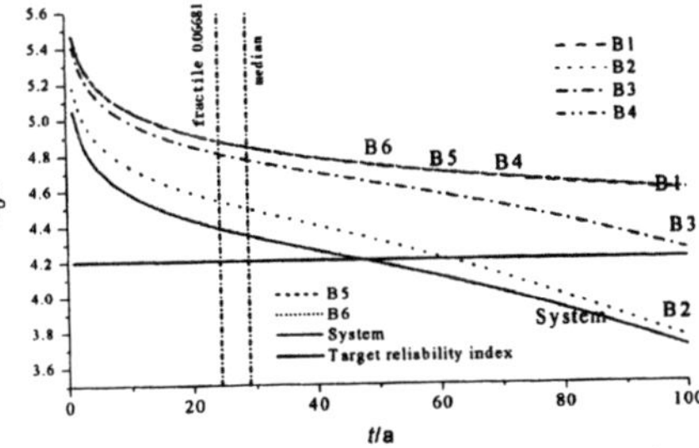

Figure 6. Time-dependent generalized reliability indexes of T-girders and deck system under concrete carbonation

Utility functions

Decisions in the face of uncertainty are based on outcomes of possible actions. The simplest outcome of an action is its expected monetary reward. However, in many situations such a monetary reward can not reasonably measure social impacts, life losses or losses of owner reputations. So called utility function is used for measuring the outcome. The utility function $U(r_w)$ is a function of the monetary reward r_w. It should satisfies simple axioms, for instance,

Axiom 1: if $r_{w1}>r_{w2}$, then $U(r_{w1}) > U(r_{w2})$; if $r_{w2}>r_{w1}$, then $U(r_{w2})> U(r_{w1})$; and if $r_{w2} = r_{w1}$, then $U(r_{w2}) = U(r_{w1})$.

Axiom 2: if a certain gain r_{w1} is equal to a gain r_{w2} with a probability P and a gain r_3 with a probability $(1-P)$ then $U(r_{w1}) = P\, U(r_{w2}) + (1-P)\, U(r_{w3})$.

An example of a convex utility for a risk avoiding owner is

$$U\left(r_w\right)=1-\exp(-Ar_w) \tag{24}$$

If $A = (\ln2)/10^6$, then $U(10^6) = 1/2$, $U(\infty) =1$. This means that a utility of even 10^{100} can not be two times as large as that of 10^6 for the owner. If $A = \ln 2\,/l_{max}$, Eq. (24) changes to

$$U\left(r_w\right)=1-\exp[-(r_w \ln 2)/l_{max}] \quad r_w<0 \tag{25}$$

where l_{max}= owner's maximum acceptable loss. Here, $U(0)=0$, $U(-l_{max})= -1$ and $U(-\infty)= -\infty$.

Optimization of bridge inspection/repair schedules by using pre-posterior decision making

Pre-posterior decision making

A decision is to choose an action with the largest expected utility. For the pre-posterior decision making, there are two decisions at any decision point: to choose an action based on existing information, or to do a trial for new data to modify a prior deteriorating model and produces a posterior model. Denote an action set by $A=\{a_1, a_2,...a_n\}$, and bridge condition random variable by S which has prior probability density $f'(s)$, a utility generated by action a_i for a bridge condition s by $U(a_i, s)$. The prior expected utility of action a_i is

$$EU'_{a_i} = \int_{s\in S} U(a_i,s)f'(s)ds \tag{26}$$

An optimum action given by the prior decision is

$$a'_{opt} = a_j \,|\,\{EU'_{a_j} = \max_{i=1,\cdots n}(EU'_{a_j})\} \tag{27}$$

Denote a measured value of bridge condition S by Y, belonging to a set Y. Denote a conditional probability density of Y by $f_{Y|S}(y|s)$. A posterior density of S for outcome y of Y is

$$f''(s) = f(s|y) = f_{Y|S}(y|s)f'(s)\Big/ \int_{s\in S} f_{Y|S}(y|s)f'(s)ds \tag{28}$$

An optimum posterior expected utility of action a_j is

$$EU''_{a_j,y} = \int_{e\in E} U(a_j,e)f''(s)ds \tag{29}$$

An optimum action a_k corresponding to the posterior density of S on the condition y is

$$a''_{opt\,y} = a_k \,|\,\{EU''_{a_k\,y} = \max_{p=1,\cdots n}(EU''_{a_p\,y})\} \tag{30}$$

Then, a conditional utility of an inspection, UI, for measured value y is

$$UI|y = \int_{s\in S} U(a_k,s)f''(s|y)ds - \int_{s\in S} U(a_j,s)f'(s)ds \tag{31}$$

An expected utility of an inspection is

$$EUI = \int_{y\in Y} UI|y\, f'(y)dy = \int_{y\in Y} UI|y dy \int_{s\in S} f_{Y|S}(y|s)f'(s)ds \tag{32}$$

$EUI <0$ means that this inspection is not needed and a decision should be based on Eq. (26).

Utility of bridge inspections

The utility of a bridge inspection is the utility of a bridge repair schedule based on the posterior deteriorating model minus that based on the prior deteriorating model, see Figure 7 with the ordinate as the time-dependent reliability and the abscissa as the time. Denote the time for first inspection based on the prior density by T_0. Assume two possible posterior deteriorating models: 1 and 2 which an inspection may lead to. The time for first repair based on the model 2 is T_2. Thus, if an inspection supports model 2 and a repair is taken at time T_0, the bridge reliability at time T_2 would be less than a given reliability limit $P_{R,min}$. The time for first repair based on the model 1 is T_1. Thus, if an inspection supports model 1 and a repair is still taken at time T_0, the bridge reliability at time T_2 is clearly higher than $P_{R,min}$. The too early repair uses resources which would be used for some more urgent projects, or makes an interest loss.

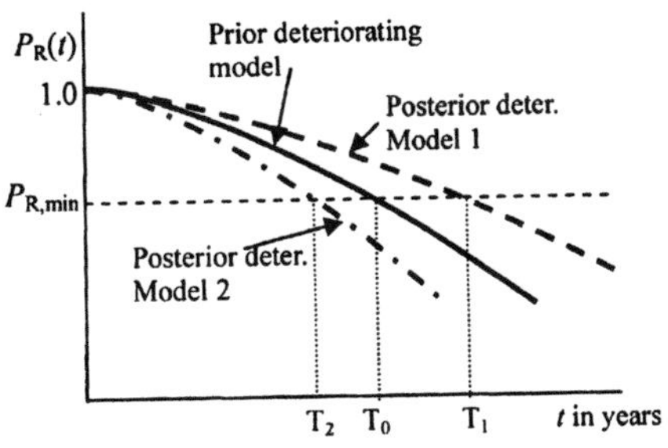

Figure. 7 Utility of bridge inspections

Examples

Baihe bridge, shown in Figs. 1 and 2, is again used as an example. The time-dependent reliabilities are given by the program TDSR. Assume that λ_C or λ_e are given by inspections and that their probability density is discretized at 10 points with equal density increments from lower corrosion rate to higher one for simplification.

For the T-girders in a design working life of 100a β_T =4.2. Denote the time-dependent reliability of complete bridge deck in a time period $[0, t]$ by $P_R(t)$, $0 < t \leq 100$. Assume that a repair action will be immediately taken as soon as $P_R(t)$ reduces to the target reliability P_{RT}. The prior decision making tells that the first repair time should be the 27th year. An inspection gives an updated distribution of λ_C or λ_e for the pre-posterior decision analysis. Each of the 10 discrete points predicts a first repair time. The corresponding pre-posterior decision analysis shows the times as the 29th, 28th, 28th, 27th, 27th, 27th, 27th, 26th, 26th and 25th years. Consider a simple repair action - replacing T- girders. Total monetary reward at time t is

$$r_w(t) = -C_I(t) - C_R(t) - C_F(t)$$

where $C_I(t) = C_{I,L}(t) + C_{I,Fix}$ and $C_R(t) = C_{R,L}(t) + C_{R,Fix}$, C_I= inspection cost, C_R= repair cost, and C_F= failure loss, $C_{R,L}(t)$, $C_{I,L}(t)$= labor costs for repair and inspection, respectively, $C_{R,Fix}$, $C_{I,Fix}$= the material and technique costs for repair and inspection, respectively. Assume the owner's maximum acceptable loss l_{max} = 500,000. The utility function Eq. (25) change to $U(r_w) = 1 - \exp[(-r_w \ln 2)/500\,000]$. $U(a_j, s)$ in Eq. (31) can be r_w or $U(r_w)$. Denote the first repair time from the ith deteriorating model by $T''_{R1,i}$ corresponding to the ith discrete point of λ_C or λ_e. $T''_{R1,10}$ is the earliest time for the first repair. The expected utility of an inspection is

$$EUI(t) = 0.1 \sum_{i=1}^{10} UI_i(t) \tag{33}$$

where UI_i= inspection utility corresponding to the ith discrete point of λ_C or λ_e. Consider all possible inspection and repair time t.

$$\text{For } t \leq T''_{R1,10}, \quad UI_i = \begin{cases} UI\left[\dfrac{C_R(T''_{R1,i})}{(1+r)^{T''_{R1,i}}} + \dfrac{C_I(t)}{(1+r)^t}\right] - UI\left[\dfrac{C_R(T'_{R1})}{(1+r)^{T'_{R1}}}\right], & T''_{R1,i} \geq T'_{R1} \\[2ex] -UI\left[\dfrac{C_R(T'_{R1}) + C_F(T'_{R1})(P_{RT} - P''_{Ri}(T'_{R1}))}{(1+r)^{T'_{R1}}}\right] + UI\left[\dfrac{C_R(T''_{R1,i})}{(1+r)^{T''_{R1,i}}} + \dfrac{C_I(t)}{(1+r)^t}\right], & T''_{R1,i} < T'_{R1} \end{cases} \tag{34}$$

where r= interest rate, T_{R1} = first repair time from the prior model, $P''_{Ri}(t)$ = time-dependent reliability from the ith posterior model, $P_{RT} = \Phi(4.2) = 0.999\ 986\ 65$.

For later times: $T'_{R1,m+1} < t < T'_{R1,m} < T_{R1}$, an expected utility of an inspection is

$$UI_i = UI\left[\frac{C_R(t)+C_F(t)(P_S - P''_{Si}(t)) + C_I(t)}{(1+r)^t}\right] - UI\left[\frac{C_R(T'_{R1})+C_F(T'_{R1})(P_S - P''_{Si}(T'_{R1}))}{(1+r)^{T'_{R1}}}\right] \quad i = m+1,\cdots,10 \quad (35)$$

Consider $C_{I,L}(t) = 50\ (1+rt)$, $C_{R,L}(t) = 3\ 000\ (1+rt)$, $C_{R,Fix} = 500$ and $C_{I,Fix} = 50$, based on data from Beijing Highway Bureau, and assume $C_F(t) = 200\ 000\ (1+rt)$ RMB, all in RMB 1000 and $r = 0.06$. Figures 8 and 9 give the expected inspection utilities and rewards and show that an inspection in the 28th year has a maximum expected utility 0.08, and a maximum expected reward RMB 24 000.

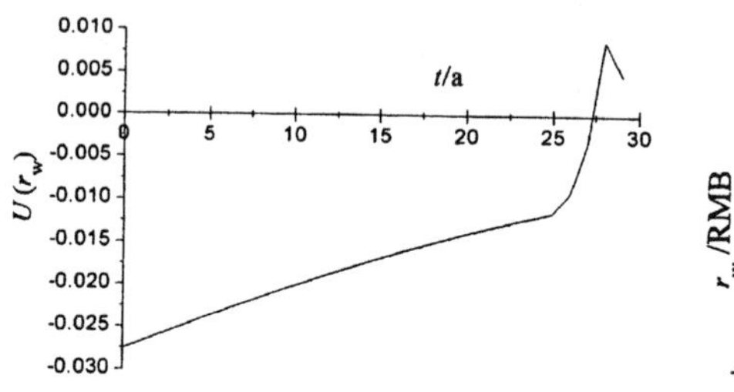

Figure 8. Expected utility of a bridge Inspection

Figure 9. Expected reward of a bridge inspection

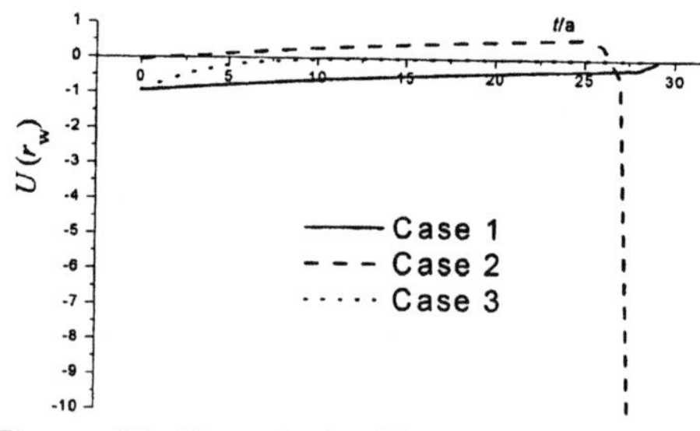

Figure 10. Expected utility curves for 3 Cases

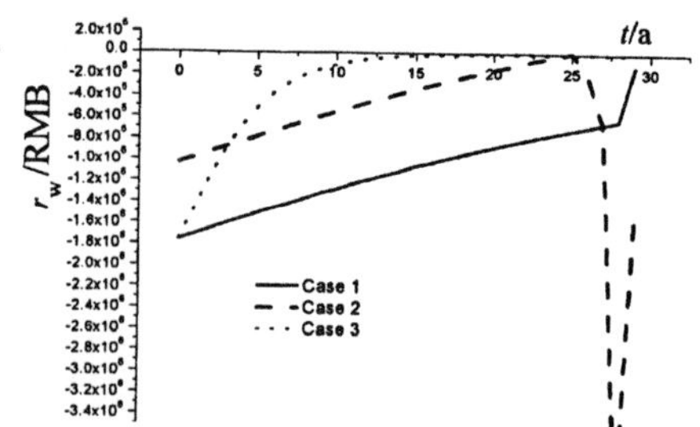

Figure 11. Expected reward curves for 3 Cases

Table 2. Three Cost cases, in RMB 1000

Cost case	$C_{I,L}(t)$	$C_{I,Fix}$	$C_{R,L}(t)$	$C_{R,Fix}$	$C_F(t)$	r
1					$8\times10^5(1+rt)$	0.06
2	$50(1+rt)$	50	$880(1+rt)$	500	$8\times10^9(1+rt)$	0.06
3					$8\times10^9(1+rt)$	0.50

Table 3. Comparison between expected utilities and rewards for 3 Cost cases

Cost case	Peak of $U(r_w)$		Peak of r_w		Necessity of inspect.
	time	value	time	value	
1	29	-0.05	29	-1.21×10^5	No
2	25	0.58	25	9 599	Yes
3	29	-8.68×10^{-5}	25	-164.8	No

An inspection reward consists of (a) an interest of repair cost during a period from repair time given by the pre-posterior model to that given by the prior model, (b) a failure loss and (c) an inspection cost. If (a) is dominant, a later repair would be better. If (c) is dominant, an earlier repair would be better. Consider three Cost cases, see Table 2, with case 1 for a less failure loss and case 3 for a larger interest rate. The expected inspection utilities and rewards for the three Cost cases are shown in Figures 10 and 11. Table 3 shows information on their peaks. The peak of inspection utility curve for case 2 is 4 years earlier then that for case 1, because of the larger failure loss for case 2. The expected inspection utility goes down rapidly after the 28th year because the failure loss becomes larger than the owner's financial capacity.

Conclusions

Pre-posterior decision making utilizes data from inspection and can give better schedules. Utility functions can respond owner's financial capacity and psychology for risk-avoiding. Time-dependent reliabilities for the ultimate limit state of RC bridge resistances due to chloride attack and concrete carbonation reduce to the target value in the 30th and 50th years, respectively, in Beijing area. Both show that actual expected working lives of typical RC bridges in Beijing are much less than the design working life of 100 years. Both corrosions make the time to concrete cover cracking due to the steel corrosions earlier than the time when time-dependent reliabilities for the ultimate limit state reduces to the target value.

Acknowledgement

This paper is a result of a National Key Research (973) Project 'Basic research on safety of major engineering project under the geo-hazard environment' numbered 2002CB4127009.

References

Ellingwood, B. & Mori, Y. 1991. Probabilistic methods for life prediction of concrete structures in nuclear power plants. *Trans., 11th Ins. Conf. on Stuct. Mech. In Reactor Technol.*, Tokyo, Japan, Vol. D: 291-296.

Fidjestoel, P. & Tuutti, K. 1998. Chloride induced corrosion in high performance concrete - lifetime versus diffusivity and receptivity. In O.E. Gjorv, (eds) 2nd *Internat. conference on Concrete under severe conditions*, Tromso; Norway, 1(2), London: 133-142.

Fontana, M.G. & Greene, N.D. 1978. *Corrosion Engineering.* NewYork, McGraw-Hill.

GB/T 50283-1999. *Unified Design Standard for Reliability of Highway Structures.* Beijing, China Jihua Press. (in Chinese)

ISO 2394. 1998. *General Principles on Reliability for Structures.* Switzerland, ISO.

Liu, T. & Weyers, R.W. 1998. Modeling the dynamic corrosion process in chloride contaminated concrete structures. *Cement and Concrete Research*, 28(3): 365-379.

Liu, Y.P. & Weyers, R.E. 1998. Modeling of the time to corrosion cracking in chloride contaminated reinforced concrete structures. *ACI Materials Journal*, 95: 675-681.

Niu, D.T. 2003. *Durability and Life Forecast of Reinforced Concrete Structures.* Beijing, China Science Press, (in Chinese)

Qin, Q. & Li, L.Y. 2005. Optimization of Inspection Schedules of RC Bridges Based on Pre-Posterior Decision Analysis and Time- Dependent Reliability. In G.. Augusti (eds), *ICOSSAR2005- Interna- tional Conference on Structural Safety and Reliability*, Rome, Italy.

Qin, Q. & Yang, X.G. 2005. Time-dependent reliability of deteriorating structures, *ICOSSAR2005-International Conference on Structural Safety and Reliability*, Rome, Italy, Millpress

Stewart, M.G. & Rosowsky, D.V. 1998. Time-dependent reliability of deteriorating reinforced concrete bridge decks. *Structural Safety*, 20(1): 91-109.

Thoft-Christensen, P. 2000. Stochastic modeling of the crack initiation time for reinforced concrete structures. In *2000 Stuctures Congress*, Philadelphia, May 8-10.

Zhao, Y.X., & Jin, W.L. 2004. Corrosion rates of reinforced steel bars when concrete cover cracking due to corrosion. *Journal of Hydraulic Engineering*, 11: 14-18. (in Chinese)

Authors index